EXACT SOLUTIONS IN THREE-DIMENSIONAL GRAVITY

This self-contained text systematically presents the determination and classification of exact solutions in three-dimensional Einstein gravity. The book explores the theoretical framework and general physical and geometrical characteristics of each class of solutions, and includes information on the researchers responsible for their discovery. Beginning with the physical character of the solutions, they are identified and ordered on the basis of their geometrically invariant properties, their symmetries, and their algebraic classifications, or from the standpoint of their physical nature; for example, electrodynamic fields, fluid, scalar field, or dilaton. Consequently, this text serves as a thorough catalogue of $(2+1)$-exact solutions to the Einstein equations coupled to matter and fields, and to the vacuum solutions of topologically massive gravity with a cosmological constant. The solutions are also examined from different perspectives, building a conceptual bridge between exact solutions of three- and four-dimensional gravities and thus providing graduates and researchers with an invaluable resource in this important topic in gravitational physics.

ALBERTO A. GARCÍA-DÍAZ is Emeritus Professor at the Center for Research and Advanced Studies of the National Polytechnic Institute, Mexico (CINVESTAV-IPN). His research throughout his career has focused on algebraic classification in four-dimensional gravity, nonlinear electrodynamics and dilaton fields.

CAMBRIDGE MONOGRAPHS ON MATHEMATICAL PHYSICS

General Editors: P. V. Landshoff, D. R. Nelson, S. Weinberg

S. J. Aarseth *Gravitational N-Body Simulations: Tools and Algorithms*†
J. Ambjørn, B. Durhuus and T. Jonsson *Quantum Geometry: A Statistical Field Theory Approach*†
A. M. Anile *Relativistic Fluids and Magneto-fluids: With Applications in Astrophysics and Plasma Physics*
J. A. de Azcárraga and J. M. Izquierdo *Lie Groups, Lie Algebras, Cohomology and Some Applications in Physics*†
O. Babelon, D. Bernard and M. Talon *Introduction to Classical Integrable Systems*†
F. Bastianelli and P. van Nieuwenhuizen *Path Integrals and Anomalies in Curved Space*†
D. Baumann and L. McAllister *Inflation and String Theory*
V. Belinski and E. Verdaguer *Gravitational Solitons*†
J. Bernstein *Kinetic Theory in the Expanding Universe*†
G. F. Bertsch and R. A. Broglia *Oscillations in Finite Quantum Systems*†
N. D. Birrell and P. C. W. Davies *Quantum Fields in Curved Space*†
K. Bolejko, A. Krasiński, C. Hellaby and M-N. Célérier *Structures in the Universe by Exact Methods: Formation, Evolution, Interactions*
D. M. Brink *Semi-Classical Methods for Nucleus-Nucleus Scattering*†
M. Burgess *Classical Covariant Fields*†
E. A. Calzetta and B.-L. B. Hu *Nonequilibrium Quantum Field Theory*
S. Carlip *Quantum Gravity in 2+1 Dimensions*†
P. Cartier and C. DeWitt-Morette *Functional Integration: Action and Symmetries*†
J. C. Collins *Renormalization: An Introduction to Renormalization, the Renormalization Group and the Operator-Product Expansion*†
P. D. B. Collins *An Introduction to Regge Theory and High Energy Physics*†
M. Creutz *Quarks, Gluons and Lattices*†
P. D. D'Eath *Supersymmetric Quantum Cosmology*†
J. Dereziński and C. Gérard *Mathematics of Quantization and Quantum Fields*
F. de Felice and D. Bini *Classical Measurements in Curved Space-Times*
F. de Felice and C. J. S Clarke *Relativity on Curved Manifolds*†
B. DeWitt *Supermanifolds, 2nd edition*†
P. G. O. Freund *Introduction to Supersymmetry*†
F. G. Friedlander *The Wave Equation on a Curved Space-Time*†
J. L. Friedman and N. Stergioulas *Rotating Relativistic Stars*
Y. Frishman and J. Sonnenschein *Non-Perturbative Field Theory: From Two Dimensional Conformal Field Theory to QCD in Four Dimensions*
J. A. Fuchs *Affine Lie Algebras and Quantum Groups: An Introduction, with Applications in Conformal Field Theory*†
J. Fuchs and C. Schweigert *Symmetries, Lie Algebras and Representations: A Graduate Course for Physicists*†
Y. Fujii and K. Maeda *The Scalar-Tensor Theory of Gravitation*†
J. A. H. Futterman, F. A. Handler, R. A. Matzner *Scattering from Black Holes*†
A. S. Galperin, E. A. Ivanov, V. I. Ogievetsky and E. S. Sokatchev *Harmonic Superspace*†
R. Gambini and J. Pullin *Loops, Knots, Gauge Theories and Quantum Gravity*†
T. Gannon *Moonshine beyond the Monster: The Bridge Connecting Algebra, Modular Forms and Physics*†
A. García-Díaz *Exact Solutions in Three-Dimensional Gravity*
M. Göckeler and T. Schücker *Differential Geometry, Gauge Theories, and Gravity*†
C. Gómez, M. Ruiz-Altaba and G. Sierra *Quantum Groups in Two-Dimensional Physics*†
M. B. Green, J. H. Schwarz and E. Witten *Superstring Theory Volume 1: Introduction*
M. B. Green, J. H. Schwarz and E. Witten *Superstring Theory Volume 2: Loop Amplitudes, Anomalies and Phenomenology*
V. N. Gribov *The Theory of Complex Angular Momenta: Gribov Lectures on Theoretical Physics*†
J. B. Griffiths and J. Podolský *Exact Space-Times in Einstein's General Relativity*†
S. W. Hawking and G. F. R. Ellis *The Large Scale Structure of Space-Time*†
F. Iachello and A. Arima *The Interacting Boson Model*†
F. Iachello and P. van Isacker *The Interacting Boson-Fermion Model*†
C. Itzykson and J. M. Drouffe *Statistical Field Theory Volume 1: From Brownian Motion to Renormalization and Lattice Gauge Theory*†
C. Itzykson and J. M. Drouffe *Statistical Field Theory Volume 2: Strong Coupling, Monte Carlo Methods, Conformal Field Theory and Random Systems*†
G. Jaroszkiewicz *Principles of Discrete Time Mechanics*

C. V. Johnson *D-Branes*†
P. S. Joshi *Gravitational Collapse and Spacetime Singularities*†
J. I. Kapusta and C. Gale *Finite-Temperature Field Theory: Principles and Applications, 2nd edition*†
V. E. Korepin, N. M. Bogoliubov and A. G. Izergin *Quantum Inverse Scattering Method and Correlation Functions*†
J. Kroon *Conformal Methods in General Relativity*
M. Le Bellac *Thermal Field Theory*†
Y. Makeenko *Methods of Contemporary Gauge Theory*†
S. Mallik and S. Sarkar *Hadrons at Finite Temperature*
N. Manton and P. Sutcliffe *Topological Solitons*†
N. H. March *Liquid Metals: Concepts and Theory*†
I. Montvay and G. Münster *Quantum Fields on a Lattice*†
P. Nath *Supersymmetry, Supergravity, and Unification*
L. O'Raifeartaigh *Group Structure of Gauge Theories*†
T. Ortín *Gravity and Strings, 2nd edition*
A. M. Ozorio de Almeida *Hamiltonian Systems: Chaos and Quantization*†
L. Parker and D. Toms *Quantum Field Theory in Curved Spacetime: Quantized Fields and Gravity*
R. Penrose and W. Rindler *Spinors and Space-Time Volume 1: Two-Spinor Calculus and Relativistic Fields*†
R. Penrose and W. Rindler *Spinors and Space-Time Volume 2: Spinor and Twistor Methods in Space-Time Geometry*†
S. Pokorski *Gauge Field Theories, 2nd edition*†
J. Polchinski *String Theory Volume 1: An Introduction to the Bosonic String*†
J. Polchinski *String Theory Volume 2: Superstring Theory and Beyond*†
J. C. Polkinghorne *Models of High Energy Processes*†
V. N. Popov *Functional Integrals and Collective Excitations*†
L. V. Prokhorov and S. V. Shabanov *Hamiltonian Mechanics of Gauge Systems*
S. Raychaudhuri and K. Sridhar *Particle Physics of Brane Worlds and Extra Dimensions*
A. Recknagel and V. Schiomerus *Boundary Conformal Field Theory and the Worldsheet Approach to D-Branes*
R. J. Rivers *Path Integral Methods in Quantum Field Theory*†
R. G. Roberts *The Structure of the Proton: Deep Inelastic Scattering*†
C. Rovelli *Quantum Gravity*†
W. C. Saslaw *Gravitational Physics of Stellar and Galactic Systems*†
R. N. Sen *Causality, Measurement Theory and the Differentiable Structure of Space-Time*
M. Shifman and A. Yung *Supersymmetric Solitons*
H. Stephani, D. Kramer, M. MacCallum, C. Hoenselaers and E. Herlt *Exact Solutions of Einstein's Field Equations, 2nd edition*†
J. Stewart *Advanced General Relativity*†
J. C. Taylor *Gauge Theories of Weak Interactions*†
T. Thiemann *Modern Canonical Quantum General Relativity*†
D. J. Toms *The Schwinger Action Principle and Effective Action*†
A. Vilenkin and E. P. S. Shellard *Cosmic Strings and Other Topological Defects*†
R. S. Ward and R. O. Wells, Jr *Twistor Geometry and Field Theory*†
E. J. Weinberg *Classical Solutions in Quantum Field Theory: Solitons and Instantons in High Energy Physics*
J. R. Wilson and G. J. Mathews *Relativistic Numerical Hydrodynamics*†

† Available in paperback

Exact Solutions in Three-Dimensional Gravity

ALBERTO A. GARCÍA-DÍAZ
Center for Research and Advanced Studies of the National Polytechnic Institute, Mexico (CINVESTAV)

✓ Cambridge University Press acknowledges contributions to Chapters 16-18 by David Chow, Christopher N. Pope and Ergin Sezgin (previously published as 'David D K Chow et al 2010 Class. Quantum Grav. 27 105001').

✓ Cambridge University Press acknowledges contributions to Chapter 19 by David Chow, Christopher N. Pope and Ergin Sezgin (previously published as 'David D K Chow et al 2010 Class. Quantum Grav. 27 105002').

CAMBRIDGE
UNIVERSITY PRESS

CAMBRIDGE
UNIVERSITY PRESS

University Printing House, Cambridge CB2 8BS, United Kingdom

One Liberty Plaza, 20th Floor, New York, NY 10006, USA

477 Williamstown Road, Port Melbourne, VIC 3207, Australia

4843/24, 2nd Floor, Ansari Road, Daryaganj, Delhi – 110002, India

79 Anson Road, #06–04/06, Singapore 079906

Cambridge University Press is part of the University of Cambridge.

It furthers the University's mission by disseminating knowledge in the pursuit of education, learning, and research at the highest international levels of excellence.

www.cambridge.org
Information on this title: www.cambridge.org/9781107147898
DOI: 10.1017/9781316556566

First published 2017

Printed in the United Kingdom by Clays, St Ives plc

A catalogue record for this publication is available from the British Library.

Library of Congress Cataloging-in-Publication Data
Names: García-Díaz, Alberto A., 1942- author.
Title: Exact solutions in three-dimensional gravity / Alberto A. García-Díaz (Center for Research and Advanced Studies of the National Polytechnic Institute, Mexico (CINVESTAV).
Other titles: Cambridge monographs on mathematical physics.
Description: Cambridge, United Kingdom ; New York, NY : Cambridge University Press, 2017. | Series: Cambridge monographs on mathematical physics
Identifiers: LCCN 2017012130 | ISBN 9781107147898 (hardback ; alk. paper) | ISBN 1107147891 (hardback ; alk. paper)
Subjects: LCSH: Gravitation–Mathematics. | Quantum gravity–Mathematics. | Einstein field equations. | Space and time.
Classification: LCC QC178 .G365 2017 | DDC 539.7/54–dc23
LC record available at https://lccn.loc.gov/2017012130

ISBN 978-1-107-14789-8 Hardback

Contents

Preface

I have been working on this book project over a period of many years in order to create a concise but comprehensive account of exact solutions in the three-dimensional Einstein theory of gravity. This theory, although tangentially related to the real gravity world, is a good model from which to extract some relevant conclusions about that world, taking advantage of simplifications due to the reduction in dimensions. Of course, after fifty years of existence, $(2+1)$ gravity has been approached from many perspectives, but, to my mind, there is still no satisfactory account of solutions of physical interest. I therefore wrote the present book on exact solutions in $(2+1)$ gravity following closely the pattern of the classic book, *Exact solutions in Einstein's field equations* with the aim of bringing to each class of solutions presented its theoretical framework, its general physical and geometrical characteristics, and references to the researchers who discovered or studied them.

I am greatly indebted to my late friends and collaborators J.F. Plebański, N. R. Sibgatullin and S. del Campo for stimulating talks. Many thanks are due to my collaborators E. Ayón-Beato, C. Campuzano, M. Cataldo, G. Gutierrez, F. Hehl, A. Macías, V. Manko, N. Mitskievich, and C. Terrero for many enlightening talks, and advice on various problems I faced when dealing with this task. This work could not have come about without a fruitful sabbatical stay at the Physics Department of the California University at Davis under the sponsorship of Professor S. Carlip, to whom I am indebted, and to UC MEXUS-CONACYT for financial support. I cordially thank my friends, particularly Y. Gurievich and M. Lopez, and colleagues from the Departamento de Física, Centro de Investigación y de Estudios Avanzados del I.P.N., for their encouragement. Many thanks to my assistant E.Vargas-Díaz for her daily support and typographical help in the final preparation of the manuscript of this book. The support of the Consejo Nacional de Ciencia y Tecnología (CONACYT), México, through various grants (at present: Grant CONACyT 178346) is acknowledged.

I really do appreciate the patience and support of my daughters, Ana Alicia and Alexandra Sofia, and sons, Sergey, Albert, Alberto-Tiko, and Alex, over all these years of writing this book.

Please do not hesitate to let me know about inconsistencies, unevenness, omissions or errors to be found in this text. Constructive comments are welcome: aagarcia@fis.cinvestav.mx.

1
Introduction

Given the large number of exact solutions that exist today in (2 + 1) Einstein gravity the purpose of the present book is to present a complete and concise list of exact solutions with emphasis on their physical and geometrical properties from the beginnings of the field in 1963 to the present, to be useful for the audience of experts and young researchers. Emphasis is given to solutions to the Einstein equations in the presence of matter and fields, for instance, point particle solutions, perfect fluids, cosmological spacetimes, dilatons, inflatons, and stringy solutions. The second part of this book deals with solutions to vacuum topologically massive gravity with a cosmological constant, as there exist three big families of spacetimes: the inhomogeneous Bianchi class of solutions, the Kundt spacetimes and the Cotton type N wave fields.

To avoid unnecessary typing, the cosmological constant is denoted by Λ, AdS spacetime stands for an asymptotically anti-de Sitter spacetime with $\Lambda < 0$, dS spacetime stands for an asymptotically de Sitter spacetime with $\Lambda > 0$, $3D$ stands for three dimensions, while $(2+1)D$ spacetime denotes $(2+1)$-dimensional spacetime, PF stands for perfect fluid, and ρ or μ denotes the fluid energy density. Occasionally we use SL for spacelike, TL for timelike, and ST to denote spacetime. On the other hand, when publications by various authors are cited, an abbreviation of their family names, including their first capital initials, are given; for instance, EEqs. and EM mean, respectively, Einstein equations and Einstein–Maxwell, MTW stands for Misner, Thorne, and Wheeler; FRW reads Friedmann–Robertson–Walker, and BTZ denotes Bañados–Teitelboim–Zanelli.

1.1 Main Features of (2 + 1) Gravity

In the early work by Giddings, Abbott, and Kuchař (GAK) (Giddings et al., 1984) it is stated that "the lowest dimension in which the Einstein Theory makes sense is $n = 3$." Consequently, bearing in mind the self-contained nature of this book, the main features of $(2+1)$-dimensional gravity are presented following the GAK pattern and essentially maintaining their wording.

1.1.1 Field Equations and Curvature Tensors

Einstein's theory of relativity, as a theory of a gravitational spacetime, can be based on two postulates which are independent of the spacetime dimensions; these postulates demand that the field equations take the form of the Einstein equations:

$$G_{\mu\nu} := R_{\mu\nu} - \frac{1}{2} R\, g_{\mu\nu} = \kappa T_{\mu\nu} - \Lambda\, g_{\mu\nu}, \tag{1.1}$$

where $G_{\mu\nu}$ is the Einstein tensor, $T_{\mu\nu}$ is the energy–momentum tensor (which, by virtue of the Bianchi identity, fulfills the energy conservation condition $T^{\mu}{}_{\nu;\mu} = 0$), Λ is a cosmological constant, and κ is a coupling constant. And second, the spacetime geometry is determined by the Riemann curvature tensor $R^{\alpha}{}_{\beta\gamma\delta}$.

The Riemann tensor in three dimensions possesses six algebraically independent components: as many as the number of independent components of the Ricci tensor; therefore, the Riemann tensor is completely determined by the Ricci tensor and the scalar curvature, namely:

$$\begin{aligned} R_{\alpha\beta\gamma\delta} &= g_{\alpha\gamma}\, R_{\beta\delta} - g_{\alpha\delta}\, R_{\beta\gamma} - g_{\beta\gamma}\, R_{\alpha\delta} + g_{\beta\delta}\, R_{\alpha\gamma} \\ &\quad -\frac{1}{2}(g_{\alpha\gamma} g_{\beta\delta} - g_{\alpha\delta}\, g_{\beta\gamma})R. \end{aligned} \tag{1.2}$$

There is no room for the Weyl conformal tensor, which is thus zero.

Due to the Einstein equations, the Riemann tensor can be expressed in terms of the Einstein tensor or, in turn, through the energy momentum tensor $T_{\alpha\gamma}$ as

$$\begin{aligned} R_{\alpha\beta\gamma\delta} &= \kappa[g_{\alpha\gamma}\, T_{\beta\delta} - g_{\alpha\delta}\, T_{\beta\gamma}] + \kappa[g_{\beta\delta}\, T_{\alpha\gamma} - g_{\beta\gamma}\, T_{\alpha\delta}] \\ &\quad -\kappa(g_{\alpha\gamma} g_{\beta\delta} - g_{\alpha\delta}\, g_{\beta\gamma})T. \end{aligned} \tag{1.3}$$

In three dimensions, the coupling constant κ is measured in units of 1/mass, and therefore defines a natural mass unit.

1.1.2 Matter Distribution Locally Curves the Spacetime

Since, in a $3D$ spacetime, the Riemann curvature tensor is expressible solely through the energy–momentum tensor (1.3), thus, in an empty spacetime, where $T_{\alpha\gamma} = 0$, the spacetime is locally flat:

$$R_{\alpha\beta\gamma\delta} = 0.$$

Therefore, the flat spacetime is the field solution to the vacuum Einstein equations $G_{\mu\nu} = 0 \rightarrow R_{\mu\nu} = 0 = R$, ($T_{\mu\nu} = 0$ and $\Lambda = 0$). Staruszkiewicz (1963), in his pioneering article, stressed this fact by writing: "three-dimensional gravitation theory is a theory without a field of gravitation; where no matter is present, space is flat. Curvature can arise only if matter or energy are present."

1.1.3 Point Particles Produce Global Effects on the Spacetime

The first publication on gravity in $(2+1)$ dimensions by Staruszkiewicz (1963) was devoted to the description of static solutions determined by point sources. Point particles move along geodesics. In the points where the particles are located there arise conical defects (conical singularities) that can be felt at infinity; the total mass in the spacetime is proportional to the deficit angle at infinity. Because the angle deficit cannot increase by 2π, the mass is bounded from above. See Chapter 2 for details.

1.1.4 Newtonian Limits

This section has to be subdivided into three subsections: first, to recall the content of the Newtonian theory; second, to reveal the existence of the Newtonian limit in the standard $(3+1)$, or $(1+(n-1))$, gravity via the weak gravitational field treatment; and finally, to show that the slow motion limit of the $(2+1)$ gravity occurs without acceleration.

Newtonian Theory of Gravity

The Newtonian theory of gravity is based on the Newtonian potential ϕ fulfilling the Poisson equation with matter density ρ,

$$\nabla^2\phi = 4\pi\, G\, \rho, \tag{1.4}$$

which generates the Newtonian field causing the accelerated motion of test particles in it:

$$\frac{d^2\, x^i}{dt^2} = -\delta^{i\,j}\partial_j\, \phi. \tag{1.5}$$

G stands for the Newton constant of gravitation.

Weak Gravitational Theory in n Dimensions

In this paragraph, starting from the n-dimensional Einstein equations for weak gravitational fields (expansion of the metric components), the equation obeyed by the weak fields, and related to the energy density, is derived. The limit of the geodesic equation of motion for test particles in the case of slow velocities and weak gravitational fields is derived and its consequences analyzed.

GAK Linearized Approach

Following GAK, the linearized Einstein equations in the Hilbert–de Donder gauge reduce to the inhomogeneous wave equation

$$\Box\, h_{\alpha\beta} = -2\kappa \left[\tau_{\alpha\beta} - \frac{\tau}{n-2} \eta_{\alpha\beta} \right]. \tag{1.6}$$

For small perturbations,

$$h_{\alpha\beta} = g_{\alpha\beta} - \eta_{\alpha\beta},\ (\eta_{\alpha\beta}) = \mathrm{diag}\,(-1, 1, 1, 1). \tag{1.7}$$

In weak fields, the linearized stresses τ_{ij} (spatial components) are negligible in comparison to the mass density $\tau_{00} = \rho$, and assuming additionally the quasi-staticity of the field and sources, one arrives from (1.6) at

$$\nabla^2\, h_{00} = -2\kappa\, \frac{n-3}{n-2}\, \tau_{00}, \tag{1.8}$$

or, identifying $h_{00} = -2\,\phi$, one gets

$$\nabla^2\, \phi = \kappa\, \frac{n-3}{n-2}\, \rho. \tag{1.9}$$

The geodesic equation

$$\frac{dx^\alpha}{ds^2} + \Gamma^\alpha{}_{\mu\nu}\, \frac{dx^\mu}{ds}\, \frac{dx^\nu}{ds} = 0, \tag{1.10}$$

for slow motion in the linearized limit, becomes

$$\frac{dx^i}{dt^2} - \frac{1}{2}\, \delta^{ij} \partial_j\, h_{00} = 0 \to \frac{dx^i}{dt^2} = -\, \delta^{ij} \partial_j\, \phi. \tag{1.11}$$

Carlip Linearized Approach
In Chapter 1 of Carlip (1998), it is stated that "general relativity in 2+1 dimensions has a Newtonian limit in which there is no force between static point masses." The starting point to establish this assessment is the approximation (1.7), and it continues with the n-dimensional field equations given in the harmonic gauge (which can always be chosen) such that:

$$\begin{aligned} &-\frac{1}{2}\eta^{\mu\nu}\partial_\mu\, \partial_\nu\, \bar{h}_{\alpha\beta} + \mathcal{O}(h^2) = \kappa T_{\alpha\beta}, \\ &\eta^{\mu\nu}\partial_\mu\, \bar{h}_{\nu\beta} = 0, \end{aligned} \tag{1.12}$$

where

$$\bar{h}_{\alpha\beta} = h_{\alpha\beta} - \frac{1}{2}\eta_{\alpha\beta}\, \eta^{\mu\nu} h_{\mu\nu} \to h_{\alpha\beta} = \bar{h}_{\alpha\beta} - \frac{1}{n-2}\eta_{\alpha\beta}\, \eta^{\mu\nu}\bar{h}_{\mu\nu}. \tag{1.13}$$

The Newtonian limit is obtained:

by setting $T_{00} = \rho$, where ρ is the mass density;
by equating to zero all other components of the stress–energy tensor;
by ignoring time derivatives;

then, identifying

$$\bar{h}_{00} = -4\phi, \tag{1.14}$$

the linearized equations (1.12) reduce to

$$\nabla^2 \phi = \frac{\kappa}{2}\, \rho. \tag{1.15}$$

In this limit, the geodesic equation (1.10) reduces to

$$\frac{d^2 x^i}{dt^2} - \frac{1}{2}\, \partial_i\, h_{00} = 0 \tag{1.16}$$

and, taking into account (1.13) and (1.14), becomes

$$\frac{d^2 x^i}{dt^2} = -2 \frac{n-3}{n-2}\, \delta^{ij} \partial_j\, \phi. \tag{1.17}$$

Discrepancies

Comparing these results reported by GAK, Giddings et al. (1984), and Carlip (1998), one notices that the numerical coefficients in the GAK equations for the limits of Newton and particle motion equations correspond to those in the limit of the geodesic motion and to the Newton equation respectively of Carlip (1998).

The Outcome of the Dilemma

A detailed derivation of the linearized Einstein theory is done in 3+1 dimensions by Tonnelat (1959), Chapter 12, §1, 2, 3, which can be extended to $(1 + (n-1))$ dimensions, practically without any changes. In this manner, using the de Donder conditions, $\sigma^\beta = \frac{1}{\sqrt{-g}} \partial_\alpha (\sqrt{-g}\, g^{\alpha\beta}) = 0$ – which allow for the existence of (isothermal) harmonic coordinates and the use of the subclass of quasi-Lorentzian coordinates in the linearization problem of the Einstein equations – one gets the Newton limit in the form (1.15). Moreover, the limit of the geodesic equation for slow motion is also established by Tonnelat (1959), which in terms of the "bar" quantities, in the terminology of MTW, Misner et al. (1973, Chapter 18), gives (1.16), and consequently, in term of the Newtonian potential, it reduces to (1.17).

Weak Gravitational Theory in (2 + 1) Dimensions

Having at hand the linearized expressions of the Einstein equations and of the geodesic equations, in any dimension n, by means of the equations (1.15) and (1.17), correspondingly, one easily recognizes that in $(2+1)$ gravity the Newtonian limit holds in the two spatial dimensions, but the Newtonian acceleration equation fails to be true; the geodesic slow motion occurs without acceleration, $\frac{d^2 x^i}{dt^2} = 0$.

1.1.5 No Geodesic Deviation for Dust

It is apparent that the geodesic deviation for neighboring moving particles has to vanish. Consider a congruence of geodesics with tangent vectors u^α, and let the separation vectors from one geodesic to another be V^μ; then the geodesic deviation equation is

$$\nabla_u \nabla_u V^\alpha = R^\alpha{}_{\beta\gamma\delta} u^\beta u^\gamma V^\delta. \tag{1.18}$$

Assume now that this congruence is modeled by a tube of dust with energy–momentum tensor

$$T^{\alpha\beta} = \rho u^\alpha u^\beta. \tag{1.19}$$

Substituting this tensor into the expression of the Riemann tensor (1.3) and contracting it with u, one gets

$$R^\alpha{}_{\beta\gamma\delta} u^\beta u^\gamma = 0, \tag{1.20}$$

which in turn implies

$$\nabla_u \nabla_u V^\alpha = 0. \tag{1.21}$$

Therefore, in a $(2+1)$ spacetime the world lines of dust do not deviate. In particular, the trajectories of point particles do not approximate to one another; in other words, there is no acceleration between them. This final observation is another effect which is present in $(2+1)$ gravity. Summarizing, in this theory there is *no action at distance:* in a $3D$ spacetime, gravitational effects do not propagate outside the matter content; test particles outside the matter region move along geodesics without experiencing acceleration and geodesic deviation.

1.1.6 No Dynamic Degrees of Freedom

In more than three dimensions, the Weyl tensor encodes the information about the Riemann curvature not caused by matter. Since in $3D$ spacetime the Weyl tensor vanishes – that is, there is no room for it – then, because curvature is only produced by matter, the gravitational field has no dynamic degrees of freedom. Another way to arrive at this absence of degrees of freedom is through a counting argument appealing to the canonical geometrodynamics, as was done in $3D$ by Giddings et al. (1984) and by Carlip (1998) in nD. Roughly speaking, in the canonical geometrodynamics, the spacetime is foliated by means of a one-parameter family of spacelike hypersurfaces, $x^\alpha = x^\alpha(x^a, t)$; thus one can define an intrinsic metric

$$g_{ab} = g_{\alpha\beta} X^\alpha_a X^\beta_b, \; X^\alpha_a := \frac{\partial x^\alpha}{\partial x^a}, \; a = 1, \ldots, n-1, \; \alpha = 1, \ldots, n-1, t,$$

and a field of unit normals U^α, to these hypersurfaces, together with the extrinsic curvatures, $K_{ab} = -U_{\alpha;\beta} X^\alpha_a X^\beta_b$. The Einstein equations are then decomposed with respect to the normal and tangential directions to the hypersurfaces; then we introduce the lapse function N, the shift vector N^a, the intrinsic metric g_{ab} and the momentum conjugate to the metric p^{ab}, defined by means of the extrinsic metric, which are elevated to the category of canonical variables. Einstein equations are then expressed in terms of all of them, with constraints: the projections $G_{\alpha\beta} U^\alpha U^\beta$, (1), $G_{\alpha\beta} U^\alpha X^\beta_a$, $(n-1)$, the evolution eq. for p: $\dot{p}^{ab} \sim G_{ab} X^\alpha_a X^\beta_b$, $n(n-1)/2$, and the evolution eq. for g: $\dot{g}_{ab}$, $n(n-1)/2$.

In n dimensions, the intrinsic metric possesses $n(n-1)/2$ components, and the conjugate momentum also has $n(n-1)/2$ components, and together the number of their components is $n(n-1)$. On the other hand, one can fix n of them by choosing n coordinates, and additionally n by the constraints; consequently, the number of degree of freedom in the canonical data is: $n(n-1)-2n = n(n-3)$. Hence, in three dimensions there is no freedom in the prescription of the initial data in the initial hypersurface.

As a consequence of this lack of degrees of freedom, there are *no gravitational waves in 3D flat spacetime*; in the terminology of Gott and Alpert (1984), there are no gravity waves in flatland; no gravitons.

1.1.7 Black Holes in $(2+1)$ Gravity

So there are no black holes in asymptotically flat spacetime; the asymptotically flat space is flat everywhere, and as such it does not allow for any solution different to the one corresponding to the Minkowskian $3D$ metric. However, let us consider a different situation from the flat asymptotic: for asymptotically anti-de Sitter $(2+1)$ spacetime there exists a black hole solution found by Bañados, Teitelboim, and Zanelli: the BTZ black hole, Bañados et al. (1992). Since that original discovery, various classes of black hole solutions, in the presence of fields and matter, have been reported in the literature.

1.1.8 Gravity in the Presence of Other Fields and Matter

$3D$ gravity in the presence of other fields is worthy of deep study; many achievements have been made since studies began. The present text is simply designed to show the existence of big families of exact solutions in three dimensions and their parallelism, if any, with those classes of the standard $4D$ gravity. For instance, one may attempt a bridge between $3D$ conformally flat metrics and nD conformally flat metrics for incompressible perfect fluids, Friedmann–Robertson–Walker cosmology, and FRW dilaton–inflaton theory, among others.

1.2 Algebraic Classification

In $4D$ gravity, to characterize adequately the gravitational field, there exist two main classifications: the Petrov classification of the Weyl conformal tensor – Petrov types of gravitational fields – and the Plebański–Pirani classification of the tensor of matter or of the traceless Ricci tensor. In (2+1) gravity the situation simplifies considerably: one is dealing with symmetric tensors for the matter from one side, and, to characterize the conformal properties of the metric, the Cotton tensor from the other.

1.2.1 Classification of the Cotton–York Tensor

The role of the conformal tensor in $(2+1)$ gravity is played by the Cotton tensor, Cotton (1899); see Stephani et al. (2003) for a more recent reference. In

$3D$ geometry, the conformal property of the space is guaranteed by the vanishing of the conformal Cotton tensor. This tensor is defined by means of the covariant derivatives of the Ricci tensor and of the scalar curvature according to

$$C^{\alpha\beta} = C^{\beta\alpha} = \eta^{\mu\nu(\alpha} \left(R^{\beta)}{}_{\mu} - \frac{1}{4} R\,\delta^{\beta)}{}_{\mu} \right)_{;\nu}, \tag{1.22}$$

where the symmetry has been introduced explicitly. Notice that the Cotton tensor is traceless:

$$C^{\alpha}{}_{\alpha} = 0. \tag{1.23}$$

To classify the Cotton tensor with respect to its eigenvalues, one has to solve a generalized eigenvalue problem:

$$\left(C^{\alpha\beta} - \lambda\, g^{\alpha\beta}\right) V_{\beta} = 0\,, \quad C^{[\alpha\beta]} = 0\,, \quad C^{\alpha\beta}\, g_{\alpha\beta} = 0\,. \tag{1.24}$$

By lowering one index, one can reformulate this task as an ordinary eigenvalue problem for the matrix $C_{\alpha}{}^{\beta}$. However, in that case, the symmetry $C^{\alpha\beta} = C^{\beta\alpha}$ is no longer present:

$$\left(C_{\alpha}{}^{\beta} - \lambda\, \delta_{\alpha}^{\beta}\right) V_{\beta} = 0\,, \quad C_{\alpha}{}^{\alpha} = 0\,. \tag{1.25}$$

Accordingly, the matrix $C_{\alpha}{}^{\beta}$ is no longer symmetric and the roots of the characteristic polynomial

$$\det\left(C_{\alpha}{}^{\beta} - \lambda\, \delta_{\alpha}^{\beta}\right) = 0 \tag{1.26}$$

may be complex. This point seems to have been overlooked by Barrow et al. (1986).

I will present a classification of $C_{\alpha}{}^{\beta}$ along the lines of Garcia–Hehl–Heineke–Macías (GHHM), García et al. (2004), where the components are referred to an orthonormal basis:

$$g = g_{\alpha\beta} dx^{\alpha}\, dx^{\beta} = \eta_{ab}\, \Theta^{a}\, \Theta^{b},\ (\eta_{ab}) = \mathrm{diag}(-1, 1, 1). \tag{1.27}$$

The trace-free condition $(1.25)_2$ reads explicitly

$$C_1{}^1 + C_2{}^2 + C_3{}^3 = 0\,. \tag{1.28}$$

Accordingly, we can eliminate $C_3{}^3$, e.g., from $(1.25)_1$. Then the secular determinant reads

$$\det \begin{vmatrix} C_1{}^1 - \lambda & C_1{}^2 & C_1{}^3 \\ -C_1{}^2 & C_2{}^2 - \lambda & C_2{}^3 \\ -C_1{}^3 & C_2{}^3 & -C_1{}^2 - C_2{}^2 - \lambda \end{vmatrix} = 0\,, \tag{1.29}$$

with the five matrix elements $C_1{}^2, C_1{}^2, C_1{}^3, C_2{}^2, C_2{}^3$. The equation to determine the eigenvalues λ amounts explicitly to

$$\lambda^3 + b\,\lambda + c = 0\,, \tag{1.30}$$

where

$$b := -(C_1{}^1)^2 - C_1{}^1 C_2{}^2 - (C_2{}^2)^2 + (C_1{}^2)^2 + (C_1{}^3)^2 - (C_2{}^3)^2\,, \tag{1.31}$$

$$c := \left[(C_1{}^1)^2 C_2{}^2 + C_1{}^1 (C_2{}^2)^2 + C_1{}^1 (C_1{}^2)^2 + C_1{}^1 (C_2{}^3)^2 \right. \\ \left. + (C_1{}^2)^2 C_2{}^2 + 2\, C_1{}^2 C_1{}^3 C_2{}^3 - (C_1{}^3)^2 C_2{}^2\right] . \tag{1.32}$$

The roots of (1.29) are given by

$$\lambda_1 = A\,, \quad \lambda_2 = -\frac{A}{2} + i\frac{\sqrt{3}}{2}B\,, \quad \lambda_3 = -\frac{A}{2} - i\frac{\sqrt{3}}{2}B\,, \tag{1.33}$$

with

$$A := \frac{D^2 - 12b}{6D}\,, \quad B := \frac{D^2 + 12b}{6D}\,, \quad D := \left(-108c + 12\sqrt{12b^3 + 81c^2}\right)^{1/3} .$$

A cubic polynomial with real coefficients has at least one real root and the complex roots have to be complex conjugates. The Petrov types, Jordan normal forms and Segré notations of the Cotton tensor read:

Table 1.2.1 *Algebraic classification of the Cotton tensor*

"Petrov" type	Jordan form	Segré notation	eigenvalues relation
I	$\begin{pmatrix} \lambda_1 & 0 & 0 \\ 0 & \lambda_2 & 0 \\ 0 & 0 & -\lambda_1 - \lambda_2 \end{pmatrix}$	[111]	$\lambda_1 \neq \lambda_2,\, \lambda_3 = -\lambda_1 - \lambda_2$
D	$\begin{pmatrix} \lambda_1 & 0 & 0 \\ 0 & \lambda_1 & 0 \\ 0 & 0 & -2\lambda_1 \end{pmatrix}$	[(11)1]	$\lambda_1 = \lambda_2 \neq 0, \lambda_3 = -2\lambda_1$
II	$\begin{pmatrix} \lambda_1 & 1 & 0 \\ 0 & \lambda_1 & 0 \\ 0 & 0 & -2\lambda_1 \end{pmatrix}$	[21]	$\lambda_1 = \lambda_2 \neq 0, \lambda_3 = -2\lambda_1$
N	$\begin{pmatrix} 0 & 1 & 0 \\ 0 & 0 & 0 \\ 0 & 0 & 0 \end{pmatrix}$	[(21)]	$\lambda_1 = \lambda_2 = \lambda_3 = 0$
III	$\begin{pmatrix} 0 & 1 & 0 \\ 0 & 0 & 1 \\ 0 & 0 & 0 \end{pmatrix}$	[3]	$\lambda_1 = \lambda_2 = \lambda_3 = 0$
O	$\begin{pmatrix} 0 & 0 & 0 \\ 0 & 0 & 0 \\ 0 & 0 & 0 \end{pmatrix}$		

This parallels exactly the Petrov classification of the Weyl tensor in four dimensions in Stephani et al. (2003). This comes about since the Weyl tensor in $4D$ is equivalent to a (complex) 3×3 trace-free matrix, as $C_\alpha{}^\beta$ in $3D$; for a similar classification of $C_{\alpha\beta}$, see Hall and Capocci (1999). A detailed derivation of the Cotton tensor in any dimension together with an account of its properties is presented here in Chapter 20; see also García et al. (2004).

1.2.2 Classification of the Energy–Momentum Tensor

The standard classification of the energy–momentum tensor T_{ab} takes advantage of its symmetry property, $T_{ab} = T_{ba}$, where the Latin letters denote the indices with respect to the orthonormal basis (1.27). For that reason the eigenvectors are found by solving the matrix equation

$$T_{ab}\,V^b = \lambda\,\eta_{ab}\,V^b, \rightarrow (T_{ab} - \lambda\,\eta_{ab})\;V^b = 0. \tag{1.34}$$

Searching the values of λ that cancel the determinant of the matrix

$$\begin{bmatrix} T_{11} + \lambda & T_{12} & T_{13} \\ T_{12} & T_{22} - \lambda & T_{23} \\ T_{13} & T_{23} & T_{33} - \lambda \end{bmatrix}, \tag{1.35}$$

namely the roots of the eigenvalue polynomial

$$\begin{aligned} &\lambda^3 + c_2\,\lambda^2 + c_1\,\lambda + c_0 = 0, \\ &c_0 := T_{11},\, T_{22}\,T_{33} - {T_{13}}^2\,T_{22} + 2\,T_{12}\,T_{23}\,T_{13} - {T_{12}}^2\,T_{33} - T_{11}\,{T_{23}}^2, \\ &c_1 := -T_{11}\,T_{22} - T_{11}\,T_{33} + {T_{12}}^2 + T_{22}\,T_{33} + {T_{13}}^2 - {T_{23}}^2, \\ &c_2 := T_{11} - T_{22} - T_{33}, \end{aligned} \tag{1.36}$$

which allows for three roots, with its possible degenerations,

$$\begin{aligned} \lambda_1 &= -\frac{c_2}{3} + \frac{1}{6}\,\sqrt[3]{Rd} - 6\,F\,\sqrt[-3]{Rd}, \\ \lambda_{2,3} &= -\frac{c_2}{3} - \frac{1}{12}\,\sqrt[3]{Rd} + 3\,F\,\sqrt[-3]{Rd} \pm \frac{1}{12}\,i\sqrt{3}\left(\sqrt[3]{Rd} + 36\,F\,\sqrt[-3]{Rd}\right), \\ Rd &:= 36\,c_1\,c_2 - 108\,c_0 - 8\,{c_2}^3 + 12\,D, \\ D &:= \sqrt{12\,{c_1}^3 - 3\,{c_1}^2{c_2}^2 - 54\,c_1\,c_2\,c_0 + 81\,{c_0}^2 + 12\,c_0\,{c_2}^3}, \\ F &= \frac{1}{3}\,c_1 - \frac{1}{9}{c_2}^2, \end{aligned} \tag{1.37}$$

one would be able to determine the eigenvectors corresponding to each root.

The nomenclature used for eigenvectors and algebraic types of tensors is borrowed from Plebański (1964): timelike, spacelike, null, and complex vectors are denoted respectively by **T**, **S**, **N**, and **Z**. For algebraic types are used the symbols: $\{\lambda_1 T, \lambda_2 S_2, \lambda_3 S_3\} \equiv \{T, S, S\}$, meaning that the first real eigenvalue λ_1 gives rise to a timelike eigenvector **T**, the second real eigenvalue λ_2 is associated with a spacelike eigenvector $\mathbf{S}_2$, and finally the third real eigenvalue λ_3 is related to a spacelike eigenvector $\mathbf{S}_3$; for the sake of simplicity I use the symbols $\{T, S, S\}$. It is clear that $\{N, N, S\}$ stands for the algebraic type allowing for two different real eigenvalues giving rise to two null eigenvectors while the third real root is associated with a spacelike eigenvector. When there are single and double real eigenvalues giving rise correspondingly to timelike and spacelike eigenvectors, the algebraic type is denoted by $\{T, 2\,S\}$; consequently, for a triple real eigenvalue, if that were the case, the types could be $\{3T\}$, $\{3N\}$, or $\{3S\}$. For a complex eigenvalue λ_Z, in general, the related eigenvectors are complex and are denoted by **Z** and $\bar{\mathbf{Z}}$ for its complex conjugate; the possible types are $\{T, Z, \bar{Z}\}$, $\{N, Z, \bar{Z}\}$, or $\{S, Z, \bar{Z}\}$.

In standard (3+1) gravity, the classification of the tensor of matter is replaced by the classification of the traceless Ricci tensor $S^{\alpha}{}_{\beta} = R^{\alpha}{}_{\beta} - \frac{1}{4}R\delta^{\alpha}{}_{\beta}$, $S^{\alpha}{}_{\alpha} = 0$; this is because the number of algebraically independent components of the Weyl tensor is 10, of the traceless Ricci tensor is 9, and of the scalar curvature is 1 – which total 20: the number of independent components of the Riemann tensor.

1.2.3 Classification of the Traceless Ricci Tensor

In the same spirit, one may classify the traceless symmetric $3D$ Ricci tensor, leaving aside the scalar curvature:

$$S^{\alpha}{}_{\beta} = R^{\alpha}{}_{\beta} - \frac{1}{3}R\delta^{\alpha}{}_{\beta},\ S^{\alpha}{}_{\alpha} = 0. \tag{1.38}$$

Since this tensor $S^{\alpha}{}_{\beta}$ shears the same properties owed by the Cotton tensor in $3D$, then established algebraic classification for the Cotton tensor extends and holds for the $S^{\alpha}{}_{\beta}$ tensor too. Using the above-proposed notation, the algebraic types for the Cotton tensor can also be given as: $\{T, S, S\}, \{T, N, N\}, \{S, Z, \bar{Z}\}$, and so on.

1.3 Brown–York Energy, Mass, and Momentum for Stationary Metrics

In this section we establish the general forms of the energy and momentum functions for metrics with non-flat anti-de Sitter asymptotics following the Brown–York approach (Brown and York Jr., 1993); see also Brown, Creighton, and Mann (Brown et al., 1994). The (2 + 1)-dimensional stationary cyclic symmetric metric to be used is given by

$$ds^2 = -N^2dt^2 + L^{-2}\,dr^2 + K^2[d\phi + W dt]^2, \tag{1.39}$$

where the structural functions N, L, K, and W depend on the variable r. The timelike vector u_μ normal to the hypersurface $\Sigma : t_\Sigma = \text{const.}$ and the spacelike vector n_μ normal to the surface $\overset{2}{B} : r_{\overset{2}{B}} = R = \text{const.}$ are given by

$$u_\mu = -N\delta^t_\mu,\ u^\mu = \frac{1}{N}\delta^\mu_t - \frac{W}{N}\delta^\mu_\phi,$$
$$n_\mu = \frac{1}{L}\delta^r_\mu,\ n^\mu = L\delta^\mu_r. \tag{1.40}$$

Therefore the projection metrics are:

$$\begin{aligned} ds^2|_{\Sigma:t=\text{const.}} &= L^{-2}\,dr^2 + K^2\,d\phi^2 = h_{ij}dx^i\,dx^j,\\ ds^2|_{\overset{2}{B}:r=R=\text{const.}} &= -N^2\,dt^2 + K^2(d\phi + W dt)^2 = \gamma_{AB}dx^A\,dx^B,\\ ds^2|_{B:t=\text{const.},r=R=\text{const.}} &= K^2\,d\phi^2 = \sigma_{\imath\jmath}dx^\imath\,dx^\jmath. \end{aligned} \tag{1.41}$$

The components of the projection tensor h are explicitly given by

$$h_{\mu\nu} = g_{\mu\nu} + u_\mu u_\nu,$$
$$h_{\mu\nu} = K^2W^2\delta^t_\mu\delta^t_\nu + WK^2(\delta^t_\mu\delta^\phi_\nu + \delta^\phi_\mu\delta^t_\nu) + L^{-2}\delta^r_\mu\delta^r_\nu + K^2\delta^\phi_\mu\delta^\phi_\nu,$$
$$h_\mu{}^\nu = W\delta^t_\mu\delta^\nu_\phi + \delta^r_\mu\delta^\nu_r + \delta^\phi_\mu\delta^\nu_\phi, h_i{}^j = \delta^j_i, \det(h_{ij}) = K^2/L^2. \tag{1.42}$$

The components of the projection tensor γ amount to

$$\gamma_{\mu\nu} = g_{\mu\nu} - n_\mu n_\nu,$$
$$\gamma_{\mu\nu} = -(N^2 - K^2W^2)\delta^t_\mu\delta^t_\nu + WK^2(\delta^t_\mu\delta^\phi_\nu + \delta^\phi_\mu\delta^t_\nu) + K^2\delta^\phi_\mu\delta^\phi_\nu,$$
$$\gamma_\mu{}^\nu = \delta^t_\mu\delta^\nu_t + \delta^\phi_\mu\delta^\nu_\phi, \det(\gamma_{AB}) = -K^2\,N^2. \tag{1.43}$$

Notice that the indices μ, ν associated with the $3D$ spacetime can be replaced by indices A, B running $0 \sim t, 3 \sim \phi$. The components of the projection tensor σ amount to

$$\sigma_{\mu\nu} = g_{\mu\nu} + u_\mu u_\nu - n_\mu n_\nu,$$
$$\sigma_{\mu\nu} = K^2W^2\delta^t_\mu\delta^t_\nu + WK^2(\delta^t_\mu\delta^\phi_\nu + \delta^\phi_\mu\delta^t_\nu) + K^2\delta^\phi_\mu\delta^\phi_\nu,\ \sigma_{\imath\jmath} = K^2\delta^\phi_\imath\delta^\phi_\jmath,$$
$$\sigma^{\mu\nu} = \delta^\mu_\phi\delta^\nu_\phi/K^2,\ \sigma^{\imath\jmath} = \delta^\imath_\phi\delta^\jmath_\phi/K^2,\ \det(\sigma_{\imath\jmath}) = K^2, \tag{1.44}$$

where $\imath$, $\jmath$ run only $3 \sim \phi$.

To evaluate extrinsic curvatures, one needs the expressions of the symmetric Christoffel symbols, which amount to

$$\Gamma^t{}_{tr} = \frac{1}{2N^2}\left(2NN_{,r} - K^2WW_{,r}\right),\ \Gamma^t{}_{r\phi} = -\frac{1}{2N^2}K^2W_{,r},$$
$$\Gamma^r{}_{tt} = L^2\left(NN_{,r} - KW^2K_{,r} - K^2WW_{,r}\right),$$
$$\Gamma^r{}_{t\phi} = -\frac{1}{2}L^2K\left(2WK_{,r} + KW_{,r}\right),$$
$$\Gamma^r{}_{rr} = -\frac{1}{L}L_{,r},\ \Gamma^r{}_{\phi\phi} = -L^2KK_{,r},$$
$$\Gamma^\phi{}_{tr} = \frac{1}{2KN^2}\left(-2WKNN_{,r} + W^2K^3W_{,r} + 2N^2WK_{,r} + N^2KW_{,r}\right),$$
$$\Gamma^\phi{}_{r\phi} = \frac{1}{2KN^2}\left(WK^3W_{,r} + 2N^2K_{,r}\right), \tag{1.45}$$

while all other components vanish. The extrinsic curvature to the hypersurface $\Sigma : t = \text{const.}$ is given by the spatial tensor $K_{\mu\nu}$, namely

$$K_{\mu\nu} = -h^\alpha_\mu\nabla_\alpha u_\nu = -h_{\mu\beta}g^{\beta\alpha}u_{\nu;\alpha}$$
$$K_{\mu\nu} = \frac{1}{2N}K^2WW_{,r}(\delta^t_\mu\delta^r_\nu + \delta^r_\mu\delta^t_\nu) + \frac{1}{2N}K^2W_{,r}(\delta^\phi_\mu\delta^r_\nu + \delta^r_\mu\delta^\phi_\nu),$$
$$K^{\mu\nu} = \frac{1}{2N}L^2W_{,r}(\delta^\mu_\phi\delta^\nu_r + \delta^\mu_r\delta^\nu_\phi),$$
$$K_\mu{}^\nu = \frac{1}{2N}L^2K^2WW_{,r}\delta^t_\mu\delta^\nu_r + \frac{1}{2N}L^2K^2W_{,r}\delta^\phi_\mu\delta^\nu_r + \frac{1}{2N}W_{,r}\delta^r_\mu\delta^\nu_\phi, \tag{1.46}$$

thus the trace of $K_{\mu\nu}$ is zero, $K^{\mu}_{\mu} = 0$.

The momentum tensor $P^{\mu\nu} = \frac{1}{2\kappa}\sqrt{\det(h_{ij})}\,[K^{\alpha}_{\alpha}\,h^{\mu\nu} - K^{\mu\nu}]$ for the hypersurface Σ becomes

$$P^{\mu\nu} = -\frac{1}{4\kappa\,N}LKW_{,r}(\delta^{\mu}_{r}\delta^{\nu}_{\phi} + \delta^{\mu}_{\phi}\delta^{\nu}_{r}), \tag{1.47}$$

while the surface momentum density vector $j_{\mu} = -2\sigma_{\mu\nu}P^{\nu\alpha}n_{\alpha}/\sqrt{\det h_{ij}}$ amounts to

$$j_{\mu} = \frac{1}{2\kappa}\frac{L}{N}K^{2}WW_{,r}\delta^{t}_{\mu} + \frac{1}{2\kappa}\frac{L}{N}K^{2}W_{,r}\delta^{\phi}_{\mu}. \tag{1.48}$$

Consequently, the surface momentum density j_a reduces in the studied case to

$$j_{\phi} = -2\frac{1}{K}\sigma_{\phi\phi}P^{\phi\,r} = \frac{1}{2\kappa}\frac{L}{N}\,K^{2}\,W_{,r} \tag{1.49}$$

modulo the additive constant related to the reference spacetime.

The energy density is evaluated by using the tensor

$$k_{\mu\nu} = -\sigma^{\alpha}_{\mu}h^{\beta}_{\alpha}n_{\lambda;\beta}h^{\lambda}_{\nu} = h_{\mu\alpha}\Theta^{\alpha\beta}h_{\beta\nu},\ \Theta_{\mu\nu} = -\gamma^{\beta}_{\mu}n_{\nu;\,\beta},$$

which amounts to

$$k_{\mu\nu} = -LKWK_{,r}[W\delta^{t}_{\mu}\delta^{t}_{\nu} + 2\delta^{\phi}_{(\mu}\delta^{t}_{\nu)}] - LKK_{,r}\delta^{\phi}_{\mu}\delta^{\phi}_{\nu}, \tag{1.50}$$

while

$$k^{\mu}{}_{\nu} = -\frac{1}{K}LWK_{,r}\delta^{\mu}_{\phi}\delta^{t}_{\nu} - \frac{1}{K}L\,K_{,r}\delta^{\mu}_{\phi}\delta^{\phi}_{\nu}. \tag{1.51}$$

Rising with $\sigma_{\phi\phi} = K^{2} = \sigma^{\phi\phi}$ one of the indexes of the component $k_{\phi\phi} = -LKK_{,r}$ of the extrinsic curvature k associated with the metric of $\overset{2}{B}$, one arrives at

$$k := k^{\imath}_{\imath} = \sigma^{\phi\phi}k_{\phi\phi} = -\frac{1}{K}LK_{,r}. \tag{1.52}$$

Therefore the energy density ϵ becomes

$$\epsilon = \frac{1}{\kappa}k|^{cl}_{0} = -\frac{1}{\kappa\,K}LK_{,r}|_{R} - \epsilon_{0}. \tag{1.53}$$

As far as the integral characteristics are concerned, the total quasilocal energy $E = \int_{B}dx\sqrt{\sigma}\epsilon = 2\pi\,K\,\epsilon$ is given by

$$E = -2\frac{\pi}{\kappa}LK_{,r}|_{R} - 2\pi\,K(R)\epsilon_{0}, \tag{1.54}$$

while the total mass related to the timelike Killing vector $\xi^{\mu} = (\frac{\partial}{\partial t})^{\mu} = \delta^{\mu}_{t}$, $M(\frac{\partial}{\partial t}) = -\int_{B}dx\sqrt{\sigma}(\epsilon\,u_{\mu} + j_{\mu})\xi^{\mu}$ amounts to

$$M(\partial/\partial t) = -2\frac{\pi}{\kappa}NLK_{,r}|_{R} - \frac{\pi}{\kappa}\frac{L}{N}K^{3}WW_{,r}|_{R} - 2\pi\,NK|_{R}\epsilon_{0}. \tag{1.55}$$

Finally, the total momentum $J(\frac{\partial}{\partial \phi}) = \int_B dx\sqrt{\sigma} j_\mu \zeta^\mu$ associated with the Killing vector $\zeta^\mu = (\frac{\partial}{\partial \phi})^\mu = \delta^\mu_\phi$, is given by

$$J(\partial/\partial\phi) = \int_0^{2\pi} d\phi\, j_\phi\, K = \frac{\pi}{\kappa}\frac{L}{N} K^3\, W_{,r}|_R,\; j_\phi = \frac{1}{2\pi}\frac{J(R)}{K(R)}. \tag{1.56}$$

Incidentally, other representations of the mass and momentum density are:

$$M(\partial/\partial t) = N(R)E(R) - W(R)J(R). \tag{1.57}$$

The extrinsic curvature $\Theta_{\mu\nu} = -\gamma^\alpha_\mu \nabla_\alpha n_\nu = -n_{\nu;\alpha}\gamma^\alpha{}_\mu$ of the surface boundary $\overset{2}{B}$ reduces to

$$\begin{aligned}\Theta_{\mu\nu} = &-L(KW^2K_{,r} + K^2WW_{,r} - NN_{,r})\delta^t_\mu\delta^t_\nu \\ &-LK(2WK_{,r} + KW_{,r})\delta^t_{(\mu}\delta^\phi_{\nu)} - LKK_{,r}\delta^\phi_\mu\delta^\phi_\nu,\end{aligned} \tag{1.58}$$

with trace Θ equal to

$$\Theta = -\frac{L}{NK}(KN_{,r} + NK_{,r}), \tag{1.59}$$

which, used in the definition of the boundary momentum

$$\pi^{\mu\nu} = -\frac{1}{2\kappa}\sqrt{-\det\gamma_{AB}}(\Theta\,\gamma^{\mu\nu} - \Theta^{\mu\nu}),$$

taking into account that $\det\gamma_{AB} = -K^2\, N^2$, gives

$$\begin{aligned}\pi^{\mu\nu} = &-\frac{L}{2\kappa\, N}K_{,r}\delta^\mu_t\delta^\nu_t + \frac{L}{2\kappa\, N}(2WK_{,r} + KW_{,r})\delta^{(\mu}_t\delta^{\nu)}_\phi \\ &+\frac{L}{2\kappa\, N\, K}(NN_{,r} - W^2KK_{,r} - K^2WW_{,r})\delta^\mu_\phi\delta^\nu_\phi.\end{aligned} \tag{1.60}$$

This tensor is used in the construction of the stress tensor

$$s^{\alpha\beta} = \frac{2}{\sqrt{\sigma}\, N}\sigma^\alpha_\mu\,\pi^{\mu\nu}\sigma^\beta_\nu - s_0{}^{\alpha\beta}. \tag{1.61}$$

1.3.1 Summary of Quasilocal Mass, Energy, and Angular Momentum

Summarizing, to evaluate quasilocal mass, energy, and angular momentum of spacetimes with asymptotics different from the flat case, in particular the anti-de Sitter, one uses the quasilocal formalism developed in Brown and York Jr. (1993); Brown et al. (1994). For a stationary cyclic symmetric metric of the form

$$g = -N(\rho)^2\, dT^2 + L(\rho)^{-2}d\rho^2 + K(\rho)^2\left[d\Phi + W(\rho)dT\right]^2,$$

the surface energy density is given by

$$\epsilon(\rho) = -\frac{L(\rho)}{\pi\, K(\rho)}\frac{d}{d\rho}K(\rho) - \epsilon_0, \tag{1.62}$$

where ϵ_0 is the reference energy density, which in the case of solutions with a negative cosmological constant $\Lambda = -1/l^2$ corresponds to the density of the anti-de Sitter spacetime, namely $\epsilon_0 = -\frac{1}{\pi\rho}\sqrt{1+\frac{\rho^2}{l^2}}$. The momentum density is determined from

$$j(\rho) = \frac{K(\rho)^2\, L(\rho)}{2\,\pi\, N(\rho)}\frac{d}{d\rho}W(\rho). \tag{1.63}$$

The integral momentum $J(\rho)$, global energy $E(\rho)$, and integral mass $M(\rho)$ are correspondingly given by

$$\begin{aligned} J(\rho) &= 2\pi\, K(\rho)\, j(\rho),\\ E(\rho) &= 2\pi\, K(\rho)\, \epsilon(\rho),\\ M(\rho) &= E(\rho)\, K(\rho) - W(\rho)\, J(\rho). \end{aligned} \tag{1.64}$$

Evaluation of these physical quantities for the studied classes of solutions is presented in the corresponding sections.

1.4 Decomposition with Respect to a Frame of Reference

Barrow, Shaw, and Tsagas (BST), Barrow et al. (2006), developed the covariant decomposition of the spacetime with respect to a family of timelike observers with worldlines tangential to the 3-velocity field u^a in analogy to the standard $(3+1)$ covariant approach to general relativity introduced by Ehlers (1961); see also Ellis and Elst (1999). This topic is included here for the sake of reference; I retain the original presentation and wording with certain minor changes; for instance, the energy density is denoted here by μ instead of ρ. Throughout this section, Latin indices run the spacetime index values.

1.4.1 Kinematics of the Frame

The 3-velocity field determines the time direction and is normalized so that $u_a\, u^a = -1$. The $2D$ space orthogonal to u^a is defined by means of the projection tensor

$$h_{ab} = g_{ab} + u_a\, u_b, \tag{1.65}$$

with properties

$$h_{ab} = h_{(ab)}, h_{ab}u^b = 0, h_{ab}h^b{}_c = h_{ac}, h^a{}_a = 2. \tag{1.66}$$

Using h_{ab} one also defines the covariant derivative operator of the $2D$ space as

$$D_a = h_a{}^b \nabla_b. \tag{1.67}$$

The irreducible kinematic variables, which describe the motion of the observers, are obtained by decomposing the covariant derivative of the 3-velocity. This splitting gives

$$\nabla_a\, u_b = \sigma_{ab} + \omega_{[ab]} + \frac{1}{2}\Theta\, h_{ab} - \dot{u}_a\, u_b, \tag{1.68}$$

where

$$\text{the shear} : \sigma_{ab} = D_{(a}u_{b)},\ \sigma_{ab}u^b = 0, \tag{1.69a}$$

$$\text{the vorticity} : \omega_{ab} = D_{[a}u_{b]},\ \omega_{ab}u^b = 0, \tag{1.69b}$$

$$\text{the area expansion} : \Theta = D^a u_a = \nabla^a\, u_a, \tag{1.69c}$$

$$\text{the acceleration} : \dot{u}_a = u^b \nabla_b\, u_a,\ \dot{u}_a\, u^a. \tag{1.69d}$$

The meaning of these quantities is explained very well in the BST paper: "The tensor

$$D_a\, u_b = h^c_a h^d_b \nabla_c u_d = \sigma_{ab} + \omega_{ab} + \frac{1}{2}\Theta\, h_{ab} \tag{1.70}$$

describes changes in the relative position of the worldlines of two neighboring observers. When the latter follow the motion of a fluid, the effect of Θ is to change the area of a given fluid element, without causing rotation or shape distortion. This scalar also defines the average scale factor a by

$$\frac{\dot{a}}{a} = \frac{1}{2}\Theta. \tag{1.71}$$

The shear monitors distortions in the element's shape that leave the area unaffected, while vorticity describes changes in its orientation under constant area and shape. The symmetric and trace-free nature of σ_{ab} ensures that it has only two independent components, while the antisymmetry of ω_{ab} guarantees that the vorticity tensor is determined by a single component. In other words, the shear and the vorticity correspond to a vector and a scalar respectively. The latter reflects the fact that the rotational axis has been reduced to a point. Defining the $2D$ permutation tensor ϵ_{ab}, with $\epsilon_{ab}u^b = 0$, the vorticity scalar is

$$\omega = \frac{1}{2}\epsilon_{ab}\omega^{ab}, \tag{1.72}$$

with $\omega_{ab} = \omega\epsilon_{ab}$. Note that $\epsilon_{ab} = \eta_{abc}u^c$, by definition, where η_{abc} is the $3D$ totally antisymmetric alternating tensor. The latter satisfies the condition $\eta_{abc}\eta^{pqr} = 3!\delta^p_{[a}\delta^q_b\delta^r_{c]}$, which ensures that $\epsilon_{ab}\epsilon^{cd} = 2h^b_{[a}\,h^d_{c]}$."

1.4.2 Perfect Fluid Referred to a Frame of Reference

Suppose that the matter source of the $3D$ gravitational field governed by the Einstein equations

$$E_{ab} := R_{ab} - \frac{1}{2}g_{ab}R + \Lambda\, g_{ab} - \kappa\, T_{ab} = 0 \tag{1.73}$$

is a perfect fluid. Then, relative to an observer moving with 3-velocity u^a, the energy–momentum tensor of the material component takes the form

$$T_{ab} = (\mu + p)\, u_a\, u_b + p\, g_{ab} = \mu\, u_a\, u_b + p\, h_{ab},\ T = T^a_a = 2p - \mu, \tag{1.74}$$

where μ is the energy density, and p is pressure. Substituting the above into the Einstein field equations (1.73), the latter read

$$R_{ab} = 2(\kappa p - \Lambda)u_a\, u_b + [\kappa(\mu - p) + 2\Lambda]h_{ab},\ R = 2\kappa(\mu - 2p) + 6\Lambda. \tag{1.75}$$

The above also provides the following auxiliary relations:

$$R_{ab}u^a\, u^b = 2(\kappa p - \Lambda),\, h_a^c h_b^d R_{cd} = [\kappa(\mu - p) + 2\Lambda]h_{ab},\ h_a^b R_{bc}u^c = 0. \tag{1.76}$$

The Bianchi identities imply that

$$\nabla^a G_{ab} = 0 \to \nabla^a T_{ab} = 0. \tag{1.77}$$

Explicitly, it reads

$$\begin{aligned} &u_b\, u_a\, \nabla^a\, \mu + (\mu + p)u_b \nabla^a\, u_a + (\mu + p)u_a \nabla^a\, u_b + h_{ab}\, \nabla^a\, p = 0 \\ &\to u_b\, u_a\, \nabla^a\, \mu + (\mu + p)\Theta\, u_b + (\mu + p)\dot{u}_b + D_b\, p = 0. \end{aligned} \tag{1.78}$$

The timelike and spacelike parts of this relation lead to the $3D$ fluid conservation laws:

$$\text{for the energy density}:\ u_a\, \nabla^a\, \mu := \dot{\mu} = -\Theta(\mu + p), \tag{1.79}$$

$$\text{for the momentum density}\ :\ (\mu + p)\dot{u}_a = -D_a p. \tag{1.80}$$

Notice that these conservation laws of a perfect fluid have the same functional form as their $4D$ counterparts.

The intrinsic curvature of the $2D$ space orthogonal to u_a is determined by projecting the Riemann tensor, (1.2), onto the $2D$ space by means of h_a^b:

$$\begin{aligned} \mathcal{R}_{abcd} &= h_a^q h_b^s h_c^f h_d^p R_{qsfp} - v_{ac}v_{bd} + v_{ad}v_{bc}, \\ v_{ab} &= D_b u_a = \sigma_{ab} + \omega_{ab} + \frac{1}{2}\Theta h_{ab}, \end{aligned} \tag{1.81}$$

where v_{ab} is the relative position tensor. Note that v_{ab} characterizes the extrinsic curvature (i.e., the second fundamental form of the space.) Assuming perfect-fluid matter, i.e., equations (1.75), the projected Riemann tensor of the $2D$ (spatial) sections reads

$$\begin{aligned} &\mathcal{R}_{abcd} = (\kappa\, \mu - \frac{1}{4}\Theta^2 + \Lambda)(h_{ac}h_{bd} - h_{ad}h_{bc}) \\ &-\frac{1}{2}\Theta\, [(\sigma_{ac} + \omega_{ac})h_{bd} + (\sigma_{bd} + \omega_{bd})h_{ac} - (\sigma_{ad} + \omega_{ad})h_{bc} - (\sigma_{bc} + \omega_{bc})h_{ad}] \\ &-(\sigma_{ac} + \omega_{ac})(\sigma_{bd} + \omega_{bd}) + (\sigma_{ad} + \omega_{ad})(\sigma_{bc} + \omega_{bc}), \end{aligned} \tag{1.82}$$

with $\mathcal{R}_{abcd} = \mathcal{R}_{[ab][cd]}$. Similarly, as in the standard $(3+1)$ gravity, the isotropic pressure does not contribute to the curvature of the space orthogonal to u^a.

Also, in the absence of anisotropy (i.e., when σ_{ab} and ω_{ab} vanish), the above reduces to

$$\mathcal{R}_{abcd} = (\kappa\,\mu - \frac{1}{4}\Theta^2 + \Lambda)(h_{ac}h_{bd} - h_{ad}h_{bc}). \tag{1.83}$$

Defining

$$\mathcal{R}_{ab} = \mathcal{R}^c{}_{acb} \tag{1.84}$$

as the local $2D$ Ricci tensor, by contracting expression (1.82) one obtains the following $3D$ analogue of the Gauss–Codacci formula

$$\mathcal{R}_{ab} = (\kappa\,\mu - \frac{1}{4}\Theta^2 + \sigma^2 - \omega^2)\,h_{ab}, \tag{1.85}$$

where the shear and rotation scalars are defined correspondingly as

$$2\sigma^2 = \sigma_{ab}\sigma^{ab},\ 2\omega^2 = \omega_{ab}\omega^{ab}. \tag{1.86}$$

In deriving the above result (1.85), the relations

$$\sigma_{c[a}\omega^c{}_{b]} = 0,\ \sigma_{c[a}\sigma^c{}_{b]} = 0,\ \omega_{c[a}\omega^c{}_{b]} = 0 \tag{1.87}$$

have been used. The former holds because $\sigma_{12}(\omega^1{}_1 + \omega^2{}_2) = 0$ (i.e., the single independent component vanishes due to the trace-free nature of the shear). Similarly, the two independent components of

$$\sigma_{c(a}\sigma^c{}_{b)} = 0$$

are also identically zero. Last, the result $\omega_{c(a}\omega^c{}_{b)} = 0$ is guaranteed by the relation $\omega_{ab} = \omega\eta_{ab}$ and the properties of η_{ab}. The absence of a skew and also of a symmetric and trace-free part from Eq. (1.84) agrees with the symmetries of the Riemann tensor in $2D$ spaces.

Finally, the trace of (1.84) leads to the curvature scalar of the spatial sections, which may also be seen as the generalized Friedmann equation for $3D$ spacetimes:

$$\mathcal{R} = \mathcal{R}^a_a = 2(\kappa\,\mu - \frac{1}{4}\Theta^2 + \sigma^2 - \omega^2 + \Lambda). \tag{1.88}$$

Besides these theoretical aspects of fluids, the paper by Barrow et al. (2006) deals with the kinematics of fluid matter, too. However, that material is outside the scope of the present work.

2

Point Particle Solutions

The first publication (Staruszkiewicz, 1963) on $(2+1)$ gravity dated back to 1963, and it was devoted to the description of static solutions determined by point sources.

In $(2+1)$ gravity, it is assumed that the spacetime is $3D$ and obeys the $3D$ Einstein equations. The flat spacetime, $R_{\alpha\beta\gamma\delta} = 0$, is the field solution to the vacuum Einstein equations

$$G_{\mu\nu} = 0 \rightarrow R_{\mu\nu} = 0 = R, \; (T_{\mu\nu} = 0 \text{ and } \Lambda = 0.)$$

In Staruszkiewicz's wording: $3D$ gravitation theory is a theory without a field of gravitation; where no matter is present, space is flat. Curvature can arise only if matter or energy are present. Point particles are the most simple material objects to be considered in $3D$; they move along geodesics, and do not interact with each other; their world-lines are lines of conical singularities.

2.1 Staruszkiewicz Point Source Solutions

This section answers the question: What does the gravitational field of a point mass look like? The source of the Einstein equations is located along a single time-like world line (a geodesic) on which curvature may be present. The spacetime is certainly static, thus for a foliation $t =$ const. the spatial hypersurface is flat except for a single point. Hence, one is dealing with a cone; or, more precisely, with a family of cones with vertexes on the world line of each point particle, if there are many. Choosing the static, circularly symmetric Schwarzschild metric

$$g = -A(r)^2 dt^2 + B(r)^2 dr^2 + r^2 d\phi^2, \; 0 \leq r \leq \infty, \; 0 \leq \phi < 2\pi, \tag{2.1}$$

the Einstein equations $R_{\mu\nu} = 0$ requires $A(r) = A_0 =$ const, and $B(r) = B_0 =$ const. Adopting a new time coordinate $\tau = A_0\, t$ one gets

$$g = -d\tau^2 + B_0^2 dr^2 + r^2 d\phi^2, \; 0 \leq r \leq \infty, \; 0 \leq \phi < 2\pi, \tag{2.2}$$

Next, accomplishing the coordinate transformation

$$r = \rho/B_0,\ \phi = \Phi\, B_0 \tag{2.3}$$

one arrives at the $(2+1)$–Minkowski metric in polar coordinates

$$g = -d\tau^2 + d\rho^2 + \rho^2 d\Phi^2,\ 0 \le \rho \le \infty,\ 0 \le \Phi < 2\pi/B_0. \tag{2.4}$$

The surface $\tau =$ const describes, because of $0 \le \Phi < 2\pi/B_0$, a cone with vertex at $\rho = 0$ and deficit angle def$\Phi := 2\pi(1 - 1/B_0)$.

In Gott and Alpert (1984), the metric (2.2) is embedded as a cone $z(r) = (B_0^2 - 1)^{1/2} r$ in the Minkowski $(3+1)$ spacetime:

$$g = -dt^2 + dz^2 + dr^2 + r^2 d\phi^2,\ \text{cone: } z(r) = (B_0^2 - 1)^{1/2} r. \tag{2.5}$$

2.1.1 Relationship Between the Deficit Angle and Mass

The deficit angle is related to the mass of the point particle. To establish this fact, let us consider a perfect fluid source with energy–momentum tensor $T_\mu^\nu = (\rho + p) u_\mu u^\nu + p\, \delta_\mu^\nu$, $u_\mu = A(r)\delta_\mu^t$. The Einstein equations for the metric (2.1) are

$$\begin{aligned} &-\frac{dB}{dr} + \kappa\, B^3 r\, \rho\,(r) = 0 \to B^{-2} = B_0 - 2\kappa \int_0^r s\, \rho(s)\, d\, s, \\ &\frac{dA}{dr} - \kappa\, p\,(r)\, B^2 A\, r = 0,\ -\frac{d^2 A}{dr^2} B + \frac{dA}{dr}\frac{dB}{dr} + \kappa\, p\,(r)\, A\, B^3 = 0. \end{aligned} \tag{2.6}$$

For dust $p = 0$, one has

$$\begin{aligned} &B^{-2} = B_0 - 2\kappa\, M(r),\ M(r) = \int_0^r s\, \rho(s)\, ds, \\ &A\,(r) = A_0, A\,(r) \to 1. \end{aligned} \tag{2.7}$$

In Cornish and Frankel (1991) one finds that for a point source $\rho = M\delta(\mathbf{r})$, and $p = 0$,

$$(\mathrm{e}^{-2\eta}) = B^{-2} = 1 - \frac{\kappa}{\pi} M,\ \kappa_3 = 2\pi\, G, \tag{2.8}$$

which results from the integration of

$$\begin{aligned} \int_0^r r\, M\, \delta(\mathbf{r}) d\, r &= \frac{1}{2\,\pi} \int_0^{2\pi} \int_0^r r\, M\, \delta(\mathbf{r}) d\, r\, d\, \phi \\ &= \frac{1}{2\,\pi} \int^x \int^y M\, \delta(x)\delta(y) d\, x\, d\, y = \frac{M}{2\,\pi}. \end{aligned}$$

Thus, one has a conical geometry

$$\begin{aligned} &g = -dt^2 + \frac{dr^2}{1 - \frac{\kappa}{\pi} M} + r^2 d\phi^2,\ r = \tilde{r}\sqrt{1 - \frac{\kappa}{\pi} M}, \phi = \tilde{\phi}/\sqrt{1 - \frac{\kappa}{\pi} M}, \\ &g = -dt^2 + d\tilde{r}^2 + \tilde{r}^2 d\tilde{\phi}^2,\ 0 \le \tilde{r} \le \infty,\ 0 \le \tilde{\phi} < 2\pi\sqrt{1 - \frac{\kappa}{\pi} M}, \end{aligned} \tag{2.9}$$

with a deficit angle proportional to the mass M

$$\text{def}\tilde{\phi} := 2\pi\left(1 - \sqrt{1 - \frac{\kappa}{\pi}M}\right) \approx \kappa M + O((\kappa M)^2), \tag{2.10}$$

which geometrically corresponds to remove a wedge of angle $\text{def}\tilde{\phi} \approx \kappa M$ with the identification of the cut edges.

This Dirac δ point source has been analyzed in detail in Cornish and Frankel (1991), where the lack of a Newtonian limit is linked to the absence of acceleration in the geodesic equations for particles initially at rest. Also one finds some lines on this single point particle solution in the appendix of Clément (1976); this article is devoted to extended particles in $2D$ space.

2.2 Staruszkiewicz Single Point Source Solution

A static cyclic symmetric (2 + 1)-dimensional metric can be given always in isotropic coordinates as

$$g = -A(x,y)^2 dt^2 + e^{2\,\eta(x,y)}\left(dx^2 + dy^2\right). \tag{2.11}$$

The Einstein equations $G_\mu^\nu = 0$ yield

$$\begin{aligned} A(x,y) &= A_0 \to 1, \\ G_0^0 &= e^{-2\eta}\left(\frac{\partial^2\eta}{\partial y^2} + \frac{\partial^2\eta}{\partial x^2}\right) = 0 \to \eta = F(x+iy) + F^*(x+iy), \end{aligned} \tag{2.12}$$

because of the flat Laplace character of the partial equation for $\eta(x,y)$. In particular, for a single point particle located at the point (x_1, y_1), one may choose

$$\eta(x,y) = \ln\left[(x-x_1)^2 + (y-y_1)^2\right]^{-\alpha}. \tag{2.13}$$

Thus, the metric (2.11) becomes

$$g = -dt^2 + \left[(x-x_1)^2 + (y-y_1)^2\right]^{-2\alpha}\left(dx^2 + dy^2\right). \tag{2.14}$$

Accomplishing the coordinate transformation $x = x_1 + r\cos\theta$, $y = y_1 + r\sin\theta$, one gets

$$g = -dt^2 + r^{-2\alpha}\left(dr^2 + r^2\,d\theta^2\right), \tag{2.15}$$

carrying out a second transformation $r = \rho^a$ one arrives at

$$\begin{aligned} g &= -dt^2 + a^2\,\rho^{2a(1-\alpha)-2} d\rho^2 + \rho^{2a(1-\alpha)}\,d\theta^2 \\ \to g &= -dt^2 + a^2\,d\rho^2 + \rho^2\,d\theta^2, \text{ for } a(1-\alpha) = 1. \end{aligned} \tag{2.16}$$

Comparing with the conical metric (2.9) one establishes that

$$a = 1/\sqrt{1 - \frac{\kappa}{\pi}M},\ \alpha = 1 - \sqrt{1 - \frac{\kappa}{\pi}M} \approx \frac{\kappa}{2\pi}M,$$

$$g = -dt^2 + d\rho^2/\sqrt{1 - \frac{\kappa}{\pi}M} + \rho^2\, d\theta^2, \tag{2.17}$$

which explicitly gives the relation between the exponent α for the single point particle solution in isotropic coordinates with a mass M:

$$\begin{aligned} g &= -dt^2 + \left[(x - x_1)^2 + (y - y_1)^2\right]^{-2\alpha} \left(dx^2 + dy^2\right), \\ \alpha &= 1 - \sqrt{1 - \frac{\kappa}{\pi}M} = \frac{\mathrm{def}\theta}{2\pi}. \end{aligned} \tag{2.18}$$

2.2.1 No Parallelism With the (3 + 1) *Schwarzschild Solution*

The Schwarzschild solution is thought of as an external gravitational field produced by a static mass. Nevertheless, it is well known that the origin of the Schwarzschild coordinates, $r = 0$, is a singularity, which, all the worse, is spacelike, and as such cannot be associated with a Dirac δ point source.

Assume for a while that the Schwarzschild mass is located at the origin $r = 0$, as it is in the book by Ortín (2004), (§7.2), p. 202, and that it follows a world-line with timelike 4-velocity u_μ. Hence, one can represent it by means of an energy–momentum δ-tensor: $T_{\mu\nu} = m\, u_\mu\, u_\nu \delta(\vec{x})$. The Einstein equations in Schwarzschild coordinates evaluated in the limit as $r \to 0$ yield not only the G_{tt} different from zero proportional to $T_{tt} = m\delta^{(3)}(r)$, but the remaining diagonal components are proportional to a δ-function; this is related to the spacelike nature of the Schwarzschild singularity.

2.3 Staruszkiewicz Two Point Sources Solution

The static cyclic (2 + 1)-dimensional metric for two point masses can be given in isotropic coordinates as

$$\begin{aligned} g &= -dt^2 + \mathrm{e}^{2\,\eta(x,y)} \left(dx^2 + dy^2\right) = -dt^2 + B(x,y)^2 \left(dx^2 + dy^2\right), \\ B\,(x,y) &= B_0{}^2 \left[(x - x_1)^2 + (y - y_1)^2\right]^{-\alpha_1/2} \left[(x - x_2)^2 + (y - y_2)^2\right]^{-\alpha_2/2}, \end{aligned} \tag{2.19}$$

where (x_1, y_1) and (x_2, y_2) stand for the spatial location of the first and second point masses m_1 and m_2, respectively, $\alpha_1 = \frac{m_1}{\kappa\pi}$ and $\alpha_2 = \frac{m_2}{\kappa\pi}$. The proper Staruszkiewicz two point sources solution corresponds to the arrangement of the masses along the real axis of z, i.e., with $y_1 = 0 = y_2$.

2.4 Deser–Jakiw–'t Hooft Static $\mathcal{N}$ Point Sources Solution

The static $\mathcal{N}$ point sources solution derived by Deser, Jackiw and 't Hooft (Deser et al., 1984) is described, in isotropic coordinates, by the metric

$$g = -dt^2 + e^{2\,\eta(x,y)}\left(dx^2 + dy^2\right) = -dt^2 + \Phi(x,y)\left(dx^2 + dy^2\right). \qquad (2.20)$$

The Einstein tensor corresponding to this metric is given by:

$$G^{\mu}_{\nu} = \delta^{\mu}_{0}\,\delta^{0}_{\nu}\,G^{0}_{0},\; G^{0}_{0} = e^{-2\eta}\left(\frac{\partial^2\eta}{\partial y^2} + \frac{\partial^2\eta}{\partial x^2}\right) = e^{-2\eta}\Delta\eta = \frac{1}{2\,\Phi}\Delta\ln\Phi,$$
$$G = G^{\mu}_{\mu} = -\frac{1}{2}R = G^{0}_{0}. \qquad (2.21)$$

For vacuum $G^{\mu}_{\nu} = 0$, the Cartesian Laplace equation allows for real solutions of the form

$$\eta = F(x+iy) + F(x-iy) = F(z) + F(z^*), \Phi = f(z)\,f(z^*), z := x+iy,$$

which made evident the flatness of the metric in this case

$$g = -dt^2 + f(z)f(z^*)\,dz\,dz^* = -dt^2 + dZ\,d\,Z^*, Z := \int^z f(z)\,dz.$$

For n static point particles, the DJ'tH solution is given by

$$\Phi(x,y) = {C_0}^2\prod_n |r - r_n|^{-\alpha_n} = {C_0}^2\prod_n\left[(x-x_n)^2 + (y-y_n)^2\right]^{-\alpha_n/2},$$
$$2\eta = \ln\Phi = -\frac{\kappa}{\pi}\sum_n m_n\ln|r - r_n| + \ln C_0^2,\; \alpha_n := \frac{\kappa}{\pi}\,m_n, \qquad (2.22)$$

where $\kappa = 8\pi\,G_E$. This choice is dictated by the δ-function character of the point sources; from the properties of the Dirac δ-function; see for instance Infeld and Plebański (1960), one has

$$n = 3 : \Delta\ln r = 2\,\pi\,\delta^2(\mathbf{r}),$$
$$n \geq 3 : \Delta\frac{1}{r^{n-2}} = -(n-2)\Omega_n\,\delta^n(\mathbf{r}), \qquad (2.23)$$

where Ω_n is equal to the area of the sphere S_n of unitary radius. Hence, for the DJ'tH solution one gets

$$\Delta\ln\Phi = -\frac{\kappa}{\pi}\sum_n m_n\Delta\ln|r - r_n| = -2\kappa\sum_n m_n\delta^2(\mathbf{r} - \mathbf{r}_n). \qquad (2.24)$$

2.4.1 *Energy and Euler Invariant*

The parameters m_k stand for the masses of the point particles. Notice that

$$\sqrt{-g}\,G^0_0 = -\frac{1}{2}\sqrt{-g}\,R = \frac{1}{2}\Delta\ln\phi, \qquad (2.25)$$

hence, the energy and Euler invariant, see Eqs. (3.1–3.2) of Deser et al. (1984) and comments therein, read

$$2\kappa E = -2\int d^2x\sqrt{-g}\,G^0_0 = \int d^2x\sqrt{-g}R = -\int d^2x\Delta\ln\Phi$$
$$= 2\kappa\int d^2x\sum_n m_n\delta^2(\mathbf{r}-\mathbf{r}_n) = 2\kappa\sum_n m_n. \tag{2.26}$$

2.4.2 Energy–Momentum Tensor for $\mathcal{N}$ Point Particles

For point particles with masses m_k located at rest at $\mathbf{r}_k$ the energy–momentum tensor may be given as

$$T_{\mu\nu} = m(x,y)\,\delta^0_\mu\delta^0_\nu, T^\mu_\nu = -\,m(x,y)\,\delta^\mu_0\delta^0_\nu, T = T^\mu_\mu = -\,m(x,y),$$

The Einstein equation $G^0_0 = \kappa T^0_0$ yields

$$\frac{1}{2\Phi}\Delta\ln\Phi = -\kappa\, m\,(x,y) \rightarrow -\frac{1}{2\Phi}2\kappa\sum_n m_n\delta^2(\mathbf{r}-\mathbf{r}_n) = -\kappa\, m\,(x,y).$$

Therefore

$$T_{\mu\nu} = \Phi^{-1}\sum_n m_n\delta^2(\mathbf{r}-\mathbf{r}_n)\,\delta^0_\mu\delta^0_\nu = \Phi^{-1}\,T_{\mu\nu\,DJtH} \tag{2.27}$$

which is a weighted version of the $T^{\mu\nu}$ of Deser et al. (1984); compare with DJ'tH(2.5).

2.5 Clément Rotating Point–Particles Solution

Clément (1985) reported two families of stationary solutions to $3D$ vacuum Einstein equations: the point particles class and a second class which suffers unsatisfactory physical interpretation. Following Clément, the stationary metric (changing the signature) is given as

$$ds^2 = -h^2(dt+\omega_i dx^i)^2 + \bar{g}_{ij}dx^i\,dx^j, \tag{2.28}$$

referring to Landau and Lifshitz (1970), where the structural functions depend on space coordinates x^i. The arbitrariness of the vector potential ω_i, via the transformation $t \rightarrow t + F(r)$, can be restricted to

$$\omega^i{}_{;i} = 0, \tag{2.29}$$

and the spatial indices are raised and lowered by means of $\bar{g}_{ij}$. The vacuum Einstein equations can be written as

$$R_{00} = 0: \; -h\,h^{;i}{}_{;i} + \frac{h^4}{4}f_{ij}f^{ij} = 0, \tag{2.30a}$$

$$R^i_0 = 0: \; -\frac{1}{2}h^{-1}\left(h^3 f^{ij}\right)_{;j} = 0, \tag{2.30b}$$

$$R^{ij} = 0 : \ -\frac{h^2}{2}\, f^{ik} f^j{}_k - h^{-1} h^{;i;j} + \bar{R}^{ij} = 0, \tag{2.30c}$$

where

$$f_{ij} = \omega_{j;i} - \omega_{i;j}, \tag{2.31}$$

and $\bar{R}^{ij}$ is constructed with the spatial metric $\bar{g}^{ij}$. In $3D$ spacetime, for the spatial metric one can choose the isotropic coordinates such that

$$\bar{g}_{ij} = \mathrm{e}^{2\,u} \delta_{ij}. \tag{2.32}$$

The equation (2.29) yields

$$\omega_{i,j} = 0 \rightarrow \omega_i = \epsilon_{ij} \partial_j \phi \rightarrow f_{ij} = -\epsilon_{ij} \Delta\phi, \tag{2.33}$$

where Δ is the $2D$ flat Laplace operator.

The equations (2.30b) leads to

$$\epsilon_{ij} \left(\mathrm{e}^{-2\,u}\, h^3\, \Delta\phi\right)_{,j} = 0 \rightarrow \Delta\phi = \lambda_0\, \mathrm{e}^{2\,u}\, h^{-3}, \tag{2.34}$$

where λ_0 is a constant. Substituting this result in (2.30a), one gets

$$h^3\, \Delta\, h = -\frac{\lambda_0^2}{2}\, \mathrm{e}^{2\,u}. \tag{2.35}$$

The trace and the traceless part of (2.30c) give

$$\mathrm{e}^{-2\,u}\, \Delta\, u = \frac{3}{4} {\lambda_0}^2\, h^{-4}, \tag{2.36}$$

$$h_{;i;j} - \frac{1}{2} h^{;k}{}_{;k} \bar{g}_{ij} = 0, \tag{2.37}$$

Introducing the complex variable $z = x + iy$, the equation (2.37) becomes

$$\frac{\partial^2}{\partial z^2} h = 2 \frac{\partial u}{\partial z} \frac{\partial h}{\partial z} \rightarrow \frac{\partial h}{\partial z}(z, \bar{z}) = \mu(\bar{z})\, \mathrm{e}^{2\,u(z,\bar{z})}, \tag{2.38}$$

where $\mu(\bar{z})$ is an arbitrary function. The most interesting class of solutions arises for $\mu(\bar{z}) = 0$; then h is a constant, which without loss of generality can be chosen equal to 1, $h = 1$. Then, the equation (2.35) implies $\lambda_0 = 0$. Consequently, the Einstein equations, (2.34) and (2.36), become

$$\Delta\phi = 0 = \Delta\, u. \tag{2.39}$$

The multicenter solutions of these equations

$$\begin{aligned} \phi &= -\frac{\kappa}{2\pi} \sum_\alpha J_\alpha \ln|z - a_\alpha| \\ u &= -\frac{\kappa}{2\pi} \sum_\alpha m_\alpha \ln|z - a_\alpha| \end{aligned} \tag{2.40}$$

(where $\kappa = 8\pi G$) lead to the metric

$$ds^2 = -\left(dt + \frac{\kappa}{2\pi}\sum_\alpha J_\alpha \frac{\epsilon_{ij}\,(x^i - a^i{}_\alpha)\,dx^j}{|\boldsymbol{r} - \boldsymbol{a}_\alpha|^2}\right)^2 + \prod_\alpha |\boldsymbol{r} - \boldsymbol{a}_\alpha|^{-\kappa m_\alpha/\pi} d\boldsymbol{r}\cdot d\boldsymbol{r}, \tag{2.41}$$

representing a system of point particles of "masses" m_α, and "spins" J_α.

In the case of a single massless spinning point particle at the origin, the above metric gives rise to

$$ds^2 = -\left(dt + \frac{\kappa}{2\pi}\,J\,d\theta\right)^2 + dr^2 + B^2\,r^2 d\theta^2, \tag{2.42}$$

therefore one recovers the rotating Deser–Jakiw–t'Hooft solution (Deser et al., 1984). An interpretation of this spacetime is given in Carlip (1998) Eqs. (3.20)–(3.25), where the coordinates $\{t, \theta\}$ are subjected to the transformations

$$t \to t' = t + \frac{\kappa\,J}{2\pi}\,\theta,\ \ \theta \to \theta' = B\,\theta,$$

therefore, the spacetime experiences a conical singularity together with a timeshift.

Of course, each one of these point particle solutions possesses a vanishing Riemann tensor as a consequence of the vacuum character of the spacetimes, hence all these multiple point particle spaces are Minkowski flat; although it may sound redundant, the Cotton tensor has zero value.

3

Dust Solutions

This chapter is devoted to the derivation of dust solutions in (2 + 1) gravity, i.e., solutions to Einstein's equations for an energy momentum tensor $T_{\alpha\beta} = \mu(x^\nu)u_\alpha u_\beta$, where u_α, $u_\alpha u^\alpha = -1$, is the dust particles' velocity.

3.1 Cornish–Frankel Dust Heaviside Function Solution

In Cornish and Frankel (1991) there were reported extended static Heaviside function dust distrinbutions of the form

$$\kappa\,\mu\,(r) = 2\pi\,G\frac{M}{\pi {R_0}^2}H(r - R_0),\ \kappa := 2\pi\,G, \tag{3.1}$$

with embedding line element

$$ds^2 = -dt^2 + dz^2 + dr^2 + r^2\,d\theta^2,$$

for the embedding function z

$$z(r) = \frac{R_0}{\sqrt{\kappa\,M/\pi}}\left[1 - \sqrt{1 - \kappa\frac{M}{\pi}\frac{r^2}{R_0^2}}\right],\ 0 \leq r \leq R_0, \tag{3.2}$$

which gives a (2 + 1) metric

$$ds^2 = -dt^2 + \left(1 - \kappa\frac{M}{\pi}\frac{r^2}{R_0^2}\right)^{-1} dr^2 + r^2\,d\theta^2,\ 0 \leq r \leq R_0,$$

with value at the border

$$z(r = R_0) = \frac{R_0}{\sqrt{\kappa\,M/\pi}}(1 - \sqrt{1 - \kappa\,M/\pi}),$$

and energy momentum tensor for dust

$$T_{\alpha\beta} = \mu(r)\delta^t_\alpha\delta^t_\beta.$$

On the other hand, the external region has an embedding function Z given by

$$Z(r) = r\frac{\sqrt{\kappa M/\pi}}{\sqrt{1-\kappa M/\pi}} - R_0\frac{1}{\sqrt{\kappa M/\pi}\sqrt{1-\kappa M/\pi}}\left[1-\sqrt{1-\kappa\frac{M}{\pi}}\right],$$
$$r \geq R_0, \tag{3.3}$$

such that

$$\begin{aligned} ds^2 &= -dt^2 + dZ^2 + dr^2 + r^2\,d\theta^2 \\ &= -dt^2 + \left(1-\kappa\frac{M}{\pi}\right)^{-1} dr^2 + r^2\,d\theta^2,\ r \geq R_0,\ Z(r=R_0) = z(r=R_0). \end{aligned}$$

with zero energy momentum tensor, although a conical singularity is present.

3.2 Giddings–Abott–Kuchař Dust Solutions

This solution has been sought for irrotational dust, which geometrically means that the world lines of the dust are hyperface forming, i.e., the field of dust velocity is tangential to the generatrices of the hypersurface. Consequently, following the Giddings–Abott–Kuchař work in Giddings et al. (1984), one can introduce a Gaussian coordinate system attached to the world lines of the dust such that the time coordinate is comoving with the dust. Since any $2D$ space is always conformally flat space, the isotropic coordinates can be used, hence the metric to deal with is

$$g = -dt^2 + e^{2\,\Omega(t,x,y)}\,\left(dx^2 + dy^2\right), \tag{3.4}$$

thus the corresponding Einstein equations for dust $T_{\lambda\nu} = \kappa\, m(t,x,y)\,\delta^t_\lambda \delta^t_\nu$, are

$$E_t^t :\rightarrow e^{-2\Omega}\left(\frac{\partial^2}{\partial y^2}\Omega + \frac{\partial^2}{\partial x^2}\Omega\right) - \left(\frac{\partial}{\partial t}\Omega\right)^2 = -\kappa\, m\,(t,x,y)\,, \tag{3.5a}$$

$$E_x^t = 0 = E_t^x :\rightarrow \frac{\partial^2}{\partial x \partial t}\Omega = 0 \rightarrow \Omega = F2(t,y) + F1(x,y)\,, \tag{3.5b}$$

$$E_y^t = 0 = E_t^y :\rightarrow \frac{\partial^2}{\partial y \partial t}\Omega = 0 \rightarrow \Omega = F3(t,x) + F1(x,y)\,, \tag{3.5c}$$

$$\begin{aligned} E_x^x &= 0 = E_y^y, \rightarrow \frac{\partial^2}{\partial t^2}\Omega + \left(\frac{\partial}{\partial t}\Omega\right)^2 = 0, \rightarrow \frac{\partial}{\partial t}(\frac{\partial}{\partial t}\Omega)^{-1} = 1 \\ \rightarrow \Omega &= \ln\left(t\,f(x,y) + g(x,y)\right). \end{aligned} \tag{3.5d}$$

One may distinguishes two branches of solutions:
a) the time-dependent class $f(x,y) \neq 0$,
b) and the static case $f(x,y) = 0$.

3.2.1 Time-Dependent Class of Dust Solutions $\Omega = \ln(tf(x,y))$

The equations (3.5b) and (3.5c), after substitution of Ω from (3.5d), yield

$$\frac{\partial}{\partial x}\frac{g}{f} = 0 = \frac{\partial}{\partial y}\frac{g}{f} \to g = c_0 f \to \Omega = \ln\left((t + c_0)\, f(x,y)\right);$$

hence, shifting the time coordinate, one has without lost of generality

$$\Omega(t,x,y) = \ln\left(t\, f(x,y)\right). \tag{3.6}$$

Substituting this $\Omega(t,x,y)$ into (3.5a) one arrives at

$$\begin{aligned} & f^{-2}(\frac{\partial^2}{\partial y^2}\ln f + \frac{\partial^2}{\partial x^2}\ln f) = 1 - \kappa t^2\, m\;(t,x,y) \\ & \to \kappa\, m\;(t,x,y) = \frac{1}{t^2}(1 - f^{-2}\nabla^2 \ln f). \end{aligned} \tag{3.7}$$

Moreover, the conservation law $T^{\mu\nu}{}_{;\nu} = 0$ gives rise to a single equation for $m(t,x,y)$, namely

$$\frac{\partial m}{\partial t} + 2\, m\frac{\partial}{\partial t}\Omega = 0 \to \frac{\partial m}{\partial t} + 2\frac{m}{t} = 0 \to m(t,x,y) = \frac{1}{t^2}\mu_0(x,y). \tag{3.8}$$

Therefore, the structural function $f(x,y)$ is the only remaining to be determined function subjected to the partial nonlinear differential equation

$$f^{-2}\nabla^2 \ln f = 1 - \kappa\, \mu_0\;(x,y)\,. \tag{3.9}$$

The Riemann tensor is equal to

$$R_{\alpha\beta\mu\nu} = R_{3232}\,\delta^3_{[\alpha}\,\delta^2_{\beta]}\delta^3_{[\mu}\delta^2_{\nu]} = \kappa\,\mu_0\,(x,y)\, f^4 t^2\,\delta^3_{[\alpha}\,\delta^2_{\beta]}\delta^3_{[\mu}\delta^2_{\nu]}, \tag{3.10}$$

$$R_{3232} = -\left[e^{-2\Omega}\left(\frac{\partial^2}{\partial y^2} + \frac{\partial^2}{\partial x^2}\right)\Omega - \left(\frac{\partial}{\partial t}\Omega\right)^2\right] f\,(x,y)^4\, t^4 = \kappa\,\mu_0 f^4 t^2. \tag{3.11}$$

The evaluation of the Cotton yields

$$(C^\alpha{}_\beta) = 2\kappa \begin{bmatrix} 0 & \frac{1}{4}\frac{\partial m}{\partial y} & -\frac{1}{4}\frac{\partial m}{\partial x} \\ -\frac{1}{4}\,e^{-2\Omega}\frac{\partial m}{\partial y} & 0 & \frac{m}{2}\frac{\partial \Omega}{\partial t} + \frac{1}{4}\frac{\partial m}{\partial t} \\ \frac{1}{4}\,e^{-2\Omega}\frac{\partial m}{\partial x} & -\frac{m}{2}\frac{\partial \Omega}{\partial t} - \frac{1}{4}\frac{\partial m}{\partial t} & 0 \end{bmatrix}, \tag{3.12}$$

thus, taking into account that $\Omega = tf(x,y)$, and $\mu = \mu_0(x,y)/t^2$, one gets

$$(C^\alpha{}_\beta) = \frac{\kappa}{2\,t^2} \begin{bmatrix} 0 & \frac{\partial}{\partial y}\mu_0(x,y) & -\frac{\partial}{\partial x}\mu_0(x,y) \\ -\,e^{-2\Omega}\frac{\partial}{\partial y}\mu_0(x,y) & 0 & 0 \\ e^{-2\Omega}\frac{\partial}{\partial x}\mu_0(x,y) & 0 & 0 \end{bmatrix}. \tag{3.13}$$

The search for its eigenvectors yields

$$\lambda_1 = 0; \mathbf{V1} = [1,0,0],\ V1_\mu = -\delta^t_\mu,\ V1^\mu V1_\mu = -1,\ \mathbf{V1} = \mathbf{T1},$$

$$\lambda_2 = i\,\mathrm{e}^{-\Omega}\sqrt{(\frac{\partial\mu_0}{\partial x})^2 + (\frac{\partial\mu_0}{\partial y})^2}; \mathbf{V2} = [1, -\mathrm{e}^{-\Omega}\frac{\partial\mu_0}{\partial y}\frac{1}{\lambda_2}, \mathrm{e}^{-\Omega}\frac{\partial\mu_0}{\partial x}\frac{1}{\lambda_2}] = \mathbf{Z},$$

$$\lambda_3 = \bar{\lambda}_2;\ \mathbf{V3} = \bar{\mathbf{Z}}, \tag{3.14}$$

therefore the corresponding tensor type is Type I: $\{T, Z, \bar{Z}\}$.

The eigenvectors **V2** and **V3** are complex conjugated, while the zero eigenvalue λ_1 gives raise to a timelike eigenvector **V1**, which points in the t-coordinate direction.

3.2.2 Static Class of Dust Solutions $\Omega = \ln g(x, y)$

The only remaining equation to be solved is the resulting from equation (3.5a) from the substitution of $\Omega = \ln g(x, y)$, namely

$$g^{-2}\nabla^2 \ln g = -\kappa\, m\,(x, y)\,. \tag{3.15}$$

Thus, for a given structural function $g(x, y)$ one determines the density $m(x, y)$, which is required to fulfill the energy conditions, among them, the positiveness of the mass. This backward way of looking at this problem is an alternative too. For instance, for a set of point sources – see the previous section 2.4 and Deser et al. (1984) – one arrives at

$$g(x, y)^2 = C_0{}^2 \prod_n |r - r_n|^{-\frac{\kappa}{\pi} m_n};\ \frac{1}{g^2}\Delta \ln g = -\kappa\, m\,(x, y)\,,$$

$$m\,(x, y) = \frac{2}{g^2}\sum_n m_n \delta^2(\mathbf{r} - \mathbf{r}_n). \tag{3.16}$$

3.3 Barrow–Shaw–Tsagas Anisotropic Dust Solution; $\Lambda = 0$

The simplest $(2+1)$ homogeneous and anisotropic Bianchi I–type universe, in the terminology of Barrow et al. (2006), is given by the spacetime metric

$$ds^2 = -dt^2 + A^2(t)dx^2 + B^2(t)dy^2. \tag{3.17}$$

For this metric, the Einstein equations for dust, with energy momentum tensor

$$T_{\alpha\beta} = \rho(t)\, u_\alpha\, u_\beta,\ u_\alpha = \delta^t_\alpha,$$

yield

$$E^t_t = 0 \rightarrow \kappa\,\rho(t) = \frac{\dot{A}\dot{B}}{AB}, \tag{3.18}$$

$$E^x_x = 0 \rightarrow \frac{d^2 B}{dt^2} = 0 \rightarrow B(t) = B_1 t + B_0, \tag{3.19}$$

$$E^y_y = 0 \to \frac{d^2 A}{dt^2} = 0 \to A(t) = A_1 t + A_0, \tag{3.20}$$

where A_i, B_i, $i = 0, 1$ are integration constants.

The area expansion Θ and the shear σ scalars are

$$2\sigma(t) = \frac{\dot{A}}{A} - \frac{\dot{B}}{B}, \tag{3.21a}$$

$$-\Theta = \frac{\dot{A}}{A} + \frac{\dot{B}}{B}. \tag{3.21b}$$

The evaluation of the dust energy density gives

$$\kappa\,\rho\,(t) = \frac{B_1\,A_1}{(B_1\,t + B_0)\;(A_1\,t + A_0)}, \tag{3.22}$$

while the shear and the area expansion scalars are

$$2\,\sigma = \frac{A_1\,B_0 - B_1\,A_0}{(A_1\,t + A_0)\,(B_1\,t + B_0)}, \tag{3.23}$$

$$-\Theta = \frac{2\,A_1\,B_1\,t + A_1\,B_0 + B_1\,A_0}{(A_1\,t + A_0)\,(B_1\,t + B_0)}. \tag{3.24}$$

Moreover, from the energy conservation law $T^{\alpha\beta}{}_{;\beta} = 0$ one establishes

$$\frac{d\rho}{dt} = -\left(\frac{\dot{A}}{A} + \frac{\dot{B}}{B}\right)\rho = \Theta\,\rho, \tag{3.25}$$

and for σ, using (3.21a) and (3.21b) and the fact that solutions arise for the field equations $\ddot{A} = 0$ and $\ddot{B} = 0$, one gets

$$\frac{d\sigma}{dt} = -\left(\frac{\dot{A}}{A} + \frac{\dot{B}}{B}\right)\sigma = \Theta\,\sigma. \tag{3.26}$$

For $\Theta = 2\dot{a}/a$, where $a(t)$ is the "geometric–mean scale factor", the solution for $a(t)$ obeys

$$2\,\frac{\dot{a}}{a} = -(\frac{\dot{A}}{A} + \frac{\dot{B}}{B}) \to a(t)^2 = \frac{\alpha_0}{A(t)B(t)}. \tag{3.27}$$

For the A(t), and B(t) above, (3.20) and (3.19), one has

$$a(t)^2 = \frac{\alpha_0}{(A_1 t + A_0)\,(B_1 t + B_0)}, \tag{3.28}$$

and for the choice

$$A(t) = A_0\frac{t}{t_0},\; B(t) = B_0\frac{\kappa\rho_0 t + 2\sigma_0}{\kappa\rho_0 t_0 + 2\sigma_0}. \tag{3.29}$$

one has

$$a(t)^2 = \frac{\alpha_0 t_0(\kappa\rho_0 t_0 + 2\sigma_0)}{A_0 B_0}\frac{1}{t\,(\kappa\rho_0 t + 2\sigma_0)} =: a_0^2\frac{1}{t\,(\kappa\rho_0 t + 2\sigma_0)}, \tag{3.30}$$

which is the inverse of the "geometric–mean scale factor" $a(t)$ reported in Barrow et al. (2006).

In the cited work is given an alternative representation of the solutions derived above following the standard integration of the Einstein equations. Using the energy conservation law and the $3D$ analogue of Raychaudhuri's equation, the dynamical equations in the case of a perfect fluid obeying a barotropic state equation $p = w\rho$, where w is the barotropic index, are given in the form of

$$\dot{\rho} = -(1+w)\Theta\,\rho, \tag{3.31a}$$

$$\dot{\Theta} + \frac{1}{2}\Theta^2 = -2\kappa\, w\,\rho - 2\sigma^2 \tag{3.31b}$$

$$\dot{\sigma} + \Theta\sigma = 0, \tag{3.31c}$$

$$\omega = 0, \tag{3.31d}$$

with the constraint

$$\frac{1}{4}\Theta^2 = \kappa\rho + \sigma^2. \tag{3.32}$$

For definitions of the area expansion, shear and vorticity, correspondingly Θ, σ, and ω; see Sec. 1.4.

To my mind, there is a discrepancy in the sign at the right-hand side of the equation (3.31a). To establish this fact, let us review the energy density conservation law:

$$u^\alpha \nabla_\alpha \rho = -(\rho+p)u^\alpha{}_{;\alpha} = -(\rho+p)\Theta, \tag{3.33}$$

since, for the u^α aligned along ∂_t

$$u^\alpha \nabla_\alpha = -\frac{d}{dt},$$

therefore

$$-\frac{d}{dt}\rho = -(\rho+p)\,\Theta \to \frac{d\rho}{dt} = (\rho+p)\,\Theta,\ \frac{d\rho}{dt} = (1+w)\rho\,\Theta. \tag{3.34}$$

The solutions of this last equation for $\Theta = 2\dot{a}/a$, and of (3.31c), are

$$\rho(t) = \rho_0\left(\frac{a(t)}{a_0}\right)^{2(1+w)}, \sigma = \sigma_0\left(\frac{a(t)}{a_0}\right)^2 \tag{3.35}$$

which certainly are inverse to the ones of Barrow et al. (2006): there, integrating (3.31a) and (3.31c) for $\Theta = 2\dot{a}/a$, one arrives at

$$\text{energy density}: \rho = \rho_0\left(\frac{a_0}{a}\right)^{2(1+w)}, \quad \text{and shear: } \sigma = \sigma_0\left(\frac{a_0}{a}\right)^2.$$

For the specific problem under consideration, the sign of Θ in Eq. (38) of Barrow et al. (2006) is opposite to the correct one calculated here for (3.21b).

Cotton Tensor
The Cotton tensor is

$$C^{\alpha}{}_{\beta} = \frac{1}{2}\frac{B_1\,A_1\,(B_1\,A_0 - A_1\,B_0)}{(B_1\,t + B_0)\,(A_1\,t + A_0)}\left[\frac{\delta^{\alpha}{}_{x}\delta^{y}{}_{\beta}}{A(t)^2} + \frac{\delta^{\alpha}{}_{y}\delta^{x}{}_{\beta}}{B(t)^2}\right]. \tag{3.36}$$

Searching for its eigenvectors, one establishes that they are

$$\lambda_1 = 0, T^{\alpha} = \delta^{\alpha}{}_{t}, T^{\alpha}\,T_{\alpha} = -1, \tag{3.37a}$$

$$\lambda_2 = \frac{1}{2}\frac{B_1\,A_1\,(B_1\,A_0 - A_1\,B_0)}{A(t)^2B(t)^2},\ S1^{\alpha} = \frac{\delta^{\alpha}{}_{x}}{A(t)} + \frac{\delta^{\alpha}{}_{y}}{B(t)}, S1^{\alpha}\,S1_{\alpha} = 2, \tag{3.37b}$$

$$\lambda_3 = -\frac{1}{2}\frac{B_1\,A_1\,(B_1\,A_0 - A_1\,B_0)}{A(t)^2B(t)^2},\ S2^{\alpha} = -\frac{\delta^{\alpha}{}_{x}}{A(t)} + \frac{\delta^{\alpha}{}_{y}}{B(t)}, S2^{\alpha}\,S2_{\alpha} = 2, \tag{3.37c}$$

where T and S are used to denote respectively timelike and spacelike vectors, therefore, this Cotton tensor is of Type I: $\{T, S, S\}$.

3.4 BST Diagonal Anisotropic Dust Solutions with Λ

Homogeneous and anisotropic $(2+1)$ universes in the presence of a cosmological constant are given by the metric

$$ds^2 = -dt^2 + A^2(t)dx^2 + B^2(t)dy^2. \tag{3.38}$$

Einstein equations for dust with Λ are

$$E_{\mu\nu} := G_{\mu\nu} + \Lambda\,g_{\mu\nu} - \kappa T_{\mu\nu} = 0,\ T_{\mu\nu} = \rho(t)\,u_{\mu}\,u_{\nu},\ u_{\mu} = \delta^{t}{}_{\mu}, \tag{3.39}$$

which for the metric (3.38) explicitly read

$$E_t^t : \kappa\,\rho\,(t) + \Lambda = \frac{\left(\frac{d}{dt}B\right)\frac{d}{dt}A}{BA}, \tag{3.40a}$$

$$E_x^x : \frac{d^2}{dt^2}B\,(t) - \Lambda\,B\,(t) = 0 \to B\,(t) = B_1\,e^{\sqrt{\Lambda}t} + B_2\,e^{-\sqrt{\Lambda}t}, \tag{3.40b}$$

$$E_y^y : \frac{d^2}{dt^2}A\,(t) - \Lambda\,A\,(t) = 0 \to A\,(t) = A_1\,e^{\sqrt{\Lambda}t} + A_2\,e^{-\sqrt{\Lambda}t}. \tag{3.40c}$$

The evaluation of the characteristic functions ρ, σ and Θ yields

$$\kappa\,\rho\,(t) = -2\,\Lambda\,\frac{(B_1\,A_2 + B_2\,A_1)}{\left(B_1\,e^{\sqrt{\Lambda}t} + B_2 e^{-\sqrt{\Lambda}t}\right)\left(A_1\,e^{\sqrt{\Lambda}t} + A_2 e^{-\sqrt{\Lambda}t}\right)}, \tag{3.41a}$$

$$\sigma(t) = \frac{\sqrt{\Lambda}\,(B_2\,A_1 - B_1\,A_2)}{\left(B_1\,e^{\sqrt{\Lambda}t} + B_2 e^{-\sqrt{\Lambda}t}\right)\left(A_1\,e^{\sqrt{\Lambda}t} + A_2 e^{-\sqrt{\Lambda}t}\right)}, \tag{3.41b}$$

$$\Theta = 2\,\frac{\sqrt{\Lambda}\left(-A_1B_1\,e^{2\sqrt{\Lambda}t} + A_2B_2\,e^{-2\sqrt{\Lambda}t}\right)}{\left(B_1\,e^{\sqrt{\Lambda}t} + B_2\,e^{-\sqrt{\Lambda}t}\right)\left(A_1\,e^{\sqrt{\Lambda}t} + A_2\,e^{-\sqrt{\Lambda}t}\right)}. \tag{3.41c}$$

3D Gödel Universes

One can bring the derived solutions a Gödel-like form by introducing the trigonometric functions, namely

$$\begin{aligned} A(t, \Lambda < 0) &= a_1 \sin \sqrt{|\Lambda|}t + a_2 \cos \sqrt{|\Lambda|}t, \\ B(t, \Lambda < 0) &= b_1 \sin \sqrt{|\Lambda|}t + b_2 \cos \sqrt{|\Lambda|}t, \end{aligned} \tag{3.42a}$$

$$\begin{aligned} A(t, \Lambda > 0) &= a_1 \sinh \sqrt{\Lambda}t + a_2 \cosh \sqrt{\Lambda}t, \\ B(t, \Lambda > 0) &= b_1 \sinh \sqrt{\Lambda}t + b_2 \cosh \sqrt{\Lambda}t. \end{aligned} \tag{3.42b}$$

3.5 BST (t, x, y)-Dependent Cosmological Solutions with Comoving Dust

In this section, anisotropic dust solutions with structural functions depending on the (t, x, y) coordinates are derived. Thus, the class of metrics of interest is of the form

$$ds^2 = -dt^2 + A(t, x, y)^2 \, dx^2 + B(t, x, y)^2 \, dy^2. \tag{3.43}$$

The Einstein equations for dust with velocity $u_\mu = \delta^t_\mu$ are

$$E_{tx} : -\left(\frac{\partial^2}{\partial x \partial t} B\right) A + \left(\frac{\partial}{\partial t} A\right) \frac{\partial}{\partial x} B = 0, \tag{3.44a}$$

$$E_{ty} : \left(\frac{\partial^2}{\partial y \partial t} A\right) B - \left(\frac{\partial}{\partial t} B\right) \frac{\partial}{\partial y} A = 0, \tag{3.44b}$$

$$E_{xx} : \frac{\partial^2}{\partial t^2} B - \Lambda \, B = 0, \tag{3.44c}$$

$$E_{yy} : \frac{\partial^2}{\partial t^2} A - \Lambda \, A = 0, \tag{3.44d}$$

with the energy density resulting from equation $E_{tt} = 0$, namely

$$\begin{aligned} [\kappa \, \rho(t, x, y) + \Lambda] A^3 B^3 &= -A^2 \, B \frac{\partial^2 A}{\partial y^2} - B^2 \, A \frac{\partial^2 B}{\partial x^2} + A^2 \, B^2 \frac{\partial B}{\partial t} \frac{\partial A}{\partial t} \\ &\quad + B^2 \frac{\partial A}{\partial x} \frac{\partial B}{\partial x} + A^2 \frac{\partial A}{\partial y} \frac{\partial B}{\partial y}, \end{aligned} \tag{3.45}$$

with structural functions depending on the three variables t, x, and y.

The equations (3.44c) and (3.44d) integrate as

$$\begin{aligned} A(t, x, y) &= A_1(x, y) \, e^{\sqrt{\Lambda}t} + A_2(x, y) \, e^{-\sqrt{\Lambda}t}, \\ B(t, x, y) &= B_1(x, y) \, e^{\sqrt{\Lambda}t} + B_2(x, y) \, e^{-\sqrt{\Lambda}t}, \end{aligned} \tag{3.46}$$

which determine the structure of the metric functions of all the classes of solutions with time dependence t via a cosmological constant Λ in terms of hyperbolic $\sinh \sqrt{\Lambda}t$ and $\cosh \sqrt{\Lambda}t$ functions for positive $\Lambda = |\Lambda|$, or through trigonometric

$\sin\sqrt{|\Lambda|}t$ and $\cos\sqrt{|\Lambda|}t$ functions for negative $\Lambda = -|\Lambda| = -1/l^2$. When replacing the exponential functions in terms of the hyperbolic, or trigonometric functions depending on the sign of the Λ, the coefficient functions in front of the sines and cosines are linear combinations of the previous coefficients multiplying the exponentials (in the negative Λ these coefficients are complex conjugated); for them, Greek symbols will be used, and one has these possibilities: for positive Λ: $\Lambda > 0$, one gets

$$\begin{aligned} A(t,x,y) &= \alpha_1(x,y)\,\sinh\sqrt{\Lambda}t + \alpha_2(x,y)\,\cosh\sqrt{\Lambda}t, \\ B(t,x,y) &= \beta_1(x,y)\,\sinh\sqrt{\Lambda}t + \beta_2(x,y)\,\cosh\sqrt{\Lambda}t, \end{aligned} \tag{3.47a}$$

while for negative Λ: $\Lambda = -|\Lambda|$, one obtains

$$\begin{aligned} A(t,x,y) &= \alpha_1(x,y)\,\sin\sqrt{|\Lambda|}t + \alpha_2(x,y)\,\cos\sqrt{|\Lambda|}t, \\ B(t,x,y) &= \beta_1(x,y)\,\sin\sqrt{|\Lambda|}t + \beta_2(x,y)\,\cos\sqrt{|\Lambda|}t. \end{aligned} \tag{3.47b}$$

Therefore, the metric structure becomes

$$\begin{aligned} ds^2 = -dt^2 &+ \left[A1(x,y)\,\mathrm{e}^{\sqrt{\Lambda}t} + A2(x,y)\,\mathrm{e}^{-\sqrt{\Lambda}t}\right]^2 dx^2 \\ &+ \left[B1(x,y)\,\mathrm{e}^{\sqrt{\Lambda}t} + B2(x,y)\,\mathrm{e}^{-\sqrt{\Lambda}t}\right]^2 dy^2, \end{aligned} \tag{3.48}$$

where Λ may be considered positive or negative. The remaining Einstein equations for the above functions read

$$E^t_x : -A2(x,y)\,\frac{\partial}{\partial x}B1(x,y) + A1(x,y)\,\frac{\partial}{\partial x}B2(x,y) = 0, \tag{3.49}$$

$$E^t_y : -B2(x,y)\,\frac{\partial}{\partial y}A1(x,y) + B1(x,y)\,\frac{\partial}{\partial y}A2(x,y) = 0. \tag{3.50}$$

Since there are only two constraints on four unknown structural functions, then one may consider two of them as arbitrary; let them be *A1* and *A2*. The most general solutions of these equations, $B1(x,y)$ and $B2(x,y)$, depending on both spatial variables x and y, are

$$B2(x,y) = B1(x,y)\,\frac{\frac{\partial A2}{\partial y}}{\frac{\partial A1}{\partial y}}, \tag{3.51}$$

$$B1(x,y) = b(y)\exp\int \frac{A1\left[\left(\frac{\partial^2 A2}{\partial y\partial x}\right)\frac{\partial A1}{\partial y} - \frac{\partial A2}{\partial y}\left(\frac{\partial^2 A1}{\partial y\partial x}\right)\right]}{\left(\frac{\partial A1}{\partial y}\right)\left[A2\left(\frac{\partial A1}{\partial y}\right) - A1\left(\frac{\partial A2}{\partial y}\right)\right]}dx. \tag{3.52}$$

Of course, certain conditions on the behavior of the derivatives of *A1* and *A2* have to be imposed.

The area expansion Θ and shear constant σ are given by

$$\Theta = -2\,\frac{\sqrt{\Lambda}\left(B1e^{2\sqrt{\Lambda}t}A1 - B2e^{-2\sqrt{\Lambda}t}A2\right)}{\left(A1e^{\sqrt{\Lambda}t} + A2e^{-\sqrt{\Lambda}t}\right)\left(B1e^{\sqrt{\Lambda}t} + B2e^{-\sqrt{\Lambda}t}\right)}, \tag{3.53}$$

$$\sigma = \frac{\left(-B1A2 + A1B2\right)\sqrt{\Lambda}}{\left(A1e^{\sqrt{\Lambda}t} + A2e^{-\sqrt{\Lambda}t}\right)\left(B1e^{\sqrt{\Lambda}t} + B2e^{-\sqrt{\Lambda}t}\right)}. \tag{3.54}$$

Assume we are working with the hyperbolic functions; then

$$\begin{aligned} A(t,x,y) &= F1(x,y)\sinh\left(\sqrt{\Lambda}t\right) + F2(x,y)\cosh\left(\sqrt{\Lambda}t\right), \\ B(t,x,y) &= G1(x,y)\sinh\left(\sqrt{\Lambda}t\right) + G2(x,y)\cosh\left(\sqrt{\Lambda}t\right), \end{aligned} \tag{3.55}$$

constrained to the pair of equations

$$G2(x,y)\,\frac{\partial}{\partial y}F1(x,y) - G1(x,y)\,\frac{\partial}{\partial y}F2(x,y) = 0, \tag{3.56a}$$

$$F1(x,y)\,\frac{\partial}{\partial x}G2(x,y) - F2(x,y)\,\frac{\partial}{\partial x}G1(x,y) = 0, \tag{3.56b}$$

thus two functions should stay arbitrary; let them be $F1(x,y)$ and $F2(x,y)$. From the (3.56a) equation one gets

$$G1(x,y) = \frac{\frac{\partial}{\partial y}F1(x,y)}{\frac{\partial}{\partial y}F2(x,y)}\,G2(x,y) \tag{3.57}$$

which, substituted into the (3.56b) and integrating for $G2$, yields

$$G2(x,y) = g_2(y)\,\exp\int\frac{F2\left[\left(\frac{\partial^2}{\partial y\partial x}F2\right)\frac{\partial}{\partial y}F1 - \left(\frac{\partial^2}{\partial y\partial x}F1\right)\frac{\partial}{\partial y}F2\right]}{\left(\frac{\partial}{\partial y}F2\right)\left[F2\frac{\partial}{\partial y}F1 - F1\frac{\partial}{\partial y}F2\right]}dx \tag{3.58}$$

with arbitrary functions $F1$ and $F2$. Notice that the integration function $g_2(y)$ can be equated to 1 by accomplishing a coordinate transformation in y.

Algebraic Classification of the Cotton Tensor

For this class of metric the Cotton tensor is given by

$$\left(C^{\alpha}{}_{\beta}\right) = \frac{\kappa}{2\,A\,B}\begin{bmatrix} 0 & A^2\frac{\partial}{\partial y}m & -B^2\frac{\partial}{\partial x}m \\ -\frac{\partial}{\partial y}m & 0 & \frac{\partial}{\partial t}\left(m\,B^2\right) \\ \frac{\partial}{\partial x}m & -\frac{\partial}{\partial t}\left(m\,A^2\right) & 0 \end{bmatrix}. \tag{3.59}$$

The secular equation for the eigenvalue S is given by

$$S^3 - \left(C^2_3\,C^3_2 + C^2_1\,C^1_2 + C^3_1\,C^1_3\right)S - C^3_1\,C^1_2\,C^2_3 - C^2_1\,C^3_2\,C^1_3 = 0,$$

where the correspondence with the coordinates is $\{1,2,3\} \to \{t,x,y\}$: for the structural functions determining this metric, none of the factors of this equation is zero, hence, in general there exist three solutions for S, such that $S_1+S_2+S_3 = 0$; consequently the Cotton tensor is of Type I.

3.5.1 BST Class 2 of Solutions

Another representation of this class of solutions has been given in the same article by Barrow et al. (2006) as the Class 2. First, the structural functions *F1* and *F2* are chosen as

$$F2\,(x,y) = \mathrm{e}^{\nu(x,y)},\ F1\,(x,y) = \mathrm{e}^{\nu(x,y)}F\,(x,y)\,, \tag{3.60}$$

which substituted into (3.56a) determines

$$G1\,(x,y) = \frac{\left[\left(\frac{\partial \nu}{\partial y}\right)F + \frac{\partial F}{\partial y}\right]}{\left(\frac{\partial \nu}{\partial y}\right)}\,G2\,(x,y)\,. \tag{3.61}$$

Entering *G1* into (3.56b) and integrating for *G2* one arrives at

$$G2 = \frac{1}{S(y)}\frac{\left(\frac{\partial \nu}{\partial y}\right)}{\left(\frac{\partial F}{\partial y}\right)}\,\mathrm{e}^{-\alpha(x,y)},\ \alpha\,(x,y) := \int^x\left(\frac{\partial F}{\partial x}\right)\frac{\left(\frac{\partial \nu}{\partial y}\right)}{\left(\frac{\partial F}{\partial y}\right)}dx. \tag{3.62}$$

Finally, the sought solution can be given as

$$\begin{aligned}
ds^2 &= -dt^2 + A\,(t,x,y)^2\,dx^2 + B\,(t,x,y)^2\,dy^2,\\
A\,(t,x,y) &= \mathrm{e}^{\nu(x,y)}\left[F\,(x,y)\sinh\left(\sqrt{\Lambda}t\right) + \cosh\left(\sqrt{\Lambda}t\right)\right],\\
B\,(t,x,y) &= \frac{\mathrm{e}^{-\alpha(x,y)}}{S(y)}\frac{\frac{\partial \nu}{\partial y}}{\frac{\partial F}{\partial y}}\left[\left(F + \frac{\frac{\partial F}{\partial y}}{\frac{\partial \nu}{\partial y}}\right)\sinh\left(\sqrt{\Lambda}t\right) + \cosh\left(\sqrt{\Lambda}t\right)\right],\\
\alpha\,(x,y) &= \int^x\left(\frac{\partial F}{\partial x}\right)\frac{\frac{\partial \nu}{\partial y}}{\frac{\partial F}{\partial y}}dx,
\end{aligned} \tag{3.63}$$

where $F(x,y)$ and $\nu(x,y)$ are arbitrary functions $\frac{\partial F}{\partial y} \neq 0$, and $S(y)$ is an integration function, which can be set equal to unity by a redefinition of the y-coordinate. The evaluation of the energy density can done by means of the expression (3.45). This general dust solution corresponds to the one given in paragraph 6.2 Class 2 in Barrow et al. (2006), through

$$\begin{aligned}
ds^2 = &-dt^2 + \mathrm{e}^{-2\nu(x,y)}\frac{[\partial_y(R(x,y,t)\mathrm{e}^{\nu(x,y)})]^2}{S^2(y)(\partial_y F(x,y))^2}\,\mathrm{e}^{-2\alpha(x,y)}dy^2\\
&+R^2(x,y,t)\mathrm{e}^{2\nu(x,y)}\,dx^2,
\end{aligned}$$

$$R(x,y,t) = \mathrm{ch}_\Lambda t + F(x,y)\mathrm{sh}_\Lambda t,$$
$$\mathrm{sh}_\Lambda t := \frac{1}{\sqrt{\Lambda}}\sinh\sqrt{\Lambda}t, \mathrm{ch}_\Lambda t := \cosh\sqrt{\Lambda}t. \quad (3.64)$$

Szekeres Solutions

Szekeres solutions arise for $\partial_x F = 0$, and

$$\mathrm{e}^{-\nu(x,y)} = A(y)x^2 + 2B(y)x + C(y).$$

3.5.2 BST Class 1 Spacetime

Apart from the general solutions above, there is a subclass of solutions arising for $A1(x,y) = a1(x)$ and $A2(x,y) = a2(x)$ depending on the single spatial variable x. In such case, the equation (3.50) is identically fulfilled, while equation (3.49) allows the solutions

$$B1(x,y) = \int a1(x)\,L(x,y)\,dx + b1(y),$$
$$B2(x,y) = \int a2(x)\,L(x,y)\,dx + b2(y), \quad (3.65)$$

where $L(x,y)$ is an arbitrary function. In this representation, the energy density amounts to

$$\kappa\,\rho = -\frac{\left(2\,\Lambda\,a2(x)\,B1 + 2\,\Lambda\,B2a1(x) + \frac{\partial}{\partial x}L\right)\mathrm{e}^{2\sqrt{\Lambda}t}}{\left(\mathrm{e}^{2\sqrt{\Lambda}t}B1 + B2\right)\left(a2(x) + \mathrm{e}^{2\sqrt{\Lambda}t}a1(x)\right)}, \quad (3.66)$$

or

$$\kappa\,\rho + \Lambda = \frac{-\frac{\partial}{\partial x}L + \Lambda\left(\mathrm{e}^{\sqrt{\Lambda}t}B1 - B2\mathrm{e}^{-\sqrt{\Lambda}t}\right)\left(\mathrm{e}^{\sqrt{\Lambda}t}a1 - a2\mathrm{e}^{-\sqrt{\Lambda}t}\right)}{\left(\mathrm{e}^{\sqrt{\Lambda}t}B1 + B2\mathrm{e}^{-\sqrt{\Lambda}t}\right)\left(\mathrm{e}^{2\sqrt{\Lambda}t}a1 + a2\mathrm{e}^{-\sqrt{\Lambda}t}\right)}, \quad (3.67)$$

to maintain the parallelism with the results of Barrow et al. (2006).

In Barrow et al. (2006) this solution was reported in the alternative form

$$ds^2 = -dt^2 + [C(x)\mathrm{sh}_\Lambda(t) + D(x)\mathrm{ch}_\Lambda(t)]^2 dx^2$$
$$+[W(x,y)\mathrm{sh}_\Lambda(t) + V(x,y)\mathrm{ch}_\Lambda(t)]^2 dy^2,$$
$$W(x,y) = \int_{x_0}^{x} C(\xi)\frac{\partial L(\xi,y)}{\partial\xi}d\xi + f(y),$$
$$V(x,y) = \int_{x_0}^{x} D(\xi)\frac{\partial L(\xi,y)}{\partial\xi}d\xi + g(y), \quad (3.68)$$

with dust energy density

$$\kappa\,\rho + \Lambda = \frac{-\partial_x^2 L(x,y) + [W\mathrm{ch}_\Lambda(t) + \Lambda\,V\mathrm{sh}_\Lambda(t)][C\mathrm{ch}_\Lambda(t) + \Lambda\,D\mathrm{sh}_\Lambda(t)]}{[W\mathrm{sh}_\Lambda(t) + V\mathrm{ch}_\Lambda(t)][C\mathrm{sh}_\Lambda(t) + D\mathrm{ch}_\Lambda(t)]},$$

where the dependence on x, and y has been omitted. Notice that the definitions of the functions $L(x, y)$ differ; $L_{(3.65)} = \partial_x L_{(3.68)}$.

3.5.3 BST Class 3 of Dust Solutions

This class of solutions arises for the choice $A1(x, y) = \alpha_0\, A2(x, y)$, which substituted into equation (3.50) yields $B1(x, y) = \alpha_0\, B2(x, y)$; for these relations the equation (3.49) is immediately satisfied. The metric amounts to

$$ds^2 = -dt^2 + (\alpha_0\, e^{\sqrt{\Lambda t}} + e^{-\sqrt{\Lambda t}})^2\, [A_2(x, y)^2 dx^2 + B_2(x, y)^2\, dy^2]. \tag{3.69}$$

Since any $2D$ space is conformally flat, one may write the spatial sector of this metric as

$$e^{2\,\phi(x,y)}(\alpha_0\, e^{\sqrt{\Lambda t}} + e^{-\sqrt{\Lambda t}})^2\, (dx^2 + dy^2). \tag{3.70}$$

Thus the spacetime for positive $\Lambda > 0$ is

$$ds^2 = -dt^2 + e^{2\,\phi(x,y)}(b_0\, \sinh\sqrt{\Lambda}t + a_0\, \cosh\sqrt{\Lambda}t)^2\, (dx^2 + dy^2), \tag{3.71a}$$

where α_0 and β_0 are constants, with energy density $\rho(t, x, y)$ given by

$$\kappa\, \rho = -\frac{\Lambda\, ({a_0}^2 - {b_0}^2) + [\frac{\partial^2}{\partial y^2}\nu\,(x, y) + \frac{\partial^2}{\partial x^2}\nu\,(x, y)]e^{-2\,\nu(x,y)}}{[b_0\, \sinh\sqrt{\Lambda}t + a_0\, \cosh\sqrt{\Lambda}t]^2} \tag{3.72}$$

and scalar area expansion

$$\Theta = -2\,\sqrt{\Lambda}\frac{a_0\, \sinh\left(\sqrt{\Lambda}t\right) + b_0\, \cosh\left(\sqrt{\Lambda}t\right)}{a_0\, \cosh\left(\sqrt{\Lambda}t\right) + b_0\, \sinh\left(\sqrt{\Lambda}t\right)}. \tag{3.73}$$

The proper BST formulation is given by

$$ds^2 = -dt^2 + e^{2\phi(x,y)}(\mu\, \mathrm{sh}_\Lambda t + \mathrm{ch}_\Lambda t)^2\, (dx^2 + dy^2), \tag{3.74}$$

with dust energy density

$$\kappa\, \rho(x, y) + \Lambda = \frac{(\mu \mathrm{ch}_\Lambda t + \Lambda \mathrm{sh}_\Lambda t)^2 - e^{-2\phi(x,y)}\nabla^2\phi(x, y)}{(\mu\, \mathrm{sh}_\Lambda t + \mathrm{ch}_\Lambda t)^2}. \tag{3.75}$$

The shear tensor and rotation tensor are zero. The FRW solution emerges when solution arises for $e^{-2\phi(x,y)}\nabla^2\phi(x, y) = \text{constant}$.

The second solution, for negative cosmological constant $\Lambda = -L^2$, is given by

$$ds^2 = -dt^2 + e^{2\,\nu(x,y)}[b_0\, \sin(L\, t) + a_0\, \cos(L\, t)]^2\, (dx^2 + dy^2), \tag{3.76}$$

with energy density

$$\kappa\, \rho = \frac{L^2({a_0}^2 + {b_0}^2) - e^{-2\,\nu(x,y)}[\frac{\partial^2}{\partial x^2}\nu\,(x, y) + \frac{\partial^2}{\partial y^2}\nu\,(x, y)]}{[a_0\, \cos(Lt) + b_0\, \sin(Lt)]^2}, \tag{3.77}$$

and scalar area expansion

$$\Theta = L\,\frac{\left(a_0{}^2 - b_0{}^2\right)\sin\left(2\,Lt\right) - 2b_0\,a_0\,\cos\left(2\,Lt\right)}{\left[a_0\,\cos\left(Lt\right) + b_0\,\sin\left(Lt\right)\right]^2}. \tag{3.78}$$

The shear tensor vanishes.

3.6 Rooman–Spindel Dust Gödel Non-Diagonal Model

Rooman and Spindel (1998) reported various families of dust models depending on one spatial coordinate, exhibiting certain properties belonging to the 4D Gödel solution.

The metric in comoving coordinates is given by Eq.(80) in Rooman and Spindel (1998) as

$$ds^2 = -[dt + Z(r)d\phi]^2 + dr^2 + Y(r)^2 d\phi^2. \tag{3.79}$$

The conservation of the energy momentum tensor $T_{\mu\nu} = (\mu + p)u_\mu u_\nu + p\,g_{\mu\nu}$ for a perfect fluid (dust) moving along the time direction ∂_t,

$$u^\mu = -\delta^\mu{}_t,\; u_\mu = \delta_\mu{}^t + Z\delta_\mu{}^\phi, \tag{3.80}$$

requires the constancy of the pressure, $\frac{dp}{dr} = 0, p(r) = p_0$.

The Einstein equation

$$G^\phi{}_t = 0 : Y\,\frac{d^2 Z}{dr^2} - \frac{dZ}{dr}\frac{dY}{dr} = 0, \tag{3.81}$$

integrates as

$$Y\left(r\right) = C_1\,\frac{d}{dr}Z\left(r\right), \tag{3.82}$$

while $G^r{}_r = 0 = (G^\phi{}_\phi)$ gives the relation

$$\kappa\,p_0 = \Lambda + \frac{1}{4C_1{}^2} = \text{constant}, \tag{3.83}$$

and the equation $G^t{}_t = 0 = (G^t{}_\phi)$ becomes

$$\frac{d^3 Z}{dr^3} + \left(\kappa\,\mu\left(r\right) + \Lambda - \frac{3}{4C_1{}^2}\right)\frac{d\,Z}{dr} = 0. \tag{3.84}$$

Although the obtained solution possesses a constant pressure throughout the entire spacetime, it is endowed with matter modeled through the energy density subject to the equation (3.84). As pointed out in Rooman and Spindel (1998), this equation can be thought of as defining the matter content for a given function $Z(r)$, or, conversely, for a given plausible matter distribution one determines the structural function $Z(r)$.

The evaluation of the fluid covariant derivative properties yields zero, i.e., the area expansion Θ, shear σ, rotation ω and acceleration vector all vanish.

As far as the algebraic type of the Cotton tensor is concerned, it is clearer and more straightforward to establish it from the alternative representation of this metric in term of the the structural function $Y(r)$. Instead of (3.82), one now uses

$$\frac{d}{dr}Z\left(r\right)=F_0\,Y\left(r\right). \tag{3.85}$$

Thus, there remains a single constraint

$$\frac{d^2}{dr^2}Y\left(r\right)+\left(\kappa\,m\left(r\right)+\Lambda-\frac{3}{4}\,{F_0}^2\right)Y\left(r\right)=0. \tag{3.86}$$

The Cotton tensor in terms of $Y(r)$ and $M(r)$, $4\,\kappa\,m\left(r\right)+{F_0}^2+4\,\Lambda=8\,M\left(r\right)$, becomes

$$Y\left(C^{\alpha}{}_{\beta}\right)=\left[\begin{array}{ccc}-2\,F_0YM-Z\frac{dM}{dr} & 0 & -3\,ZF_0YM-\left(Y^2+Z^2\right)\frac{dM}{dr}\\ 0 & F_0YM & 0\\ \frac{dM}{dr} & 0 & Z\frac{dM}{dr}+F_0YM\end{array}\right], \tag{3.87}$$

from which it is evident that the roots for the eigenvalues could be real or complex conjugated, fulfilling the traceless condition $\lambda_1+\lambda_2+\lambda_3=0$, therefore the algebraic types could be Type I, and Type I_Z.

In the next paragraph, for the case of incompressible dust $m(r)=m_0$, the eigenvectors of the Cotton tensor are shown explicitly.

Rooman–Spindel Dust

If the pressure is set equal to zero, $p=0$, thus $C_1^2=-\frac{1}{4\Lambda}$, then one is dealing with an AdS, $\Lambda<0$, dust star, with metric

$$ds^2=-[dt+Z(r)d\phi]^2+dr^2+\left(C_1\,\frac{d\,Z}{dr}\right)^2d\phi^2, \tag{3.88}$$

and energy density subjected to

$$\frac{d^3\,Z}{dr^3}+\left(\kappa\,\mu\left(r\right)+4\Lambda\right)\frac{d\,Z}{dr}=0. \tag{3.89}$$

Incompressible Dust Gödel Models

A quite particular class of solution emerges under the constancy of the energy density, $\mu=\mu_0$. The energy equation becomes

$$\frac{d^3}{dr^3}Z\left(r\right)+M_0\,\kappa\,\frac{d}{dr}Z\left(r\right)=0, \tag{3.90}$$

where

$$\kappa M_0=\kappa\mu_0-\frac{3}{4\,{C_1}^2}+\Lambda=\kappa\mu_0+4\Lambda,\ \Lambda<0. \tag{3.91}$$

allowing the general solutions

$$Z(r) = C_O + C_2 \sin\left(\sqrt{\kappa M_O}\, r\right) + C_3 \cos\left(\sqrt{\kappa M_O}\, r\right),$$
$$M_0 > 0, a^{-2} = \kappa M_0, \tag{3.92a}$$

or

$$Z(r) = C_O + C_2 \sinh\left(\sqrt{\kappa\, M_O}\, r\right) + C_3 \cosh\left(\sqrt{\kappa\, M_O}\, r\right),$$
$$M_0 < 0,\ a^{-2} = -\kappa M_0. \tag{3.92b}$$

Following Lubo et al. (1999), see Eq. (95), these solutions can be given as

$$ds^2 = -dt^2 + dr^2 + \gamma^2 \left(\sinh^2(\frac{r}{a}) - a^2c^2(\kappa + \cosh(\frac{r}{a}))^2\right) d\phi^2$$
$$-2ac\gamma(\kappa + \cosh(\frac{r}{a}))dt\, d\phi, \tag{3.93a}$$
$$ds^2 = -dt^2 + dr^2 + \gamma^2 \left(\mathrm{e}^{(\frac{2r}{a})} - a^2c^2(\kappa + \mathrm{e}^{(\frac{r}{a})})^2\right) d\phi^2$$
$$-2ac\gamma(\kappa + \mathrm{e}^{(\frac{r}{a})})dt\, d\phi, \tag{3.93b}$$
$$ds^2 = -dt^2 + dr^2 + \gamma^2 \left(\cosh^2(\frac{r}{a}) - a^2c^2(\kappa + \sinh(\frac{r}{a}))^2\right) d\phi^2$$
$$-2ac\gamma(\kappa + \sinh(\frac{r}{a}))dt\, d\phi, \tag{3.93c}$$

Locally, these metrics are mutually equivalent; for coordinate transformations relating to them, see Rooman and Spindel (1998); for low density stars, $\kappa\mu < -4\Lambda$, they are determined by restricting the domain of r to $r < r_{boundary}$, hence they correspond to different non-diffeomorphic subsets of a larger space on which r is unrestricted.

On the other hand, the metric of the high density star $\kappa\mu > -4\Lambda$ is given by

$$ds^2 = -dt^2 + dr^2 + \gamma^2 \left(\sin^2(\frac{r}{a}) - a^2c^2(\kappa + \cos(\frac{r}{a}))^2\right) d\phi^2$$
$$-2ac\gamma(\kappa + \cosh(\frac{r}{a}))dt\, d\phi. \tag{3.94}$$

For the physical interpretation of the above solutions one has to return to the original Gödel works.

The Cotton tensor (3.87), for this incompressible case

$$m(r) = m_0 \to M(r) = M_0; 2F_O M_0 = C_O,$$

amounts to

$$(C^\alpha{}_\beta) = \begin{bmatrix} -C_O & 0 & -\frac{3}{2}\, Z\, C_O \\ 0 & \frac{1}{2}\, C_O & 0 \\ 0 & 0 & \frac{1}{2}\, C_O \end{bmatrix},$$

thus, the eigenvalues are: a single $\lambda_1 = -C_0$ and a double one $\lambda_2 = C_0/2$ with eigenvectors

$$\begin{aligned}
\lambda_1 &= -C_0, V1^\alpha = \delta^\alpha{}_t, V1^\alpha V1_\alpha = -1,\\
\lambda_2 &= \frac{C_0}{2}, V2^\alpha = -Z\, V_3 \delta^\alpha{}_t + V_2 \delta^\alpha{}_x + V_3 \delta^\alpha{}_y, V^\alpha V_\alpha = V_2{}^2 + V_3{}^2 Y^2,\\
\lambda_3 &= \frac{C_0}{2}, V3^\alpha = -Z\, v_3 \delta^\alpha{}_t + v_2 \delta^\alpha{}_x + v_3 \delta^\alpha{}_y, V^\alpha V_\alpha = v_2{}^2 + v_3{}^2 Y^2,
\end{aligned} \tag{3.95}$$

in particular, for the components

$$v_1 = V_1, v_2 = -\frac{V_3{}^2 Y^2}{V_2}, v_3 = V_3,$$

the vectors $\boldsymbol{V}2$ and $\boldsymbol{V}3$ are orthogonal. Consequently, the Cotton tensor for these Gödel kinds of solutions is of Type I: $\{T, S, S\}$.

4

A Shortcut to (2+1) Cyclic Symmetric Stationary Solutions

From a general cyclic symmetric stationary metric of the $(2+1)$-gravity with a negative $\Lambda = -1/l^2$ a shortcut in the derivation of the three families of cyclic solutions is given. There arise: two branches for the BTZ black hole solution, the stationary Coussaert–Henneaux (CH) solution, the cyclic CH $SO(2) \times SO(2)$ cosmological metric, and additionally the static AdS class with its sub-branches: the one-parametric AdS solution, called the AdS static BTZ black hole, and the AdS naked singularity metric.

4.1 Cyclic Symmetric Stationary Solutions in Canonical Coordinates

Within (2+1)-dimensional Einstein theory, the general form of a stationary cyclic symmetric line element – allowing for a timelike Killing vector $\boldsymbol{\partial_t}$ and a spacelike Killing vector field $\boldsymbol{\partial_\phi}$ – can be given as

$$g = -\frac{F(r)}{H(r)}dt^2 + \frac{dr^2}{F(r)} + H(r)(d\phi + W(r)dt)^2. \tag{4.1}$$

For this metric, the Einstein equations $E_\mu{}^\nu := G_\mu{}^\nu - \delta_\mu{}^\nu/l^2 = 8\pi\, T_\mu{}^\nu$, with negative cosmological constant, in the vacuum case $T_\mu{}^\nu = 0$, yield:

$$E_\phi{}^t = \frac{1}{2}\frac{d}{dr}(H^2\frac{dW}{dr}) = 0 \rightarrow \frac{dW}{dr} = \frac{J}{H^2}, \tag{4.2}$$

$$E_t{}^t + E_\phi{}^\phi + 2\,E_r{}^r = \frac{1}{2}\left(\frac{d^2F}{dr^2} - \frac{8}{l^2}\right) = 0 \rightarrow F(r) = \frac{4}{l^2}r^2 + F_1 r + F_0, \tag{4.3}$$

$$E_t{}^t - E_r{}^r - W\,E_\phi{}^t = \frac{F}{2H}\frac{d^2H}{dr^2} = 0 \rightarrow H(r) = H_1 r + H_0. \tag{4.4}$$

The integration of (4.2), assuming $H_1 \neq 0$, gives

$$W(r) = -\frac{J}{H_1}\frac{1}{H(r)} + W_0 = -\frac{J}{H_1}\frac{1}{H_1 r + H_0} + W_0. \tag{4.5}$$

Without loss of generality one can set $W_0 = 0$. Moreover, the substitution of the above structural functions into the remaining Einstein equations gives rise to a condition on the integration constants, which can be solved, for instance, for F_0

$$F_0 = \frac{H_0}{H_1}F_1 + \frac{J^2}{H_1^2} - \frac{4}{l^2}\frac{H_0^2}{H_1^2} \tag{4.6}$$

Therefore the metric tensor components become

$$\begin{aligned} g_{tt} &= -\frac{1}{H_1}\left(\frac{4}{l^2}r + F_1 - 4\frac{H_0}{l^2 H_1}\right), \\ 1/g_{rr} &= \frac{4}{l^2}r^2 + F_1\, r + \frac{J^2}{H_1^2} + \frac{H_0}{H_1}\left(F_1 - 4\frac{H_0}{l^2 H_1}\right), \\ g_{t\,\phi} &= -\frac{J}{H_1},\ g_{\phi\,\phi} = H_1\, r + H_0. \end{aligned} \tag{4.7}$$

4.1.1 Bañados–Teitelboim–Zanelli Solution in Canonical Polar ρ Coordinates

Accomplishing in (4.1), with structural functions determined by (4.3)–(4.5) and constraint (4.6), the linear transformations and re-parameterizations

$$\begin{aligned} &t \to t\sqrt{H_1},\ \phi \to 2\phi/\sqrt{H_1},\ r \to \rho^2/4 - H_0/H_1,\ 4F(\rho)/\rho^2 \to f(\rho), \\ &J \to J_0\, H_1/4, F_1 \to \frac{8}{l^2}\frac{H_0}{H_1} - M, \end{aligned} \tag{4.8}$$

one arrives at the standard BTZ form presented in Bañados et al. (1992)

$$\begin{aligned} g &= -f(\rho)dt^2 + \frac{d\rho^2}{f(\rho)} + \rho^2\left(d\phi - 2\frac{J_0}{\rho^2}dt\right)^2, \\ f(\rho) &= \frac{\rho^2}{l^2} - M + 4\frac{J_0^2}{\rho^2}. \end{aligned} \tag{4.9}$$

For a detailed analysis see Bañado et al. (1993). This representation of the BTZ metric in polar circumference coordinate $ds_{\bigodot} = \rho\, d\phi$ is useful when one considers, for instance, the vacuum limit of the metrics describing electric fields.

Brown–York Quasilocal Characteristics of BTZ

The Brown–York quasilocal characterization of stationary (2+1) metrics is given in detail in Section 1.3 and in the summary 1.3.1. Using these quantities, the energy–momentum characterization for the metric (4.9) amount to: momentum

$$\begin{aligned} j_\phi &= \frac{J_0}{2\pi\rho},\ j_\phi \approx \frac{J_0}{2\pi\rho}, \\ J_\phi &= J_0,\ J_\phi \approx J_0, \end{aligned} \tag{4.10a}$$

energy density

$$\begin{aligned}\epsilon(\rho, M, \epsilon_0) &= -\frac{\sqrt{f(\rho)}}{\pi\rho} - \epsilon_0,\ \epsilon(\rho, M, 0) \approx -\frac{1}{\pi l} + \frac{l}{2\pi}\frac{M}{\rho^2},\\ \epsilon(\rho, M, \epsilon_0(M_0)) &\approx \frac{l}{2\pi}\frac{M - M_0}{\rho^2},\\ \epsilon_0(M_0) &:= -\frac{\sqrt{\rho^2/l^2 - M_0}}{\pi\,\rho},\ \epsilon_0(M_0) \approx -\frac{1}{\pi\, l} + \frac{l\, M_0}{2\pi\,\rho^2},\end{aligned} \tag{4.10b}$$

total quasi-local energy

$$\begin{aligned}&E(\rho, M, \epsilon_0) = -2\sqrt{f(\rho)} - 2\pi\rho\epsilon_0,\\ &E(\rho, M, 0) \approx -2\frac{\rho}{l} + \frac{l}{\rho}M, E(\rho, M, \epsilon_0(M_0)) \approx \frac{l}{\rho}(M - M_0),\end{aligned} \tag{4.10c}$$

total quasi-local mass

$$\begin{aligned}&M(\rho, M, \epsilon_0) = -2\frac{\rho^2}{l^2} + 2M - 2\pi\rho\sqrt{f(\rho)}\epsilon_0,\ M(\rho, M, 0) = -2\frac{\rho^2}{l^2} + 2M,\\ &M(\rho, M, \epsilon_0(M_0)) \approx M - M_0,\end{aligned} \tag{4.10d}$$

where $\approx$ denotes "approximate value of the considered quantity in the limit as $\rho \to \infty$."

As we see, these quantities depend on the choice of the referential energy density ϵ_0: the referential $\epsilon_0 = 0$ can be thought of as the energy corresponding to the flat Minkowski spacetime, while $\epsilon_0(M_0) = -\frac{\sqrt{\rho^2/l^2 - M_0}}{\pi\,\rho}$ corresponds to energy density of the static BTZ metric – a common choice.

4.1.2 BTZ Solution Counterpart

On the other hand, accomplishing in (4.1) the following linear transformations and re-parametrization

$$\begin{aligned}&t \to t\sqrt{H_1}l,\ \phi \to \frac{2}{l\sqrt{H_1}}\Phi,\ r \to \rho^2/4 - \frac{l^2}{4}F_1 + \frac{H_0}{H_1},\ 4F(\rho)/\rho^2 \to f(\rho),\\ &J \to J_0\, H_1, F_1 \to \frac{8}{l^2}\frac{H_0}{H_1} - M,\end{aligned} \tag{4.11}$$

one arrives at the BTZ solution in the form

$$\begin{aligned}g &= -\rho^2\left(dt + \frac{J_0}{2\rho^2}d\phi\right)^2 + \frac{d\rho^2}{f(\rho)} + f(\rho)d\phi^2\\ &= -\rho^2\frac{f(\rho)}{\rho^2/l^2 + M}dt^2 + \frac{d\rho^2}{f(\rho)} + (\rho^2/l^2 + M)[d\phi - \frac{J_0/2}{\rho^2/l^2 + M}dt]^2,\\ f(\rho) &= \frac{\rho^2}{l^2} + M + \frac{J_0^2}{4\rho^2}.\end{aligned} \tag{4.12}$$

It becomes apparent that this metric form is just another real cut of the metric (4.9) when subjecting it to the complex transformations $t \to i\,\phi$ and $\phi \to i\,t$. This representation of the BTZ metric counterpart is useful when one considers, for example, the vacuum limit of the metrics describing magnetic fields.

Moreover, by accomplishing in the above metric the transformation of the radial coordinate

$$t \to t/l,\ \rho \to l\,\sqrt{\rho^2/l^2 - M_0},\ \phi \to l\,\phi, \tag{4.13}$$

one gets the standard BTZ metric (4.9).

Brown–York Quasilocal Characteristics of BTZ Counterpart

As far as to the energy–momentum characteristics are concerned, for the second form of the metric (4.12) one has

$$\begin{aligned}
j_\phi &= -\frac{J_0}{2\,l^2\,\pi\sqrt{\rho^2/\,l^2 + M}},\ j_\phi \approx -\frac{J_0}{2\,l\,\pi\rho},\\
J_\phi &= -\frac{J_0}{l^2},\ J_\phi \approx -\frac{J_0}{l^2},\\
\epsilon(\rho, M, \epsilon_0) &= -\frac{1}{l^2\pi}\frac{\sqrt{f(\rho)}}{\rho^2/\,l^2 + M} - \epsilon_0,\ \epsilon(\rho, M, 0) \approx -\frac{1}{\pi l} + \frac{l}{2\pi}\frac{M}{\rho^2},\\
E(\rho, M, \epsilon_0) &= -\frac{2}{l^2}\rho\frac{\sqrt{f(\rho)}}{\sqrt{\rho^2/\,l^2 + M}} - 2\pi\,\epsilon_0\sqrt{\rho^2/\,l^2 + M},\\
E(\rho, M, 0) &\approx -2\frac{\rho}{l^2},\ E(\rho, M, \epsilon_{0c}(M_0)) \approx \frac{1}{\rho}(M - M_0),\\
M(\rho, M, \epsilon_0) &= -\frac{2}{l^2}\rho^2 - 2\pi\,\rho\,\epsilon_0\,\sqrt{f(\rho)},\ M(\rho, M, 0) = -2\frac{\rho^2}{l^2},\\
M(\rho, M, \epsilon_{0c}(M_0)) &\approx M - M_0,\\
\epsilon_{0c}(M_0) &:= -\frac{\rho}{\pi\,l^2\,\sqrt{\rho^2/\,l^2 + M_0}},\\
\epsilon_{0c}(M_0) &\approx -\frac{1}{\pi\,l} + \frac{l\,M_0}{2\pi\,\rho^2} \approx \epsilon_0(M_0).
\end{aligned} \tag{4.14}$$

4.1.3 Coussaert–Henneaux Metrics

There exists the branch of solutions with constant $H(r) = H_0 = J\,l/2$, thus the integration of (4.2) yields $W(r) = 4r/(Jl^2)$, and therefore the resulting metric can be given as

$$\begin{aligned}
g &= -2\frac{F(r)}{J\,l}dt^2 + \frac{dr^2}{F(r)} + \frac{J\,l}{2}\left(d\phi + \frac{4\,r}{J\,l^2}dt\right)^2,\\
F(r) &= \frac{4}{l^2}r^2 + F_1\,r + F_0,
\end{aligned} \tag{4.15}$$

which, by changing $t/\sqrt{Jl/2} \to t$, $\phi\sqrt{Jl/2} \to \phi$, can be brought to the form

$$\begin{aligned} g &= -F(r)dt^2 + \frac{dr^2}{F(r)} + \left(d\phi + \frac{2\,r}{l}dt\right)^2, \\ F(r) &= \frac{4}{l^2}r^2 + F_1\,r + F_0 = \frac{4}{l^2}[(r + \frac{l^2}{8}F_1)^2 \pm A_\pm^2], \\ A_\pm^2 &:= \pm\frac{l^2}{4}(F_0 - \frac{l^2}{16}F_1^2). \end{aligned} \tag{4.16}$$

Coussaert–Henneaux Stationary Metric

Subjecting (4.16) to the transformations

$$\begin{aligned} t &= \frac{l^2}{4\,A_+}\tilde{t},\; r = -\frac{l^2}{2}F_1 + A_+\,\sinh\tilde{r},\; \phi = -\frac{l^3\,F_1}{4\,A_+}\tilde{t} + \frac{l}{2}\tilde{\phi}, \\ A_+ &:= \frac{l}{2}\sqrt{F_0 - \frac{l^2}{16}F_1^2},\; F(r) \to \frac{4}{l^2}A_+^2\,\cosh^2\tilde{r} \end{aligned} \tag{4.17}$$

one gets the standard Coussaert and Henneaux (1994b) metric; see also Ayón–Beato et al. (2004),

$$g = \frac{l^2}{4}\left[-d\tilde{t}^2 + 2\,\sinh\tilde{r}d\tilde{t}\,d\tilde{\phi} + d\tilde{\phi}^2 + d\tilde{r}^2\right]. \tag{4.18}$$

Time-Dependent $SO(2)\times SO(2)$ Cyclic Metric

Moreover, if one were allowing the structural function $F(r)$ to range over negative values, which takes place in the case of $A_-^2 = -\frac{l^2}{4}(F_0 - l^2\,F_1^2)$, with lower negative sign in $F(r)$ from (4.16), subjecting (4.16) to the transformations

$$\begin{aligned} t &= \frac{l^2}{4\,A_-}\theta,\; r = -\frac{l^2}{8}F_1 + A_-\,\sin\tau,\; \phi = \frac{l^3\,F_1}{16\,A_-}\theta + \frac{l}{2}\Phi, \\ A_- &:= \frac{l}{2}\sqrt{\frac{l^2}{16}F_1^2 - F_0},\; F(r) \to -\frac{4}{l^2}A^2\,\cos^2\tau, \end{aligned} \tag{4.19}$$

one arrives at a time-dependent (τ-coordinate) spacetime

$$g = \frac{l^2}{4}\left[d\theta^2 + d\Phi^2 + 2\,\sin\tau d\theta\,d\Phi - d\tau^2\right]. \tag{4.20}$$

This spacetime is the self-dual spacetime with isometry $SO(2)\times SO(2)$; see equation (34) of Ayón–Beato et al. (2004) and details therein.

4.2 Static AdS Black Hole

Working with the canonical metric (4.1) the Einstein equations in the static case, $W(r) = 0$, simplify considerably. In principle, one obtains the anti-de Sitter static black hole solution from the BTZ metric simply by setting $J = 0$, or following step by step the procedure exhibited above in the stationary case.

4.2.1 Static BTZ Solution

In polar coordinates the static BTZ solution assumes the form

$$g = -f(\rho, M_0)dt^2 + \frac{d\rho^2}{f(\rho, M_0)} + \rho^2\, d\phi^2,$$

$$f(\rho, M_0) := \frac{\rho^2}{l^2} - M_0, \tag{4.21}$$

where the mass parameter is now denoted by M_0.
In these coordinates the energy characteristics acquire the form

$$\epsilon(\rho, M_0, \epsilon_0) = -\frac{1}{\pi\rho}\sqrt{f(\rho, M_0)} - \epsilon_0,\ \ \epsilon(\rho, M_0, 0) \approx -\frac{1}{\pi l} + \frac{l}{2\pi}\frac{M_0}{\rho^2},$$

$$E(\rho, M_0, \epsilon_0) = -2\sqrt{f(\rho, M_0)} - 2\pi\, \rho\, \epsilon_0,\ \ E(\rho, M_0, 0) \approx -2\frac{\rho}{l} + \frac{l}{\rho}M_0,$$

$$M(\rho, M_0, \epsilon_0) = -2f(\rho, M_0) - 2\pi\rho\sqrt{f(\rho, M_0)}\,\epsilon_0,$$

$$M(\rho, M_0, 0) \approx -2\frac{\rho^2}{l^2} + 2M_0. \tag{4.22}$$

Introducing the energy density function $\epsilon(\rho, M_0)$,

$$\epsilon(\rho, M_0)\ := -\frac{1}{\pi\rho}\sqrt{f(\rho, M_0)} \tag{4.23}$$

one brings the above functions to the forms

$$\epsilon(\rho, M_0, \epsilon_0) = \epsilon(\rho, M_0) - \epsilon_0,$$

$$E(\rho, M_0, \epsilon_0) = 2\pi\rho\,[\epsilon(\rho, M_0) - \epsilon_0]\,,$$

$$M(\rho, M_0, \epsilon_0) = \sqrt{f(\rho, M_0)}E(\rho, M_0, \epsilon_0). \tag{4.24}$$

Choosing the Base Energy Density; AdS Spacetime

For $M_0 = -1$ one obtains the standard anti-de Sitter solution in polar coordinates

$$g = -f(\rho)dt^2 + \frac{d\rho^2}{f(\rho)} + \rho^2\, d\phi^2,$$

$$f(\rho) = \frac{\rho^2}{l^2} + 1, \tag{4.25}$$

with energy density

$$\epsilon(\rho, -1, 0) = \epsilon_{AdS}(\rho) := -\frac{1}{\pi\rho}\sqrt{\frac{\rho^2}{l^2} + 1},\ \ \epsilon_{AdS}(\rho) \approx -\frac{1}{\pi l} - \frac{l}{2\pi\,\rho^2},$$

$$\epsilon(\rho, -1, \epsilon_0)\ =\ \epsilon_{AdS}(\rho) - \epsilon_0, \tag{4.26}$$

therefore, for the referential energy density choice $\epsilon_0 = \epsilon_{AdS}(\rho)$, the mass–energy functions of the anti-de Sitter become zero:

$$\epsilon(\rho, -1, \epsilon_{AdS}) = 0,\ E(\rho, -1, \epsilon_{AdS}(\rho)) = 0, M(\rho, -1, \epsilon_{AdS}(\rho)) = 0.$$

This is one of the plausible choices for the base energy density of the anti-de Sitter spacetime; see Brown et al. (1994) in the paragraph below Eq. (4.12). Another possibility takes place for the naked singularity solution with $M_0 = 0$,

$$g = -f(\rho,0)dt^2 + \frac{d\rho^2}{f(\rho,0)} + \rho^2\, d\phi^2,\ f(\rho,0) = \frac{\rho^2}{l^2}, \tag{4.27}$$

with energy characteristics

$$\begin{aligned} \epsilon(\rho,0) &:= -\frac{1}{\pi\, l},\ \epsilon(\rho,0,\epsilon_0) = -\frac{1}{\pi\, l} - \epsilon_0, \\ E(\rho,0,\epsilon_0) &= 2\pi\,\rho\,\epsilon(\rho,0,\epsilon_0),\ M(\rho,0,\epsilon_0) = \frac{\rho}{l}\, E(\rho,0,\epsilon_0). \end{aligned} \tag{4.28}$$

For the particular referential energy density $\epsilon_0 = -\frac{1}{\pi\, l}$, all energy characteristics of this naked anti-de Sitter solution vanish,

$$\epsilon(\rho,0,\epsilon_0 = -\frac{1}{\pi\, l}) = 0, E(\rho,0,\epsilon_0 = -\frac{1}{\pi\, l}) = 0, M(\rho,0,\epsilon_0 = -\frac{1}{\pi\, l}) = 0,$$

in agreement with Brown et al. (1994), Eq. (4.12).

Limits of the Mass–Energy Functions of the Static BTZ Black Hole

For the static BTZ metric (4.21), with M_0 parameter, the energy characteristics at infinity, $\rho \to \infty$, become:
for the "naked" referential $\epsilon_0 = -\frac{1}{\pi\, l}$

$$\begin{aligned} \epsilon\left(\rho, M_0, \epsilon_0 = -\frac{1}{\pi\, l}\right) &\approx \frac{M_0\, l}{2\pi\rho^2}, \\ E\left(\rho, M_0, \epsilon_0 = -\frac{1}{\pi\, l}\right) &\approx \frac{M_0\, l}{\rho}, \\ M\left(\rho, M_0, \epsilon_0 = -\frac{1}{\pi\, l}\right) &\approx M_0, \end{aligned} \tag{4.29}$$

while for the AdS referential $\epsilon_0 = \epsilon_{AdS}$

$$\begin{aligned} \epsilon(\rho, M_0, \epsilon_{AdS}) &\approx \frac{l}{2\pi\rho^2}(1+M_0), \\ E(\rho, M_0, \epsilon_{AdS}) &\approx \frac{l}{\rho}(1+M_0), \\ M(\rho, M_0, \epsilon_{AdS}) &\approx 1 + M_0. \end{aligned} \tag{4.30}$$

These results can be gathered in a more compact form by introducing the base energy density equipped with the discrete parameter $m = -1$, and 0,

$$\epsilon_0(\rho, m) =:= -\frac{1}{\pi\rho}\sqrt{\frac{\rho^2}{l^2} - m},\ \epsilon_0(\rho,m) \approx -\frac{1}{\pi l} + \frac{l\, m}{2\pi\,\rho^2}, \tag{4.31}$$

as

$$\begin{aligned}\epsilon(\rho, M_0, \epsilon_0(\rho, m)) &\approx \frac{l}{2\pi\rho^2}(M_0 - m),\\ E(\rho, M_0, \epsilon_0(\rho, m)) &\approx \frac{l}{\rho}(M_0 - m),\\ M(\rho, M_0, \epsilon_0(\rho, m)) &\approx M_0 - m.\end{aligned} \tag{4.32}$$

4.2.2 Static AdS Solution Counterpart

Setting $J_0 = 0$ in the metric (4.12), one arrives at the static BTZ metric counterpart

$$g = -\rho^2 dt^2 + \frac{d\rho^2}{(\rho^2/l^2 + M_0)} + (\rho^2/l^2 + M_0)d\phi^2. \tag{4.33}$$

In these coordinates the energy characteristics are given by

$$\begin{aligned}\epsilon(\rho, M_0, \epsilon_0) &= -\frac{\rho}{\pi l^2}\frac{1}{\sqrt{\rho^2/l^2 + M_0}} - \epsilon_0,\ \epsilon(\rho, M_0, 0) \approx -\frac{1}{\pi l} + \frac{l}{2\pi}\frac{M_0}{\rho^2},\\ E(\rho, M_0, \epsilon_0) &= -2\frac{\rho}{l^2} - 2\pi\,\epsilon_0\sqrt{\rho^2/l^2 + M_0},\ E(\rho, M_0, 0) \approx -2\frac{\rho}{l^2},\\ M(\rho, M_0, \epsilon_0) &= -2\frac{\rho^2}{l^2} - 2\pi\rho\,\epsilon_0\sqrt{\rho^2/l^2 + M_0},\ M(\rho, M_0, 0) \approx -2\frac{\rho^2}{l^2}.\end{aligned} \tag{4.34}$$

One could define the base energy density as the one corresponding to the black hole limit of the anti-de Sitter spacetime or that associated with the naked singularity; both cases can be handled by introducing the discrete parameter $m = 0, -1$, and defining the base energy density function $\epsilon(\rho, m)$,

$$\epsilon_0(\rho, m) := -\frac{\rho}{\pi\, l^2}\frac{1}{\sqrt{\rho^2/l^2 + m}},\ \epsilon_0(\rho, m) \approx -\frac{1}{\pi l} + \frac{l}{2\pi}\frac{m}{\rho^2}. \tag{4.35}$$

For such a choice of the base energy density, the mass–energy characteristics of the proper anti-de Sitter solution counterpart, as it should be, vanish:

$$\epsilon(\rho, m, \epsilon_0(\rho, m)) = 0,\ E(\rho, m, \epsilon_0(\rho, m)) = 0,\ M(\rho, m, \epsilon_0(\rho, m)) = 0.$$

The limits at spatial infinity $\rho \to \infty$ of the energy quantities are

$$\begin{aligned}\epsilon(\rho, M_0, \epsilon_0(\rho, m)) &\approx \frac{l}{2\pi\rho^2}(M_0 - m)\\ E(\rho, M_0, \epsilon_0(\rho, m)) &\approx \frac{1}{\rho}(M_0 - m),\\ M(\rho, M_0, \epsilon_0(\rho, m)) &\approx M_0 - m.\end{aligned} \tag{4.36}$$

By accomplishing in metric (4.33) the transformations (4.13) one gets the anti-de Sitter black hole metric (4.21). The presence of powers of l in these transformations explains the appearance of different powers in the limits at spatial infinity $\rho \to \infty$ of the energy–momentum characteristics for the BTZ and AdS black hole solutions compared with the corresponding quantities of their solution counterparts.

4.3 Symmetries of the Stationary and Static Cyclic Symmetric BTZ Metrics

Although it is known that the BTZ solution possesses two Killing vectors – the timelike symmetry along the time coordinate and the spacelike symmetry along the orbits of the periodic angular variable – in my opinion, some comments on this respect can be added to clarify how the number of six Killing vectors solutions for the BTZ metric structure reduces to the quoted two. In this framework, the six symmetries of the anti-de Sitter space with parameter M_0, $AdS(M_0)$, are derived; the static BTZ allowing for *time* + *polar* coordinates possesses a timelike and one 2π-periodic circular symmetries. For the anti-de Sitter space with parameter $M_0 = -1$, denoted simply by AdS, there are six symmetries: time, circular, and four boots symmetries.

The study of the symmetries of the stationary and static cyclic symmetric BTZ families and AdS classes of solutions starts with the stationary metric for the standard BTZ solution

$$g = -F(r)^2\, dt^2 + \frac{dr^2}{F(r)^2} + r^2\,[d\phi + W(r)dt]^2\,,$$
$$F(r)^2 = \frac{r^2}{l^2} - M + \frac{J^2}{4r^2},\ W(r) = -\frac{J}{2r^2}. \tag{4.37}$$

The covariant components of the Killing vectors V_α are denoted by $v\alpha$, namely

$$V_1(t, r, \phi) = v1,\ V_2(t, r, \phi) = v2,\ V_3(t, r, \phi) = v3. \tag{4.38}$$

The Killing equations $EQ_{\mu\nu} := V_{\mu;\nu} + V_{\nu;\mu} = 0$ amount explicitly to

$$EQ11 = \frac{\partial}{\partial t} v1 - \frac{rF(r)^2}{l^2} v2 = 0, \tag{4.39a}$$

$$2EQ12 = \frac{\partial}{\partial r} v1 + \frac{\partial}{\partial t} v2 - \frac{J}{l^2 r F(r)^2}\, v3 - 2\,\frac{r}{l^2 F(r)^2} v1 = 0, \tag{4.39b}$$

$$2EQ13 = \frac{\partial}{\partial \phi} v1 + \frac{\partial}{\partial t} v3 = 0, \tag{4.39c}$$

$$EQ22 = F(r)\frac{\partial}{\partial r} v2 + v2\,\frac{d}{dr} F(r) \to v2 = \frac{F_1(t,\phi)}{2lF(r)} = 0, \tag{4.39d}$$

$$EQ23 = \frac{\partial}{\partial \phi} v2 + \frac{\partial}{\partial r} v3 + \frac{J}{rF(r)^2}\, v1 + 2\,\frac{(Ml^2 - r^2)}{rl^2 F(r)^2} v3 = 0, \tag{4.39e}$$

$$EQ33 = \frac{\partial}{\partial \phi} v3 + r\, F(r)^2\, v2 = 0, \tag{4.39f}$$

where $F_i(t, \phi)$, $i = 1, 2, 3$, are integration functions.

Isolating $v1$ from (4.39e) in terms of $v2$ and $v3$ and their derivatives, one gets

$$r l^2 J v1 = 2\, r \left(r^2 - M l^2\right) v3 - r^2\, l^2\, F(r)^2 \left(\frac{\partial}{\partial \phi} v2 + \frac{\partial}{\partial r} v3\right). \tag{4.40}$$

Next, substituting $v1$ from above and the first integral of $v2$ from (4.39d) into equation (4.39b), one arrives at a linear second-order equation for $v3$ with integrals

$$v3 = \frac{r^2}{2} F_2(t, \phi) + F_3(t, \phi) - \frac{1}{2} \frac{r\, l\, F(r)}{M^2 l^2 - J^2} \left(J \frac{\partial}{\partial t} + M \frac{\partial}{\partial \phi}\right) F_1(t, \phi) \tag{4.41}$$

which, substituted together with $v2$ from (4.39d) into Eq. (4.40) for $v1$, gives

$$\begin{aligned} v1 = {} & 2 \frac{\left(r^2 - M l^2\right)}{J l^2} F_3(t, \phi) - \frac{1}{4} J\, F_2(t, \phi) \\ & + \frac{1}{2} \frac{r\, l\, F(r)}{\left(l^2 M^2 - J^2\right)} \left(\frac{J}{l^2} \frac{\partial}{\partial \phi} + M \frac{\partial}{\partial t}\right) F_1(t, \phi). \end{aligned} \tag{4.42}$$

The dependence of the Killing vector components on the r variable has been established; it remains still to determine their dependence on the t and ϕ variables hiding in the $F_1(t, \phi)$, $F_2(t, \phi)$ and $F_3(t, \phi)$ functions. Substituting the expressions of $v1$ from (4.42), $v3$ from (4.41), and $v2$ from (4.39d), one arrives at the independent equations

$$F_1 J^2 + l^2 J \frac{\partial^2}{\partial t \partial \phi} F_1 + l^2 \left(\frac{\partial^2}{\partial \phi^2} F_1\right) M - F_1 M^2 l^2 = 0, \tag{4.43a}$$

$$F_1 J^2 + l^2 J \frac{\partial^2}{\partial t \partial \phi} F_1 + M \left(\frac{\partial^2}{\partial t^2} F_1\right) l^4 - F_1 M^2 l^2 = 0, \tag{4.43b}$$

with integral

$$F_1 = C_1 e^{\frac{\sqrt{Ml - J}(l\phi + t)}{l^{3/2}}} + C_2 e^{\frac{\sqrt{Ml + J}(l\phi - t)}{l^{3/2}}} + C_3 e^{-\frac{\sqrt{Ml + J}(l\phi - t)}{l^{3/2}}} + C_4 e^{-\frac{\sqrt{Ml - J}(l\phi + t)}{l^{3/2}}}. \tag{4.44}$$

Furthermore, there have to be solved constraints on $F_2(t, \phi)$ and $F_3(t, \phi)$, namely

$$\begin{aligned} & \frac{\partial}{\partial t} F_3 = 0, \frac{\partial}{\partial \phi} F_3 = 0, F_3(t, \phi) = J\, C_6 = \text{const.}, \\ & \frac{\partial}{\partial t} F_2 = 0, \frac{\partial}{\partial \phi} F_2 = 0, F_2(t, \phi) = C_5 = \text{const.} \end{aligned} \tag{4.45}$$

where the integration constants are denoted through C_i, $i = 1, \ldots, 6$. Finally, the covariant Killing vector components of the BTZ solution are

$$V_1 = \frac{1}{2} \frac{r\,F(r)}{l^{3/2}\sqrt{M^2l^2 - J^2}} \left(C_1 \sqrt{Ml + J} \exp(\frac{\sqrt{Ml - J}}{l^{3/2}}(l\phi + t)) \right.$$
$$- C_2 \sqrt{Ml - J} \exp(\frac{\sqrt{J + Ml}}{l^{3/2}}(l\phi - t))$$
$$+ C_3 \sqrt{Ml - J} \exp(-\frac{\sqrt{J + Ml}}{l^{3/2}}(l\phi - t))$$
$$\left. - C_4 \sqrt{Ml + J} \exp(-\frac{\sqrt{Ml - J}}{l^{3/2}}(l\phi + t)) \right)$$
$$-\frac{J}{4} C_5 - 2 \frac{(Ml^2 - r^2)}{l^2} C_6, \tag{4.46a}$$

$$V_2 = \frac{1}{2lF(r)} \left(C_1 \exp(\frac{\sqrt{Ml - J}}{l^{3/2}}(l\phi + t)) + C_2 \exp(\frac{\sqrt{J + Ml}}{l^{3/2}}(l\phi - t)) \right.$$
$$\left. + C_3 \exp(-\frac{\sqrt{J + Ml}}{l^{3/2}}(l\phi - t)) + C_4 \exp(-\frac{\sqrt{Ml - J}}{l^{3/2}}(l\phi + t)) \right), \tag{4.46b}$$

$$V_3 = -\frac{1}{2l^{1/2}} \frac{r\,F(r)}{\sqrt{M^2l^2 - J^2}} \left(C_1 \sqrt{Ml + J} \exp(\frac{\sqrt{Ml - J}}{l^{3/2}}(l\phi + t)) \right.$$
$$+ C_2 \sqrt{Ml - J} \exp(\frac{\sqrt{Ml + J}}{l^{3/2}}(l\phi - t))$$
$$- C_3 \sqrt{Ml - J} \exp(-\frac{\sqrt{Ml + J}}{l^{3/2}}(l\phi - t))$$
$$\left. - C_4 \sqrt{Ml + J} \exp(-\frac{\sqrt{Ml - J}}{l^{3/2}}(l\phi + t)) \right) + \frac{r^2}{2} C_5 + J\,C_6. \tag{4.46c}$$

These expressions allow one to determine the Killing vector $\mathbf{k_i}$ associated with its corresponding integration constant C_i for each of the possible i, by means of $V_\mu = \sum_{i=1}^{6} k_{i\mu}$, where $k_{i\mu} = C_i V_{i\mu}$, for each fixed value of i. The contravariant vectors' components ${k_i}^\mu$, $\boldsymbol{\partial}_{k_i} = {k_i}^\mu \frac{\partial}{\partial x^\mu}$, are derived from the relationship $V^\mu = V_\nu g^{\nu\mu} = \sum_{i=1}^{6} k_{i\nu} g^{\nu\mu} = \sum_{i=1}^{6} {k_i}^\mu = \sum_{i=1}^{6} C_i V_i^\mu$. Explicitly, these contravariant Killing vectors' components ${k_i}^\mu$, $\boldsymbol{\partial}_{k_i} = {k_i}^\mu \frac{\partial}{\partial x^\mu} = C_i V_i^\mu$, $i = 1, \ldots, 6$ are given below. The reason to include the integration constants in the definitions of the Killing vectors ${k_i}^\mu$ is related to the domain of definition of the spatial coordinates; in the case of the existence of a periodic coordinate some Killing vectors vanish, which can be easily achieved by setting certain structural constant equal to zero. Explicitly, these Killing vectors are:

$$C_1,\ \boldsymbol{\partial}_{k_1};\ {k_1}^\mu = C_1 \exp\left(\frac{\sqrt{Ml - J}}{l^{3/2}}(l\phi + t)\right) \left[\frac{1}{4} \frac{Jl - 2r^2}{\sqrt{Ml - J}} \frac{1}{r\,F(r)} \frac{1}{l^{3/2}}, \right.$$
$$\left. \frac{1}{2}\frac{F(r)}{l}, -\frac{1}{4} \frac{Jl + 2r^2 - 2Ml^2}{\sqrt{Ml - J}} \frac{1}{l^{5/2}} \frac{1}{r\,F(r)} \right], \tag{4.47a}$$

$$C_2,\ \boldsymbol{\partial}_{k_2};\ {k_2}^{\mu} = C_2 \exp\left(\frac{\sqrt{J+Ml}\,(l\phi - t)}{l^{3/2}}\right)\left[\frac{1}{4}\frac{l\,J+2\,r^2}{\sqrt{J+Ml}}\frac{1}{l^{3/2}}\frac{1}{r\,F(r)},\right.$$
$$\left.\frac{1}{2}\frac{F(r)}{l}, \frac{1}{4}\frac{(l\,J+2\,Ml^2-2\,r^2)}{\sqrt{J+Ml}}\frac{1}{l^{5/2}}\frac{1}{r\,F(r)}\right], \tag{4.47b}$$

$$C_3,\ \boldsymbol{\partial}_{k_3};\ {k_3}^{\mu} = C_3 \exp\left(-\frac{\sqrt{J+Ml}}{l^{3/2}}(l\phi - t)\right)\left[-\frac{1}{4}\frac{l\,J+2\,r^2}{\sqrt{J+Ml}}\frac{1}{l^{3/2}}\frac{1}{r\,F(r)},\right.$$
$$\left.\frac{1}{2}\frac{F(r)}{l}, -\frac{1}{4}\frac{(l\,J+2\,Ml^2-2\,r^2)}{\sqrt{J+Ml}}\frac{1}{l^{5/2}}\frac{1}{r\,F(r)}\right], \tag{4.47c}$$

$$C_4,\ \boldsymbol{\partial}_{k_4};\ {k_4}^{\mu} = C_4 \exp\left(-\frac{\sqrt{M\,l-J}\,(l\phi + t)}{l^{3/2}}\right)\left[-\frac{1}{4}\frac{J\,l-2\,r^2}{\sqrt{M\,l-J}}\frac{1}{l^{3/2}}\frac{1}{r\,F(r)},\right.$$
$$\left.\frac{1}{2}\frac{F(r)}{l}, \frac{1}{4}\frac{J\,l+2\,r^2-2\,Ml^2}{\sqrt{M\,l-J}}\frac{1}{l^{5/2}}\frac{1}{r\,F(r)}\right], \tag{4.47d}$$

$$C_5,\ \boldsymbol{\partial}_{k_5};\ {k_5}^{\mu} = C_5[0,0,1/2], \tag{4.47e}$$

$$C_6,\ \boldsymbol{\partial}_{k_6};\ {k_6}^{\mu} = C_6[-2,0,0]. \tag{4.47f}$$

For completeness, the list of the independent Killingian commutators is given:

$$\partial_{[k_6}\partial_{k_1]} = -2C_6\frac{\sqrt{Ml-J}}{l^{3/2}}\partial_{k_1}, \partial_{[k_6}\partial_{k_3]} = -2C_6\frac{\sqrt{Ml+J}}{l^{3/2}}\partial_{k_3},$$
$$\partial_{[k_6}\partial_{k_4]} = 2C_6\frac{\sqrt{Ml-J}}{l^{3/2}}\partial_{k_4}, \partial_{[k_6}\partial_{k_2]} = 2C_6\frac{\sqrt{Ml+J}}{l^{3/2}}\partial_{k_2}, \tag{4.48a}$$
$$\partial_{[k_5}\partial_{k_1]} = C_5\frac{1}{2}\frac{\sqrt{Ml-J}}{l^{1/2}}\partial_{k_1}, \partial_{[k_5}\partial_{k_3]} = -C_5\frac{1}{2}\frac{\sqrt{Ml+J}}{l^{1/2}}\partial_{k_3},$$
$$\partial_{[k_5}\partial_{k_4]} = -C_5\frac{1}{2}\frac{\sqrt{Ml-J}}{l^{1/2}}\partial_{k_4}, \partial_{[k_5}\partial_{k_2]} = C_5\frac{1}{2}\frac{\sqrt{Ml+J}}{l^{1/2}}\partial_{k_2}, \tag{4.48b}$$
$$\partial_{[k_1}\partial_{k_4]} = -\frac{C_1C_4}{2C_6}\frac{1}{l^{5/2}\sqrt{Ml-J}}\partial_{k_6} + \frac{C_1C_4}{C_5}\frac{2}{l^{7/2}\sqrt{Ml-J}}\partial_{k_5}, \tag{4.48c}$$
$$\partial_{[k_3}\partial_{k_2]} = -\frac{C_3C_2}{2C_6}\frac{1}{l^{5/2}\sqrt{Ml+J}}\partial_{k_6} - \frac{C_3C_2}{C_5}\frac{2}{l^{7/2}\sqrt{Ml+J}}\partial_{k_5}. \tag{4.48d}$$

All anti-de Sitter metrics for coordinates $\{t, \rho, \phi\}$ – merely names – ranging

$$-\infty \leq t \leq \infty, -\infty \leq \rho \leq \infty, -\infty \leq \phi \leq \infty$$

allows for six symmetries, i.e., six Killing vectors. All these spaces in these coordinates are maximally symmetric spaces.

Another is the situation if the spatial coordinates are constrained to range

$$0 \leq \rho \leq \infty, 0 \leq \phi \leq 2\pi,$$

in such case ρ and ϕ are polar coordinates with ϕ being the angular coordinate with period 2π. Since the expressions of four of the Killing vector fields depending on ϕ do not exhibit the angular symmetry in 2π, invariance under the change

$\phi \to \phi + 2\pi$, therefore there is no room for the corresponding symmetries and the integration constants associated with those vectors ought to be zero. Consequently the metric with positive M allowing for polar angular coordinate, and only that, possesses only two Killing vectors, ∂_t and ∂_ϕ (two symmetries: the time translation and the 2π-periodic angular rotation). This spacetime is known as the stationary BTZ black hole. For more about this, see also Ayón–Beato et al. (2004).

A similar situation takes place in the case of the static anti-de Sitter metric. By setting the rotation parameter equal to zero, $J = 0$, the above expressions (4.48) give the Killing vectors for the static anti–de Sitter spacetime. Again, in the case of the coordinates restricted to ranges $0 \leq \rho \leq \infty$, $0 \leq \phi \leq 2\pi$, the static anti-de Sitter metric allows only for two Killing vectors: $\boldsymbol{\partial}_t$, and $\boldsymbol{\partial}_\phi$, i.e., the time translation and the 2π-periodic angular rotation; otherwise, when there are six constants, the space is maximally symmetric.

4.3.1 Symmetries of the AdS Metric for Negative M, $M = -\alpha^2$

If M is negative, one can equate it to $-\alpha^2$. Moreover, instead of complex exponential function, it will be better to use trigonometric sine and cosine functions. Thus, one can give the Killing vector components as $V = V_\mu dx^\mu = C_a V_{a\mu} dx^\mu$

$$V_1 = \frac{r\sqrt{\alpha^2 l^2 + r^2}}{\alpha\, l^3}\left[-C_1 \sin(\alpha\,\phi)\cos\left(\frac{\alpha\, t}{l}\right) + C_2 \sin(\alpha\,\phi)\sin\left(\frac{\alpha\, t}{l}\right)\right.$$
$$\left. - C_3 \cos(\alpha\,\phi)\cos\left(\frac{\alpha\, t}{l}\right) + C_4 \cos(\alpha\,\phi)\sin\left(\frac{\alpha\, t}{l}\right)\right]$$
$$+ C_5(\alpha^2 l^2 + r^2), \tag{4.49}$$

$$V_2 = \sqrt{\alpha^2 l^2 + r^2}\left[C_1 \sin(\alpha\,\phi)\sin\left(\frac{\alpha\, t}{l}\right) + C_2 \sin(\alpha\,\phi)\cos\left(\frac{\alpha\, t}{l}\right)\right.$$
$$\left. + C_3 \cos(\alpha\,\phi)\sin\left(\frac{\alpha\, t}{l}\right) + C_4 \cos(\alpha\,\phi)\cos\left(\frac{\alpha\, t}{l}\right)\right], \tag{4.50}$$

$$V_3 = \frac{r\sqrt{\alpha^2 l^2 + r^2}}{\alpha\, l^2}\left[C_1 \cos(\alpha\,\phi)\sin\left(\frac{\alpha\, t}{l}\right) + C_2 \cos(\alpha\,\phi)\cos\left(\frac{\alpha\, t}{l}\right)\right.$$
$$\left. - C_3 \sin(\alpha\,\phi)\sin\left(\frac{\alpha\, t}{l}\right) - C_4 \sin(\alpha\,\phi)\cos\left(\frac{\alpha\, t}{l}\right)\right] + C_6 r^2. \tag{4.51}$$

This anti-de Sitter metric (cosmological constant negative, $\Lambda = -1/l^2$) for the coordinates $\{t, \rho, \phi\}$ – merely names – ranging $-\infty \leq t \leq \infty$, $-\infty \leq \rho \leq \infty$; $-\infty \leq \phi \leq \infty$ allows for six symmetries, i.e., six Killing vectors. For these ranges of determination of the coordinates, the space is maximally symmetric.

If the spatial coordinates are restricted to range

$$0 \leq \rho \leq \infty,\; 0 \leq \phi \leq 2\pi,$$

and α is set equal to unity, $\alpha = 1 = -M$, then in such case ρ and ϕ become polar coordinates with ϕ being the angular coordinate with period 2π. This spacetime – the (proper) anti-de Sitter space (with $M = -1$) – allows for six symmetries, and as such it is maximally symmetric.

In this chapter, in the framework of the (2+1)-dimensional Einstein theory with cosmological constant different families of exact solutions for cyclic symmetric stationary (static) metrics in the presence of a negative cosmological constant have been derived. Specific branches of solutions in the general case are determined via a straightforward integration. In this systematic approach, all known cyclic symmetric solutions of the considered class are properly identified.

5

Perfect Fluid Static Stars; Cosmological Solutions

The purpose of this chapter is to determine the static circularly (cyclic) symmetric spacetimes coupled to perfect fluids via a straightforward integration of the Einstein equations. The structural functions of the metric depend on the energy density, which remains in general arbitrary. Spacetimes for fluids fulfilling linear (barotropic) and polytropic state equations are explicitly derived. By the way, we demonstrate here that the incompressible perfect fluid solution is the only conformally flat (with vanishing of the Cotton tensor) circularly symmetric solution. Since there is no boundary to determine a zero-pressure surface, all these solutions fall in the cosmological category. The perfect fluid is characterized by a fluid velocity u^μ and energy–momentum tensor

$$T_{\mu\nu} = (p+\mu)u_\mu\, u_\nu + p g_{\mu\nu},\; u_\mu\, u^\mu = -1, \tag{5.1}$$

where μ and p are the energy density and the pressure; throughout this text, the energy density will be denoted with μ or ρ, as are commonly used in hydrodynamics.

5.1 Static Circularly Symmetric Fluid Solutions

The static circularly symmetric metric commonly is chosen in curvature coordinates

$$g = -\mathrm{e}^{2\,\nu(r)}dt^2 + \mathrm{e}^{2\,\lambda(r)}\,dr^2 + r^2\,d\theta^2, \tag{5.2}$$

therefore, the Einstein equations are

$$\begin{aligned} E_t^t : \frac{d}{dr}\lambda\,(r) &= \kappa\, r\,\mu\,(r)\,\mathrm{e}^{2\,\lambda(r)}, \rightarrow \frac{d}{dr}\mathrm{e}^{-2\,\lambda(r)} = -2\kappa\, r\mu\,(r) \rightarrow \\ \mathrm{e}^{-2\,\lambda(r)} &= C_0 - 2\int^r \kappa\, r\mu\,(r)\; dr = C_0 - \frac{\kappa}{\pi}\, M\,(r)\,, \\ M\,(r) &:= 2\pi \int^r r\,\mu\,(r)\; dr, \end{aligned} \tag{5.3a}$$

$$E_r^r : \frac{d}{dr}\nu(r) - \kappa r e^{2\lambda(r)} p(r) = 0 \to \nu(r) = \kappa \int^r \frac{r p(r)\, dr}{C_0 - \frac{\kappa}{\pi} M(r)}, \tag{5.3b}$$

$$E_\theta^\theta : \frac{d^2\nu}{dr^2} + \left(\frac{d\nu}{dr}\right)^2 - \frac{d\nu}{dr}\frac{d\lambda}{dr} - \kappa p e^{2\lambda} = 0. \tag{5.3c}$$

By the way, the substitution of the derivatives of λ from (5.3a) and ν from (5.3b) into equation (5.3c) yields the energy momentum conservation equation (5.4) given below. The conservation law $T^{\mu\nu}{}_{;\nu} = 0$ gives rise to a single equation equation after Oppenheimer and Volkoff (1939), namely

$$\frac{dp}{dr} = -\frac{d\nu}{dr}(p+\mu) \to \frac{dp}{dr} = -\kappa \frac{rp}{C_0 - \kappa M(r)/\pi}(p+\mu). \tag{5.4}$$

In hydrodynamics, usually a state equation – a dependence between the pressure and the energy – has to be provided.

Moreover, an integral quantity arise from the combination of (5.3b) and (5.3c), $(E_\theta^\theta - E_r^r/r)e^{\nu(r)}$, namely

$$\frac{d}{dr}\left(\frac{e^{-\lambda}}{r}\frac{de^\nu}{dr}\right) = 0. \tag{5.5}$$

5.1.1 Cotton Tensor Types

The Cotton tensor for this perfect fluid is

$$C^\mu{}_\nu = -\frac{\kappa}{4} e^{-\lambda} \frac{d\mu(r)}{dr}(r\, e^{-\nu} \delta_t^\mu \delta_\nu^\theta - \frac{e^\nu}{r} \delta_\theta^\mu \delta_\nu^t). \tag{5.6}$$

Since the Cotton vanishes for constant density $\mu(r) = \mu_0$ then this kind of solution is conformally flat. Therefore, the incompressible perfect fluid ($\mu(r) = \mu_0$) is conformally flat.

The search for its eigenvectors yields

$$\lambda_1 = 0; \mathbf{V1} = [0, V^2, 0],\ V1_\mu = V^2 g_{rr}\delta^r_\mu,\ V1^\mu V1_\mu = (V^2)^2 g_{rr},\ \mathbf{V1} = \mathbf{S1},$$

$$\lambda_2 = i\frac{\kappa}{4}e^{-\lambda}\frac{d\mu(r)}{dr}; \mathbf{V2} = [V^1, 0, V^3 = -\frac{ie^\nu V^1}{r}],\ \mathbf{V2} = \mathbf{Z},$$

$$\lambda_3 = -i\frac{\kappa}{4}e^{-\lambda}\frac{d\mu(r)}{dr}; \mathbf{V3} = [V^1, 0, V^3 = \frac{ie^\nu V^1}{r}],\ \mathbf{V3} = \bar{\mathbf{Z}}, \tag{5.7}$$

consequently the corresponding tensor type is Type I: $\{S, Z, \bar{Z}\}$.

5.2 Incompressible Static Star

For this class of solutions, the corresponding spacetimes are conformally flat; the Cotton becomes Type O. The easiest case to handle is the fluid with constant density $\mu(r) = \mu_0$, which integrates completely:

$$\begin{aligned}
e^{-2\,\lambda(r)} &= C_0 - r^2\kappa\,\mu_0,\\
\frac{d}{dr}p\,(r) &= -\frac{\kappa\,p\,(r)\,r\,(\mu_0 + p\,(r))}{C_0 - r^2\kappa\,\mu_0} \rightarrow p(r) = \frac{B_0\mu_0\sqrt{C_0 - r^2\kappa\,\mu_0}}{1 - B_0\sqrt{C_0 - r^2\kappa\,\mu_0}},\\
\nu\,(r) &= \nu_0 + \ln\left(1 - B_0\sqrt{C_0 - r^2\kappa\,\mu_0}\right),
\end{aligned} \tag{5.8}$$

where C_0 and B_0 are integration constants which one fixes via boundary conditions.

5.2.1 Collas Static Star with Constant Density μ_0

In Collas (1977), the search for a perfect fluid solution was posed in the paragraph: *B Interior solutions*. The condition that the component g_{rr}, at the boundary of the mass distribution r_0, has to be 1, as for the exterior Minkowski metric, gives rise to the Eq. (37) in Collas (1977), namely

$$e^{-2\,\lambda(r)} = C_0 - 2\int^{r_0} \kappa\, r\mu\,(r)\; dr = 1 \tag{5.9}$$

from which one determines C_0, which in the particular case of constant density μ_0 results in $C_0 = \kappa\mu_0 r_0^2 = 8\pi\mu_0 r_0^2$. The pressure is expressible as

$$p(r) = B\mu_0\frac{[1 + 8\pi\mu_0(r_0^2 - r^2)]^{1/2}}{[1 + 8\pi\mu_0 r_0^2]^{1/2} - B[1 + 8\pi\mu_0(r_0^2 - r^2)]^{1/2}}, \tag{5.10}$$

comparable with Eq. (40) of Collas (1977). Thus the requirement of vanishing pressure at the boundary of the fluid ball fails to be fitted, hence the author concludes that it is not possible to have a fluid with constant density μ and nonzero pressure $p \neq 0$.

This dilemma found a solution in the forthcoming formulation for the constant density μ_0 and variable bounded pressure $p(r)$ branch of metrics, but not allowing for a match to the exterior Minkowski spacetime.

5.2.2 Giddings–Abott–Kuchař Static Star with μ_0

In the work by Giddings et al. (1984) a nontrivial perfect fluid solution with constant density and variable pressure is found. The first requirement to fulfill was the vanishing of the pressure at some border R, which yields to $C_0 = R^2\kappa\,\mu_0$, therefore

$$p(r) = \frac{B_0\mu_0\sqrt{\kappa\,\mu_0}\,\sqrt{R^2 - r^2}}{1 - B_0\sqrt{\kappa\,\mu_0}\,\sqrt{R^2 - r^2}}. \tag{5.11}$$

Next, the finite value of the pressure $p_c = p(r = 0)$ at the center $r = 0$ of the ball fixes the constant B_0 to be

$$B_0\sqrt{\kappa\,\mu_0}R = p_c/(\mu_0 + p_c),$$

consequently

$$p(r) = p_c\,\mu_0 \frac{\sqrt{R^2 - r^2}}{R(p_c + \mu_0) - p_c\sqrt{R^2 - r^2}}$$

$$= p_c \frac{\sqrt{1 - (r/R)^2}}{1 + \frac{p_c}{\mu_0} - \frac{p_c}{\mu_0}\sqrt{1 - (r/R)^2}}, \tag{5.12a}$$

$$\mathrm{e}^{-2\,\lambda(r)} = \kappa\,\mu_0(1 - r^2) = \kappa\,\mu_0 R^2(1 - (r/R)^2) = g^{rr}, \tag{5.12b}$$

$$\nu\,(r) = \nu_0 + \ln\left(\frac{\mu_0}{\mu_0 + p_c}\right) + \ln\left(1 + \frac{p_c}{\mu_0} - \frac{p_c}{\mu_0}\sqrt{1 - (r/R)^2}\right). \tag{5.12c}$$

Notice that at this boundary also $g^{rr}(r_b) \to 0$, $g_{rr}(r_b) \to \infty$, which means that one is facing a coordinate singularity there.

Accomplishing the coordinate transformations

$$r\,' = \frac{r}{R\sqrt{\kappa\,\mu_0}},\ \theta\,' = R\sqrt{\kappa\,\mu_0}\,\theta,\ \nu_0 = -\ln\frac{\mu_0}{\mu_0 + p_c}, \tag{5.13}$$

one arrives, dropping primes, at the GAK metric, Eq. (118) of Giddings et al. (1984):

$$g = -\left(1 + \frac{p_c}{\mu_0} - \frac{p_c}{\mu_0}\sqrt{1 - r^2\kappa\,\mu_0}\right)^2 dt^2 + \frac{1}{1 - r^2\kappa\,\mu_0}\,dr^2 + r^2\,d\theta^2$$

$$p\,(r) = \frac{p_c\mu_0\sqrt{1 - r^2\kappa\,\mu_0}}{p_c + \mu_0 - p_c\sqrt{1 - r^2\kappa\,\mu_0}},\ \mu(r) = \mu_0,\ M(r) = \pi\,\mu_0\,r^2, \tag{5.14}$$

where $p\,(r)$, $\mu\,(r)$, $m\,(r)$ denote correspondingly the pressure, the mass density, and the integral mass. The range of the circular coordinate θ has been restored to be $0 \leq \theta \leq 2\pi$. In this formulation of the fluid solution with constant μ, as before, the center is located at $r = 0$, $p(r = 0) = p_c$, and the vanishing of the pressure requires $r_b = 1/\sqrt{\kappa\,\mu_0}$, at which the metric component $g_{rr}(r_b) \to \infty$, *therefore the matching to an exterior metric, what ever it may be, is not possible; the continuity of the metric components is one of the main requirements to be satisfied.* The solution with μ_0 is simply a cosmological one.

As was stated before, the solutions with constant energy density μ_0 are Cotton zero, i.e., conformally flat as in the $(3+1)$ Schwarzschild–Kottler case.

5.2.3 Cornish–Frankel Static Star with μ_0

The incompressible perfect fluid, $\mu(r) = \mu_0$, has been treated by various authors: Collas (1977), Giddings et al. (1984), and again by Cornish and Frankel (1991) in their *Example C* with the solution in the form

$$g = -\left[1+\frac{p_c}{\mu_0}\left(1-\left(1-(\frac{r}{R})^2\right)^{1/2}\right)\right]^2 dt^2 + \left[1-(\frac{r}{R})^2\right]^{-1} dr^2 + r^2 d\theta^2,$$

$$p(r) = p_c\left[1-(\frac{r}{R})^2\right]^{1/2}\left[1+\frac{p_c}{\mu_0}\left(1-\left(1-(\frac{r}{R})^2\right)^{1/2}\right)\right]^{-1}. \tag{5.15}$$

The pressure vanishes on the boundary R of the star, but on this circumference the metric exhibits a coordinate singularity, therefore the fluid ball blows up in these confines.

5.3 Cornish–Frankel Static Polytropic Solutions

Cornish and Frankel (1991) successfully treated static polytropic perfect fluid cases for the metric (5.2). The starting point in deriving this class of fluids is the Oppenheimer–Volkoff pressure equation (5.4) together with the mass–energy–density definition (5.3a). Isolating $C_0 \pi - \kappa M(r)$ from (5.4), one has

$$C_0 \pi - \kappa M(r) = -\frac{\pi r \kappa p(r)(\mu(r) + p(r))}{\frac{d}{dr} p(r)}, \tag{5.16}$$

which, differentiated with respect to r, yields

$$r(2p-\mu) + \frac{p(\mu+p)}{\frac{dp}{dr}} + \frac{r p \frac{d\mu}{dr}}{\frac{dp}{dr}} - \frac{r p(\mu+p)\frac{d^2 p}{dr^2}}{\left(\frac{dp}{dr}\right)^2} = 0. \tag{5.17}$$

For the family of polytropic fluids, the pressure and energy density can be given as

$$p(r) = p_0 \Theta(r)^{n+1}, \; \mu(r) = \mu_0 \Theta(r)^n, \tag{5.18}$$

where n is the polytropic index, not necessarily an integer. Moreover, one may impose one condition on the pressure and energy at the center of the fluid ball, namely

$$p(r=0) = p_0 \Theta(0)^{n+1}, \; \mu(r=0) = \mu_0, \Theta(r=0) = 1. \tag{5.19}$$

A second condition, such as the value of $\Theta(r = R) = 0$, where R denotes the boundary value of r, is not needed; on the contrary, if this were the case, then a relation between the boundary value R and the center values of the pressure and the energy density would arise. In any case, at the boundary R a relation (a transcendent one) between $\Theta(R)$, μ_0, p_0, and R will arise; see below.

The above equation (5.17), in terms of Θ, can be rearranged into the form

$$d\left[\ln(r) - \ln\left(\frac{d}{dr}\Theta\right) - (n+1)\ln(\Theta) + (2n+3)\ln(\mu_0 + p_0\,\Theta)\right] = 0.$$

Therefore, one arrives at

$$\frac{d\Theta}{dr} = B_O\, r \frac{(\mu_0 + p_0\,\Theta)^{2\,n+3}}{\Theta^{n+1}}$$
$$\rightarrow \frac{B_O\, r^2}{2} = \int \frac{\Theta^{n+1}}{(\mu_0 + p_0\,\Theta)^{2\,n+3}} d\Theta + A_O. \tag{5.20}$$

The total integral is given in terms of polynomials of Θ, namely

$$-\int \frac{\Theta^{n+1}}{(\mu_0 + p_0\,\Theta)^{2\,n+3}} d\Theta = \frac{1}{{\mu_0}^{n+1}\,{p_0}^{n+2}} \sum_{k=0}^{n+1} S_{kn}, \tag{5.21}$$

$$S_{kn}(\Theta(r)) = (-1)^k \frac{(n+1)!\,{\mu_0}^k}{k!\,(n+1-k)!\,(n+1+k)} \frac{1}{(p_0\,\Theta + \mu_0)^{n+1+k}}. \tag{5.22}$$

Hence, evaluating A_O at $r = 0$, $\Theta(r = 0) = 1$, one arrives at

$$-\frac{B_O}{2} r^2 = \frac{1}{{\mu_0}^{n+1}\,{p_0}^{n+2}} \sum_{k=0}^{n+1} \left[S_{kn}(\Theta(r)) - S_{kn}(\Theta = 1)\right]. \tag{5.23}$$

At this stage, let us see if one can fix the value of the integration constant B_O. The evaluation of

$$\mathrm{e}^{-2\lambda(r)} = C_O - \frac{\kappa}{\pi} M(r) = -\frac{\pi\, r\kappa\, p\,(\mu + p)}{\frac{dp}{dr}} = -\frac{\kappa r}{n+1} (\mu_0 + p_0\,\Theta)\,\Theta^{n+1} \frac{dr}{d\Theta}$$
$$= -\frac{1}{B_O} \frac{\kappa}{n+1} \frac{\Theta^{2n+2}}{(\mu_0 + p_0\,\Theta)^{2n+2}} = \frac{1}{g_{rr}},$$
$$g_{\Theta\Theta} = -\frac{1}{B_O} \frac{n+1}{\kappa r^2} \frac{1}{(\mu_0 + p_0\,\Theta)^{2n+2}}, \tag{5.24}$$

therefore, the positiveness of the spatial metric components requires B_O to be negative, $B_O < 0$, say $B_O = -2$. The integration of $\nu(r)$, (5.3b) is straightforward, $\exp(2\nu) = N_0^2\,(\mu_0 + p_0\,\Theta)^{-2\,n-2}$.

Changing Θ to a more conventional radial coordinate $\rho \equiv \Theta$, the CF solution in the (t, ρ, ϕ) coordinates can be given as

$$ds^2 = -(\mu_0 + p_0\,\rho)^{-2\,n-2}\, dt^2 + \frac{n+1}{2\kappa\, r^2} \frac{d\rho^2}{(\mu_0 + p_0\,\rho)^{2n+2}} + r^2\, d\phi^2, \tag{5.25}$$

$$r(\rho)^2 := \frac{1}{{\mu_0}^{n+1}\,{p_0}^{n+2}} \sum_{k=0}^{n+1} \left[S_{kn}(\rho) - S_{kn}(\rho = 1)\right], \tag{5.26}$$

$$S_{kn}(\rho) = (-1)^k \frac{(n+1)!\,{\mu_0}^k}{k!\,(n+1-k)!\,(n+1+k)} \frac{1}{(\mu_0 + p_0\,\rho)^{n+1+k}}, \tag{5.27}$$

$$p(\rho) = p_0\rho^{n+1}, \mu(\rho) = \mu_0\rho^n. \tag{5.28}$$

Since this solution does not allow for a zero-pressure circle $p(\rho) = 0$ except at $\rho = 0$, this kind of solution is a cosmology that extends over the whole space.

As far as to the mass is concerned, let us return to the equations (5.24), with $B_0 = -2$,

$$C_0 - \frac{\kappa}{\pi} M(r) = \frac{1}{2n+1} \frac{\kappa\, \Theta^{2n+2}}{(\mu_0 + p_0\, \Theta)^{2n+2}} = \frac{1}{g_{rr}}. \tag{5.29}$$

From this equation one concludes that at the origin $r = 0$, and at the boundary $r = R$ one correspondingly has

$$C_0 = \frac{\kappa}{\pi} M(r=0) + \frac{1}{2n+1} \frac{\kappa}{(\mu_0 + p_0)^{2n+2}}, \tag{5.30}$$

$$C_0 - \frac{\kappa}{\pi} M(r=R) = \frac{1}{2n+1} \frac{\kappa\, \Theta(R)^{2n+2}}{(\mu_0 + p_0 \Theta(R))^{2n+2}} = \frac{1}{g_{rr}(R)}. \tag{5.31}$$

From the second condition, it becomes clear that the choice $\Theta(R) = 0$, made in Cornish and Frankel (1991), is not an adequate one since in that case $g^{rr}(R) = 0$, $g_{rr}(R) \to \infty$–a black fluid ball-plate. Moreover, as was been mentioned above, the condition $\Theta(R) = 0$ impose a dependence of the form $R(\mu_0, p_0)$, which is also unwelcome. Consequently, it is better to adopt as the mass within the radius r the quantity

$$M(\Theta) = M(0) + \frac{\pi}{2\,(\mu_0 + p_0)^{2n+2}\,(n+1)} - \frac{\Theta(r)^{2n+2}\,\pi}{2\,(\mu_0 + p_0\,\Theta(r))^{2n+2}\,(n+1)}.$$

In particular, the mass inside the ball with $r = R$, is given by

$$M(r=R) = M(0) + \frac{\pi}{2} \frac{1}{n+1} \frac{1}{(\mu_0 + p_0)^{2n+2}} - \frac{\pi}{2} \frac{1}{n+1} \frac{\Theta(R)^{2n+2}}{(\mu_0 + p_0 \Theta(R))^{2n+2}}.$$

On the other hand, at this boundary $r = R$, the metric components are finite if $\Theta(R) \neq 0$, and $\mu_0 + p_0\Theta(R) \neq 0$; see (5.31). Recall that $\Theta(R)$ is a root of the algebraic polynomial equation of degree $n + k + 1$ for $\Theta(R)$ arising from

$$R^2 = \frac{1}{{\mu_0}^{n+1}\,{p_0}^{n+2}} \sum_{k=0}^{n+1} \left[S_{kn}(\Theta(R)) - S_{kn}(\Theta(0) = 1)\right].$$

This is a complete analytic solution, for any polytropic index n and has no parallel comparing with $(3 + 1)$-dimensional case, where only few polytropic solutions are known.

5.3.1 Static Star with a Stiff Matter $p(r) = \mu(r)$

The integration of the Einstein equations yields

$$e^{2\nu(r)} = \frac{1}{2\kappa\, C_2} e^{-C_1 r^2/2},\; e^{2\lambda(r)} = -\frac{C_1}{2\kappa\, C_2} e^{-C_1 r^2/2},\; p(r) = C_2\, e^{C_1 r^2/2},$$
$$p(0) = p_0 = C_2,\; C_1 = -2\kappa\, C_2 = -2\kappa\, p_0,$$

hence, this simple conformally flat solution can be given as

$$d\,s^2 = -e^{\kappa\, p_0\, r^2}\, dt^2 + e^{\kappa\, p_0\, r^2}\, dr^2 + r^2\, d\theta^2,$$
$$p(r) = p_0 e^{-\kappa\, p_0\, r^2}, \tag{5.32}$$

from which its cosmological character becomes apparent.

5.3.2 Static Star with Pure Radiation $p = \mu(r)/2$

The static radiation or ultra-relativistic particles solution $p = \mu(r)/2$ is given by the metric (corrected) and pressure of the form

$$g = -\left[1 - (\frac{r}{R})^2\right]^{-2} dt^2 + \left[1 - (\frac{r}{R})^2\right]^{-4} dr^2 + r^2 d\theta^2,$$
$$p(r) = \gamma \left[1 - (\frac{r}{R})^2\right]^3, \gamma = \frac{2}{\kappa\, R^2}. \tag{5.33}$$

In Cornish and Frankel (1991), (*Example B*), there is a misprint in the component $g_{tt} = -\left[1 - (\frac{r}{R})^2\right]^2$, instead of the correct one of (5.33) with power -2.

6

Static Perfect Fluid Stars with Λ

The purpose of this chapter is to determine all static circularly (cyclic) symmetric spacetimes with a cosmological constant coupled to perfect fluids with and without zero-pressure surfaces via a straightforward integration of the Einstein equations. The structural functions of the metrics depend on the energy density, which remains in general arbitrary. Spacetimes for fluids fulfilling linear (barotropic) and polytropic state equations are explicitly derived; they contain, among other families, stiff matter, incoherent radiation, and non-relativistic degenerate fermions. By the way, we demonstrate here that the incompressible perfect fluid solution is the only conformally flat – in the sense of the vanishing of the Cotton tensor – circularly symmetric solution.

In Section 6.1, the Einstein equations for the static circularly symmetric (2+1) metric with a cosmological constant coupled to a perfect fluid solution with variable density ρ and pressure p are exhibited and integrated; in this chapter we are using ρ to denote the energy density instead of the μ used in the previous chapters. Section 6.2 is devoted to representing this whole class of spacetimes in a canonical coordinate system. For a given equation of state of the form $p = p(\rho)$, certain particular families of perfect fluid solutions are derived; as concrete examples, the subcases of fluids obeying the barotropic law $p = \gamma\, \rho$ in Section 6.3, and those fluids subjected to a polytropic law $p = C\, \rho^{\gamma}$ in Section 6.4, are derived. Moreover, the incompressible fluid, $\rho =$ const., is the only conformally flat static circularly symmetric solution coupled to perfect fluids. In Section 6.5, from the Oppenheimer–Volkoff equation certain properties of the studied solutions are established: for positive pressure p and positive density ρ, a microscopically stable fluid possesses a monotonically decreasing energy density; and conversely, these results are in close relation with the ones reported by Cruz and Zanelli (1995), which unfortunately contains various misprints.

In Section 6.6, to facilitate the comparison of the interior Schwarzschild (3+1) solution with cosmological constant, the incompressible perfect fluid (2+1) solution is derived. With this aim in mind, we search for an adequate representation of the corresponding structural functions and related quantities of these $(3+1)$ and $(2+1)$ spacetimes. A comparison table is presented. Therefore, by dimensional reduction from the interior Schwarzschild with Λ solution, one can obtain the incompressible perfect fluid (2+1) solution.

6.1 Equations for a (2+1) Static Perfect Fluid Metric

As far as we know, in most of the publications dealing with the search for perfect fluid solutions in (2+1) gravity – see for instance Cornish and Frankel (1991); Cruz and Zanelli (1995) – in the integration process the energy–momentum conservation, i.e., the Oppenheimer–Volkoff equation, has been used as a clue to obtain the desired results. By contrast, we prefer to solve directly the corresponding Einstein equations; in such case the energy–momentum conservation equations trivially hold.

The line element of static circularly symmetric (2+1) spacetimes, in coordinates $\{t, r, \theta\}$, is given by

$$ds^2 = -N(r)^2 dt^2 + \frac{dr^2}{G(r)^2} + r^2 d\theta^2. \tag{6.1}$$

The Einstein equations with cosmological constant for a perfect fluid energy–momentum tensor T_{ab}:

$$\begin{aligned} G_{ab} &= R_{ab} - \frac{1}{2} g_{ab} R = \kappa T_{ab} - \Lambda g_{ab}, \\ T_{ab} &= (p+\rho) u_a u_b + p g_{ab},\ u_a = -N \delta^t_a, \end{aligned} \tag{6.2}$$

for the metric (6.1) explicitly amount to

$$\begin{aligned} G_{tt} &= \frac{N^2}{2r}\frac{dG^2}{dr} = -N^2(\kappa\rho + \Lambda), \\ G_{rr} &= -\frac{1}{rN}\frac{dN}{dr} = -\frac{1}{G^2}(\kappa p - \Lambda), \\ G_{\theta\theta} &= -\frac{r^2}{N}\left[G^2\frac{d^2N}{dr^2} + \frac{1}{2}\frac{dN}{dr}\frac{dG^2}{dr}\right] = -r^2(\kappa p - \Lambda). \end{aligned} \tag{6.3}$$

Notice that the combination of the Einstein equations $r^2\, G(r)^2 G_{rr} - G_{\theta\theta} = 0$, for $N(r) \neq$ const., gives rise to an important equation, namely

$$G^2 r^2 \frac{d^2N}{dr^2} + r\frac{dN}{dr}\left(\frac{r}{2}\frac{dG^2}{dr} - G^2\right) = 0, \longrightarrow \frac{d}{dr}\left(\frac{G}{r}\frac{dN}{dr}\right) = 0, \tag{6.4}$$

which will be extensively used throughout this paper.

The energy conservation yields the Oppenheimer–Volkoff equation (Oppenheimer and Volkoff, 1939; see subsection 6.5):

$$N\frac{dp}{dr} + [p(r) + \rho(r)]\frac{dN}{dr} = 0, \tag{6.5}$$

which used in the Cotton tensor yields to

$$C^{\alpha}{}_{\beta} = \frac{\kappa}{2} F \frac{d\rho}{dr} \left[-\frac{r}{N} \delta^{\alpha}{}_{t} \, \delta^{\phi}{}_{\beta} + \frac{N}{r} \delta^{\alpha}{}_{\phi} \, \delta^{\phi}{}_{\beta} \right]. \tag{6.6}$$

Thus the eigenvectors are

$$\lambda_1 = 0, \ \boldsymbol{S}; \lambda_2 = \frac{\kappa}{2} \, i \, F \frac{d\rho}{dr}, \ \boldsymbol{Z}, \lambda_3 = \bar{\lambda}_2, \ \bar{\boldsymbol{Z}}. \tag{6.7}$$

Therefore the Cotton tensor is of Type I: $\{S, Z, \bar{Z}\}$. The incompressible fluid, $\rho(r) = \rho_0 =$ constant, determines a conformally flat spacetime; Type O. Therefore, we conclude that the incompressible perfect fluid $\rho(N) =$ const. solution for static circularly symmetric spacetimes is unique. This result can be stated as a theorem: the incompressible perfect fluid solution is the only conformally flat static circularly symmetric spacetime for a perfect fluid source with or without cosmological constant. The similarity of this theorem with the corresponding one formulated for Schwarzschild metric, see García (1988), is noteworthy. We shall return to this spacetime in Section 6.6.

6.1.1 General Perfect Fluid Solution with Variable $\rho(r)$

In this section, we derive the most general static circularly symmetric solution via a straightforward integration of the Einstein equation with Λ for a perfect fluid following the approach presented in García and Campuzano (2003). It is easy to establish that the structural functions $G(r)$ and $N(r)$ can be integrated in quadratures.

Integrating the G_{tt}–(6.3), one arrives at

$$G(r)^2 = -\Lambda r^2 - 2\kappa \int_0^r r\rho(r)dr \equiv C - \Lambda r^2 - 2\kappa \int^r r\rho(r)dr, \tag{6.8}$$

where C is an integration constant in which we have incorporated the constant value of the integral at the lower integration limit $r = 0$, thus the remaining integral depends on the upper integration limit r; we use the r–notation for the upper integration limit as well as to denote the integration variable in the integrand because this notation is unambiguous; this convention will be used hereafter. From the second relation of (6.4), one obtains

$$\frac{dN}{dr} = n_1 \frac{r}{G(r)} \tag{6.9}$$

therefore

$$N(r) = n_1 \int_0^r \frac{r}{G(r)} dr \equiv n_0 + n_1 \int^r \frac{r}{G(r)} dr \tag{6.10}$$

and the evaluation of the pressure $p(r)$ yields

$$\kappa p(r) = \frac{1}{N(r)} \left(n_1 \, G(r) + \Lambda \, N(r) \right). \tag{6.11}$$

The metric (6.1), with $G(r)$ from (6.8), and $N(r)$ from (6.10), determines the general static circularly symmetric (2+1) solution of the Einstein equations (6.3) with Λ for a perfect fluid, characterized by a pressure given by (6.11), and an arbitrary density $\rho(r)$. The fluid-velocity is aligned along the timelike Killing direction ∂_t. To deal with realistic matter distributions one has to impose positivity conditions on the density, $\rho > 0$, and the pressure, $p > 0$, requiring additionally $\rho > p$.

For a finite distributed fluid, the pressure p becomes zero at the boundary, say $r = a$; this value of the radial coordinate r is determined as solution of the equation $p(r) = 0$.

For non-vanishing cosmological constant, assuming that the values of the structural functions at the boundary $r = a$ are $N(a)$ and $G(a)$, the vanishing at $r = a$ of the pressure $p(r)$, given by (6.11), requires $n_1 = -\Lambda N(a)/G(a)$, hence

$$\kappa\, p(r) = \frac{\Lambda}{N(r)G(a)} \left(N(r)\, G(a) - N(a)\, G(r)\right). \tag{6.12}$$

If one is interested in matching the obtained perfect fluid metric with a vacuum metric with cosmological constant Λ, the plausible choice at hand is the anti-de Sitter metric, with $\Lambda = -1/l^2$, for which $G(a) = N(a) = \sqrt{-M_\infty + a^2/l^2}$ at the boundary $r = a$. Incidentally, for a cosmological constant different from zero, there is no room for dust. The zero character of the pressure would yield the vanishing of the density, and consequently the metric reduces to the (anti-)de Sitter spacetime.

For vanishing cosmological constant, the expression of the pressure (6.11) is

$$\kappa p(r) = n_1 \frac{G(r)}{N(r)}, \tag{6.13}$$

from which it becomes apparent that the corresponding solution represents a cosmological spacetime; there is no a surface of vanishing pressure.
For vanishing Λ and zero pressure, the situation slightly changes: the function N becomes a constant, and the corresponding metric can be written as

$$ds^2 = -dt^2 + \frac{dr^2}{C - 2\kappa \int^r r\rho(r)dr} + r^2 d\theta^2, \tag{6.14}$$

for any density function ρ. Of course, the choice of ρ is restricted by physically reasonable matter distributions.

6.2 Canonical Coordinate System $\{t, N, \theta\}$

In this section we show that an alternative formulation of our general solution can be achieved in coordinates $\{t, N, \theta\}$. Indeed, from (6.9) for the derivative of the function N, in which we are including (without any loss of generality) the constant n_1, $N/n_1 \longrightarrow N, n_1\, t \longrightarrow t$, one obtains

$$\frac{dN}{r} = \frac{dr}{G(r)}, \tag{6.15}$$

hence

$$r^2 = C_0 + 2\int^N G\,dN. \tag{6.16}$$

To derive G as a function of the new variable N, one uses the G_{tt}–(6.3) in the form of

$$G\,dG = -(\kappa\,\rho + \Lambda)r\,dr = -(\kappa\,\rho + \Lambda)GdN, \tag{6.17}$$

therefore, integrating one gets

$$G(N) = C_1 - \Lambda\,N - \kappa\int^N \rho(N)\,dN. \tag{6.18}$$

Substituting this function G into the expression of r one obtains

$$H(N) := r^2 = C_0 + 2\,C_1\,N - \Lambda\,N^2 - 2\,\kappa\int^N\int^N \rho(S)\,dSdN. \tag{6.19}$$

Finally, our metric in the new coordinates $\{t, N, \theta\}$ amounts to

$$ds^2 = -N^2dt^2 + \frac{dN^2}{H(N)} + H(N)\,d\theta^2, \tag{6.20}$$

which is characterized by pressure

$$p(N) = \frac{C_1}{\kappa}\frac{1}{N} - \frac{1}{N}\int^N \rho(N)\,dN \tag{6.21}$$

and an arbitrary energy density $\rho(N)$ depending on the variable N, both functions p and ρ have to be positive.

The metric (6.20) together with the function H from (6.19) give an alternative representation of our general solution. This representation will be used to derive particular solutions for a given state equation of the form $p = p(\rho)$. In this approach the expression of the pressure (6.21) play a central role.

6.3 Perfect Fluid Solutions for a Barotropic Law $p = \gamma\,\rho$

Although in the above section we provide the general solution to the posed question of finding all solutions for circularly symmetric static metrics in (2+1)-gravity coupled to a perfect fluid in the presence of the cosmological constant, from the physical point of view, even in this lower-dimensional spacetime, it is of interest to analyze certain specific cases: for instance, the solution corresponding to a fluid obeying the barotropic law $p = \gamma\,\rho$, or the more complicated case of a polytropic law $p = \rho^\gamma$.

The barotropic law establishes a linear relation between pressure and energy density of the form

$$p(N) = \gamma\,\rho(N). \tag{6.22}$$

Substituting $p(N)$ from Eq. (6.23) into this relation, one gets

$$\frac{C_1}{\kappa} - \int^N \rho(N)\,dN = \gamma\,N\,\rho(N). \tag{6.23}$$

Differentiating this equation with respect to the variable N, one obtains

$$\frac{d}{dN}(N\,\rho) + \frac{1}{\gamma N}\,(N\,\rho) = 0, \tag{6.24}$$

which has as general integral

$$\rho(N) = C_2\frac{\gamma-1}{\gamma^2}N^{-\frac{\gamma+1}{\gamma}}, \tag{6.25}$$

where C_2 is an integration constant. Since we arrived at the above simple linear equation (6.24) through differentiation, then one has to replace the obtained result into the relation (6.23), or equivalently into (6.22), to see if there arises any constraint from it:

$$p(N) = \frac{C_1}{\kappa}\frac{1}{N} + \gamma\;\rho(N) = \gamma\;\rho(N) \longrightarrow C_1 = 0, \tag{6.26}$$

in such manner, we establish that the constant C_1 vanishes. Replacing the function $\rho(N)$ from (6.25) into the expression of $H(N)$, (6.19), and accomplishing the integration one arrives at

$$H(N) = C_0 - \Lambda N^2 + C_2 N^{(\gamma-1)/\gamma} = r^2. \tag{6.27}$$

Thus, the metric for a perfect fluid fulfilling a barotropic state equation in coordinates $\{t, N, \theta\}$ is given by

$$ds^2 = -N^2dt^2 + \frac{dN^2}{C_0 - \Lambda N^2 + 2\kappa C_2 N^{\frac{\gamma-1}{\gamma}}} + (C_0 - \Lambda N^2 + 2C_2\kappa N^{\frac{\gamma-1}{\gamma}})d\theta^2. \tag{6.28}$$

To give this solution in terms of the radial variable r, one has to be able to solve the algebraic equation, in general a transcendent one, for $N = N(r)$. Within this class of solutions merits mention the pure radiation $\gamma = 1/2$, and the stiff matter $\gamma = 1$ where the speed of sound equals the speed of light.

6.4 Perfect Fluid Solutions for a Polytropic Law $p = C\rho^\gamma$

This section is devoted to the derivation of all solutions obeying the polytropic law

$$p = C\,\rho^\gamma. \tag{6.29}$$

Using again the expression of $p(N)$ from (6.21), the above polytropic relation can be written as

$$\frac{C_1}{\kappa} - \int^N \rho(N)\, dN = C\, N\, \rho^\gamma(N). \tag{6.30}$$

Differentiating with respect to N, one obtains

$$-\rho = C\,\frac{d}{dN}(N\,\rho^\gamma) \tag{6.31}$$

which, by introducing the auxiliary function $Z := N^{1/\gamma}\,\rho$, can be written as

$$d(Z^{\gamma-1}) + \frac{1}{C} d(N^{(\gamma-1)/\gamma}) = 0 \to d\left[(\rho^{\gamma-1} + \frac{1}{C})\,N^{(\gamma-1)/\gamma}\right] = 0, \tag{6.32}$$

therefore, integrating

$$\left(\rho^{\gamma-1} + \frac{1}{C}\right)\,N^{(\gamma-1)/\gamma} = B\,C^{-(\gamma-1)/\gamma}, \tag{6.33}$$

where B is an integration constant. Consequently, the general integral of this equation can be given as

$$\rho = C^{\frac{-1}{\gamma}}\,N^{\frac{-1}{\gamma}}\left[B - C^{\frac{-1}{\gamma}}\,N^{\frac{\gamma-1}{\gamma}}\right]^{\frac{1}{\gamma-1}}. \tag{6.34}$$

Entering this ρ into the equation (6.30), taking into account that the integral of the density ρ amounts to

$$\int^N \rho(N)\, dN = -\int^N d\left[B - C^{\frac{-1}{\gamma}}\,N^{\frac{\gamma-1}{\gamma}}\right]^{\frac{\gamma}{\gamma-1}}, \tag{6.35}$$

one arrives at

$$p(N) = \frac{n_1}{\kappa}\frac{C_1}{N} + C\,\rho^\gamma = C\,\rho^\gamma \longrightarrow C_1 = 0. \tag{6.36}$$

Considering that the first integral of ρ is given by (6.35), the expression of the structural function $H(N)$ becomes

$$H(N) = r^2 = C_0 - \Lambda\,N^2 + 2\,\kappa\int^N \left[B - C^{\frac{-1}{\gamma}}\,N^{\frac{\gamma-1}{\gamma}}\right]^{\frac{\gamma}{\gamma-1}} dN. \tag{6.37}$$

Notice that the integral can be expressed in terms of hypergeometric functions, hence

$$\begin{aligned} H(N) &= C_0 - \Lambda\,N^2 + 2\,\kappa\,B^{\gamma/(\gamma-1)}\,N\,F\Big([\frac{\gamma}{\gamma-1}, -\frac{\gamma}{\gamma-1}], [\frac{\gamma}{\gamma-1}+1], \\ &\quad N^{(\gamma-1)/\gamma} C^{-1/\gamma}\, {C_1}^{-1}\Big). \end{aligned} \tag{6.38}$$

Summarizing, in the case of a polytropic equation of state $p = C\,\rho^\gamma$, the metric is given by

$$ds^2 = -N^2dt^2 + \frac{dN^2}{C_0 - \Lambda\, N^2 + 2\kappa \int^N \left[B - C^{\frac{-1}{\gamma}}\, N^{\frac{\gamma-1}{\gamma}}\right]^{\frac{\gamma}{\gamma-1}} dN}$$

$$+ \left(C_0 - \Lambda\, N^2 + 2\kappa \int^N \left[B - C^{\frac{-1}{\gamma}}\, N^{\frac{\gamma-1}{\gamma}}\right]^{\frac{\gamma}{\gamma-1}} dN \right) d\phi^2. \quad (6.39)$$

while the pressure and density are:

$$p = \frac{1}{N}\left[B - C^{\frac{-1}{\gamma}}\, N^{\frac{\gamma-1}{\gamma}}\right]^{\frac{\gamma}{\gamma-1}},\ \rho = C^{\frac{-1}{\gamma}}\, N^{\frac{-1}{\gamma}}\left[B - C^{\frac{-1}{\gamma}}\, N^{\frac{\gamma-1}{\gamma}}\right]^{\frac{1}{\gamma-1}}. \quad (6.40)$$

Within this class of solutions, the nonrelativistic degenerate fermions $\gamma = 2$, and the nonrelativistic matter and radiation $\gamma = 3/2$, are worthy of mention. Incidentally, the study of static circularly symmetric cosmological (absence of the cosmological constant) spacetimes, coupled to perfect fluids fulfilling the polytropic law was accomplished in Cornish and Frankel (1991).

6.5 Oppenheimer–Volkoff Equation

Although when Einstein equations have been fulfilled the Bianchi identities (energy–momentum conservation) trivially hold, it is of interest to establish certain properties arising from the Oppenheimer–Volkoff equation in Oppenheimer and Volkoff (1939): see for instance Cruz and Zanelli (1995) in (2+1) gravity. An alternative derivation of this equation consists in differentiating with respect to r the Einstein G_{rr}–equation (6.3); this yields

$$\kappa \frac{dp}{dr} = \frac{1}{rN}\frac{dN}{dr}\left(\frac{dG^2}{dr} - \frac{G^2}{r}\right) + \frac{G^2}{rN}\left(\frac{d^2N}{dr^2} - \frac{1}{N}(\frac{dN}{dr})^2\right), \quad (6.41)$$

substituting the second derivative $\frac{d^2N}{dr^2}$ from (6.4), and the first derivative $\frac{dN}{dr}$ from the G_{rr}-equation into(6.41) one arrives at the Oppenheimer–Volkoff equation:

$$\frac{dp}{dr} = -\frac{r}{G^2}(\kappa p - \Lambda)(\rho + p). \quad (6.42)$$

At the circle of vanishing pressure $p(a) = 0$ at $r = a$ one obtains that the pressure gradient amounts to

$$\frac{dp}{dr}|_{r=a} = \frac{\Lambda a}{G(a)^2}\rho(a). \quad (6.43)$$

Since inside the circle the pressure is positive, $p(r < a) > 0$, hence at the circle $r = a$ the pressure gradient has to be non-positive; consequently the cosmological constant ought to be negative, $\Lambda = -1/l^2 < 0$. We shall continue to use Λ instead of $-1/l^2$, keeping in mind that Λ is a negative constant.

The definition of the mass contained in the circle of radius a is given by

$$M := 2\pi \int_0^a \rho(r)\, r\, dr, \quad (6.44)$$

and since the metric components $g_{rr} = 1/G(r)^2$ has to be positive in the domain of definition of the solution, then there exits an upper limit for the mass, namely

$$M \leq \frac{\pi}{\kappa}(C - \Lambda\, a^2). \tag{6.45}$$

Matter is said to be microscopically stable if $dp/d\rho \geq 0$, which is equivalent to the statement that the speed of sound is less than the velocity of light. Since (6.41) can be written as

$$\frac{dp}{d\rho} = -\frac{r}{G^2}(\kappa\, p - \Lambda)(\rho + p)/\frac{d\rho}{dr}, \tag{6.46}$$

one concludes that for a microscopically stable fluid with positive pressure p and positive density ρ, this density is monotonically decreasing $d\rho/dr < 0$.

For our general solution in coordinates $\{t, N, \theta\}$, metric (6.20), from the expression (6.21) for the pressure, one establishes

$$\frac{dp}{d\rho} = -\frac{1}{N}(\rho + p)/\frac{d\rho}{dN}, \tag{6.47}$$

therefore the density is monotonically decreasing $d\rho/dN < 0$ if the matter is microscopically stable $dp/d\rho \geq 0$, and conversely.

Moreover, our fluids, fulfilling the barotropic state equation $p = \gamma\rho$, $\gamma > 0$, as well as those ones obeying the polytropic law $p = C\rho^\gamma$, $C > 0, \gamma > 0$, are microscopically stable fluids.

6.6 Perfect Fluid Solution with Constant Density

As has been established above, for $\rho =$ const. the Cotton tensor vanishes, and consequently the corresponding conformally flat space is unique; we shall refer to it as the incompressible perfect fluid.

In this section, it is shown that one can achieve a full correspondence of the metrics and structural functions for incompressible perfect fluids in (2+1) and (3+1) gravities. By an appropriate choice of the constant densities and cosmological constants, via a dimensional reduction (freezing of one of the spatial coordinates of the (3+1) spacetime), one obtains the (2+1) metric structure from the (3+1) solution. To achieve the intended purpose, the conformally flat static spherically symmetric perfect fluid (3+1) solution with cosmological constant is presented in a form which allows comparison with the incompressible static circularly symmetric perfect fluid with Λ–term of the (2+1) gravity.

In the canonical coordinate system $\{t, N, \theta\}$, for $\rho = \rho_0$, the metric, the expression of the function H, which in its turn establishes the relation to the radial coordinate r, and the pressure are given by:

$$ds^2 = -N^2 dt^2 + \frac{dN^2}{H} + H d\theta^2, \tag{6.48}$$

$$H = C_0 + 2C_1 N - (\Lambda + \kappa\rho_0)N^2 =: r^2, \tag{6.49}$$

$$p = -\rho_0 + \frac{C_1}{\kappa}\frac{1}{N}. \tag{6.50}$$

This unfamiliar-looking solution can be given in terms of the radial variable r by expressing N as function of r, $N = N(r)$.

Having in mind the comparison of the (2+1) incompressible perfect fluid with its (3+1) relative – the Schwarzschild interior solution – we shall derive it from the very beginning by integrating the Einstein equations (6.3) in coordinates $\{t, r, \theta\}$.

For $\rho = \text{const.}$, the integral of (6.8) gives

$$G(r) = \sqrt{C - (\kappa\rho + \Lambda)r^2}. \tag{6.51}$$

Substituting $G(r)$ from (6.51) into (6.10), one obtains

$$N(r) = n_0 - \frac{n_1}{\Lambda + \kappa\rho} G(r), \tag{6.52}$$

which can be written as $N(r) = C_1 + C_2\, G(r)$.

The evaluation of pressure $p(r)$ from (6.3) yields

$$\kappa p(r) = \frac{1}{(\kappa\rho + \Lambda)N(r)} \left[n_1 \kappa\rho\, G(r) + n_0 \Lambda(\kappa\rho + \Lambda)\right]. \tag{6.53}$$

This pressure has to vanish at the boundary $r = a$, which imposes a relation on the constants: $n_0 = -n_1\, \kappa\rho G(a)/[\Lambda(\kappa\rho + \Lambda)]$, where $G(a)$ is the value of the function $G(r)$ at the boundary, i.e., $G(a)$ is equal to the external value for the $G(r)$ corresponding with the vacuum solution plus Λ. A similar comment applies to $N(a)$. Replacing n_0 in (6.52), the function $N(r)$ becomes

$$N(r) = -\frac{n_1}{\Lambda(\kappa\rho + \Lambda)} \left[\kappa\rho G(a) + \Lambda G(r)\right]. \tag{6.54}$$

Evaluating $N(r)$ at $r = a$, establishes that $n_1 = -\Lambda\, N(a)/G(a)$. Consequently, $N(r)$ amounts to

$$N(r) = \frac{N(a)}{G(a)(\kappa\rho + \Lambda)} \left[\kappa\rho\, G(a) + \Lambda\, G(r)\right]. \tag{6.55}$$

Substituting n_0, n_1, and $N(r)$ into (6.53), one gets

$$p(r) = \rho\,\Lambda\, \frac{G(a) - G(r)}{\kappa\rho G(a) + \Lambda G(r)}. \tag{6.56}$$

Summarizing, the (2+1) metric for an incompressible perfect fluid is given by

$$ds^2 = -N(r)^2 dt^2 + \frac{dr^2}{G(r)^2} + r^2 d\theta^2 \tag{6.57}$$

with structural functions $G(r)$ from (6.51) and $N(r)$ from (6.55) and characterized by a density $\rho = \text{const.}$ and pressure p given by (6.56).

6.6.1 (3+1) Static Spherically Symmetric Perfect Fluid Solution

In this subsection we review the main structure of the interior perfect fluid solution in the presence of the cosmological constant Λ – the interior Schwarzschild metric with Λ – for the $(3+1)$ static spherically symmetric metric of the form

$$ds^2 = -N(r)^2 dt^2 + \frac{dr^2}{G(r)^2} + r^2\left(d\theta^2 + \sin^2\theta d\phi^2\right). \tag{6.58}$$

The Einstein equations with cosmological constant for perfect fluids for the metric (6.58) explicitly amount to

$$\begin{aligned}
G_{tt} &= \frac{N^2}{r^2}\left(r\frac{dG^2}{dr} + G^2 - 1\right) = -N^2(\kappa\rho + \Lambda),\\
G_{rr} &= -\frac{1}{G^2 N r^2}\left(2rG^2\frac{dN}{dr} - N + NG^2\right) = -\frac{1}{G^2}(\kappa p - \Lambda),\\
G_{\theta\theta} &= -\frac{r}{N}\left(G^2\frac{dN}{dr} + \frac{1}{2}N\frac{dG^2}{dr} + rG^2\frac{d^2N}{dr^2} + \frac{r}{2}\frac{dN}{dr}\frac{dG^2}{dr}\right) = -r^2(\kappa p - \Lambda),\\
G_{\phi\phi} &= \sin^2\theta\ G_{\theta\theta}.
\end{aligned} \tag{6.59}$$

In what follows we shall omit the dependence on r of the functions we are dealing with, except when needed for clarity. The Einstein equations for a perfect fluid energy–momentum tensor in four dimensions have the same form as the ones in three dimensions, except for the modifications due to the change of dimensionality; for instance, the expressions of the scalar curvatures R are different. Because of the corresponding equations can be found in textbooks, we do not exhibit them here explicitly.

Since we are interested in conformally flat solutions, we require the vanishing of the conformal Weyl tensor, which for static spherically symmetric perfect fluids fulfills the following equation

$$\frac{d}{dr}\left(\frac{G^2 - 1}{r^2}\right) = 0 \longrightarrow G(r) = \sqrt{1 + c_0 r^2}. \tag{6.60}$$

On the other hand, from the equation G_{tt}, one arrives at

$$G(r) = \sqrt{1 - \frac{1}{3}(\kappa\rho + \Lambda)r^2}, \tag{6.61}$$

therefore, comparing with (6.60), one has $c_0 = -(\kappa\rho + \Lambda)/3, \longrightarrow \rho = \text{const}$. Hence, the resulting solution corresponds to an incompressible perfect fluid; see Adler et al. (1965).

Moreover, from $(r^2G^2G_{rr} - G_{\theta\theta}) = 0$, taking into account the form of the function G from (6.61), the general expression of $N(r)$ is

$$N(r) = C_1 + C_2 G(r). \tag{6.62}$$

The evaluation of the pressure p, from G_{rr} Einstein equation, yields

$$\kappa p(r) = \frac{1}{3N(r)} \left[C_1 (2\Lambda - \kappa\rho) - 3C_2\, \kappa\, \rho\, G(r) \right]. \tag{6.63}$$

where $G(r)$ and $N(r)$ are determined in (6.61) and (6.62), respectively.

This result can be stated in the form of a generalization of the Gürses and Gürsey (1975) theorem to the case of Λ: the only conformally flat spherically symmetric static solution to the Einstein equations with cosmological constant for a perfect fluid is given by the metric (6.58) with structural functions $G(r)$ and $N(r)$ defined respectively by (6.61) and (6.62). Moreover, replacing in the metric (6.58) $\sin^2\theta$ by $\sinh^2\theta$, and θ^2, one obtains correspondingly the pseudo-spherical and flat branches of the solutions.

The constants C_1 and C_2 are determined through the values of structural functions at the boundary $r = a$, where the pressure vanishes, $p(r = a) = 0$; they are:

$$C_1 = 3\, C_2 \kappa\rho \frac{G(a)}{2\Lambda - \kappa\rho}, \quad C_2 = \frac{N(a)}{2G(a)} \frac{2\Lambda - \kappa\rho}{\Lambda + \kappa\rho}, \tag{6.64}$$

where, $G(a)$ is the value of the function $G(r)$ at the boundary $r = a$, i.e., $G(a)$ is equal to the external value of $G(r)$ corresponding to the vacuum plus Λ solution. A similar comment applies to $N(a)$. We shall return to this point at the end of this section.

Substituting the expressions of C_1 and C_2 into Eq. (6.62), one has

$$N(r) = \frac{N(a)}{2G(a)(\kappa\rho + \Lambda)} \left[3\kappa\rho G(a) + (2\Lambda - \kappa\rho) G(r) \right]. \tag{6.65}$$

Replacing C_1, C_2 and the above expression of $N(r)$ into (6.63), one gets

$$p(r) = \rho(2\Lambda - \kappa\rho) \frac{G(a) - G(r)}{3\kappa\rho G(a) + (2\Lambda - \kappa\rho) G(r)}. \tag{6.66}$$

For the external Schwarzschild with Λ solution, known also the Kottler solution, Stephani et al. (2003); Kottler (1918), the functions $N(r)$ and $G(r)$ are equal one to another, $N(r) = G(r)$, namely

$$N(r) = G(r) = \sqrt{1 - \frac{2m}{r} - \frac{\Lambda}{3} r^2}, \text{ for } r \geq a. \tag{6.67}$$

Evaluating the mass contained in the sphere of radius a for a constant density $\kappa\rho$, one obtains $2m = \kappa\rho a^3/3$, therefore

$$N(r) = G(r) = \sqrt{1 - \frac{\kappa\rho}{3} \frac{a^3}{r} - \frac{\Lambda}{3} r^2}, \text{for } r \geq a, \tag{6.68}$$

consequently at $r = a$, one has

$$N(a) = G(a) = \sqrt{1 - \frac{\kappa\rho + \Lambda}{3} a^2}. \tag{6.69}$$

In the limit of vanishing cosmological constant, $\Lambda = 0$, one arrives at the interior Schwarzschild solution.

6.6.2 Comparison Table

A comparison table of perfect fluid solutions with constant ρ is given:

Table 6.1 *Incompressible perfect fluid solutions*

(3 + 1) solution	(2 + 1) solution
$ds^2 = -N^2dt^2 + \frac{dr^2}{G^2} + r^2d\Omega^2$ $G^2 = 1 - \frac{1}{3}(\kappa\rho + \Lambda)r^2$ $N = \frac{1}{2(\kappa\rho+\Lambda)}\frac{N(a)}{G(a)}[3\kappa\rho G(a)$ $+(2\Lambda - \kappa\rho)G(r)]$ $p = \rho\frac{(2\Lambda-\kappa\rho)(G(a)-G(r))}{3\kappa\rho G(a)+(2\Lambda-\kappa\rho)G(r)}$	$ds^2 = -N^2dt^2 + \frac{dr^2}{G^2} + r^2d\phi^2$ $G^2 = C - (\kappa\rho + \Lambda)r^2$ $N = \frac{1}{(\kappa\rho+\Lambda)}\frac{N(a)}{G(a)}[\kappa\rho G(a) + \Lambda G(r)]$ $+\Lambda G(r)]$ $p = \rho\Lambda\frac{G(a)-G(r)}{\kappa\rho G(a)+\Lambda G(r)}$
Kottler:	(anti-)de Sitter:
$G(a)^2 = 1 - \frac{2m}{r} - \frac{1}{3}\Lambda r^2;$ $2m = \kappa\rho a^3/3$ $N(a) = G(a)$	$G(a)^2 = -M_\infty - \Lambda a^2;$ $C = \kappa\rho a^2 - M_\infty > 0$ $N(a) = G(a)$
$2\Lambda_4 - \kappa_4\rho_4 \rightarrow 6\Lambda_3, \kappa_4\rho_4 \rightarrow 2\kappa_3\rho_3$ $G_4(r) \rightarrow G_3(r), N_4(r) \rightarrow N_3(r), \kappa_4 p_4(r) \rightarrow 2\kappa_3 p_3(r)$	

Comparing the structure corresponding to perfect fluid solutions with constant ρ in (3+1) gravity with the structure of the (2+1) incompressible perfect fluid solution one arrives at the following correspondence: $2\Lambda_4 - \kappa_4\rho_4 \rightarrow 6\Lambda_3$, $3\kappa_4\rho_4 \rightarrow 6\kappa_3\rho_3$, which yields $G_4(r) \rightarrow G_3(r), N_4(r) \rightarrow N_3(r)$, $\kappa_4 p_4(r) \rightarrow 2\kappa_3 p_3(r)$. Remembering that in (2+1) gravity there is no Newtonian limit; the choice of κ_3 is free, thus by selecting κ_3 appropriately one can achieve that $p_4(r) \rightarrow p_3(r)$ and $\rho_4 \rightarrow \rho_3$.

From this comparison table one can easily conclude that the (2+1) incompressible perfect fluid can be derived from the Schwarzschild interior metric by a simple dimensional reduction: freezing one of the spatial coordinates, say $\theta = \pi/2$, in the (3+1) solution, one obtains the corresponding (2+1) spacetime.

Since we accomplished a scaling transformation of the r-coordinate, accompanied with the inverse scaling of the angular coordinate ϕ, one may argue that a conical singularity could arise; one may overcome this problem by saying that the angular coordinate should be fixed once one brings the canonical form to the (2+1) metric with $G_3(r) = \sqrt{1 - (\kappa\rho + \Lambda)r^2}$.

In this chapter all perfect fluid solutions for the static circularly symmetric spacetime have been derived. The general solution is presented in the standard coordinate system $\{t, r, \theta\}$, and alternatively, in a system – the canonical one – with coordinates $\{t, N, \theta\}$. From the physical point of view, particularly interesting are those fluids fulfilling the linear (barotropic) equation of state, $p = \gamma\rho$, as well as those subjected to the polytropic law $p = \rho^{\gamma}$; both families are derived in details from our general metric referred to the coordinate system $\{t, N, \theta\}$. Therefore, the derived solutions describe, among other things, stiff matter, pure radiation, incoherent radiation, nonrelativistic degenerate fermions, etc. The incompressible perfect fluid solution with cosmological constant of the (2+1) gravity is singled out among all static circularly spacetimes as the only conformally flat space – its Cotton tensor vanishes – sharing this conformally flatness property with its (3+1) relative, the Schwarzschild interior perfect (incompressible) fluid solution with Λ; a comparison table for these incompressible fluids is included.

7
Hydrodynamic Equilibrium

Under Buchdahl's conditions on the behavior of the density and the pressure for regular fluid static stars in the presence of a cosmological constant, the bounds of the mass are determined in any dimension. Here we work out the n-dimensional case because of its parallelism with the $3D$ case.

For $(2+1)$-dimensional perfect fluid stars in hydrodynamic equilibrium there are no bounds on the mass, except for their positiveness; the metric for a constant density distribution is derived, and its matching with the external static solution with a negative cosmological constant is accomplished. In the $(d \geq 4)$-dimensional case the existence of bounds for the mass is established. The metric for a constant density is derived and its matching with the external static solution is carried out. Some mistakes in previous works on the topic are pointed out.

7.1 Generalized Buchdahl's Theorem

The main objective of this section is to establish that a generalized Buchdahl's theorem, Buchdahl (1959), holds in $d \geq 4$-dimensions. For lower dimensions, $d < 4$, there is no room for bounds of the mass distribution.

Theorem 7.1 *If a perfect fluid distribution fulfills the conditions:*

- *it is described by a one-parameter state equation* $p = p(\mu)$,
- *the density is positive,* $\mu > 0$, *and monotonically decreasing,* $\frac{d\mu}{dr} < 0$,
- *it is microscopically stable,* $\frac{dp}{d\mu} \geq 0 \rightarrow \frac{dp}{dr} \leq 0$,

in d dimensions, there is a bound on the density given by

$$\frac{M(R)}{R^{d-3}} \leq \frac{(d-2)^2}{(d-1)^2} + \frac{\sqrt{(d-2)^4 - 2\,\Lambda\, R^2\,(d-1)\,(d-2)}}{(d-1)^2} \geq \frac{2m_{skt}}{R^{d-3}}, \tag{7.1}$$

where the subscript skt stands for Schwarzschild–Kottler–Tangherlini; see Kottler (1918), Tangherlini (1963). For vanishing cosmological constant one gets

$$\frac{1}{R^{d-3}}M(R) \leq 2\,\frac{(d-2)^2}{(d-1)^2} \geq \frac{2}{R^{d-3}}m_{skt}. \tag{7.2}$$

Moreover, for four dimensions, $d=4$, this inequality reduces just to the well-known expression

$$\frac{1}{R}M(R) \leq \frac{8}{9} \geq \frac{2}{R}m_{skt}. \tag{7.3}$$

In (2+1) dimensions there is no bound on the mass.

7.2 Stellar Equilibrium in $(2+1)$ Dimensions with Λ

The work Cruz and Zanelli (1995) is devoted to the study of the equilibrium for static perfect fluid solutions with cosmological constant, generalizing in this manner the GAK analysis. Unfortunately, their work suffers from various misprints; see García (2014). Hence, we shall address the topic from the very beginning.

The Einstein's equations for a static $(2+1)$ metric in curvature coordinates

$$g = -e^{2\,\nu(r)}dt^2 + e^{2\,\lambda(r)}\,dr^2 + r^2 d\theta^2 \tag{7.4}$$

for a perfect fluid in the presence of a cosmological constant, $T_{\mu\nu} = (p(r) + \mu(r))u_\mu u_\nu + p(r)\,g_{\mu\nu}, u_\alpha = e^{\nu(r)}\delta^t_\alpha$, can be given as

$$\begin{aligned} E^\beta{}_\alpha &:= R^\beta{}_\alpha - \frac{1}{2}\delta^\beta{}_\alpha R - \kappa T^\beta{}_\alpha + \Lambda\delta^\beta{}_\alpha, \\ T^\beta{}_\alpha &= -\rho(r)\delta^\beta{}_t\delta^t{}_\alpha + p(r)\,\delta^\beta{}_r\delta^r{}_\alpha + p(r)\,\delta^\beta{}_\theta\delta^\theta{}_\alpha, \end{aligned} \tag{7.5}$$

explicitly

$$\begin{aligned} E^1_1 &= 0 : \frac{d}{dr}\lambda = r\,(\kappa\,\mu + \Lambda)e^{2\,\lambda} \to e^{-2\,\lambda(r)} = C_0 - \frac{\kappa}{\pi}\,m\,(r)\,, \\ m\,(r) &:= 2\pi\int^r r\,(\mu + \Lambda/\kappa)\,dr = \int^r 2\pi\,r\,\mu\,dr + \frac{\pi}{\kappa}\Lambda\,r^2 =: M(r) + \frac{\pi}{\kappa}\Lambda\,r^2, \end{aligned} \tag{7.6a}$$

$$E^2_2 = 0 : \frac{d\nu}{dr} - re^{2\,\lambda(r)}\,(\kappa\,p - \Lambda) = 0 \to \nu = \int^r \frac{(\kappa p - \Lambda)\,r\,dr}{C_0 - \frac{\kappa}{\pi}\,m\,(r)}, \tag{7.6b}$$

$$E^3_3 = 0 : \frac{d^2}{dr^2}\nu + \left(\frac{d}{dr}\nu\right)^2 - \left(\frac{d}{dr}\nu\right)\frac{d}{dr}\lambda + (\Lambda - \kappa\,p)\,e^{2\,\lambda} = 0. \tag{7.6c}$$

The substitution of the derivative $\frac{d\nu}{dr}$ from (7.6*b*) into (7.6*c*) yields the same equation arising from the energy–momentum conservation law $T^{\alpha\beta}{}_{;\beta} = 0$, namely

$$\frac{dp}{dr} = -\,(\mu + p)\,\frac{d\nu}{dr} = -\mathrm{e}^{2\,\lambda} r\,(\kappa\,p - \Lambda)\,(\mu + p)$$

$$\rightarrow \frac{d\,p}{dr} = -\frac{r\,(\kappa\,p - \Lambda)\,(\mu + p)}{C_0 - \kappa\,M\,(r)\,/\pi - \Lambda\,r^2}. \tag{7.7}$$

On the other hand, the substitution of $p(r)$ from (7.6*b*) into (7.6*c*) gives rise to

$$\frac{d^2\nu}{dr^2} + \left(\frac{d\nu}{dr}\right)^2 - \left(\frac{d\nu}{dr}\right)\frac{d\lambda}{dr} - \frac{1}{r}\frac{d\nu}{dr} = 0, \tag{7.8}$$

which is a first-order equation for $N(r) := \frac{d\nu}{dr}$, which can be written in a very simple form by introducing the functions

$$\xi(r) := r\,\mathrm{e}^{\lambda(r)},\ Z := N/\xi, \tag{7.9}$$

namely

$$\frac{d}{dr}\left(\frac{N}{\xi}\right) + (\frac{N}{\xi})^2\xi = 0 \rightarrow dZ^{-1} = \xi\,dr \rightarrow Z^{-1} = C_1 + \int_0^r \xi(r)dr. \tag{7.10}$$

The equation for ν becomes

$$d\nu = Z\xi\,dr = \frac{\xi\,dr}{C_1 + \int_0^r \xi(r)dr} \rightarrow \nu(r) = \ln\,[C_1 + \int_0^r \xi(r)dr] + \ln C_2/2,$$

$$\nu\,(r) = \ln\left(\int_0^r \mathrm{e}^{\lambda(r)} r dr + C_1\right) + \ln C_2/2, \tag{7.11}$$

the const $C_2/2 \rightarrow 1$ by scaling the time coordinate.

Substituting this integral in (7.6*b*) one obtains the pressure

$$p\,(r) = \frac{1}{\kappa\,\mathrm{e}^{\lambda(r)}\left(\int_0^r \mathrm{e}^{\lambda(r)} r dr + C_1\right)} + \frac{\Lambda}{\kappa}. \tag{7.12}$$

The integration of the equation for $\lambda\,(r)$ gives

$$\mathrm{e}^{\lambda(r)} = 1/\sqrt{C_0 - \frac{\kappa}{\pi}\int_0^r (2\pi\,\mu(\tilde{r})\,\tilde{r}\,d\tilde{r}) - \Lambda\,r^2},$$

$$\lambda\,(r) = -1/2\,\ln\left(C_0 - \frac{\kappa}{\pi}\,M\,(r) - \Lambda\,r^2\right),\ \frac{d}{dr}M\,(r) = 2\,\pi\,\mu\,(r)\,r, \tag{7.13}$$

consequently

$$\mathrm{e}^{\nu(r)} = C_1 + \int_0^r [r'\,dr'/\sqrt{C_0 - \frac{\kappa}{\pi}\int_0^{r'} (2\pi\,\mu(\tilde{r})\,\tilde{r}\,d\tilde{r}) - \Lambda\,r'^2}]. \tag{7.14}$$

Finally, the pressure results in

$$\kappa\,p\,(r) = \frac{\sqrt{C_0 - \frac{\kappa}{\pi}\int_0^r (2\pi\,\mu(\tilde{r})\,\tilde{r}\,d\tilde{r}) - \Lambda\,r^2}}{C_1 + \int_0^r [r'\,dr'/\sqrt{C_0 - \frac{\kappa}{\pi}\int_0^{r'} (2\pi\,\mu(\tilde{r})\,\tilde{r}\,d\tilde{r}) - \Lambda\,r'^2}]} + \Lambda. \tag{7.15}$$

This relation determines the pressure p through the energy density μ in a functional manner: if p were expressed by a state equation of the form $p = p(\mu)$, the equation (7.15) gives rise to an integral differential equation for the energy as function of the variable r.

The pressure has to vanish at boundary r_b (perimeter) of the circle, $p(r_b) = 0$, where the mass function $M(r)$ determines the total mass of the fluid $M(r_b)$ on the circle. Because the metric signature has to be preserved throughout the whole spacetime, the positiveness of g_{rr} imposes an upper bound on the value of the total mass, namely

$$M(r_b) \leq \frac{\pi}{\kappa}(C_0 - \Lambda r_b^2). \tag{7.16}$$

7.2.1 Cruz–Zanelli Existence of Hydrostatic Equilibrium for $\Lambda \leq 0$

Cruz and Zanelli (1995) established that: a perfect fluid in hydrostatic equilibrium (pressure monotonically decreasing),

$$\mu(r \leq r_b) > 0,\ p(r \leq r_b) > 0,\ \frac{dp}{dr}|_{r\leq r_b} < 0,$$

is only possible for $\Lambda \leq 0$.

The condition on Λ follows from the energy–momentum conservation equation (7.7)

$$\frac{d}{dr}p\,(r) = -\frac{r\,(\kappa\,p - \Lambda)\,(\mu + p)}{C_0 - \kappa\,M\,(r)\,/\pi - \Lambda\,r^2},$$

which evaluated at the boundary yields

$$\frac{dp}{dr}|_{r=r_b} = \frac{r_b\Lambda}{C_0 - \Lambda\,r_b^2 - \kappa\,M\,(r_b)\,/\pi}\mu(r_b),$$

which for $\mu(r) \geq 0$, and $M(r_b)$ fulfilling (7.16) is non-positive only if $\Lambda \leq 0$. Moreover, since for $\Lambda \leq 0$ the right-hand side of

$$\frac{d}{dr}p\,(r) = -\,r(\kappa\,p - \Lambda)\,(p + \mu)\,g_{rr}$$

is always negative, then $p\,(r)$ is a decreasing function such that

$$p\,(r=0) = p_c > p\,(r_b) = 0.$$

7.2.2 No Buchdahl's Inequality in $(2+1)$ Hydrostatics

The existence of a Buchdahl-like bound on the mass density is based, in four and more dimensions, on an inequality arising from the energy conservation law, the Oppenheimer–Volkoff equation, expressed in terms of the metric structural functions, see (7.27). In three dimensions the corresponding equation (7.8), to which (7.27) reduces for $d = 3$, does not allow for an inequality; it integrates in quadratures, (7.11). Therefore, in three dimensions there is no Buchdahl–like bound on the mass density.

7.2.3 Static Star with Constant Density μ_0 and $\Lambda = -1/l^2 \leq 0$

The static star with uniform density μ_0 is characterized by mass and pressure given respectively as:

$$M(r) = \pi\,\mu_0\,r^2, m(r) = \pi\,\mu_0\,r^2 + \frac{\pi}{\kappa}\Lambda\,r^2,$$
$$p(r) = \mu_0\frac{\sqrt{C_0 - \frac{\kappa}{\pi}m(r)} - \sqrt{C_0 - \frac{\kappa}{\pi}m(r_b)}}{-\sqrt{C_0 - \frac{\kappa}{\pi}m(r)} - \frac{\kappa}{\Lambda}\mu_0\sqrt{C_0 - \frac{\kappa}{\pi}m(r_b)}}. \tag{7.17}$$

A star of uniform density in hydrostatic equilibrium ($\Lambda = -1/l^2$) possesses central mass and pressure of the form

$$M(0) = 0, m(0) = 0,$$
$$p_c = \mu_0\frac{\sqrt{C_0} - \sqrt{C_0 - \frac{\kappa}{\pi}m(r_b)}}{-\sqrt{C_0} + \kappa\,l^2\mu_0\sqrt{C_0 - \frac{\kappa}{\pi}m(r_b)}}, \tag{7.18}$$

at the boundary $r = r_b$

$$M(r_b) = \pi\,r_b^2\mu_0 = \mu_0\,S_{\odot}, p(r_b) = 0,$$

while for $r_b = \sqrt{\frac{C_0}{\kappa\,\mu_0}}\sqrt{1 + \frac{1}{\kappa\,l^2\,\mu_0^2}}$ the pressure becomes infinity, $p \to \infty$, and the mass equates to

$$M = \frac{\kappa\,C_0}{\pi}(1 + \frac{1}{\kappa\,l^2\,\mu_0}), (m = \frac{\kappa\,C_0}{\pi}(1 - \frac{1}{\kappa^2\,l^4\,\mu_0^2})).$$

The evaluation of $e^{2\nu}$ or the metric component $g_{tt} = -e^{2\nu}$ yields

$$e^{2\nu} = \left(\frac{\kappa\,\mu_0\,l^2\sqrt{C_0 - \frac{\kappa}{\pi}m(r_b)} - \sqrt{C_0 - \frac{\kappa}{\pi}m(r)}}{\kappa\,\mu_0\,l^2 - 1}\right)^2. \tag{7.19}$$

which is different compared with the expression determined by Eq. (23) of Cruz and Zanelli (1995).

The external solution to which the uniform fluid solution can be matched is the static anti-de Sitter metric with parameter M_0, known also as the static BTZ solution

$$g = -(-M_0 + \frac{r^2}{l^2})dt^2 + dr^2/(-M_0 + \frac{r^2}{l^2}) + r^2 d\phi^2 \tag{7.20}$$

the continuity at the boundary r_b of the metric for the fluid is achieved

$$e^{2\nu(r_b)} = e^{-2\lambda(r_b)} = C_0 + \frac{r_b^2}{l^2} - \frac{\kappa}{\pi}M = -M_0 + \frac{r_b^2}{l^2}$$
$$\to \kappa = \pi,\; M(r_b) = M = M_0 - C_0. \tag{7.21}$$

The static perfect fluid solution with $\Lambda = -1/l^2$ exhibits an event horizon at $r_h = \frac{C\,l^2}{\kappa\,\mu_0\,l^2-1}$.

7.3 Buchdahl Theorem in *d* Dimensions

The static spherical symmetric metric in curvature Schwarzschild coordinates for a d-dimensional spacetime is given by

$$\begin{aligned} g &:= -\mathrm{e}^{2\,\nu(r)}dt^2 + \mathrm{e}^{2\,\lambda(r)}dr^2 + r^2\,d\Omega^2_{d-2}, \\ d\Omega^2_{d-2} &= d\chi_2^2 + \sin^2(\chi_2)d\chi_3^2 + \ldots + \sin^2(\chi_2)\sin^2(\chi_3)\ldots\sin^2(\chi_{d-2})\,d\chi^2_{d-1}. \end{aligned} \tag{7.22}$$

In the presence of a perfect fluid and a cosmological constant Λ, the Einstein equation are:

$$\begin{aligned} E^\beta{}_\alpha &:= R^\beta{}_\alpha - \frac{1}{2}\delta^\beta{}_\alpha R - \kappa T^\beta{}_\alpha + \Lambda\delta^\beta{}_\alpha = 0, \\ T^\beta{}_\alpha &= -\rho(r)\delta^\beta{}_0\delta^0{}_\alpha + p(r)\,\delta^\beta{}_1\delta^1{}_\alpha + \cdots + p(r)\,\delta^\beta{}_{d-1}\delta^{d-1}{}_\alpha. \end{aligned} \tag{7.23}$$

The time component $E_{t\,t}$ of the Einstein equations yields

$$\kappa\,\rho\,(r) + \Lambda = (d-2)\,\mathrm{e}^{-2\,\lambda(r)}\left[\frac{1}{r}\frac{d\lambda\,(r)}{dr} - \frac{(d-3)}{2r^2} + \frac{(d-3)}{2r^2}\mathrm{e}^{2\,\lambda(r)}\right], \tag{7.24}$$

while the radial Einstein equation component $E_{r\,r}$ gives rise to

$$\kappa\,p\,(r) - \Lambda = (d-2)\,\mathrm{e}^{-2\,\lambda(r)}\left[\frac{1}{r}\frac{d\nu\,(r)}{dr} + \frac{(d-3)}{2r^2} - \frac{(d-3)}{2r^2}\mathrm{e}^{2\,\lambda(r)}\right]. \tag{7.25}$$

The remaining Einstein equations $E_{d-2\,d-2}$ related to angular components yield

$$\begin{aligned} \mathrm{e}^{2\,\lambda}\,(\kappa\,p\,(r) - \Lambda) = {} & r^2\frac{d^2\nu}{dr^2} + \left(\frac{d\nu}{dr}\right)^2 r^2 + r\,(d-3)\,\frac{d\nu}{dr} - r^2\frac{d\lambda}{dr}\frac{d\nu}{dr} \\ & -\frac{d-3}{2}\left(\mathrm{e}^{2\lambda}(d-4) - (d-4) + 2\,r\frac{d\lambda}{dr}\right). \end{aligned} \tag{7.26}$$

Replacing the expression of $p(r)$ from (7.25) into (7.26) one gets a second-order equation for $\nu(r)$, namely

$$\frac{d^2\nu}{dr^2} + \left(\frac{d\nu}{dr}\right)^2 - \frac{1}{r}\frac{d\nu}{dr} - \left(\frac{d\nu}{dr}\right)\frac{d\lambda}{dr} = (d-3)\left[\frac{1}{r}\frac{d\lambda}{dr} - \frac{\mathrm{e}^{2\lambda}}{r^2} + \frac{1}{r^2}\right]. \tag{7.27}$$

This equation can be rewritten as

$$r\,\mathrm{e}^{(\lambda-\nu)}\,\frac{d}{dr}\left(\frac{1}{r}\mathrm{e}^{-\lambda}\,\frac{d\mathrm{e}^{\nu}}{dr}\right) = -\,(d-3)\,\mathrm{e}^{2\lambda}\left[\frac{r}{2}\frac{d}{dr}\left(\frac{1}{r^2}\,\mathrm{e}^{-2\lambda}\right) + \frac{1}{r^2}\right]. \tag{7.28}$$

The right-hand side of (7.28), containing λ and its derivative, can be significatively simplified by using the solution of the density equation (7.24) for λ, namely

$$e^{-2\lambda} = 1 - 2\frac{r^2\Omega(r)}{d-2} + 2C_1\frac{r^{-d+3}}{d-2} - 2\frac{\Lambda r^2}{(d-1)(d-2)}, \tag{7.29}$$

where the function $\Omega(r)$ is defined by

$$\Omega = \kappa r^{1-d}\int r^{d-2}\rho(r)\,dr. \tag{7.30}$$

Therefore, one may write (7.28) as

$$r\,e^{-\nu}\frac{d}{dr}\left(\frac{1}{r}e^{-\lambda}\frac{de^{\nu}}{dr}\right) = \frac{d-3}{d-2}r\,e^{\lambda}\left(\frac{d}{dr}\Omega(r) + C_1\frac{(d-1)}{r^d}\right). \tag{7.31}$$

To bring (7.31) to the Buchdahl's form one introduces the variables

$$\zeta := e^{\nu},\; y := e^{-\lambda},\; d\xi = 2\frac{r}{y}dr;\; x = r^2; \tag{7.32}$$

consequently one gets

$$\frac{d^2\zeta}{d\xi^2} = \frac{1}{2}\frac{d-3}{d-2}\zeta\left(\frac{1}{y}\frac{d}{d\xi}\Omega + \frac{C_1}{2}\frac{d-1}{r^{1+d}}\right) = \frac{1}{2}\frac{d-3}{d-2}\zeta\left(\frac{d}{dx}\Omega + \frac{C_1}{2}\frac{d-1}{x^{(1+d)/2}}\right). \tag{7.33}$$

To have solutions with regular centers one imposes certain conditions on the density and on the metric components; therefore, one requires, for $d > 4$, C_1 to be zero: $C_1 = 0$. Thus for any dimension equal to or greater than four, the leading term in g^{rr} is 1.

The three-dimensional case, $d = 3$, is singled out, since in this case the leading term is a constant $1 + C_1 = M$ related to the mass for negative cosmological constant. Moreover, the second-order equation equates to zero, $\frac{d^2\zeta}{d\xi^2} = 0$. Consequently, its solution becomes

$$\frac{d^2\zeta}{d\xi^2} = 0 \to \zeta = A_1\,\xi + A_0 \simeq e^{\nu} = A_0 + A_1\int_0^r 2re^{\lambda}dr, \tag{7.34}$$

where it has been taking into account the definitions given in (7.32). Hence, there is no room for a restriction on the mass via a Buchdahl procedure.

7.3.1 Buchdahl's Inequalities

Under the conditions of the Lagrange mean value theorem – see, for instance, Korn and Korn (1961)[1] – the integral of $\frac{d\zeta}{d\xi}$ yields

$$\begin{aligned}(\xi - 0)\frac{d\zeta}{d\xi}(\bar{\xi}) &\overset{\text{m.v.}}{=} \int_0^{\xi}\frac{d\zeta}{d\xi}d\xi = [\zeta(\xi) - \zeta(0)]\\ \to \frac{d\zeta}{d\xi}(\xi_{m.v.}) &= \frac{\zeta(\xi) - \zeta(0)}{\xi}, 0 < \bar{\xi} < \xi,\end{aligned} \tag{7.35}$$

[1] Lagrange mean value theorem. If $f(x)$ is continuous in $[a, b]$:
i) and differentiable on (a, b), then in (a, b) there exists a number X such that $f(b) - f(a) = f'(X)(b - a)$,
ii) then on (a, b) there exists a number X such that $\int_a^b f(x)dx = f(X)(b - a)$.

where $\xi_{m.v.}$ is the value of ξ where the mean value Lagrange theorem holds; let us call it the "mean value" ξ coordinate. Moreover, since $\frac{d^2\zeta}{d\xi^2} \leq 0 \to \frac{d\zeta}{d\xi}(\xi_0) \geq \frac{d\zeta}{d\xi}(\xi)$, $\xi_0 < \xi$, and considering that $\zeta(\xi)$ is a positive (increasing) function, then

$$\zeta(\xi_0) < \zeta(\xi),\, \xi_0 < \xi \to \frac{d\zeta}{d\xi}(\bar{\xi} > \xi_{m.v.}) \leq \frac{\zeta(\xi) - \zeta(0)}{\xi}, 0 < \xi_{m.v.} < \bar{\xi} < \xi. \tag{7.36}$$

Because of the positiveness of ξ and $\zeta(\xi)$ and its increasing property implying $\frac{d\zeta}{d\xi} \geq 0$, one can write

$$\frac{d\zeta}{d\xi}(\bar{\xi}) \leq \frac{\zeta(\xi)}{\xi}, 0 < \xi_{m.v.} < \bar{\xi} < \xi. \tag{7.37}$$

In various publications, the inequality referred to as the one due to the mean value theorem is

$$\frac{d\zeta}{d\xi}(\xi) \leq \frac{\zeta(\xi)}{\xi}; \tag{7.38}$$

see, for instance, Straumann (1984), paragraph 6.6.3, Ponce de Leon and Cruz (2000), Mak and Harko (2000), and Zarro (2009). This last inequality is treated as a function of ξ in the whole range of its variability, without any comment on the restrictions imposed by the mean value theorem. The writing of (7.38) in that way is an abuse of typing, although it would be considered correct if one should have in mind that ξ stands for the $\bar{\xi}$ at the left-hand side of this inequality, which should be very doubtful. Whatever the case, it is essentially misleading.

In terms of the original variables, the above inequality reads

$$\frac{1}{2\bar{r}} \mathrm{e}^{-\lambda(\bar{r})}\, \mathrm{e}^{\nu(\bar{r})} \frac{d\nu}{dr}(\bar{r}) \leq \frac{1}{2\int_0^r r\mathrm{e}^{\lambda} dr} \mathrm{e}^{\nu(r)}. \tag{7.39}$$

Isolating from (7.25) the derivative $\frac{d\nu}{dr}$ and substituting in the middle of the resulting equation the expression for $\mathrm{e}^{-2\lambda}$ from (7.29), one gets

$$\frac{1}{r}\frac{d}{dr}\nu = \mathrm{e}^{2\lambda}\left[\frac{\kappa}{d-2}p\,(r) + \frac{d-3}{d-2}\Omega\,(r) - \frac{2}{(d-1)\,(d-2)}\Lambda\right], \tag{7.40}$$

which substituted into (7.39) yields the inequality

$$\frac{1}{2}\mathrm{e}^{\lambda(\bar{r})+\nu(\bar{r})}\left[\frac{\kappa\, p\,(\bar{r})}{d-2} + \frac{d-3}{d-2}\Omega\,(\bar{r}) - \frac{2\,\Lambda}{(d-1)\,(d-2)}\right]\int_0^r 2\,r\mathrm{e}^{\lambda} dr \leq \mathrm{e}^{\nu(r)}. \tag{7.41}$$

The actual problem of handling the inequality (7.41) is hidden in the evaluation of the integral $\xi(r) = \int_0^r 2\,r\mathrm{e}^{\lambda} dr$. To avoid dealing with specific models, i.e., to give particular densities $\rho(r)$, one can follow the Buchdahl procedure consisting in replacing $\Omega(r)$ by $\Omega(R)$ and use the resulting inequality and the arising integral expressed in terms of elementary functions, namely, radicals. Since $\rho(r)$ is assumed to be a monotonically decreasing function, then $\frac{d\rho}{dr} \leq 0$, therefore Ω is a decreasing function too; consequently

$$\frac{d\Omega}{dr} \leq 0 \rightarrow \Omega(r) > \Omega(R),\ \mathrm{e}^{\lambda(\Omega(R),r)} \leq \mathrm{e}^{\lambda(\Omega(r),r)},$$
$$\int_0^r \mathrm{e}^{\lambda(r)} dr^2 \geq \int_0^r \mathrm{e}^{\lambda(\Omega(R),r)} dr^2 \geq \int_0^{\bar{r}} \mathrm{e}^{\lambda(\Omega(R),r)} dr^2, 0 < \bar{r} \leq r \leq R. \tag{7.42}$$

Thus, the inequality (7.41) can be rewritten as

$$\frac{1}{2}\mathrm{e}^{\lambda(\bar{r})}\left[\frac{\kappa\, p(\bar{r})}{d-2} + \frac{d-3}{d-2}\Omega(\bar{r}) - \frac{2}{(d-1)(d-2)}\Lambda\right]\int_0^{\bar{r}} 2\, r\mathrm{e}^{\lambda(\Omega(R),r)} dr \leq 1. \tag{7.43}$$

The integration of $\xi(r)$ for $\Omega(R)$ is straightforward:

$$\xi(\Omega(R), r) = \int_0^r \frac{dr^2}{\sqrt{1 - 2\,\frac{r^2}{d-2}\omega(R)}} = -\frac{d-2}{\omega(R)}\Delta^{1/2}(r)|_0^r,$$
$$\Delta := 1 - 2\,\frac{r^2}{d-2}\omega(R),\ \omega(R) := \Omega\,(R) + \frac{\Lambda}{(d-1)},$$
$$\omega(R) = \frac{1}{R^{d-1}}\int_0^R \kappa\, r^{d-2}\rho\,(r)\, dr + \frac{\Lambda}{d-1} =: \frac{M(R)}{R^{d-1}} + \frac{\Lambda}{d-1}. \tag{7.44}$$

At the boundary $r = R$ the fluid metric has to be matched to the external vacuum with Λ solution known as the d-dimensional Schwarzschild–Kottler–Tangherlini (SKT) solution, which is given by the metric (7.22) with structural functions

$$\mathrm{e}^{-2\lambda(r)} = \mathrm{e}^{2\nu(r)} = 1 - 2\,\frac{r^2}{d-2}\left(\frac{2m}{r^{d-1}} + \frac{\Lambda}{d-1}\right) = 1 - 2\,\frac{r^2}{d-2}\,\omega_{skt}(r),$$
$$\omega_{skt}(r) := \frac{2m}{r^{d-1}} + \frac{\Lambda}{d-1}. \tag{7.45}$$

At the hypersphere boundary $r = R$ one has

$$\mathrm{e}_{skt}^{-\lambda(R)} = \mathrm{e}_{skt}^{\nu(R)} = \mathrm{e}_{pf}^{-\lambda(R)} = \mathrm{e}_{pf}^{\nu(R)} = \sqrt{1 - 2\,\frac{R^2}{d-2}\,\omega(R)} = \Delta^{1/2}(R),$$
$$\omega_{skt}(R) = \omega_{pf}(R) = \omega(R). \tag{7.46}$$

Thus, at $r = R$, taking into account that $p(R) = 0$ and that

$$\xi(\Omega(R), R) = -\frac{d-2}{\omega(R)}\Delta^{1/2}(r)|_0^R = \frac{d-2}{\omega(R)}(1 - \Delta^{1/2}(R))$$

the inequality (7.43) becomes

$$\frac{1}{2}\frac{d-2}{\omega(R)}\left[\frac{d-3}{d-2}\omega\,(R) - \frac{1}{d-2}\Lambda\right]\left(1 - \Delta^{1/2}(R)\right) \leq \Delta^{1/2}(R). \tag{7.47}$$

Isolating $\Delta^{1/2}(R)$ one arrives at

$$\Delta^{1/2}(R) \geq \frac{(d-3)\omega\,(R) - \Lambda}{(d-1)\omega\,(R) - \Lambda}$$
$$\rightarrow \Delta(R) = 1 - 2\frac{r^2}{d-2}\omega(R) \geq \frac{[(d-3)\omega\,(R) - \Lambda]^2}{[(d-1)\omega\,(R) - \Lambda]^2}. \tag{7.48}$$

Hence, for vanishing Λ one obtains the well-known inequality $\Delta(R, \Lambda = 0) \geq \frac{(d-3)^2}{(d-1)^2}$, which, for dimension four, $d = 4$, yields Buchdahl's inequality, $\Delta \geq \frac{1}{9}$.

The algebraic equation for ω is a cubic equation with one of the roots being zero, $\omega = 0$; the remaining quadratic equation amounts to

$$\omega^2 - 2\,\frac{\left[(d-2)^2 + \Lambda\,R^2\,(d-1)\right]}{R^2\,(d-1)^2}\omega + \frac{\Lambda\,\left[2\,(d-2) + \Lambda\,R^2\right]}{R^2\,(d-1)^2} = 0, \tag{7.49}$$

with roots $\omega_\pm$ where $\omega_+ \geq \omega(R)$ is given by

$$\omega(R) \leq \frac{(d-2)^2 + \Lambda\,R^2\,(d-1)}{R^2\,(d-1)^2} + \frac{\sqrt{(d-2)^4 - 2\,\Lambda\,R^2\,(d-1)\,(d-2)}}{R^2\,(d-1)^2}. \tag{7.50}$$

Taking into account the definition of $\omega(R)$, (7.44), through Ω and M, one finally obtains the main inequality:

$$\frac{M(R)}{R^{d-3}} \leq \frac{(d-2)^2}{(d-1)^2} + \frac{\sqrt{(d-2)^4 - 2\,\Lambda\,R^2\,(d-1)\,(d-2)}}{(d-1)^2} \geq \frac{2m_{skt}}{R^{d-3}}. \tag{7.51}$$

For vanishing cosmological constant one gets

$$\frac{M(R)}{R^{d-3}} \leq 2\,\frac{(d-2)^2}{(d-1)^2} \geq \frac{2}{R^{d-3}}m_{skt}. \tag{7.52}$$

Moreover, for the fourth dimension, $d = 4$, this inequality reduces just to the well-known expression $\frac{1}{R}M(R) \leq \frac{8}{9} \geq \frac{2}{R}m_{skt}$.

7.3.2 Constant Density Solution

In this subsection the static perfect fluid regular solution with constant density and a cosmological constant is derived and its matching to the external d-dimensional SKT solution is accomplished.

The family of static perfect fluid with constant density and Λ is unique: if the integral mass density Ω (7.30) is constant Ω_0, then $\rho =$ constant too;

$$r^{d-1}\Omega_0 = \kappa\int r^{d-2}\rho\,(r)\,dr \rightarrow \Omega_0\frac{d}{dr}r^{d-1} = \kappa r^{d-2}\rho\,(r)$$
$$\rightarrow \rho\,(r) = \frac{1}{(d-1)\kappa}\Omega_0 = \rho_0. \tag{7.53}$$

The integral of $e^{-2\lambda(r)}$, for metrics which are regular at the origin of the coordinates, yields

$$e^{-2\lambda(r)} = 1 - 2\frac{\kappa\,\rho_0\,r^2}{(d-1)(d-2)} - 2\frac{\Lambda\,r^2}{(d-1)(d-2)}. \tag{7.54}$$

As far as the integral for $\nu(r)$ from (7.27) is concerned, that equation becomes

$$\frac{d}{dr}\left(\frac{1}{r}e^{-\lambda}\frac{de^{\nu}}{dr}\right) = 0 \to e^{\nu} = C_0 + C_1\int_0^r re^{\lambda}dr \tag{7.55}$$

which, substituting e^{λ} and integrating, can be given as

$$e^{\nu(r)} = C_0 - \frac{C_1}{2}\frac{(d-1)(d-2)}{\kappa\,\rho_0 + \Lambda}e^{-\lambda(r)}. \tag{7.56}$$

By substituting $e^{\lambda(r)}$ and $e^{\nu(r)}$ and its derivatives into (7.25) one obtains the expression of the pressure $p(r)$,

$$\begin{aligned} p(r) = & \frac{2\,C_1\,(d-2)\,(\kappa\,\rho_0 + \Lambda)\,e^{-\lambda(r)}}{\kappa\,\left[2\,C_0\,(\kappa\,\rho_0 + \Lambda) - C_1\,(d-1)\,(d-2)\,e^{-\lambda(r)}\right]} \\ & + \frac{2\Lambda}{\kappa\,(d-1)} - \rho_0\frac{d-3}{d-1}. \end{aligned} \tag{7.57}$$

At the boundary $r_b = R$ of the fluid distribution, the pressure has to vanish, $p(R) = 0$; this condition determines one of the constants, namely

$$C_1 = 2\frac{C_0}{\kappa\,\rho_0}\frac{(\kappa\,\rho_0 + \Lambda)\left[\kappa\rho_0(d-3) - 2\,\Lambda\right]}{(d-2)\,(d-1)^2}e^{\lambda(R)}, \tag{7.58}$$

therefore the pressure vanishing at the boundary can be written as

$$p = \rho_0\frac{(\kappa\,d\rho_0 - 3\,\kappa\,\rho_0 - 2\,\Lambda)\left(e^{\lambda(R)} - e^{\lambda(r)}\right)e^{-\lambda(r)}}{(d-1)\,\kappa\,\rho_0 - (\kappa\,d\rho_0 - 3\,\kappa\,\rho_0 - 2\,\Lambda)\,e^{\lambda(R)-\lambda(r)}}. \tag{7.59}$$

Matching to the External SKT Solution

The external solution is the d–dimensional Schwarzschild–Kottler–Tangherlini solution, which is given by the metric (7.22) with structural functions

$$e^{-2\lambda(r)} = e^{2\nu(r)} = 1 - 2\frac{\Lambda\,r^2}{(d-1)(d-2)} - 4\frac{mr^{-d+3}}{d-2}. \tag{7.60}$$

From the continuity of the metric components across the fluid edge one gets:

$$e^{-2\lambda_f(R)} = e^{-2\lambda_{skt}(R)} \to m = \frac{\kappa\,\rho_0}{2}\frac{R^{d-1}}{d-1}, \tag{7.61}$$

and

$$\begin{aligned} & e^{2\nu_f(R)} = e^{2\nu_{skt}(R)} \to e^{\nu(R)} = C_0 - \frac{C_1}{2}\frac{(d-1)(d-2)}{\kappa\,\rho_0 + \Lambda}e^{-\lambda(R)} = e^{-\lambda(R)} \\ & \to C_0 = \left(1 + \frac{C_1}{2}\frac{(d-1)(d-2)}{\kappa\,\rho_0 + \Lambda}\right)e^{-\lambda(R)}. \end{aligned} \tag{7.62}$$

Taking into account the expression (7.58) for C_1 arising from the vanishing of the pressure at the border $r_b = R$, one arrives at

$$C_0 = \frac{\kappa \rho_0 (d-1)}{2(\kappa \rho_0 + \Lambda)} e^{-\lambda(R)},$$
$$C_1 = \frac{\kappa \rho_0 (d-3) - 2\Lambda}{(d-2)(d-1)} e^{-\lambda(R)}. \tag{7.63}$$

Finally, the metric components of a regular static solution in d-dimensions for a perfect fluid with constant density ρ_0 in the presence of a cosmological constant Λ allowing for a matching on the hypersphere R with the external Schwarzschild–Kottler–Tangherlini solution can be given as

$$1/g_{rr} = e^{-2\lambda(r)} = 1 - 2\frac{\kappa \rho_0 r^2}{(d-1)(d-2)} - 2\frac{\Lambda r^2}{(d-1)(d-2)},$$
$$-g_{tt} = e^{2\nu(r)} = e^{-2\lambda(R)} \left[1 + \frac{1}{2}\frac{\kappa \rho_0 (d-3) - 2\Lambda}{\kappa \rho_0 + \Lambda} \left(e^{-\lambda(R)} - e^{-\lambda(r)}\right)\right]^2,$$
$$p = \rho_0 \frac{[\kappa \rho_0 (d-3) - 2\Lambda]\left(e^{\lambda(R)} - e^{\lambda(r)}\right)}{(d-1)\kappa \rho_0 - [\kappa \rho_0 (d-3) - 2\Lambda] e^{\lambda(R)-\lambda(r)}} e^{-\lambda(r)}, \tag{7.64}$$

where the behavior at the frontier R becomes apparent. The event horizon, if there is any, is located at $r_h = \sqrt{\frac{(d-1)(d-2)}{2(\kappa \rho_0 + \Lambda)}}$.

8

Stationary Circularly Symmetric Perfect Fluids with Λ

In the theory of black holes, an important place is assigned to the problem of their interior: understanding it as the matter that generates the external gravitational field. Commonly, the matter is modeled through a perfect fluid. It is worth recalling that in $(3+1)$ gravity the interior solution to the Schwarzschild black hole is modeled by an interior Schwarzschild perfect fluid solution with constant energy density, but the question still remains open as to what is the interior solution for the rotating Kerr black hole? In $(2+1)$ gravity one may ask the same question with respect to the rotating BTZ black hole. As we shall see in this chapter, there is an interior solution to the BTZ black hole, modeled by a perfect fluid with constant energy density. Moreover, we succeeded in deriving perfect fluid solutions, such that their fluid velocity possesses differential rotation. On the other hand, rigidly rotating exact perfect fluid solutions derived by Rooman and Spindel (1998) and Lubo et al. (1999) are also reported. Those papers deal with stationary circularly symmetric gravitational sources of the perfect fluid type, with a cosmological constant with focus on restrictions on the physical parameters of the solutions due to the matching conditions between the interior and exterior geometries. In particular, it is established there that finite sources and absence of closed timelike curves privilege negative values of the cosmological constant. Moreover, for stationary configurations, the field equations for constant energy densities have been explicitly solved; if, additionally, the pressure vanishes, interior Gödel-like stars arise.

It is worth pointing out that the literature on stationary perfect fluid solutions of (2+1) gravity is rather scarce; among it one may cite Rooman and Spindel (1998); Lubo et al. (1999); García (2004); Cataldo (2004); Gürses (1994); Obukhov (2003). From Section 8.1 to Section 8.2.5 I follow the presentation given in García (2004), while for the remaining two sections I follow Rooman and Spindel (1998), and Lubo et al. (1999).

8.1 Stationary Differentially Rotating Perfect Fluids

The main goal of this section is the derivation of interior solutions modeled through a differentially rotating perfect fluid in the presence of a cosmological constant for the stationary circularly symmetric (2+1) metric

$$ds^2 = -N(r)^2 dt^2 + \frac{dr^2}{F(r)^2} + r^2 \left(d\phi + W(r)dt\right)^2 . \tag{8.1}$$

It will be demonstrated that this class of stationary circularly symmetric (2+1) spacetimes coupled to a perfect fluid are determined completely by an arbitrary function $N(r)$; for a given function $N(r)$ one determines the remaining function $F(r)$, the energy density, and the pressure. Various families of solutions will be explicitly given.

The Einstein equations to be considered are

$$R_{\alpha\beta} - \frac{R}{2}\, g_{\alpha\beta} + g_{\alpha\beta}\,\Lambda = \kappa T_{\alpha\beta},$$

where the perfect fluid energy–momentum tensor is given by

$$\begin{aligned} T_{\alpha\beta} &= (p+\rho)\, u_\alpha u_\beta + p\, g_{\alpha\beta}, \\ u^\alpha &= \frac{1}{\sqrt{N^2 - r^2\,(W+\Omega)^2}}\,(\delta_t{}^\alpha + \Omega\,\delta_\phi{}^\alpha)\,. \end{aligned} \tag{8.2}$$

The function $\Omega(r)$ describes the differential rotation property of the fluid; if $d\Omega(r) \neq 0$, one is dealing with a differentially rotating perfect fluid, while if $\Omega(r) = \text{const.}$, the fluid is rigidly rotating. κ stands for the (2+1) gravitational constant, $\rho(r)$ and $p(r)$ denote respectively the perfect fluid energy density and isotropic pressure.

The independent Einstein equations for the metric (8.1), coupled to a perfect fluid and a cosmological constant Λ, are:
the pressure

$$\kappa p\,(r) - \Lambda = \frac{F^2}{4r\,N^2}\left(r^3\,(\dot W)^2 + 4\,N\,\dot N\right), \tag{8.3a}$$

the energy density

$$\begin{aligned} \rho\,(r) + \Lambda = {} & \left[r\left(\frac{W}{N}\right)^2 - \frac{1}{r}\right] F\dot F - rF^2\left(\frac{W}{N}\right)^2 \frac{\dot N}{N} \\ & - \frac{r^2}{4}\left(\frac{F}{N}\right)^2 (\dot W)^2 + r\left(\frac{F}{N}\right)^2 \left[\frac{\dot F}{F} - \frac{\dot N}{N}\right] (\Omega^2 + 2\,W\,\Omega), \end{aligned} \tag{8.3b}$$

the equation for the metric function W

$$EQWd := \ddot W + \frac{3}{r}\dot W + \left[\frac{\dot F}{F} - \frac{\dot N}{N}\right]\left(\dot W + \frac{2}{r}W + \frac{2}{r}\Omega\right) = 0, \tag{8.3c}$$

and the equation for the metric function N

$$EQNd = \ddot{N} + \left(\frac{\dot{F}}{F} - r\left(\frac{W}{N}\right)^2 - \frac{1}{r}\right)\dot{N} + r\frac{\dot{F}}{F}\frac{W^2}{N} - r^2\frac{(\dot{W})^2}{N}$$
$$+\frac{r}{N}\left[\frac{\dot{F}}{F} - \frac{\dot{N}}{N}\right](\Omega^2 + 2\,W\,\Omega) = 0. \tag{8.3d}$$

Incidentally, replacing $(\Omega^2 + 2\,W\,\Omega)$ from (8.3d) in (8.3b), one gets ρ in terms of the metric functions only,

$$\kappa\rho\,(r) + \Lambda = -F^2\left(\frac{\ddot{N}}{N} - \frac{\dot{N}}{rN} - \frac{3}{4}\frac{r^2}{N^2}\dot{W}^2\right) - F\dot{F}\left(\frac{\dot{N}}{N} + \frac{1}{r}\right). \tag{8.3e}$$

Therefore, we have established the following result:

The class of stationary circularly symmetric (2+1) solutions coupled to a differentially rotating perfect fluid, $d\Omega/dr \neq 0$, and a cosmological constant is defined by: the metric (8.1), the isotropic pressure (8.3a), the fluid density (8.3b), or equivalently (8.3e), and the structural functions: $W(r)$, and $N(r)$ fulfilling the differential equations (8.3c), and (8.3d), respectively, for the four unknown functions N, F, W, and Ω. A constraint on these functions is provided by a state equation – a relation between the pressure p and the energy density ρ, thus still there is a freedom in one function; one may think of equation (8.3c) as defining the differential rotation function Ω and replace it into the equation (8.3d) for N; in this manner the constraints occur in terms of the metric functions. Some mistakes contained in the expression for Ω reported in García (2004) are here corrected.

8.2 Garcia Stationary Rigidly Rotating Perfect Fluids

The main goal of this section is the derivation of interior solutions modeled through a rigidly rotating perfect fluid in the presence of a cosmological constant for the stationary circularly symmetric (2+1) metric (8.1). It will be demonstrated that this class of stationary circularly symmetric (2+1) spacetimes coupled to a perfect fluid is determined completely by an arbitrary function $N(r)$; for a given function $N(r)$ one determines the remaining function $F(r)$, the energy density, and the pressure. Various families of solutions will be explicitly given. The perfect fluid energy–momentum tensor associated to the Einstein equations is given by

$$T_{\alpha\beta} = (p+\rho)\,u_\alpha u_\beta + p\,g_{\alpha\beta},\; u^\alpha = \frac{1}{\sqrt{N^2 - r^2\,W^2}}\,(\delta_t{}^\alpha)\,. \tag{8.4}$$

Therefore, the explicit expressions of the Einstein equations are derived from the previous ones (8.3) by setting the differential rotation equal to zero, $\Omega = 0$, hence the pressure

$$\kappa p\,(r) - \Lambda = \frac{F^2}{4r\,N^2}\left(r^3\,(\dot{W})^2 + 4\,N\,\dot{N}\right), \tag{8.5a}$$

the energy density

$$\rho\,(r) + \Lambda = \left[r\,\left(\frac{W}{N}\right)^2 - \frac{1}{r}\right] F\,\dot{F} - rF^2\left(\frac{W}{N}\right)^2\frac{\dot{N}}{N} - \frac{r^2}{4}\left(\frac{F}{N}\right)^2\dot{W}^2, \tag{8.5b}$$

the equation for the metric function W

$$EQW := \ddot{W} + \frac{3}{r}\dot{W} + \left[\frac{\dot{F}}{F} - \frac{\dot{N}}{N}\right]\left(\dot{W} + \frac{2}{r}W\right) = 0, \tag{8.5c}$$

and the equation for the metric function N

$$EQN := \ddot{N} + \left(\frac{\dot{F}}{F} - r\left(\frac{W}{N}\right)^2 - \frac{1}{r}\right)\dot{N} + r\frac{\dot{F}}{F}\frac{W^2}{N} - r^2\frac{\dot{W}^2}{N}. \tag{8.5d}$$

Another expression for ρ is given by (8.3e). Certainly, the energy–momentum conservation equations are fulfilled modulo the equations arising from the Einstein equations. Therefore, a state equation $\rho = \rho(p)$ gives rise to a third equation for the three unknown structural functions $F(r)$, $W(r)$, and $N(r)$. A second possibility of looking at the Einstein–perfect fluid set of equations consists in a particular choice of one of the structural functions, say for instance W, solving for a second structural function, N or F, and substituting this last relation into the energy and the pressure; one is then left with two functions expressed in terms of one unknown function and its derivatives. An explicit example of this approach is given in the next subsection.

The Cotton tensor for this class of metrics is of the general form

$$(C^\alpha{}_\beta) = \begin{bmatrix} c^1_1 & 0 & c^1_3 \\ 0 & c^2_2 & 0 \\ c^3_1 & 0 & -(c^1_1 + c^2_2) \end{bmatrix}, \tag{8.6}$$

in the standard correspondence $\{t, r, \phi\} \to \{1, 2, 3\}$. The eigenvalues of this tensor are

$$\begin{aligned} \lambda_1 &= c^2_2 \to \boldsymbol{S}1, \\ \lambda_{2,3} &= -\frac{1}{2}c^2_2 \pm \frac{1}{2}\sqrt{(c^2_2 + 2\,c^1_1)^2 + 4\,c^3_1\,c^1_3}, \end{aligned} \tag{8.7}$$

Depending on the radical the roots $\lambda_{2,3}$ could be complex, hence the type of the Cotton tensor would be Type I: $\{S, Z, \bar{Z}\}$. This kind of metric allows for conformally flat solutions too.

8.2.1 Rigidly Rotating Perfect Fluid Solution with $W(r) = J/(2r^2)$

Incidentally, the $W(r) = W_1/r^2$ singles out a unique class of rigidly rotating perfect fluid solutions. In fact, the equation EQW (8.5c) allows for a unique solution for W independent of the functions F and N; requiring

$$\begin{aligned}\frac{d^2W}{dr^2} + \frac{3}{r}\frac{dW}{dr} = 0, \rightarrow W = \frac{W_1}{r^2},\ W_0 = 0,\\ \frac{dW}{dr} + \frac{2}{r}W = 0, \rightarrow W = \frac{W_1}{r^2},\ W_1 = -J/2,\ \text{BTZ-like rotation.}\end{aligned} \tag{8.8}$$

Hence, the stationary circularly symmetric (2+1) metric to be studied in this subsection for the derivation of interior solutions modeled through a rigidly rotating perfect fluid with a unit 4 velocity

$$u^\alpha = 2\frac{r}{\sqrt{4\,r^2\,N^2 - J^2}}\,\delta_t{}^\alpha,$$

in the presence of a cosmological constant Λ amounts to

$$ds^2 = -N(r)^2dt^2 + \frac{dr^2}{F(r)^2} + r^2\left(d\phi - \frac{J}{2r^2}dt\right)^2. \tag{8.9}$$

This class of stationary (2+1) spacetimes coupled to a perfect fluid are determined completely by an arbitrary function $N(r)$; for a given function $N(r)$ one determines the remaining function $F(r)$, the energy density, and the pressure.

The equation EQN (8.5d), once substituted $W = -J/(2\,r^2)$, can be written as

$$EQN: \frac{d}{dr}\left(\left(4\,r^3\,\dot{N}\,N + J^2\right)\frac{F}{Nr^4}\right) = 0, \tag{8.10}$$

therefore

$$F(r) = F_0\,\frac{N\,r^4}{4\,r^3N\,\dot{N} + J^2} \rightarrow \dot{N} = \frac{1}{4}\frac{rF_0}{F} - \frac{1}{4}\frac{J^2}{r^3N}, \tag{8.11}$$

where dots denote derivatives with respect to the radial coordinate r. Substituting $W = -J/(2r^2)$ and using $\dot{N}$ from (8.11) into the expression (8.5a) for the pressure, one obtains

$$\kappa\,p(r) - \Lambda = \frac{F_0}{4}\frac{F}{N}. \tag{8.12}$$

The expression of the energy density for $W = -J/(2r^2)$ from (8.5b) yields

$$\rho\,(r)\,\kappa + \Lambda = -\frac{1}{4}\frac{J^2\left(r\,\dot{N} + N\right)F^2}{r^4N^3} + \frac{1}{4}\frac{\left(J^2 - 4\,r^2\,N^2\right)}{N^2r^3}F\,\dot{F}, \tag{8.13}$$

which, modulo the function $F(r)$ from (8.11), can be written entirely in terms of N and its derivatives as

$$\kappa\rho\,(r) + \Lambda = -\frac{F_0^2\, r^4}{4\left(4r^3 N\dot{N} + J^2\right)^3}\Big[4r^4 N\ddot{N}(J^2 - 4r^2\, N^2) + 4r^3\dot{N}\left(J^2\, r\,\dot{N} + 4r^2\, N^3 + J^2\, N\right) + 16\, J^2\, r^2\, N^2 - 3J^4\Big]. \quad (8.14)$$

Consequently, the class of rigidly rotating fluid stationary circularly symmetric solutions for the metric (8.9) are determined through an arbitrary function $N(r)$; for any given function $N(r)$ one determines the metric function $F(r)$ from (8.11), the isotropic pressure (8.12), and the fluid density (8.14).

Various families of solutions will be explicitly given in the forthcoming subsections. The particular interior García (2004) solution with constant energy density matchable to the exterior BTZ black hole will also be derived.

8.2.2 Garcia Interior Solution with Constant Energy Density

In this section the incompressible branch of solutions is studied in detail. The equation (8.14) for $\rho = \rho_0$ to determine $N(r)$ amounts to

$$(\kappa\,\rho_0 + \Lambda)\left(4\,r^3 N\,\dot{N} + J^2\right)^3 = \frac{F_0^2}{4} r^4\Big[4r^4 N\,\ddot{N}\,(4r^2 N^2 - J^2) - 4J^2 r^3\,\dot{N}\,(r\dot{N} + N) - 16r^2 N^2 (J^2 + r^3 N\dot{N}) + 3J^4\Big]. \quad (8.15)$$

We search $N(r)$ by means of an auxiliary function $Q(r)$ as

$$N(r)^2 = \frac{J^2}{4r^2} + Q(r), \quad (8.16)$$

which, substituted in (8.15), yields a simple nonlinear equation

$$2F_0^2 rQ\frac{d^2Q}{dr^2} - 2F_0^2 Q\frac{dQ}{dr} - 8(\kappa\,\rho_0 + \Lambda)\left(\frac{dQ}{dr}\right)^3 - F_0^2 r\left(\frac{dQ}{dr}\right)^2 = 0. \quad (8.17)$$

Introducing a new independent variable $x = r^2$, the above equation can be rewritten as

$$Q\frac{d^2Q}{dx^2} - 8\frac{\kappa\,\rho_0 + \Lambda}{F_0^2}\left(\frac{dQ}{dx}\right)^3 - \frac{1}{2}\left(\frac{dQ}{dx}\right)^2 = 0, \quad (8.18)$$

which can be brought to the form

$$\frac{d(\frac{dQ}{dx})}{\frac{dQ}{dx}} - \frac{d(\frac{1}{2} + 8\frac{\kappa\,\rho_0+\Lambda}{F_0^2}\frac{dQ}{dx})}{\frac{1}{2} + 8\frac{\kappa\,\rho_0+\Lambda}{F_0^2}\frac{dQ}{dx}} = \frac{dQ}{2Q}, \quad (8.19)$$

and allows the first integral

$$\frac{|\frac{dQ}{dx}|}{|\frac{1}{2} + 8\frac{\kappa\,\rho_0+\Lambda}{F_0^2}\frac{dQ}{dx}|} = \sqrt{\frac{|Q|}{|Q_0|}}, \quad (8.20)$$

where vertical bars stand for absolute values. The above relation can be put into the differential form

$$d\left(B_0 - 4\frac{\kappa\,\rho_0 + \Lambda}{F_0^2 Q_0}\,x\right) = d\left(1 - 8\frac{\kappa\,\rho_0 + \Lambda}{F_0^2}\sqrt{\frac{|Q|}{|Q_0|}}\right)^2; \tag{8.21}$$

hence, integrating one obtains

$$Q(r) = \epsilon_Q Q_0 \left(\frac{F_0^2}{8(\kappa\,\rho_0 + \Lambda)}\right)^2 (1 - \epsilon K(r))^2,$$

$$K(r) := \sqrt{B_0 - 4\epsilon_Q \frac{\kappa\,\rho_0 + \Lambda}{F_0^2 Q_0} r^2}, \tag{8.22}$$

where $\epsilon_Q = \pm$ and $\epsilon = \pm$ assume their signs $\pm$ independently; of course one could choice $\epsilon_Q Q_0$ as a single constant $\tilde{Q}$; nevertheless, as we shall see in the forthcoming treatment, it is more convenient to consider Q_0 as a positive constant, proportional to $q_0^2 := Q_0\left(\frac{F_0^2}{8(\kappa\,\rho_0+\Lambda)}\right)^2$, and the signs switch to ϵ.

Summarizing, with all generality, the structural functions can be given as

$$W(r) = -\frac{J^2}{2r^2}, \tag{8.23}$$

$$N(r)^2 = \frac{1}{4}\frac{J^2}{r^2} + \epsilon_N q_0^2 \left[1 - \epsilon K(r)\right]^2, \tag{8.24}$$

$$F(r) = 4\epsilon\frac{\kappa\,\rho_0 + \Lambda}{F_0}\frac{K(r)}{1 - \epsilon K(r)} N(r), \tag{8.25}$$

$$K(r) := \sqrt{B_0 - \frac{1}{16}\frac{F_0^2}{(\kappa\,\rho_0 + \Lambda) q_0^2} r^2}, \tag{8.26}$$

where $q_0^2 := Q_0\left(\frac{F_0^2}{8(\kappa\,\rho_0+\Lambda)}\right)^2$, $\epsilon = \pm$, and $\epsilon_N = \pm$.

The pressure amounts to

$$\kappa p(r) = \frac{\epsilon\,\kappa\,\rho_0 K(r) + \Lambda}{1 - \epsilon K(r)}. \tag{8.27}$$

From this expression one establishes that at certain value $r = r_{\text{zp}}$, where the subscript zp denotes zero pressure, the pressure may vanish for $\epsilon = 1$, $\Lambda = -1/l^2$, and $\epsilon = -1$, $\Lambda = 1/l^2$; one has to keep in mind these sign conditions when extracting square roots; for instance, $\sqrt{\Lambda^2} = \sqrt{(-\epsilon\Lambda)^2} = -\epsilon\Lambda$. The vanishing of the numerator of $p(r)$, $K(r_{\text{zp}}) = -\epsilon\,\Lambda/(\kappa\,\rho_0)$, yields

$$B_0 = \epsilon_N \frac{F_0^2}{16(\kappa\,\rho_0 + \Lambda) q_0^2} r_{\text{zp}}^2 + \frac{\Lambda^2}{\kappa^2 \rho_0^2}. \tag{8.28}$$

Therefore, for vanishing pressure at the circle r_{zp}, the structural functions $N(r)$ and $F(r)$ amount to

$$N(r)^2 = \frac{J^2}{4r^2} + \epsilon_N q_0^2 \left[1 - \epsilon K(r, r_{\text{zp}})\right]^2, \tag{8.29}$$

$$F(r) = 4\epsilon \frac{\kappa \rho_0 + \Lambda}{F_0} \frac{K(r, r_{\text{zp}})}{1 - \epsilon K(r, r_{\text{zp}})} N(r), \tag{8.30}$$

$$K(r, r_{\text{zp}}) = \sqrt{\frac{\Lambda^2}{\kappa^2 \rho_0^2} + \epsilon_N \frac{F_0^2}{16(\kappa \rho_0 + \Lambda) q_0^2} (r_{\text{zp}}^2 - r^2)}. \tag{8.31}$$

Moreover, at the radius r_{zp}, the functions $N(r)$ and $F(r)$ become

$$N(r_{\text{zp}})^2 = \frac{1}{4} \frac{J^2}{r_{\text{zp}}^2} + \epsilon_N q_0^2 \frac{(\kappa \rho_0 + \Lambda)^2}{\kappa^2 \rho_0^2}, \tag{8.32}$$

$$F(r_{\text{zp}}) = -\frac{4\Lambda}{F_0} N(r_{\text{zp}}). \tag{8.33}$$

On the other hand, the vacuum solution for the metric (8.9) is determined by the structural functions

$$N(r)^2 = F(r)^2 = \frac{1}{4} \frac{J^2}{r^2} - \Lambda r^2 - M, \tag{8.34}$$

which in the case of negative cosmological constant $\Lambda = -1/l^2$ becomes the well-known BTZ black hole solution given by the metric (8.9) with

$$N(r)^2 = F(r)^2 = \frac{1}{4} \frac{J^2}{r^2} + \frac{r^2}{l^2} - M. \tag{8.35}$$

Comparing (8.34) with (8.32) at r_{zp}, one arrives at

$$q_0^2 = \epsilon_N \frac{\kappa^2 \rho_0^2}{(\kappa \rho_0 + \Lambda)^2} (-\Lambda r_{\text{zp}}^2 - M), \tag{8.36}$$

and consequently we have the following sub-branches of solutions:

A)$\epsilon_N = 1$,
A1: $\Lambda = -1/l^2$, $q_0^2 = \frac{\kappa^2 \rho_0^2}{(\kappa \rho_0 - 1/l^2)^2} (\frac{r_{\text{zp}}^2}{l^2} - M)$, $r_{\text{zp}}^2/l^2 - M > 0$,
A2: $\Lambda = 1/l^2$, $q_0^2 = -\frac{\kappa^2 \rho_0^2}{(\kappa \rho_0 + 1/l^2)^2} (\frac{r_{\text{zp}}^2}{l^2} + M)$, $r_{\text{zp}}^2/l^2 + M < 0$,

B) $\epsilon_N = -1$,
B1:$\Lambda = -1/l^2$,$q_0^2 = -\frac{\kappa^2 \rho_0^2}{(\kappa \rho_0 - 1/l^2)^2} (\frac{r_{\text{zp}}^2}{l^2} - M)$, $r_{\text{zp}}^2/l^2 - M < 0$,
B2: $\Lambda = 1/l^2$, $q_0^2 = \frac{\kappa^2 \rho_0^2}{(\kappa \rho_0 + 1/l^2)^2} (\frac{r_{\text{zp}}^2}{l^2} + M)$, $r_{\text{zp}}^2/l^2 + M > 0$.

Substituting $F_0 = -4\epsilon \Lambda$, which in turn yields $F(r_{\text{zp}}) = \epsilon N(r_{\text{zp}})$, and q_0 from (8.36) into (8.29–8.31), one obtains the expressions of the structural functions satisfying matching conditions with external gravitational fields in the presence of a cosmological constant of both signs:

$$N(r)^2 = \frac{1}{4}\frac{J^2}{r^2} - \frac{\Lambda\, r_{\rm zp}^2 + M}{(\kappa\,\rho_0 + \Lambda)^2}\kappa^2\,\rho_0^2\left[1 - \epsilon K(r, r_{\rm zp})\right]^2, \tag{8.37}$$

$$F(r) = -\frac{\kappa\,\rho_0 + \Lambda}{\Lambda}\,\frac{K(r, r_{\rm zp})}{1 - \epsilon K(r, r_{\rm zp})}\,N(r), \tag{8.38}$$

$$K(r, r_{\rm zp}) = -\epsilon\frac{\Lambda}{\kappa\,\rho_0}\sqrt{1 - \frac{\kappa\,\rho_0 + \Lambda}{\Lambda\, r_{\rm zp}^2 + M}(r_{\rm zp}^2 - r^2)}, \tag{8.39}$$

accompanied by the pressure

$$p(r) = \rho_0\,\Lambda\,\frac{1 - \sqrt{1 - \frac{\kappa\,\rho_0 + \Lambda}{\Lambda\, r_{\rm zp}^2 + M}(r_{\rm zp}^2 - r^2)}}{\kappa\,\rho_0 + \Lambda\sqrt{1 - \frac{\kappa\,\rho_0 + \Lambda}{\Lambda\, r_{\rm zp}^2 + M}(r_{\rm zp}^2 - r^2)}}. \tag{8.40}$$

8.2.3 Interior Perfect Fluid Solution to the BTZ Black Hole

One of the most interesting interior solutions of the studied class, which matches with the exterior BTZ black hole, arises for $\Lambda = -1/l^2$, and is determined by the metric (8.9) with structural functions

$$N(r)^2 = \frac{1}{4}\frac{J^2}{r^2} - \frac{M - r_{\rm zp}^2/l^2}{(\kappa\,\rho_0 - 1/l^2)^2}\left(\kappa\,\rho_0 - \frac{1}{l^2}\mathcal{K}(r, r_{\rm zp})\right)^2, \tag{8.41}$$

$$F(r) = (\kappa\,\rho_0 - 1/l^2)\,\frac{\mathcal{K}(r, r_{\rm zp})}{\kappa\,\rho_0 - \frac{1}{l^2}\mathcal{K}(r, r_{\rm zp})}\,N(r), \tag{8.42}$$

$$\mathcal{K}(r, r_{\rm zp}) := \sqrt{1 - \frac{\kappa\,\rho_0 - 1/l^2}{M - r_{\rm zp}^2/l^2}(r_{\rm zp}^2 - r^2)}, \tag{8.43}$$

This solution is characterized by the pressure

$$p(r) = -\frac{\rho_0}{l^2}\,\frac{1 - \sqrt{1 - \frac{\kappa\,\rho_0 - 1/l^2}{M - r_{\rm zp}^2/l^2}(r_{\rm zp}^2 - r^2)}}{\kappa\,\rho_0 - \frac{1}{l^2}\sqrt{1 - \frac{\kappa\,\rho_0 - 1/l^2}{M - r_{\rm zp}^2/l^2}(r_{\rm zp}^2 - r^2)}}. \tag{8.44}$$

and a constant energy density ρ_0.

8.2.4 Alternative Parametrization

By introducing new constants c_1, and c_2 through

$$q_0 = \frac{c_1}{4},\; B_0 = 1 - 16\frac{c_2}{c_1^2},\; F_0 = -\epsilon\, 4\Lambda \tag{8.45}$$

one brings the structural functions $N(r)$ and $F(r)$, (8.24) and (8.25), to the form

$$N(r)^2 = \frac{1}{4}\frac{J^2}{r^2} + \frac{\epsilon_N}{16}\left(c_1 + 4\frac{\Lambda}{\sqrt{\kappa^2\,\rho_0^2 - \Lambda^2}}\,K(r)\right)^2,$$

$$F(r) = 4\epsilon(\kappa\,\rho_0 + \Lambda)\,\frac{K(r)}{c_1\sqrt{\kappa^2\,\rho_0^2 - \Lambda^2} + 4\Lambda K(r)}\,N(r),$$

$$K(r) := -\epsilon\frac{\sqrt{\kappa^2\,\rho_0^2 - \Lambda^2}}{4\Lambda}\sqrt{c_1^2 - 16c_2 - \epsilon_N\,\frac{16\Lambda^2}{\kappa\,\rho_0 + \Lambda}r^2}, \tag{8.46}$$

accompanied by the isotropic pressure,

$$\kappa\,p(r) = \Lambda\,\frac{c_1\sqrt{\kappa^2\,\rho_0^2 - \Lambda^2} - 4\kappa\rho_0 K(r)}{c_1\sqrt{\kappa^2\,\rho_0^2 - \Lambda^2} + 4\Lambda K(r)}, \tag{8.47}$$

where $\epsilon = \pm$, and $\epsilon_N = \pm$ independently. As far as the behavior of the pressure is concerned, one may establish that for r running from the value $r_{\mathrm{press}\uparrow\infty} = \sqrt{-\epsilon_N\,c_2(\kappa\,\rho_0 + \Lambda)/\Lambda^2}$, at which point the denominator of the pressure p (8.47) becomes equal to zero, up to the value r_{zp} at which the numerator of the pressure vanishes, $K(r_{\mathrm{zp}}) = c_1\,\sqrt{\kappa^2\,\rho_0^2 - \Lambda^2}/(4\kappa\rho_0)$, the pressure is a positive function of the radial coordinate, which decreases from a very large value at $r_{\mathrm{press}\uparrow\infty}$ to zero at the boundary circle r_{zp}, at which

$$N(r_{\mathrm{zp}})^2 = F(r_{\mathrm{zp}})^2 = \frac{1}{4}\frac{J^2}{r_{\mathrm{zp}}^2} + \epsilon_N\,\frac{c_1^2}{16}\,\frac{(\kappa\,\rho_0 + \Lambda)^2}{\kappa^2\,\rho_0^2},$$

$$c_1 = 4\frac{\kappa\,\rho_0}{\sqrt{\kappa^2\,\rho_0^2 - \Lambda^2}}\sqrt{c_2 + \epsilon_N\,\frac{\Lambda^2}{\kappa\,\rho_0 + \Lambda}r_{\mathrm{zp}}^2}.$$

Taking into account the existence of a zero-pressure circle $r = r_{\mathrm{zp}}$, the structural functions from (8.46) can be given as

$$N(r)^2 = \frac{1}{4}\frac{J^2}{r^2} + \frac{\epsilon_N}{\kappa^2\,\rho_0^2 - \Lambda^2}\left[\kappa\,\rho_0\,K(r_{\mathrm{zp}}) + \Lambda\,K(r)\right]^2,$$

$$F(r) = \epsilon\,(\kappa\,\rho_0 + \Lambda)\,\frac{K(r)}{\kappa\,\rho_0\,K(r_{\mathrm{zp}}) + \Lambda K(r)}\,N(r),$$

$$K(r) := \sqrt{c_2 + \frac{\epsilon_N\,\Lambda^2\,r_{\mathrm{zp}}^2}{\kappa\,\rho_0 + \Lambda} + (\kappa\,\rho_0 - \Lambda)\,(r_{\mathrm{zp}}^2 - r^2)}, \tag{8.48}$$

together with

$$p(r) = -\rho_0\,\Lambda\,\frac{K(r) - K(r_{\mathrm{zp}})}{\kappa\,\rho_0\,K(r_{\mathrm{zp}}) + \Lambda\,K(r)}, \tag{8.49}$$

where $K(r_{\mathrm{zp}}) = K(r = r_{\mathrm{zp}})$, $c_1 = 4\epsilon\,\kappa\,\rho_0 K(r_{\mathrm{zp}})/(\kappa^2\,\rho_0^2 - \Lambda^2)$, meaningful for positive and negative values of the cosmological constant Λ and the parameter M.

Notice that the vacuum solution arises as a limiting case of the above interior metric functions, (8.49), for $\rho_0 = 0$ accompanied with the identification $c_2 \to -\epsilon_N M$.

Comparing (8.34) with the expression (8.49), one establishes that the fulfillment of the matching conditions, $N_{\rm ext}(r_{\rm zp}) = N_{\rm fluid}(r_{\rm zp})$, together with $F_{\rm ext}(r_{\rm zp}) = F_{\rm fluid}(r_{\rm zp})$, is guaranteed by

$$K(r_{\rm zp})^2 = -\epsilon_N \frac{\kappa\,\rho_0 - \Lambda}{\kappa\,\rho_0 + \Lambda}(\Lambda\, r_{\rm zp}^2 + M).$$

Substituting $K(r_{\rm zp})$ into $K(r)$ from (8.48), one has

$$K(r) = K(r_{\rm zp})\sqrt{1 - \frac{\kappa\,\rho_0 - \Lambda}{\Lambda\, r_{\rm zp}^2 + M}(r_{\rm zp}^2 - r^2)}, \tag{8.50}$$

and hence the structural functions assume the form given in (8.37)–(8.39).

8.2.5 Barotropic Rotating Perfect Fluids Without Λ

The equations (8.13) and (8.14), for the linear state equation $p = (\gamma - 1)\rho$ without Λ, gives rise to a single differential equation

$$\begin{aligned}&4r^4 N\left(4r^2N^2 - J^2\right)\ddot{N} + 4r^4\left[4\,(1-\gamma)\,r^2\,N^2 - J^2\right](\dot{N})^2\\&-4r^3 N\left[4r^2N^2 - (1-2\gamma)J^2\right]\dot{N} + J^2\left[(4-\gamma)J^2 - 16\,r^2\,N^2\right] = 0\end{aligned} \tag{8.51}$$

with nontrivial solution

$$N(r)^2 = \frac{J^2}{4r^2} + \frac{1}{4}\left(c_2\,r^2 - c_1\right)^{\frac{2}{2-\gamma}}. \tag{8.52}$$

Substituting the above $N(r)$ into (8.11), one determines of the function $F(r)$

$$F(r) = F_0 \frac{2-\gamma}{2\,c_2}\, N(r)\,\left(c_2\,r^2 - c_1\right)^{-\frac{\gamma}{2-\gamma}}, \tag{8.53}$$

and finally the explicit expressions of the pressure and the energy density are achieved:

$$\kappa\,p = F_0^2\,\frac{2-\gamma}{8\,c_2}\,r^2\,\left(c_2\,r^2 - c_1\right)^{-\frac{\gamma}{2-\gamma}},\ \rho = \frac{1}{\gamma - 1}p. \tag{8.54}$$

It is clear from the energy conditions that $0 < \gamma < 2$. These perfect fluid solutions can be thought of as cosmological solutions which extend to infinity. At $r_{\rm zp} = \sqrt{c_1/c_2}$ both the pressure and $F(r)$ vanish.

8.3 Lubo–Rooman–Spindel Rotating Perfect Fluids

In this section the field equations for a stationary circularly symmetric metric in the presence of a perfect fluid as gravitational source are established; see Lubo et al. (1999), assuming (anti-)de Sitter exterior geometries. The metric in coordinates $\{t, r, \phi\}$ with structural functions $\rho(r)$, $Z(r)$, $B(r)$, and $Y(r)$ is given as

$$ds^2 = -(\rho(r)dt + Z(r)d\phi)^2 + B(r)^2dr^2 + Y(r)^2d\phi^2 = -{\theta^0}^2 + {\theta^1}^2 + {\theta^2}^2,$$
$$\theta^0 = \rho(r)dt + Z(r)d\phi,\ \theta^1 = B(r)dr,\ \theta^2 = Y(r)d\phi, \tag{8.55}$$

where θ^a, $a = 0, 1, 2$ is the orthonormal triad basis. Comparing with the corresponding metric of Lubo et al. (1999), the function $T(r)$ has been replaced by $\rho(r)$ here; the reason for proceeding in this way is due to the convenience of a variable change to the spatial radial coordinate $\rho(r)$. For the fluid energy density here is reserved $\mu(r)$.

The Einstein tensor triad components are $G^a{}_b = R^a{}_b - R\delta^a{}_b/2 = \kappa T^a{}_b - \delta^a{}_b\Lambda$ are

$$G^0{}_0 = \frac{1}{BY}\left(\frac{Y'}{B}\right)' - \frac{1}{4}\left[\frac{\rho}{BY}\left(\frac{Z}{Y}\right)'\right]^2, \tag{8.56a}$$

$$G^1{}_1 = \frac{\rho' Y'}{B^2\rho Y} + \frac{1}{4}\left[\frac{\rho}{BY}\left(\frac{Z}{Y}\right)'\right]^2, \tag{8.56b}$$

$$G^2{}_2 = \frac{1}{B\rho}\left(\frac{\rho'}{B}\right)' + \frac{1}{4}\left[\frac{\rho}{BY}\left(\frac{Z}{Y}\right)'\right]^2, \tag{8.56c}$$

$$G^2{}_0 = -G^0{}_2 = \frac{1}{2}\frac{1}{B\rho^2}\left[\frac{\rho^3}{BY}\left(\frac{Z}{\rho}\right)'\right]'. \tag{8.56d}$$

The fluid one-form is chosen as linear combination os the Killing vectors $\frac{\partial}{\partial t}$ and $\frac{\partial}{\partial \phi}$, namely

$$\mathbf{u} = \cosh\nu(r)\theta^0 + \sinh\nu(r)\theta^2,$$
$$u_\alpha = \rho\,\cosh\nu\,\delta^t_\alpha + (Z\,\cosh\nu + Y\,\sinh\nu)\,\delta^\phi_\alpha,\ u_\alpha u^\alpha = -1. \tag{8.57}$$

The energy–momentum perfect fluid tensor

$$T_{\mu\nu} = (p + \mu)u_\mu\, u_\nu + p g_{\mu\nu}, \tag{8.58}$$

where μ and p are the energy density and the pressure, referred to the orthonormal basis possesses the components

$$T^0{}_0 = -\mu\,(r)\cosh^2\nu - p\,(r)\sinh^2\nu,$$
$$T^2{}_0 = -T^0{}_2 = \cosh\nu\ \sinh\nu\ (\mu\,(r) + p\,(r)),$$
$$T^1{}_1 = p(r),\ T^2{}_2 = \mu\,(r)\sinh^2\nu + p\,(r)\cosh^2\nu, \tag{8.59}$$

throughout this section, the energy density will be denoted with μ instead of σ used in Lubo et al. (1999). The energy–momentum conservation $T^{\alpha\beta}{}_{;\beta} = 0$ yields

$$p' = -\frac{\mu + p}{\rho Y}\left[\cosh^2\nu\,\rho' Y - \sinh^2\nu\,\rho Y' - \cosh\nu\ \sinh\nu\,\rho^2\left(\frac{Z}{\rho}\right)'\right]. \tag{8.60}$$

8.3.1 Equations for Rigidly Rotating Fluids

Following Lubo et al. (1999), the stellar models to be analyzed are those with $\mathbf{u}$ aligned along θ^0, i.e., for $\nu = 0$, leaving out of consideration the differentially rotating case. Thus the components of the energy momentum tensor (8.59) become

$$T^0{}_0 = -\mu(r)\,,\, T^2{}_0 = -T^0{}_2 = 0,$$
$$T^1{}_1 = p(r) = T^2{}_2, \tag{8.61}$$

together with

$$p' = -\frac{\rho'}{\rho}(\mu + p). \tag{8.62}$$

The equation $G^2{}_0 = 0$ yields

$$\left[\frac{\rho^3}{BY}\left(\frac{Z}{\rho}\right)'\right]' = 0 \to \frac{\rho^3}{BY}\left(\frac{Z}{\rho}\right)' = -2\beta = \text{constant}, \tag{8.63}$$

and from $G^2{}_2 = G^1{}_1$ one gets

$$\frac{\rho' Y'}{BY} = \left(\frac{\rho'}{B}\right)' \simeq d\ln(Y) = d\ln(\frac{\rho'}{B}) \to Y(r) = y_0\,\frac{\rho'}{B}, y_0 \to 1, \tag{8.64}$$

which substituted into (8.63) gives

$$\frac{\rho^3}{\rho'}\left(\frac{Z}{\rho}\right)' = -2\beta \simeq d(\frac{Z}{\rho}) = \beta\, d\,\rho^{-2} \to Z(r) = \alpha\rho(r) + \frac{\beta}{\rho(r)}. \tag{8.65}$$

Therefore, the metric for a rigidly rotating fluid in all generality can be given as

$$ds^2 = -(\rho\, dt + Z\, d\phi)^2 + B^2 dr^2 + \left(\frac{\rho'}{B}\right)^2 d\phi^2,$$
$$Z(r) = \alpha\,\rho(r) + \frac{\beta}{\rho(r)},\; Y(r) = \frac{\rho'}{B}, \tag{8.66}$$

where α, β, and B are integration constants, with the energy and pressure fulfilling

$$-(\kappa\mu(r) + \Lambda) = \frac{1}{BY}\left(\frac{Y'}{B}\right)' - \frac{1}{4}\left[\frac{\rho}{BY}\left(\frac{Z}{Y}\right)'\right]^2,$$
$$\kappa p(r) - \Lambda = \frac{\rho' Y'}{B^2\rho Y} + \frac{1}{4}\left[\frac{\rho}{BY}\left(\frac{Z}{Y}\right)'\right]^2. \tag{8.67}$$

The energy conservation $p' = -\frac{\rho'}{\rho}(\mu + p)$ does not provide new conditions; for the given μ and p, it becomes an identity.

8.3.2 Garcia Representation of Stationary Perfect Fluid Solutions

A further transformation based on (8.64), $B(r)dr = y_0 d\rho/Y =: \frac{1}{\rho\, A(\rho)} d\rho$, brings the metric to the form

$$ds^2 = -(\rho\, dt + Z(\rho)\, d\phi)^2 + \frac{1}{\rho^2 A(\rho)^2} d\rho^2 + \rho^2\, A(\rho)^2\, d\phi^2,$$
$$Z(\rho) = \alpha\, \rho + \beta/\rho, \tag{8.68}$$

with energy density

$$\kappa\, \mu\, (\rho) + \Lambda = 3\, \frac{\beta^2}{\rho^4} - A^2 - 2\, \rho\, \frac{dA^2}{d\rho} - \frac{\rho^2}{2} \frac{d^2 A^2}{d\rho^2}, \tag{8.69}$$

and pressure

$$\kappa\, p\, (\rho) - \Lambda = \frac{\beta^2}{\rho^4} + A^2 + \frac{\rho}{2} \frac{dA^2}{d\rho}. \tag{8.70}$$

Providing specific state equations $\mu = \mu(p)$ one gets different families of stationary fluid solutions, evidently in a non-familiar spatial polar coordinate.

8.3.3 Barotropic Class of Solutions $p = \gamma\, \mu$

For a linear dependence between pressure and density – the barotropic state equation – of the form

$$p = \gamma\, \mu \tag{8.71}$$

one arrives at linear equation allowing for the general solution

$$A(\rho)^2 = -2\, \rho^{-\frac{\gamma+1}{\gamma}}\, \gamma\, C_1\, (\gamma - 1)^{-1} - \Lambda + \frac{\beta^2}{\rho^4} + 2\, \frac{C_2\, \gamma}{\rho^2\, (\gamma - 1)}. \tag{8.72}$$

Radiation solution $p = \mu/2$

Radiation is determined for $\gamma = 1/2$, thus

$$A(\rho)^2 = -\Lambda - 2\, \frac{C_2}{\rho^2} + 2\, \frac{C_1}{\rho^3} + \frac{\beta^2}{\rho^4}. \tag{8.73}$$

8.3.4 Constant Density Stationary Solution; $p = p(r)$, $\mu = \mu_0$

For constant mass energy density, one can obtain analytic solutions in $(2+1)D$. Integrating the equation (8.69) for $\mu = \mu_0 =$ constant with respect to A^2, one gets

$$A\, (\rho)^2 = -\Lambda - M_0\, \kappa + \frac{\beta^2}{\rho^4} + 2\, \frac{C_1}{\rho^2} - 2\, \frac{C_2}{\rho}, \tag{8.74}$$

which, substituted into (8.70), gives the pressure

$$\kappa p = -\kappa \mu_0 - \frac{C_1}{\rho}. \tag{8.75}$$

The transformation to the radial coordinate is achieved by solving

$$g_{\phi\phi} = \rho^2 A^2 - Z^2 = r^2, \tag{8.76}$$

which explicitly reads

$$\begin{aligned} r^2 &= -a\rho^2 - 2\,C_2\,\rho - 2\,\alpha\,\beta + 2\,C_1,\; a := \Lambda + \alpha^2 + \kappa\,\mu_0, \\ (\rho + \frac{C_2}{a})^2 &= -\frac{r^2}{a} + \frac{2\,(C_1 - \alpha\,\beta)\,a + {C_2}^2}{a^2}. \end{aligned} \tag{8.77}$$

8.3.5 Lubo–Rooman–Spindel Perfect Fluids $u = \theta^0$ and $g_{rr} = 1$

In this paragraph interior solutions containing, among others, one parameter family of Gödel-like geometries, are derived. All the obtained solutions are causally well behaved; some are nevertheless physically unacceptable; their angular momentum is too large compared to their mass, and such solutions would lead to naked causal singularities in case of collapse unless centrifugal forces were to prohibit them from evolving into black holes.

The choice of the radial coordinate in the stationary metric studied in Rooman and Spindel (1998) – see Eqs. (80) and (81) – is such that $g_{rr} = B(r) = 1$, therefore the metric becomes

$$ds^2 = -[\rho\,dt + (\alpha\,\rho + \beta/\rho)d\phi]^2 + dr^2 + {\rho'}^2 y_0^2 d\phi^2. \tag{8.78}$$

When regularity conditions at the origin $r = 0$ are imposed, then the following relations arises

$$\beta = -\alpha, \rho'(0) = 0, C/2 = \alpha\,\rho''(0) = \omega, \rho(0) \to 1, \tag{8.79}$$

equivalent to Eq. (1.8) of RS. Hence

$$ds^2 = -\rho^2\,dt^2 + dr^2 + \alpha^2 \left(\frac{{\rho'}^2}{\omega^2} - (\rho - \frac{1}{\rho})^2\right) d\phi^2 - 2\alpha(\rho^2 - 1)dt\,d\phi. \tag{8.80}$$

Moreover, the matching to the external AdS metric requires additionally

$$L_\infty = -\omega,\; \omega^2 < -\Lambda.$$

8.3.6 LBR Rotating Perfect Fluid with μ_0

Assuming in the metric (8.55) the metric components $g_{tt} = -1 = -g_{rr}$, one integrates its equations to arrive at a first-order equation for $\rho(r)$, namely

$$\left(\frac{d\,\rho}{d\,r}\right)^2 = \omega^2(\frac{1}{\rho^2} - 1) + \kappa(p(0) + \mu_0)(\rho - 1) - (\kappa\mu_0/2 + \Lambda)(\rho^2 - 1), \tag{8.81}$$

which integrates in terms of elliptic functions. In Rooman and Spindel (1998) additional considerations dealing with the angular momentum and quasi local mass values are stressed.

8.3.7 Rooman–Spindel Rotating Fluid Model; $g_{tt} = -1 = -g_{rr}$

There is a sub-branch of stationary fluid solutions for the metric (8.55) with metric component $g_{tt} = -1$. The metric in the gauge $g_{rr} = B(r) = 1$ becomes

$$ds^2 = -[dt + Z(r)d\phi]^2 + dr^2 + Y(r)^2 d\phi^2. \tag{8.82}$$

The equation $G^2{}_0 = 0$, (8.56d), yields

$$Y(r) = C_1 \frac{d}{dr} Z(r), \tag{8.83}$$

while (8.67) gives the pressure in the form of a relation

$$\kappa\, p(r) = \Lambda + \frac{1}{4C_1{}^2} = \text{constant}. \tag{8.84}$$

The energy fulfills

$$\frac{d^3 Z}{dr^3} + \left(\kappa\,\mu(r) + \Lambda - \frac{3}{4C_1{}^2}\right)\frac{dZ}{dr} = 0. \tag{8.85}$$

As pointed out in Rooman and Spindel (1998), this equation can be thought of as defining the matter content for a given function $Z(r)$, or, conversely, for a given plausible matter distribution one determines the structural function $Z(r)$. The star possesses a boundary if the pressure is set equal to zero, $p = 0$, thus $\Lambda = -\frac{1}{4C_1^2} < 0$, hence for this AdS dust star

$$\frac{d^3 Z}{dr^3} + (\kappa\,\mu(r) + 4\Lambda)\frac{dZ}{dr} = 0.$$

The incompressible dust was treated in detail in Section 3.6.

9

Friedmann–Robertson–Walker Cosmologies

This chapter is devoted to the derivation of exact solutions to the FRW model in (3+1) and (2+1) dimensions and to establish their correspondence following García, Cataldo and del Campo, García et al. (2003). It is widely known that any (3+1) FRW perfect fluid fulfilling a barotropic equation of state of the form $p(\rho) = (\gamma - 1)\,\rho$, under algebraic transformation rules of the parameters, can be transformed into its (2+1) FRW counterpart. It is noteworthy that the physical content of the solutions depends on the dimensionality in which they are viewed. Using four-dimensional terminology, for a vanishing cosmological constant, the (2+1) analog of the (3+1) dust ($\gamma_4 = 1$), $p_4 = 0$ is a radiation-dominated universe ($\gamma_3 = 3/2$), $p_3 = \rho_3/2$. Conversely, for a (3+1) radiation-dominated universe ($\gamma_4 = 4/3$), $p_4 = \rho_4/3$, one finds that the (2+1) counterpart is the stiff matter ($\gamma_3 = 2$), $p_3 = \rho_3$. Moreover, the analog of the (3+1) de Sitter spacetime is the (2+1) de Sitter spacetime, $\gamma_4 = 0 = \gamma_3$, with equation of state $p = -\rho = \text{const.}$ in both cases. The family of FRW barotropic solutions are derived in details. Moreover, the full description of collapsing dust is given for: the Mann–Ross dust FRW solution with Λ, Section 9.4, and the Gidding–Abbott–Kuchař dust FRW solution, Section 9.4.5.

9.1 Einstein Equations for FRW Cosmologies

The starting point in the study of FRW cosmologies begins with the homogeneous and isotropic metric equipped with a scale factor $a(t)$ (function depending on time), and a constant k, the curvature index, related to the curvature character of the three-space (closed, flat, open).

9.1.1 Einstein Equations for (3+1) FRW Cosmology

In (3+1)-dimensional gravity, the metric of the FRW model is given by

$$ds^2 = -dt^2 + a(t)^2 \left(\frac{dr^2}{1 - kr^2} + r^2\, d\theta^2 + r^2 \sin^2\theta d\phi^2 \right), \tag{9.1}$$

where, as usual, $a(t)$ is the scale factor, and $k = -1, 0, 1$. The FRW metric is conformally flat, i.e., its Weyl tensor vanishes everywhere in the domain of definition of the spacetime. The scale factor $a(t)$ of the metric (9.1) is governed by equations modeled in terms of the perfect fluid energy density ρ_4, the matter isotropic pressure p_4, and a cosmological constant Λ, if present: the energy–momentum tensor is given by

$$T_{\mu\nu} = (\rho_4(t) + p_4(t))\, u_\mu u_\nu + p_4(t)\, g_{\mu\nu},\ u_\mu = -\delta^t_\mu,$$

hence, the Einstein equations

$$R_{\mu\nu} - \frac{R}{2}\, g_{\mu\nu} + \Lambda\, g_{\mu\nu} = \kappa\, T_{\mu\nu},$$

amounts explicitly to

$$E^t{}_t : \frac{3}{a^2}\left(\dot{a}^2 + k\right) = \Lambda_4 + \kappa_4\, \rho, \tag{9.2a}$$

$$E^i{}_i : 2\, a\, \ddot{a} + \dot{a}^2 + k + a^2\, (\kappa\, p_4 - \Lambda_4) = 0$$
$$\rightarrow a\, \ddot{a} + \frac{a}{6}\, (\kappa\, \rho_4 + 3\, \kappa\, p_4 - 2\, \Lambda_4) = 0, \tag{9.2b}$$

$$\frac{d}{dt}\left(\rho_4\, a^3\right) + p_4\, \frac{da^3}{dt} = 0 \equiv T^\mu{}_{\nu;\mu} = 0. \tag{9.2c}$$

The equation (9.2c) represents the conservation of the matter content: $T^\mu{}_{\nu;\mu} = 0$. It results also from the substitution of $\dot{a}$ from (9.2a) into $\ddot{a}$ of (9.2b); thus only two of the equations of this system are independent.

9.1.2 Einstein Equations for (2+1) FRW Cosmology

In (2+1)-dimensional gravity, the analogous to the FRW metric (9.1) is given by

$$ds^2 = -dt^2 + a(t)^2 \left(\frac{dr^2}{1 - kr^2} + r^2 d\theta^2\right). \tag{9.3}$$

This metric is characterized by the vanishing of the Cotton tensor, and hence, from the $3D$ point of view, this conditions leads to the conformal flatness of the (2+1) FRW metric. Moreover, the metric (9.3) can be considered as a dimensional reduction of the metric (9.1) for any fixed value of the azimuthal angle ϕ.

The corresponding (2+1) Einstein equations for the fluid, $T_{\mu\nu} = (\rho_3(t) + p_3(t)) u_\mu u_\nu + p_3(t)\, g_{\mu\nu}$, $u_\mu = -\delta^t_\mu$, are

$$\frac{\dot{a}^2 + k}{a^2} = \kappa_3 \rho_3 + \Lambda_3, \tag{9.4a}$$

$$\ddot{a} + a\, (\kappa\, p_3 - \Lambda_3) = 0, \tag{9.4b}$$

$$\frac{d}{dt}\left(\rho\, a^2\right) + p\, \frac{da^2}{dt} = 0. \tag{9.4c}$$

Substituting $\dot{a}$ from (9.4a) into (9.4b) one arrives at (9.4c), which is the equation arising from the energy conservation $T^{\mu}{}_{\nu;\mu} = 0$.

For dust, the pressure vanishes, $p = 0$, and the conservation of the energy–momentum tensor implies that

$$3+1 : \rho\, a^3 = \rho_0\, a_0^3 = \text{const.},$$

$$2+1 : \rho\, a^2 = \rho_0\, a_0^2 = \text{const.},$$

where ρ_0 is the initial density and a_0 the value of the scale factor at the beginning of the time counting.

9.2 Barotropic Perfect Fluid FRW Solutions

In this section, by a straightforward integration of the field equations in the case of a perfect fluid, fulfilling linear state equations of the form $p + \rho = \gamma\, \rho$, the general solutions in (3+1) and (2+1) spacetimes are derived.

9.2.1 Barotropic Perfect Fluid (3 + 1) Solutions

In (3+1) FRW spaces, for linear equations of state

$$p_4 = (\gamma_4 - 1)\, \rho_4, \tag{9.5}$$

the equation (9.2c) amounts to

$$d \ln \rho_4 + 3\gamma_4 d \ln a = 0, \tag{9.6}$$

and hence its general integral is

$$\rho_4 = \rho_{40}\, a^{-3\gamma_4}, \tag{9.7}$$

where ρ_{40} is an integration constant. The integral of (9.2a) is given by

$$t - t_0 = \epsilon_a \int \frac{a^{\frac{3}{2}\gamma_4 - 1} da}{\sqrt{\frac{\Lambda_4}{3} a^{3\gamma_4} - k a^{2(\frac{3}{2}\gamma_4 - 1)} + \frac{\kappa_4}{3}\rho_{40}}}, \tag{9.8}$$

where $\epsilon_a = \pm 1$.

From (9.8), it becomes apparent that one can not, in general, express in terms of elementary functions t as function of a.

Nevertheless, for $\Lambda_4 = 0$ and arbitrary γ_4, the above integral is given in terms of hypergeometric functions, namely

$$t(a) = t_0 + \frac{\epsilon_a}{\sqrt{\rho_{40}\kappa_4/3}} \frac{a^{\frac{3}{2}\gamma_4}}{\frac{3}{2}\gamma_4} F\left(\left[\frac{1}{2}, \frac{\frac{3}{2}\gamma_4}{3\gamma_4 - 2}\right], \left[\frac{\frac{9}{2}\gamma_4 - 2}{3\gamma_4 - 2}\right], k\frac{a^{\frac{3}{2}\gamma_4 - 1}}{\rho_{40}\kappa_4/3}\right). \tag{9.9}$$

(3 + 1) FRW Solution $a(t) \sim t$

For $\gamma_4 = 2/3$, $p_4 = -\rho_4/3$, the scale factor amounts to

$$a(t) = \epsilon_a \sqrt{\frac{\kappa_4}{3}\rho_{40} - k}\; t + a_0, \tag{9.10}$$

where a_0 is a constant of integration. This solution is quite unphysical but it is included for reason of comparison with the linear dependent $a(t)$ of the $(2+1)$ case.

(3 + 1) de Sitter Cosmological Solution

On the other hand, for $\gamma_4 = 0$, $p_4 = -\rho_4 = -\rho_{40} = \text{const.}$, i.e., de Sitter spacetime, the scale factor is

$$a(t) = \frac{e^{-\epsilon_a C_0(t-t_0)}}{2C_0}\left[k + e^{2\epsilon_a C_0(t-t_0)}\right], \; C_0 = \sqrt{\frac{\kappa_4}{3}\rho_{40}}. \tag{9.11}$$

9.2.2 Barotropic Perfect Fluid (2 + 1) Solutions

For (2+1) FRW cosmology with linear state equations of the form $p_3 = (\gamma_3 - 1)\rho_3$, the dynamical field equation (9.4c) possesses as general integral

$$\rho_3 = \rho_{30} a^{-2\gamma_3}, \tag{9.12}$$

where ρ_{30} is a constant of integration. The integral of t, from (9.4a), amounts to

$$t - t_0 = \epsilon_a \int \frac{a^{\gamma_3 - 1} da}{\sqrt{\Lambda_3 a^{2\gamma_3} - k a^{2(\gamma_3-1)} + \kappa_3 \rho_{30}}}. \tag{9.13}$$

For $\Lambda_3 = 0$, the integral (9.13) can be written in terms of hypergeometric functions

$$t(a) = t_0 + \frac{\epsilon_a}{\sqrt{\kappa_3\rho_{30}}}\frac{a^{\gamma_3}}{\gamma_3} F\left(\left[\frac{1}{2}, \frac{\gamma_3}{2\gamma_3 - 2}\right], \left[\frac{3\gamma_3 - 2}{2\gamma_3 - 2}\right], k\frac{a^{2\gamma_3-2}}{\kappa_3\rho_{30}}\right). \tag{9.14}$$

Saslaw, Collas, Gidding–Abbott–Kuchař, and Cornish–Frankel Dust Solution, $a(t) \sim t$

Moreover, for zero cosmological constant $\Lambda = 0$, $\gamma_3 = 1$, and $p_3 = 0$, the scale factor can be given as

$$a(t) = \epsilon_a \sqrt{\kappa_3 \rho_{30} - k}\; t + a_0, \tag{9.15}$$

where a_0 is an integration constant.

In Collas (1977), Eq.(58)–Eq.(64), the dust-filled open universe linearly expands in time, therefore there is no deceleration, and consequently it satisfies the Hubble law, which has been derived in detail for the first time to our knowledge; although the work by Saslaw (1977), published in September 1977 (the same month of the Collas publication), also contains the same results.

This kind of solution was also reported in Giddings et al. (1984), Eq.(77). In the paragraph prior to this equation, it is pointed out "that in structure the equation for a radiation-dominated Universe in n dimensions is identical to that for a dust-filled Universe in $n+1$ dimensions."

Later, in 1991, this kind of cosmology was reported in Cornish and Frankel (1994), Eq.(4.5–4.6).

(2 + 1) de Sitter Cosmological Solution

On the other hand, for $\gamma_3 = 0$, consequently $p_3 = -\rho_3 = -\rho_{30} = \text{const.}$, i.e., one is dealing with the de Sitter metric, the scale factor amounts to

$$a(t) = \frac{e^{-\epsilon_a C_0 (t-t_0)}}{2C_0}\left[k + e^{2\epsilon_a C_0 (t-t_0)}\right], \; C_0 = \sqrt{\kappa_3 \rho_{30}}. \tag{9.16}$$

Saslaw (1977), in the paragraph following equation (6), wrote the sentence: "If $\Lambda \neq 0$, one can similarly compare models for 2+1 and 3+1 dimensional universes. For $p = 0$ – dust – one obtains $a = a_0 \exp t/\tau$, which is the analogous of the de Sitter model."

Saslaw, Cornish–Frankel Radiation-Dominated Universes; $\rho_3 = 2p_3$

In Saslaw (1977), after Eq. (6), it is stated that, in $(3+1)$ dimensions, the dust-filled universe $k = 0$, $a(t) \sim t^{2/3}$ corresponds, in $(2+1)$ dimensions, to the radiation-dominated universe with $\rho \sim a^{-3}$.

In Cornish and Frankel (1994), Eq.(4.4) and Eq.(5.20), the solutions for expanding universes in the case of radiation are given explicitly as:

$$k = 0 : t = \frac{2}{3}\frac{1}{\sqrt{2GM_0 a_0}}\, a^{3/2}, \tag{9.17a}$$

$$k = 1 : t = 2GM_0 a_0 \left(\arcsin A^{1/2} - A^{1/2}(1-A)^{1/2}\right), \tag{9.17b}$$

$$k = -1 : t = 2GM_0 a_0 \left(A^{1/2}(1+A)^{1/2} - \sinh^{-1} A^{1/2}\right), \tag{9.17c}$$

where $A = \frac{a}{\sqrt{2GM_0 a_0}}$.

9.2.3 Comparison Between (3+1) and (2+1) Barotropic Solutions

Assuming that the time t and the scale factor function $a(t)$ are structurally invariant functions for the studied (3+1) and (2+1) metrics, let us compare the integral solution t given by (9.8) with the one defined by (9.13); by accomplishing parameter scaling transformations

$$\frac{\kappa_4}{3} \to \kappa_3, \; \frac{\Lambda_4}{3} \to \Lambda_3, \; 3\gamma_4 \to 2\gamma_3 \tag{9.18}$$

in t from (9.8), one arrives at the integral t of (9.13); the equivalence in the opposite direction holds too. Thus the time t as function of the variable a is a structurally invariant function. One reaches the same conclusion when dealing with the hypergeometric function representation of t determined by (9.9) and (9.14). On the other hand, if one were able to express the scale factor a as a function of the variable t, then, via parameter scaling, one would arrive at the structurally invariant character of the function $a(t)$.

We have established in this way that any FRW cosmology, filled with a perfect fluid fulfilling a linear state equation, determined in (3+1) dimensions, can be reduced to its (2+1) counterpart by using the correspondence (9.18); the converse statement holds, too.

Moreover, considering (3+1) and (2+1) FRW cosmologies as independent entities, dominant energy conditions for fluids: $\rho \geq 0$ and $-\rho \leq p \leq \rho$, have to hold on their own account in (2+1) and (3+1) dimensions. Therefore, the (3+1) dimensional state parameter γ_4 has to fulfill the condition $0 \leq \gamma_4 \leq 2$, while independently the (2+1) dimensional state parameter γ_3 has to range the values $0 \leq \gamma_3 \leq 2$.

On the other hand, assuming that the considered spacetimes are in the correspondence (9.18), one arrives at restrictions for the values one can assign to the state parameters γ, namely:

$$0 \leq \gamma_3 \leq 2, \qquad \text{and} \qquad 0 \leq \gamma_4 \leq \frac{4}{3}.$$

Thus, the class of (3+1) perfect fluid cosmologies which participates in the correspondence with the whole family of (2+1) perfect fluid cosmologies is more narrow compared with the whole (3+1) perfect fluid cosmology; (3+1) cosmology with $\frac{4}{3} < \gamma_4 \leq 2$ are out of the comparison scheme. Hence, thinking in terms of dimensionally reduced spaces, a perfect fluid FRW solution given in (3+1) dimensions, which can be reduced to its (2+1) cosmological counterpart, possesses state parameters given in the above-specified ranges. From this point of view, using the four-dimensional terminology, for vanishing cosmological constants $\Lambda_4 = \Lambda_3 = 0$, the (2+1) analog of the (3+1) dust ($\gamma_4 = 1$), $p_4 = 0$, is a radiation-dominated universe ($\gamma_3 = 3/2$), $p_3 = \rho_3/2$. Conversely, for a (3+1) radiation-dominated universe ($\gamma_4 = 4/3$), $p_4 = \rho_4/3$, one finds that the (2+1) counterpart is the stiff matter ($\gamma_3 = 2$), $p_3 = \rho_3$. Moreover, the (3+1) de Sitter spacetime coincides with the (2+1) de Sitter spacetime, $\gamma_4 = 0 = \gamma_3$, with equation of state $p = -\rho = \text{const}$.

9.3 Polytropic Perfect Fluid FRW Solutions

In what follows, solutions for perfect fluids subjected to polytropic state equations $p = \alpha\, \rho^{\gamma}$ are derived.

9.3.1 Polytropic Perfect Fluid (3 + 1) *Solutions*

Under the polytropic state equation

$$p_4 = \alpha_4 {\rho_4}^{\gamma_4}, \tag{9.19}$$

the equation (9.2c) becomes

$$\frac{d\rho_4}{\rho_4 + \alpha_4 \rho_4{}^{\gamma_4}} + 3d \ln a = 0. \tag{9.20}$$

hence

$$d \ln \left[(\alpha_4 + \rho_4^{(-\gamma_4+1)}) a^{3(-\gamma_4+1)} \right] = 0 \rightarrow \rho_4^{\nu_4} = A_{40} a^{-3\nu_4} - \alpha_4, \tag{9.21}$$

where $\nu_4 = 1 - \gamma_4$.

Moreover, the integral of t, (9.2a), amounts to

$$t - t_0 = \int \frac{da}{\sqrt{\frac{\Lambda_4}{3} a^2 - k + \frac{\kappa_4}{3} a^2 \left[A_{40}\, a^{-3\nu_4} - \alpha_4 \right]^{1/\nu_4}}}. \tag{9.22}$$

9.3.2 Polytropic Perfect Fluid (2 + 1) Solutions

In (2+1) cosmology one encounters, as expected, a similar treatment for the polytropic case, $p_3 = \alpha_3 \rho_3{}^{\gamma_3}$, which is determined by:
the fluid energy density

$$\rho_3^{\nu_3} = A_{30} a^{-2\nu_3} - \alpha_3, \tag{9.23}$$

and the time variable t

$$t - t_0 = \int \frac{da}{\sqrt{\Lambda_3 a^2 - k + \kappa_3 a^2 \left[A_{30}\, a^{-2\nu_3} - \alpha_3 \right]^{1/\nu_3}}}. \tag{9.24}$$

9.3.3 Comparison Between (3+1) and (2+1) Polytropic Solutions

Comparing the expressions of coordinate time t, which is assumed to have the same meaning in both dimensions, as well as the scale factor $a(t)$, one arrives at: $\frac{\Lambda_4}{3} \rightleftharpoons \Lambda_3$, $\frac{\kappa_4}{3} \rightleftharpoons \kappa_3$, but one can not establish a relation between γ's, or equivalently for ν's, to reproduce the energy densities functions from one another. At most one conclude, due to the structural invariance of $a(t)$, that

$$\frac{1}{a(t)} = \left(\frac{1}{A_4} (\rho_4^{\nu_4} + \alpha_4) \right)^{\frac{1}{3\nu_4}} = \left(\frac{1}{A_3} (\rho_3^{\nu_3} + \alpha_3) \right)^{\frac{1}{2\nu_3}}. \tag{9.25}$$

Hence, for perfect fluids fulfilling the polytropic state equation $p = \alpha \rho^{(1-\nu)}$, we have no relations between exponential factors ν_d alone.

9.4 Mann–Ross Collapsing Dust FRW Solutions with Λ

In this section, following Mann and Ross (1993), the standard cosmological constant is subjected to the change: $\Lambda \rightarrow -\Lambda_{mr}$, thus for asymptotically AdS spaces $\Lambda_{mr} = 1/l^2$, while for asymptotically dS spaces $\Lambda_{mr} = -1/l^2$

The Einstein equations for a perfect fluid with a cosmological constant for the $(2+1)$ Friedmann–Robertson–Walker metric (9.3),

$$ds^2 = -dt^2 + a(t)^2 \left(\frac{dr^2}{1-kr^2} + r^2 d\phi^2 \right),$$

are given by (9.4).

For dust, which is the case to be treated from now on, the pressure vanishes, $p = 0$, and the conservation of the energy–momentum tensor implies that

$$\rho\, a(t)^2 = \rho_0\, a_0^2 = \text{constant}, \tag{9.26}$$

where ρ_0 is the initial density and a_0^2 is the value of the scale factor at the beginning of the time. The remaining equations are:

$$\frac{d^2a}{dt^2} + \Lambda_{mr}\, a = 0, \quad \left(\frac{da}{dt}\right)^2 + k - \kappa\,\rho_0\, a_0^2 + \Lambda_{mr}\, a^2 = 0. \tag{9.27}$$

9.4.1 Cosmological dS–FRW Solution

In the case of negative $\Lambda_{mr} = -1/l^2$, i.e., positive standard cosmological constant Λ, the Einstein equations (9.27) are solved by

$$a(t) = a_0 \cosh(\sqrt{-\Lambda_{mr}}\,t) + \frac{\dot a_0}{\sqrt{-\Lambda_{mr}}} \sinh(\sqrt{-\Lambda_{mr}}\,t),\ \dot a_0 = \frac{da}{dt}|_{t=t_0},$$
$$\dot a_0^2 = \kappa\,\rho_0\, a_0^2 - k - \Lambda_{mr}\, a_0^2;\ \kappa\,\rho_0\, a_0^2 - k - \Lambda_{mr}\, a_0^2 \geq 0. \tag{9.28}$$

For a monotonically increasing scale factor, $\dot a(t) < 0$, in particular, $\dot a(t = t_0) = -\sqrt{\kappa\,\rho_0 a_0^2 - k - \Lambda_{mr}\, a_0^2}$.

If the collapse occurs, it takes place at time $t = t_c$ when the scale factor vanishes, $a(t_c) = 0$, i.e., for

$$\begin{aligned} t_c &= -\frac{1}{\sqrt{-\Lambda_{mr}}} \text{arctanh}\left[\frac{a_0\sqrt{-\Lambda_{mr}}}{\dot a_0}\right] \\ &= -\frac{1}{\sqrt{-\Lambda_{mr}}} \text{arctanh}\left[\frac{a_0\sqrt{-\Lambda_{mr}}}{\sqrt{\kappa\,\rho_0\, a_0^2 - k - \Lambda_{mr}\, a_0^2}}\right]. \end{aligned} \tag{9.29}$$

The exterior metric has to be the de Sitter one, with negative parameter $M < 0$ to have the correct metric signature; see (9.32) below.

9.4.2 Asymptotically AdS–FRW Dust Solution

In the case of positive $\Lambda_{mr} = 1/l^2$, i.e., negative standard cosmological constant Λ, the Einstein equations (9.27) are satisfied by

$$a(t) = a_0 \cos(\sqrt{\Lambda_{mr}}\, t) + \frac{\dot{a}_0}{\sqrt{\Lambda_{mr}}} \sin(\sqrt{\Lambda_{mr}}\, t),\ \dot{a}_0 = \frac{da}{dt}|_{t=t_0},$$
$$\dot{a}_0 = \sqrt{\kappa\, \rho_0\, a_0^2 - k - \Lambda_{mr}\, a_0^2};\ \kappa\, \rho_0\, a_0^2 - k - \Lambda_{mr}\, a_0^2 \geq 0, \text{ real } a(t). \tag{9.30}$$

At time of the collapse, $t = t_c$, this solution has to vanish, $a(t_c) = 0$, hence

$$t_c = \frac{1}{\sqrt{\Lambda_{mr}}} \arctan\left[\frac{a_0 \sqrt{\Lambda_{mr}}}{\sqrt{\kappa\, \rho_0\, a_0^2 - k - \Lambda_{mr}\, a_0^2}}\right]. \tag{9.31}$$

9.4.3 Matching the AdS–FRW Dust to the Static BTZ

To match the dust solution of the FRW metric to the exterior static BTZ solution endowed with mass and a negative cosmological constant $\Lambda = -1/l^2 = -\Lambda_{mr}$, one has to fit the continuity conditions of the metrics and the extrinsic curvature tensors. The static BTZ can be given in the exterior coordinates $\{T, R, \phi\}$ as

$$g = -(\Lambda_{mr} R^2 - M) dT^2 + \frac{dR^2}{\Lambda_{mr} R^2 - M} + R^2 d\phi^2. \tag{9.32}$$

The boundary of the dust distribution is taken to be at r_b, and correspondingly at $R(t)$ in the exterior coordinates. On the boundary edge, the solutions ought to fulfill the continuity conditions, i.e., the vanishing of the jump of the metric $g_{\mu\nu}$, following the Israel (1966) notation: $\overset{MR}{g}_{\mu\nu} \equiv \overset{(-)}{g}_{\mu\nu}$, $\overset{BTZ}{g}_{\mu\nu} \equiv \overset{(+)}{g}_{\mu\nu}$, and of the extrinsic curvature $K_{\mu\nu}$:

$$[g_{\mu\nu}] := \overset{(-)}{g}_{\mu\nu}(r_b) - \overset{(+)}{g}_{\mu\nu}(R(t)) = 0, \tag{9.33a}$$

$$[K_{\mu\nu}] = 0. \tag{9.33b}$$

From $\overset{(-)}{g}(r_b) - \overset{(+)}{g}(R(t)) = 0$, and $\overset{(-)}{g}_{\phi\phi}(r_b) = \overset{(+)}{g}_{\phi\phi}(R(t))$ one gets correspondingly

$$-1 = -(\Lambda_{mr} R(t)^2 - M)\left(\frac{dT}{dt}\right)^2 + \frac{1}{\Lambda_{mr} R(t)^2 - M}\left(\frac{dR}{dt}\right)^2,$$
$$r_b\, a(t) = R(t). \tag{9.34}$$

The latter condition $R(t) = r_b\, a(t)$ means that in the exterior coordinates the position R of the boundary is equal to the proper distance $r_b\, a(t)$ from the origin to the dust edge. The first condition yields

$$\frac{dT}{dt} = \frac{\sqrt{\Lambda_{mr} R(t)^2 - M + \dot{R}(t)^2}}{\Lambda_{mr} R(t)^2 - M}, \tag{9.35}$$

on the dust edge; the over-dot denotes $\frac{d}{dt}$.

The initial conditions $a_0 = 1$, and $\dot{a}_0 = 0$ represent a ball of dust with initial radius $R(t_0) = r_b\, a(t_0) = r_b$ initially at rest $\dot{R}(t)|_{t_0} = r_b\, \dot{a}(t)|_{t_0} = 0$ in the exterior coordinates.

9.4.4 Determination of K_{ij}

The full determination of the $\overset{(\pm)}{K}_{ij}$ is accomplished here in detail. The standard definition of the extrinsic curvature tensor, see Eisenhart (1966), is used

$$\overset{(\pm)}{K}_{ij} = N_\alpha \mathrm{e}_i{}^\nu\, \nabla_\nu \mathrm{e}_j{}^\alpha = -\mathrm{e}_i{}^\nu\, \mathrm{e}_j{}^\alpha\, \nabla_\nu\, N_\alpha,\ \ N_\alpha \mathrm{e}_i{}^\alpha = 0, \tag{9.36}$$

where $\mathrm{e}_i{}^\nu$ are the components of the tangent vector ∂_i to the coordinate curve ξ^i defined on the surface, N_α is the normal vector to the surface, and as such is orthogonal to $\mathrm{e}_i{}^\nu$. Operationally, it is recommendable to use the directional description $\mathrm{e}_i{}^\nu \frac{\partial}{\partial x^\nu} = \frac{\partial}{\partial \xi^i}$, thus

$$\overset{(\pm)}{K}_{ij}= N_\alpha \left(\frac{\partial}{\partial \xi^i} \mathrm{e}_j{}^\alpha + \mathrm{e}_i{}^\nu\, \mathrm{e}_j{}^\sigma \Gamma^\alpha{}_{\nu\sigma} \right) = -\mathrm{e}_j{}^\alpha \left(\frac{\partial}{\partial \xi^i} N_\alpha - \mathrm{e}_i{}^\nu\, N_\sigma \Gamma^\sigma{}_{\nu\alpha} \right), \tag{9.37}$$

which is close to the Israel formulation.

Extrinsic Curvature for the Interior AdS–FRW Dust Metric

The boundary surface is located at $\{r = r_b,\ t = t,\ \phi = \phi\}$, consequently the tangent and normal vectors can be chosen as

$$\mathrm{e}_t{}^\nu = \delta_t{}^\mu,\ \mathrm{e}_\phi{}^\nu = \frac{1}{r_b\, a(t)} \delta_\phi{}^\mu,\ N_\alpha = \frac{a(t)}{\sqrt{1 - k\, r_b^2}} \delta^r{}_\alpha. \tag{9.38}$$

The components of the extrinsic curvature are

$$\overset{(-)}{K}_{tt} = N_\alpha\, \Gamma^\alpha{}_{tt} = \frac{a(t)}{\sqrt{1 - k\, r_b^2}} \Gamma^t{}_{tt} = 0,\ \overset{(-)}{K}_{t\phi}= 0, \tag{9.39}$$

$$\overset{(-)}{K}_{\phi\phi} = \frac{1}{r_b^2\, a(t)^2} N_\alpha\, \Gamma^\alpha{}_{\phi\phi} = \frac{1}{r_b^2\, a(t)\, \sqrt{1 - k\, r_b^2}} \Gamma^r{}_{\phi\phi} = -\frac{\sqrt{1 - k\, r_b^2}}{r_b\, a(t)}. \tag{9.40}$$

Extrinsic Curvature for Exterior Static BTZ Metric

For the exterior metric, the tangent and normal vectors to the surface $T(\lambda)$, $R(\lambda)$, $\phi = \phi$ are chosen as

$$\mathrm{e}_\lambda{}^\mu = T_{,\lambda} \delta^\mu_T + R_{,\lambda} \delta^\mu_R,\ \mathrm{e}_\phi{}^\mu = 1/R \delta^\mu_\phi, \tag{9.41}$$

$$\mathrm{N}_\mu = -R_{,\lambda} \delta^T_\mu + T_{,\lambda} \delta^R_\mu,\ \mathrm{e}_\lambda{}^\mu \mathrm{N}_\mu = 0 = \mathrm{e}_\phi{}^\mu \mathrm{N}_\mu, \tag{9.42}$$

Thus the extrinsic curvature tensor components are

$$\overset{(+)}{K}_{\lambda\lambda}= \mathrm{N}_\mu \mathrm{e}_\lambda{}^\nu \nabla_\nu \mathrm{e}_\lambda{}^\mu = -\mathrm{e}_\lambda{}^\nu \mathrm{e}_\lambda{}^\mu \nabla_\nu \mathrm{N}_\mu,\ \overset{(+)}{K}_{\phi\phi}= \mathrm{N}_\mu \mathrm{e}_\phi{}^\nu \nabla_\nu \mathrm{e}_\phi{}^\mu. \tag{9.43}$$

For $\overset{(+)}{K}_{\phi\phi}$ one has

$$\overset{(+)}{K}_{\phi\phi} = \mathrm{N}_\mu \Gamma^\mu{}_{\phi\phi}/R^2 = -T_{,\lambda} \frac{\Lambda_{mr} R^2 - M}{R}, \tag{9.44}$$

where the non-vanishing Christoffel symbols for the external metric are

$$\Gamma^T{}_{TR} = \frac{\Lambda_{mr} R}{\Lambda_{mr} R^2 - M} = -\Gamma^R{}_{RR},\ \Gamma^R{}_{TT} = \Lambda_{mr} R(\Lambda_{mr} R^2 - M),$$
$$\Gamma^R{}_{\phi\phi} = -R(\Lambda_{mr} R^2 - M),\ \Gamma^\phi{}_{R\phi} = \frac{1}{R}. \tag{9.45}$$

For $\overset{(+)}{K}_{\lambda\lambda}$ one has

$$\overset{(+)}{K}_{\lambda\lambda} = \mathrm{N}_\mu\, \partial_\lambda \mathrm{e}_\lambda{}^\mu + \mathrm{N}_\mu \mathrm{e}_\lambda{}^\nu \mathrm{e}_\lambda{}^\sigma \Gamma^\mu{}_{\nu\sigma}. \tag{9.46}$$

Consider first

$$\mathrm{N}_\mu\, \partial_\lambda \mathrm{e}_\lambda{}^\mu = \mathrm{N}_\mu\, \partial_\lambda (T_{,\lambda}\delta^\mu_T + R_{,\lambda}\delta^\mu_R) = (-R_{,\lambda}\delta^T_\mu + T_{,\lambda}\delta^R_\mu)\,(T_{,\lambda\lambda}\delta^\mu_T + R_{,\lambda\lambda}\delta^\mu_R)$$
$$= -T_{,\lambda\lambda}R_{,\lambda} + R_{,\lambda\lambda}T_{,\lambda}. \tag{9.47}$$

On the other hand,

$$\mathrm{N}_\mu \mathrm{e}_\lambda{}^\nu\, \mathrm{e}_\lambda{}^\sigma\, \Gamma^\mu{}_{\nu\sigma} = -R_{,\lambda}\left(T_{,\lambda}T_{,\lambda}\Gamma^T{}_{TT} + 2\,T_{,\lambda}R_{,\lambda}\Gamma^T{}_{TR} + R_{,\lambda}R_{,\lambda}\,\Gamma^T{}_{RR}\right)$$
$$+T_{,\lambda}\left(T_{,\lambda}T_{,\lambda}\Gamma^R{}_{TT} + 2\,T_{,\lambda}R_{,\lambda}\Gamma^R{}_{TR} + R_{,\lambda}R_{,\lambda}\,\Gamma^R{}_{RR}\right). \tag{9.48}$$

Substituting the Christoffel symbols one arrives at

$$\mathrm{N}_\mu \mathrm{e}_\lambda{}^\nu (\mathrm{e}_\lambda{}^\sigma\, \Gamma^\mu{}_{\nu\sigma}) = \Lambda_{mr} R(\Lambda_{mr} R^2 - M)(T_{,\lambda})^3 - 3\,T_{,\lambda}(R_{,\lambda})^2 \frac{\Lambda_{mr} R}{\Lambda_{mr} R^2 - M}. \tag{9.49}$$

Setting $\lambda = t$, and using $T_{,t}$ from (9.35), together with

$$T_{,tt} = \frac{\dot{R}\ddot{R}}{(\Lambda_{mr} R^2 - M)\sqrt{\Lambda_{mr} R^2 - M + \dot{R}^2}}$$
$$-\frac{\Lambda_{mr} R\dot{R}\left[\Lambda_{mr} R^2 - M + 2\dot{R}^2\right]}{(\Lambda_{mr} R^2 - M)^2\sqrt{\Lambda_{mr} R^2 - M + \dot{R}^2}} \tag{9.50}$$

one arrives at

$$\overset{(+)}{K}_{tt} = \frac{\partial}{\partial R}\sqrt{\Lambda_{mr} R^2 - M + \dot{R}^2}. \tag{9.51}$$

Therefore, since on the boundary $r = r_b$ the smoothness of $K_{\phi\phi}$ from (9.40) and (9.44) yields

$$-\frac{\sqrt{1 - k\,r_b^2}}{r_b\, a(t)} = -T_{,\lambda}\frac{\Lambda_{mr} R^2 - M}{R} = -\frac{\sqrt{\Lambda_{mr} R^2 - M + \dot{R}^2}}{R}, \tag{9.52}$$

where it has been used $T_{,\lambda} = T_{,t}$ from (9.35). Since on the boundary

$$R(t) = r_b\, a(t),\ \dot{R}(t) = r_b\, \dot{a}(t), \tag{9.53}$$

hence, one arrives at the condition

$$\Lambda_{mr}\, {r_b}^2 a(t)^2 - M + {r_b}^2 \dot{a}(t)^2 = 1 - k\, {r_b}^2, \tag{9.54}$$

isolating M one gets

$$M = (k + \Lambda_{mr} a(t)^2 + \dot{a}(t)^2)\, r_b^2 - 1 = \kappa\, r_b^2\, a_0^2\, \rho_0 - 1 \tag{9.55}$$

where (9.27) was used in agreement with the results reported in Mann–Ross.

The existence of an event horizon in the static BTZ black hole around the collapsing dust, $R_h = \sqrt{M\Lambda_{mr}}$, $M > 0$, requires $\rho_0 > 1/(\kappa\, r_b^2\, a_0^2)$.

9.4.5 Gidding–Abbott–Kuchař Dust FRW Solution

In Giddings et al. (1984), §7, one finds the (2+1) dynamical dust solution without a cosmological constant, analogous to the Oppenheimer–Snyder collapsing dust of $(3 + 1)$ dimensions; see Oppenheimer and Snyder (1939).

This case is contained in the above-derived solution. Setting the cosmological constant equal to zero in (9.27), one gets

$$\begin{aligned} &\frac{d^2 a}{dt^2} = 0 \to a(t) = \dot{a}_0\, t + a_0, \\ &\left(\frac{da}{dt}\right)^2 + k - \kappa\, \rho_0\, a_0^2 = 0 \to \dot{a}_0^2 + k - \kappa\, \rho_0\, a_0^2 = 0 \\ &\to \dot{a}_0 = \pm\sqrt{\kappa\, \rho_0\, a_0^2 - k}. \end{aligned} \tag{9.56}$$

This scalar factor $a(t)$ can also be obtained from (9.30) as a limit as Λ goes to zero. The collapse, if any, occurs at $a(t_c) = 0$, i.e.,

$$t_c = -a_0/\dot{a}_0 = a_0/\sqrt{\kappa\, \rho_0\, a_0^2 - k},$$

where $\dot{a}_0 = -\sqrt{\kappa\, \rho_0\, a_0^2 - k}$ has been chosen to have $t_c > 0$, assuming $a_0 > 0$; the collapse has to take place after the initial time $t = 0$.

The exterior metric (9.32) for vanishing Λ becomes a flat metric with conical singularity

$$g = -C\, dT^2 + \frac{dR^2}{C} + R^2 d\phi^2,\ C := -M > 0. \tag{9.57}$$

Choosing the dust edge at $r = r_b$, with equation $R(t)$, the matching conditions (9.33a) and (9.33b) of the FRW metric (9.3) to the metric (9.57) are those of (9.34), (9.35), (9.52) and (9.51) for $\Lambda = 0$, namely

$$R(t) = r_b\, a(t) \tag{9.58a}$$

$$\frac{dT}{dt} = \frac{\sqrt{C + \dot{R}(t)^2}}{C}, \tag{9.58b}$$

$$-\overset{(-)}{K}_{\phi\phi} = \frac{\sqrt{1 - k\, r_b^2}}{r_b\, a(t)} = \dot{T}\frac{C}{R} = -\overset{(+)}{K}_{\phi\phi}, \tag{9.58c}$$

$$\overset{(-)}{K}_{tt} = 0 = \frac{d}{d\,R}\sqrt{C + \dot{R}(t)^2} = \overset{(+)}{K}_{tt}, \tag{9.58d}$$

Using $\dot{T}$ from (9.58b) in (9.58c), taking into account that $R(t) = r_b\, a(t)$, one arrives at

$$1 - k\, r_b^2 = C + \dot{R}(t)^2, \tag{9.59}$$

which, isolating C, considering that $R(t) = r_b\, a(t)$ and the derivative $\dot{a}(t)$ from (9.56), yields

$$C = 1 - (k + \dot{a}(t)^2)\, r_b^2 = 1 - \kappa\, r_b^2\, a_0^2\, \rho_0 > 0, \tag{9.60}$$

in agreement with Mann and Ross (1993), Eq. (32).

10

Dilaton–Inflaton Friedmann–Robertson–Walker Cosmologies

The purpose of this chapter is to provide a new insight on (2+1) and (3+1) Friedmann–Robertson–Walker (FRW) cosmologies by establishing a bridge between them. In order to achieve this goal, I shall begin with a comparison of the dynamical equations corresponding to (2+1) and (3+1) FRW spacetimes coupled to matter perfect fluid sources, scalar field (inflaton, dilaton) fields, and cosmological constants. A (2+1) FRW spacetime may be considered as a dimensional reduction of the associated (3+1) FRW spacetime, arising as result of the freezing (constant value assignation) of the azimuthal angle (in spherical coordinates) of this last (3+1) space. A similar approach has been applied successfully by Cataldo, del Campo and Garcia, Cataldo et al. (2001), to the (3+1) Plebański–Carter[A] metric – see Plebański (1975) and Carter (1968) – to derive the (2+1) BTZ black hole solution.

It is shown that FRW cosmological models coupled to a single scalar field and to a perfect fluid fitting a wide class of matter perfect fluid state equations, determined in (3+1) dimensional gravity, can be related to their (2+1) cosmological counterparts, and vice versa, by using simple algebraic transformations relating gravitational constants, state parameters, perfect fluid and scalar field characteristics. It should be pointed out that the demonstration of these relations for the scalar fields and potentials does not require the fulfillment of any state equation for the scalar field energy density and pressure. As far as the perfect fluid is concerned, one has to demand the fulfillment of state equations of the form $p+\rho=\gamma\, f(\rho)$. If the considered cosmologies contain the inflaton field alone ϕ, then any (3+1) scalar field cosmology possesses a (2+1) counterpart, and vice versa.

Notice that one is tacitly assuming that coordinates remain the same ones for both (3+1) and (2+1) FRW metrics. It is notable that these spacetimes are both **conformally flat**, i.e., correspondingly their **Weyl and Cotton tensors vanish.** By associated (corresponding) spacetimes we mean spaces that belong to a specific family: for instance, spaces fulfilling a (linear) barotropic state equation, or those fitting a polytropic law. Moreover, (2+1) FRW solutions to a barotropic perfect fluid state equation are in correspondence with

(3+1) FRW cosmological solutions to a barotropic perfect fluid state equation, but with different values of the state parameters. Among other considerations, maintaining the (3+1) terminology, the (2+1) counterpart of the (3+1) radiation is the stiff matter. By a straightforward integration of the corresponding field equations, wide families of solutions are derived, which are used to check independently the fulfillment of the theorem. Special attention is addressed to solutions associated to inflaton scalar field cosmologies, for instance, to Barrow–Saich, and Barrow–Burd–Lancaster–Madsen solutions.

The outline of the present chapter is as follows:
In Section 10.1 we briefly review the Einstein field equations for (3+1) and (2+1) FRW models and demonstrate a theorem. In Section 10.2 we derive single scalar field solutions to barotropic state equations. Section 10.3 deals with spatially flat FRW solutions to barotropic state equations for both matter and scalar fields simultaneously present. In Section 10.4 we derived single scalar field spatially flat FRW solutions to state equations of the form $p_\phi + \rho_\phi = \Gamma \rho_\phi{}^\beta$. In Section 10.5, single scalar field spatially flat FRW solutions for a given scale factor $a(t)$ are determined.

Some conventions are followed: Latin and Greek letters with the subscript 0 denote constants, for instance, A_0, t_0 or $\rho_{\phi_{30}}$, although constants of common use, such as gravitational constant κ, cosmological constant Λ, and those appearing in state equations, γ, Γ, and β, will be typed without any subscript. When extracting the square root of a quantity, say m, the $\pm$ sign will be denoted by ϵ_m. In the derivation of solutions practically the same pattern will be followed. Moreover, the conventional perfect fluid description of the field equations is widely used to derive most of the solutions presented here. Abbreviations for perfect fluid (PF) and scalar field (dilaton) are used.

10.1 Equations for a FRW Cosmology with a Perfect Fluid and a Scalar Field

In this section, the field equations for isotropic homogeneous FRW models filled with a perfect fluid and a single scalar field ϕ minimally coupled to gravity with a self-interacting potential $V(\phi)$ are explicitly given. A theorem relating (3+1) and (2+1) solutions is demonstrated.

Throughout this chapter, the energy density will be denoted by $\rho(t,r)$ instead of μ to correspond with earlier publications.

10.1.1 Einstein Equations for (3+1) FRW Dilaton Cosmology

In (3+1) dimensional gravity, the metric of the FRW model is given by

$$ds^2 = dt^2 - a(t)^2 \left(\frac{dr^2}{1 - kr^2} + r^2 d\Omega^2 \right), \tag{10.1}$$

where, as usual, $d\Omega^2 := d\theta^2 + \sin^2\theta d\phi^2$, $a(t)$ is the scale factor, and $k = -1, 0, 1$ denotes the curvature index. This metric is conformally flat, i.e., its Weyl tensor vanishes everywhere in the domain of definition of the spacetime. The scale factor $a(t)$ of the metric (10.1) is governed by equations modeled in terms of the perfect fluid energy density ρ_4, the matter isotropic pressure p_4, the scalar field ϕ, the self-interacting potential $V(\phi)$, and a cosmological constant Λ, if present.

Standard Formulation of the (3+1) Field Equations

These dynamical equations are:

$$3\frac{\dot{a}^2 + k}{a^2} = \kappa_4(\rho_4 + \frac{1}{2}\dot{\phi_4}^2 + V_4) + \Lambda_4, \tag{10.2}$$

$$\dot{\rho_4} + 3\frac{\dot{a}}{a}(\rho_4 + p_4) = 0, \tag{10.3}$$

$$\ddot{\phi_4} + 3\frac{\dot{a}}{a}\dot{\phi_4} + \frac{d}{d\phi_4}V(\phi_4) = 0. \tag{10.4}$$

Perfect Fluid Formulation of the (3+1) Field Equations

In the conventional perfect fluid notation to describe the scalar field, one defines the energy density and the pressure associated with the scalar field correspondingly as:

$$\rho_{\phi_4} = \frac{1}{2}\dot{\phi_4}^2 + V_4, \tag{10.5}$$

$$p_{\phi_4} = \frac{1}{2}\dot{\phi_4}^2 - V_4. \tag{10.6}$$

In terms of these quantities, the (3+1) dynamical equations are:

$$3\frac{\dot{a}^2 + k}{a^2} = \kappa_4(\rho_4 + \rho_{\phi_4}) + \Lambda_4, \tag{10.7}$$

$$\dot{\rho_4} + 3\frac{\dot{a}}{a}(\rho_4 + p_4) = 0, \tag{10.8}$$

$$\dot{\rho}_{\phi_4} + 3\frac{\dot{a}}{a}(\rho_{\phi_4} + p_{\phi_4}) = 0, \tag{10.9}$$

The (10.3) represents the conservation of the matter content, while (10.4) corresponds to the energy conservation of the scalar field.

10.1.2 Einstein Equations for (2+1) FRW Cosmology

In (2+1)-dimensional gravity, the analogous to the FRW metric (10.1) is given by

$$ds^2 = dt^2 - a(t)^2\left(\frac{dr^2}{1 - kr^2} + r^2 d\theta^2\right). \tag{10.10}$$

This metric is characterized by the vanishing of the Cotton tensor, and hence, from the $3D$ point of view, it is conformally flat. Moreover, the metric (10.10) can be considered as a dimensional reduction of the metric (10.1) for any fixed value of the azimuthal angle ϕ.

Standard Formulation of the (2+1) Field Equations

The corresponding (2+1) Einstein equations are

$$\frac{\dot{a}^2 + k}{a^2} = \kappa_3(\rho_3 + \frac{1}{2}\dot{\phi}_3^{\ 2} + V_3) + \Lambda_3, \tag{10.11}$$

$$\dot{\rho}_3 + 2\frac{\dot{a}}{a}(\rho_3 + p_3) = 0, \tag{10.12}$$

$$\ddot{\phi}_3 + 2\frac{\dot{a}}{a}\dot{\phi}_3 + \frac{d}{d\phi_3}V(\phi_3) = 0. \tag{10.13}$$

Perfect Fluid Formulation of the (2+1) Field Equations

Introducing the scalar field density and pressure defined through

$$\rho_{\phi_3} = \frac{1}{2}\dot{\phi}_3^{\ 2} + V_3, \tag{10.14}$$

$$p_{\phi_3} = \frac{1}{2}\dot{\phi}_3^{\ 2} - V_3, \tag{10.15}$$

the (2+1) dynamical equations become

$$\frac{\dot{a}^2 + k}{a^2} = \kappa_3(\rho_3 + \rho_{\phi_3}) + \Lambda_3, \tag{10.16}$$

$$\dot{\rho}_3 + 2\frac{\dot{a}}{a}(\rho_3 + p_3) = 0, \tag{10.17}$$

$$\dot{\rho}_{\phi_3} + 2\frac{\dot{a}}{a}(\rho_{\phi_3} + p_{\phi_3}) = 0. \tag{10.18}$$

It is apparent that the field equations for metrics (10.1) and (10.10) are different because of the difference in dimensions. Nevertheless, one may assume that the time coordinate t remains the same in both (2+1) and (3+1) dimensions. Moreover, one also may assume that the scale factor $a(t)$ is a structurally invariant function depending on t and certain constants; by structural invariance we mean that under dimensional reduction the function $a(t)$ maintains its form with respect to the t variable as well as its dependence on the constants involved. The extension of this concept to functions depending on other variables is straightforward.

The main result of this section can be formulated as a theorem.

10.1.3 Correspondence Between (3+1) and (2+1) Solutions

Theorem 10.1 *Assuming that the time coordinate t and the scale factor $a(t)$ are structurally invariant functions in both (2+1) and (3+1) dimensional FRW cosmologies coupled, in each dimension, to a single scalar field and to a perfect fluid subjected to state equations $p+\rho=\gamma f(\rho)$, where $f(\rho)$ is a structurally invariant functions, then the constants and structural functions of these cosmologies are related according to the following rules:*

$$\text{parameter scaling: } \frac{\kappa_4}{3} \rightleftharpoons \kappa_3, \ \frac{\Lambda_4}{3} \rightleftharpoons \Lambda_3, 3\gamma_4 \rightleftharpoons 2\gamma_3,$$
$$\text{function scaling: } \rho_4 \rightleftharpoons \rho_3, \ \sqrt{\frac{3}{2}}\phi_4 \rightleftharpoons \phi_3,$$
$$V_4 - \frac{1}{4}\dot{\phi}_4^{\,2} \to V_3, \ V_3 + \frac{1}{6}\dot{\phi}_3^{\,2} \to V_4. \tag{10.19}$$

Proof Considering that the time coordinate t as well the scale factor $a(t)$ remain unchanged, comparing (10.2) and (10.11) one has

$$\frac{\dot{a}^2+k}{a^2} = \frac{\kappa_4}{3}(\rho_4 + \frac{1}{2}\dot{\phi}_4^{\,2} + V_4) + \frac{\Lambda_4}{3} = \kappa_3(\rho_4 + \frac{1}{2}\dot{\phi}_3^{\,2} + V_3) + \Lambda_3 \Rightarrow$$
$$\frac{\kappa_4}{3} \rightleftharpoons \kappa_3, \ \frac{\Lambda_4}{3} \rightleftharpoons \Lambda_3, \rho_4 \rightleftharpoons \rho_3, \tag{10.20}$$

together with

$$\frac{1}{2}\dot{\phi}_4^{\,2} + V_4 \rightleftharpoons \frac{1}{2}\dot{\phi}_3^{\,2} + V_3, \sim \rho_{\phi_4} \rightleftharpoons \rho_{\phi_3}. \tag{10.21}$$

Next, assuming that in each space the state equation for matter is of the form $p+\rho=\gamma f(\rho)$, where $f(\rho)$ is a structurally invariant function, i.e., it is a form-invariant function as viewed from the spaces under consideration, the matter conservation equations yield

$$\frac{da}{a} = -\frac{1}{3\gamma_4}\frac{d\rho_4}{f(\rho_4)} = -\frac{1}{2\gamma_3}\frac{d\rho_3}{f(\rho_3)}, \tag{10.22}$$

hence, because of by assumption $f(\rho_4) \rightleftharpoons f(\rho_3)$, one has

$$\ln\frac{a}{a_0} = -\frac{1}{3\gamma_4}\int^{\rho_4}\frac{d\rho}{f(\rho)} = -\frac{1}{2\gamma_3}\int^{\rho_3}\frac{d\rho}{f(\rho)}, \tag{10.23}$$

therefore

$$3\gamma_4 \rightleftharpoons 2\gamma_3. \tag{10.24}$$

To establish the remaining relationships on scalar fields ϕ and potentials $V(\phi)$ we rewrite (10.4) and (10.13) correspondingly as:

$$(3+1): \ \frac{d}{dt}\frac{1}{2}\dot{\phi}_4^{\,2} + \frac{d}{dt}V(\phi_4) + 3\frac{\dot{a}}{a}\dot{\phi}_4^{\,2} = 0,$$
$$(2+1): \ \frac{d}{dt}\frac{1}{2}\dot{\phi}_3^{\,2} + \frac{d}{dt}V(\phi_3) + 2\frac{\dot{a}}{a}\dot{\phi}_3^{\,2} = 0. \tag{10.25}$$

Starting from the (3+1) equation, assuming again that t and $a(t)$ remain unchanged under dimensional reduction, we shall establish the transformations of ϕ_4 and V_4 to recover the (2+1) equation. First, one has

$$(1-\frac{1}{3})\frac{d}{dt}\frac{1}{2}\left(\sqrt{\frac{3}{2}}\dot{\phi}_4\right)^2+\frac{d}{dt}V(\phi_4)+2\frac{\dot{a}}{a}\left(\sqrt{\frac{3}{2}}\dot{\phi}_4\right)^2=0, \tag{10.26}$$

which rewrites as

$$\frac{d}{dt}\frac{1}{2}\left(\sqrt{\frac{3}{2}}\dot{\phi}_4\right)^2+\frac{d}{dt}[V(\phi_4)-\frac{1}{4}\dot{\phi}_4{}^2]+2\frac{\dot{a}}{a}\left(\sqrt{\frac{3}{2}}\dot{\phi}_4\right)^2=0. \tag{10.27}$$

Comparing with the corresponding (2+1) equation, one establishes

$$\sqrt{\frac{3}{2}}\phi_4\rightarrow\phi_3,\ V_4-\frac{1}{4}\dot{\phi}_4{}^2\rightarrow V_3. \tag{10.28}$$

The inverse transformations read:

$$\sqrt{\frac{2}{3}}\phi_3\rightarrow\phi_4, V_3+\frac{1}{6}\dot{\phi}_3{}^2\rightarrow V_4. \tag{10.29}$$

Finally, we demonstrate that (10.21) does not contribute with an independent relation; in fact, since by definition $\rho_{\phi_4}=\frac{1}{2}\dot{\phi}_4{}^2+V_4$, replacing the correspondence above, (10.28), one obtains $\rho_{\phi_4}=\frac{1}{2}\dot{\phi}_3{}^2+V_3=:\rho_{\phi_3}$. □

Corollary *In (2+1) and (3+1) dimensional FRW cosmologies, with cosmological constants, coupled to a single scalar field, under the assumption of invariance of the time coordinate as well as the scale factors in both (3+1) and (2+1) spaces, the constants and structural functions are related according to the following rules:*

$$\frac{\kappa_4}{3}\rightleftharpoons\kappa_3,\ \frac{\Lambda_4}{3}\rightleftharpoons\Lambda_3,$$
$$\sqrt{\frac{3}{2}}\phi_4\rightleftharpoons\phi_3,\ V_4-\frac{1}{4}\dot{\phi}_4^2\rightarrow V_3,\ V_3+\frac{1}{6}\dot{\phi}_3^2\rightarrow V_4. \tag{10.30}$$

The proof follows immediately from the theorem above. Therefore, any (3+1) inflationary cosmology possesses a (2+1) partner and conversely.

Nevertheless, scalar fields, modeled conventionally in terms of fluid quantities, by definition fulfill $p_\phi+\rho_\phi=\dot{\phi}^2\neq 0$. Thus, one can think of this last equation as a state equation for the scalar field and assume the existence of relations of the form $p_\phi+\rho_\phi=\Gamma F(\rho_\phi)$. If one were assuming the structural invariant property of functions $F(\rho_\phi)$, because of the equations $\dot{\phi}_4^2=\Gamma_4 F(\rho_{\phi_4})$ and $\dot{\phi}_3^2=\Gamma_3 F(\rho_{\phi_3})$, one could straightforwardly determine the relationship between Γs, namely

$$3\Gamma_4\rightleftharpoons 2\Gamma_3. \tag{10.31}$$

In the next sections, cosmologies with scalar fields subjected to state equations $p_\phi + \rho_\phi = \Gamma F(\rho_\phi)$, for which (10.31) holds, are derived. The advantage of using these equations resides in the uniqueness of the derived solutions. For such branches of solutions one tacitly assumes that the above condition (10.31) is fulfilled.

For matter perfect fluids, dominant energy conditions require that $\rho \geq 0$, and $-\rho < p < \rho$, therefore determining this kind of solution, one has additionally to take care of the fulfillment of this inequality in each spacetime, no matter its dimension. The validity of this physical requirement is assumed to hold beyond four dimensions.

10.2 Single Scalar Field to Linear State Equations; $\Lambda = 0$

By integrating the dynamical equations for a vanishing cosmological constant in the case of a single scalar field, described in the conventional fluid formulation, fulfilling linear state equations $p_\phi + \rho_\phi = \gamma\, \rho_\phi$, the general solutions for (3+1) and (2+1) FRW spacetimes are derived. The following two subsections are devoted to general non-flat ($k \neq 0$) cosmologies, while the third one deals with spatially flat ($k = 0$) FRW spacetimes. By simple comparison one establishes that the conditions stated in the theorem hold.

10.2.1 (2+1) Solutions for a Scalar Field

The derivation of solutions for a scalar field, modeled through perfect fluid quantities, obeying a linear state equation $p_{\phi_3} = (\Gamma_3 - 1)\rho_{\phi_3}$, is given in some detail.

The equation for ρ_{ϕ_3} can be written as

$$\frac{d}{da}\rho_{\phi_3} + \frac{2}{a}(\rho_{\phi_3} + p_{\phi_3}) = 0, \tag{10.32}$$

thus, for the considered linear state equation, one obtains

$$\rho_{\phi_3} = \rho_{\phi_{30}}\, a^{-2\Gamma_3}, \quad \rho_{\phi_{30}} = \text{const.} \tag{10.33}$$

Moreover, (10.16) yields

$$\dot{a}^2 = \kappa_3 \rho_{\phi_3} a^2 - k = \kappa_3 \rho_{\phi_{30}}\, a^{-2(\Gamma_3 - 1)} - k, \tag{10.34}$$

hence

$$t - t_0 = \epsilon_a \int \frac{a^{\Gamma_3 - 1}}{\sqrt{\kappa_3 \rho_{\phi_{30}} - k\, a^{2(\Gamma_3 - 1)}}} da, \tag{10.35}$$

which can be given in terms of hypergeometric functions as

$$t(a) = t_0 + \frac{\epsilon_a}{\sqrt{\kappa_3 \rho_{\phi_{30}}}} \frac{a^{\Gamma_3}}{\Gamma_3} F\left(\left[\frac{1}{2}, \frac{\Gamma_3}{2\Gamma_3 - 2}\right], \left[\frac{3\Gamma_3 - 2}{2\Gamma_3 - 2}\right], k\frac{a^{2\Gamma_3 - 2}}{\kappa_3 \rho_{\phi_{30}}}\right). \tag{10.36}$$

Adding (10.14) and (10.15) one gets $(d\phi_3/dt)^2 = \rho_{\phi_3} + p_{\phi_3}$, which can be written as

$$\dot{a}^2 \left(\frac{d\phi_3}{da}\right)^2 = \Gamma_3 \rho_{\phi_3}. \tag{10.37}$$

Substituting in the above equation $\dot{a}^2$ from (10.34), and the expression of ρ_{ϕ_3} from (10.33), one obtains

$$d\phi_3 = \epsilon_a \epsilon_\phi \sqrt{\rho_{\phi 30}} \frac{\sqrt{\Gamma_3}}{\sqrt{\kappa_3 \rho_{\phi_{30}}} - k a^{2\Gamma_3 - 2}} \frac{da}{a}, \tag{10.38}$$

which, by introducing the variable $z = a^{\Gamma_3 - 1}$, amounts to

$$d\phi_3 = -\frac{\epsilon_a \epsilon_\phi \sqrt{\Gamma_3}}{(\Gamma_3 - 1)\sqrt{\kappa_3}} d\left[\ln\left(\sqrt{\kappa_3 \rho_{\phi_{30}}} + \sqrt{\kappa_3 \rho_{\phi_{30}} - k z^2}\right) - \ln z\right]. \tag{10.39}$$

Integrating this last expression one arrives at

$$\phi_3 - \phi_{30} = -\epsilon_a \epsilon_\phi \frac{\sqrt{\Gamma_3}}{\sqrt{\kappa_3}} \frac{1}{\Gamma_3 - 1} \left[\ln\left(\sqrt{\kappa_3 \rho_{\phi_{30}}} + \sqrt{\kappa_3 \rho_{\phi_{30}} - k a^{2(\Gamma_3 - 1)}}\right) - \ln a^{(\Gamma_3 - 1)}\right], \tag{10.40}$$

where ϕ_{30} is an integration constant. Moreover, introducing

$$C_3 = \epsilon_a \epsilon_\phi \sqrt{\kappa_3}(\Gamma_3 - 1)/\sqrt{\Gamma_3},$$

the expression of a in terms of ϕ_3 amounts to

$$a^{(\Gamma_3 - 1)} = 2\sqrt{\kappa_3 \rho_{\phi 30}} \frac{\exp\left(-C_3(\phi_3 - \phi_{30})\right)}{k + \exp\left(-2C_3(\phi_3 - \phi_{30})\right)}. \tag{10.41}$$

On the other hand, subtracting the (10.15) from (10.14), one has $2V_3 = 2\rho_{\phi_3} - \dot{\phi}_3^2 = (2 - \Gamma_3)\rho_{\phi_3}$, which explicitly becomes

$$V_3 = \frac{2 - \Gamma_3}{2} \rho_{\phi_{30}} \left[2\sqrt{\kappa_3 \rho_{\phi 30}} \frac{\exp\left(-C_3(\phi_3 - \phi_{30})\right)}{k + \exp\left(-2C_3(\phi_3 - \phi_{30})\right)}\right]^{-2\frac{\Gamma_3}{\Gamma_3 - 1}}. \tag{10.42}$$

10.2.2 (3+1) Solutions for a Scalar Field

The general solution for a scalar field, fulfilling a linear equation of state $p_{\phi_4} = (\Gamma_4 - 1)\rho_{\phi_4}$, is explicitly given by:
the scalar field density

$$\rho_{\phi_4} = \rho_{\phi_{40}} a^{-3\Gamma_4}, \quad \rho_{\phi_{40}} = \text{const.}, \tag{10.43}$$

the time variable t as function of the scale factor a

$$t - t_0 = \epsilon_a \int \frac{a^{\frac{3}{2}\Gamma_4 - 1}}{\sqrt{\frac{\kappa_4}{3}\rho_{\phi_{40}} - k\, a^{2(\frac{3}{2}\Gamma_4 - 1)}}} da, \tag{10.44}$$

which, in terms of hypergeometric functions, can be expressed as

$$t = t_0 + \frac{\epsilon_a}{\sqrt{\frac{\kappa_4}{3}\rho_{\phi_{40}}}} \frac{a^{\frac{3}{2}\Gamma_4}}{\frac{3}{2}\Gamma_4} F\left(\left[\frac{1}{2}, \frac{\frac{3}{2}\Gamma_4}{3\Gamma_4 - 2}\right], \left[\frac{\frac{9}{2}\Gamma_4 - 2}{3\Gamma_4 - 2}\right], k\frac{a^{3\Gamma_4 - 2}}{\frac{\kappa_4}{3}\rho_{\phi_{40}}}\right), \tag{10.45}$$

the scalar field

$$\phi_4 = -\frac{\epsilon_a \epsilon_\phi \sqrt{\Gamma_4}}{\sqrt{\kappa_4/3}} \frac{1}{\frac{3}{2}\Gamma_4 - 1} \left[\ln\left(\sqrt{\frac{\kappa_4}{3}\rho_{\phi_{40}}} + \sqrt{\frac{\kappa_4}{3}\rho_{\phi_{40}} - k a^{2(\frac{3}{2}\Gamma_4 - 1)}}\right) - \ln a^{(\frac{3}{2}\Gamma_4 - 1)}\right] + \phi_{40}, \tag{10.46}$$

where ϕ_{40} is an integration constant. Introducing $C_4 = \epsilon_a \epsilon_\phi \sqrt{\frac{\kappa_4}{3}}(\frac{3}{2}\Gamma_4 - 1)/\sqrt{\Gamma_4}$, the expression of the function $a(\phi_4)$ amounts to

$$a^{(\frac{3}{2}\Gamma_4 - 1)} = 2\sqrt{\frac{\kappa_4}{3}\rho_{\phi_{40}}} \frac{\exp\left(-C_4(\phi_4 - \phi_{40})\right)}{k + \exp\left(-2C_4(\phi_4 - \phi_{40})\right)}. \tag{10.47}$$

Finally, the potential V_4 can be expressed as

$$V_4 = \frac{2 - \Gamma_4}{2}\rho_{\phi_{40}} \left[2\sqrt{\frac{\kappa_4}{3}\rho_{\phi_{40}}} \frac{\exp\left(-C_4(\phi_4 - \phi_{40})\right)}{k + \exp\left(-2C_4(\phi_4 - \phi_{40})\right)}\right]^{-6\frac{\Gamma_4}{3\Gamma_4 - 2}}. \tag{10.48}$$

It is clear that these (2+1) and (3+1) solutions for a single scalar field fulfill the requirements of the theorem (corollary) and the condition (10.31).

10.2.3 Slow Roll Spatially Flat FRW Solutions

This subsection is devoted to the derivation of (3+1) and (2+1) inflationary solutions. A comparison with the existing solutions in the literature is carried out.

Lucchin–Matarrese Power Law (3+1) Solution

In (3+1) inflationary theory, one of the most important solutions for a single scalar field was found by Lucchin and Matarrese (1985); see also Liddle and Lyth (2000), which gives rise to power law inflation. In the conventional scalar field representation, this unique solution arises for a linear state equation

$$p_{\phi_4} + \rho_{\phi_4} = \frac{2}{3\alpha}\rho_{\phi_4} =: \Gamma_4 \rho_{\phi_4}. \tag{10.49}$$

Therefore, the equation for ρ_{ϕ_4}, (10.9), reads

$$d\ln \rho_{\phi_4} + \frac{2}{\alpha} d\ln a = 0, \tag{10.50}$$

with solution

$$\rho_{\phi_4} = \rho_{\phi_{40}} (a/a_0)^{-2/\alpha}, \tag{10.51}$$

where the constant a_0 has been introduced for further convenience; notice that the dynamical equations, for $k = 0$ and $\Lambda_4 = 0$, are invariant under the change $a \to a/a_0$, thus without lost of generality one may set $a_0 = 1$. Substituting this expression of ρ_{ϕ_4} into (10.7), which yields

$$d(a/a_0)^{1/\alpha} = 1/\alpha\sqrt{\kappa_4\rho_{\phi_{40}}/3}\,dt, \tag{10.52}$$

thus its integral becomes

$$a = a_0(\kappa_4/\alpha^2\rho_{\phi_{40}}/3)^{\alpha/2}t^\alpha. \tag{10.53}$$

Equating $\rho_{\phi_{40}}$ to $3\alpha^2/\kappa_4$, one arrives at the power law

$$a(t) = a_0 t^\alpha. \tag{10.54}$$

Next, the equation for ϕ_4 becomes

$$\dot{\phi}_4 = \epsilon_\phi\sqrt{2\alpha/\kappa_4}\,t^{-1}, \tag{10.55}$$

with general solution

$$\phi_4 - \phi_{40} = \epsilon_\phi\sqrt{2\frac{\alpha}{\kappa_4}}\ln t. \tag{10.56}$$

Finally, the evaluation of $V_4 = \rho_{\phi_4} - \dot{\phi}_4/2$ yields the slow roll the self-interacting potential

$$V_{\phi_4} = (3\alpha - 1)\frac{\alpha}{\kappa_4}t^{-2} = (3\alpha - 1)\frac{\alpha}{\kappa_4}e^{-\epsilon_\phi\sqrt{2\frac{\kappa_4}{\alpha}}(\phi_4 - \phi_{40})}. \tag{10.57}$$

Power Law (2+1) Solution

In (2+1) cosmology one may construct the analog to the inflaton solution via the correspondence (10.19) and the relation (10.31). In this way one obtains the (2+1) power law slow roll inflaton solution:
power law scale factor

$$a(t) = a_0 t^\alpha, \tag{10.58}$$

the homogeneous scalar field

$$\phi_3 - \phi_{30} = \epsilon_\phi\sqrt{\frac{\alpha}{\kappa_3}}\ln t, \tag{10.59}$$

slow roll the self-interacting potential

$$V_{\phi_3} = (2\alpha - 1)\frac{\alpha}{2\kappa_3}t^{-2} = (2\alpha - 1)\frac{\alpha}{2\kappa_3}e^{-2\epsilon_\phi\sqrt{\frac{\kappa_3}{\alpha}}(\phi_3 - \phi_{30})}. \tag{10.60}$$

In the conventional scalar field representation, this solution corresponds to a linear state equation of the form $p_{\phi_3} + \rho_{\phi_3} = \frac{1}{\alpha}\rho_{\phi_3} =: \Gamma_3\rho_{\phi_3}$, with $\rho_{\phi_3} = \frac{\alpha^2}{\kappa_3}t^{-2}$.

Cruz–Martínez (2+1) Flat FRW Solution

Cruz and Martínez (2000) have obtained a solution which describes a (2+1) flat FRW cosmology determined through:

$$a(t) = \left(t_0 + \epsilon_a \Gamma_3 \sqrt{\kappa_3}\, t\right)^{1/\Gamma_3}, \tag{10.61}$$

$$\phi_3(t) - \phi_{30} = \frac{1}{\sqrt{\kappa_3 \Gamma_3}} \ln\left(t_0 + \epsilon_a \Gamma_3 \sqrt{\kappa_3}\, t\right), \tag{10.62}$$

$$V(\phi_3) = \frac{2 - \Gamma_3}{2}\, e^{-2\sqrt{\kappa_3 \Gamma_3}(\phi_3 - \phi_{30})}. \tag{10.63}$$

It is clear from the comparison with the (2+1) inflaton solution exhibited above that the Cruz–Martínez solution is a slightly different parametrization of the inflaton solution.

10.3 Spatially Flat FRW Solutions for Barotropic Perfect Fluid and Scalar Field

The derivation of the general solutions for spatially flat ($k = 0$) FRW spacetimes filled simultaneously with matter and scalar field, modeled by two perfect fluids – one related to matter and the second one related to the scalar field – is presented in some details.

10.3.1 Spatially Flat FRW (3+1) Solutions ; $\gamma_4 \neq 2\Gamma_4$

For a (3+1) spatially flat FRW spacetime, the equations for ρ_4 and ρ_{ϕ_4} can be written as

$$\begin{aligned} \frac{d}{da}\rho_4 + \frac{3}{a}(\rho_4 + p_4) &= 0, \\ \frac{d}{da}\rho_{\phi_4} + \frac{3}{a}(\rho_{\phi_4} + p_{\phi_4}) &= 0, \end{aligned} \tag{10.64}$$

thus, for linear state equations

$$\begin{aligned} \rho_4 + p_4 &= \gamma_4 p_4 \\ \rho_{\phi_4} + p_{\phi_4} &= \Gamma_4 p_{\phi_4} \end{aligned} \tag{10.65}$$

their integrals are

$$\rho_4 = \rho_{40} a^{-3\gamma_4}, \ \rho_{40} = \text{const.}, \ \rho_{\phi_4} = \rho_{\phi_{40}} a^{-3\Gamma_4}, \ \rho_{\phi_{40}} = \text{const.} \tag{10.66}$$

On the other hand, (10.7) for $k = 0$ yields

$$\left(\frac{\dot a}{a}\right)^2 = \frac{\kappa_4}{3}\left(\rho_{40} a^{-3\gamma_4} + \rho_{\phi_{40}} a^{-3\Gamma_4}\right), \tag{10.67}$$

which, by introducing the new variable $z = a^{3\gamma_4/2}$, can be written as

$$\dot{z} = \frac{3}{2}\epsilon_a \gamma_4 \sqrt{\kappa_4/3}\sqrt{\rho_{40} + \rho_{\phi_{40}}\, z^{2\frac{(\gamma_4 - \Gamma_4)}{\gamma_4}}}. \tag{10.68}$$

Incidentally, the simplest solutions to (10.68) arise for $\gamma_4 = \Gamma_4$, and $\gamma_4 = 2\Gamma_4$, and amount to

$$a^{3\gamma_4/2} = \frac{1}{2}\epsilon_a \gamma_4 \sqrt{3\kappa_4(\rho_{40} + \rho_{\phi_{40}})}(t - t_0), \tag{10.69}$$

$$a^{3\gamma_4/2} = z = \frac{1}{\rho_{\phi_{40}}}\left(\frac{3}{16}\gamma_4^2 \rho_{\phi_{40}}^2 \kappa_4 (t - t_0)^2 - \rho_{40}\right). \tag{10.70}$$

the first scale factor, (10.69), gives rise to the slow roll power law inflationary solution, while the last one, (10.70), yields the proper (3+1) Barrow and Saich (1993) solution; see below.

For the general case $\gamma_4 \neq 2\Gamma_4$, the integral of (10.68), is given in terms of hypergeometric functions by

$$t - t_0 = \frac{2}{3}\frac{\epsilon_a}{\gamma_4\sqrt{\kappa_4/3\rho_{40}}}\, a^{3\gamma_4/2} F\left(\left[\frac{1}{2}, \frac{\gamma_4}{2\gamma_4 - 2\Gamma_4}\right], \left[\frac{3\gamma_4 - 2\Gamma_4}{2\gamma_4 - 2\Gamma_4}\right], -\frac{\rho_{\phi_{40}}}{\rho_{40}} a^{3(\gamma_4 - \Gamma_4)}\right). \tag{10.71}$$

Adding (10.5) and (10.6) one gets $(d\phi_4/dt)^2$, which can be written as

$$\dot{a}^2\left(\frac{d\phi_4}{da}\right)^2 = \Gamma_4 \rho_{\phi_4}. \tag{10.72}$$

Substituting above $\dot{a}^2$ from (10.67), and the expression of ρ_{ϕ_4} from (10.66), one obtains

$$\phi_4 - \phi_{40} = \epsilon_a \epsilon_\phi \frac{\sqrt{\Gamma_4}}{\sqrt{\kappa_4/3}}\frac{2}{3(\gamma_4 - \Gamma_4)} \ln\left(\sqrt{\rho_{\phi_{40}}}\, a^{\frac{3}{2}(\gamma_4 - \Gamma_4)} + \sqrt{\rho_{40} + \rho_{\phi_{40}}\, a^{3(\gamma_4 - \Gamma_4)}}\right), \tag{10.73}$$

hence, introducing $C_4 = \epsilon_a \epsilon_\phi \frac{\sqrt{\kappa_4/3}}{\sqrt{\Gamma_4}}\frac{3}{2}(\gamma_4 - \Gamma_4)$, the expression of a in terms of ϕ amounts to

$$a^{\frac{3}{2}(\gamma_4 - \Gamma_4)} = \frac{1}{2\sqrt{\rho_{\phi_{40}}}}\left(e^{C_4(\phi_4 - \phi_{40})} - \rho_{40} e^{-C_4(\phi_4 - \phi_{40})}\right). \tag{10.74}$$

Since $2V_4 = 2\rho_{\phi_4} - \dot{\phi}_4^2 = (2 - \Gamma_4)\rho_{\phi_4}$, then V_4 as function of ϕ is given by

$$V_4 = \frac{2 - \Gamma_4}{2}\rho_{\phi_{40}}\left[\frac{1}{2\sqrt{\rho_{\phi_{40}}}}\left(e^{C_4(\phi_4 - \phi_{40})} - \rho_{40} e^{-C_4(\phi_4 - \phi_{40})}\right)\right]^{-2\frac{\Gamma_4}{\gamma_4 - \Gamma_4}} \tag{10.75}$$

10.3.2 Spatially Flat FRW (2+1) Solutions; $\gamma_3 \neq 2\Gamma_3$

In the (2+1) case, for linear state equations $\rho_3 + p_3 = \gamma_3 p_3$, $\rho_{\phi_3} + p_{\phi_3} = \Gamma_3 p_{\phi_3}$, the integrals of the dynamical equations are:
the energy densities

$$\rho_3 = \rho_{30} a^{-2\gamma_3}, \; \rho_{30} = \text{const.}, \; \rho_{\phi_3} = \rho_{\phi_{30}} a^{-2\Gamma_3}, \; \rho_{\phi_{30}} = \text{const.}, \tag{10.76}$$

the time variable t, given in terms of hypergeometric functions depending on the argument a, is given by

$$t - t_0 = \frac{\epsilon_a}{\gamma_3 \sqrt{\kappa_3 \rho_{30}}} a^{\gamma_3} F\left(\left[\frac{1}{2}, \frac{\gamma_3}{2\gamma_3 - 2\Gamma_3}\right], \left[\frac{3\gamma_3 - 2\Gamma_3}{2\gamma_3 - 2\Gamma_3}\right], -\frac{\rho_{\phi_{30}}}{\rho_{30}} a^{2(\gamma_3 - \Gamma_3)}\right), \tag{10.77}$$

the scalar field

$$\phi_3 - \phi_{30} = \epsilon_a \epsilon_\phi \frac{1}{\sqrt{\kappa_3}} \frac{\sqrt{\Gamma_3}}{\gamma_3 - \Gamma_3} \ln\left(\sqrt{\rho_{\phi_{30}}}\, a^{\gamma_3 - \Gamma_3} + \sqrt{\rho_0 + \rho_{\phi_{30}} a^{2(\gamma_3 - \Gamma_3)}}\right), \tag{10.78}$$

and finally the potential

$$V_3 = \frac{2 - \Gamma_3}{2} \rho_{\phi_{30}} \left[\frac{1}{2\sqrt{\rho_{\phi_{30}}}} \left(e^{C_3(\phi_3 - \phi_{30})} - \rho_0 e^{-C_3(\phi_3 - \phi_{30})}\right)\right]^{-2\frac{\Gamma_3}{\gamma_3 - \Gamma_3}}, \tag{10.79}$$

where it as been introduced the constant $C_3 = \epsilon_a \epsilon_\phi \frac{\sqrt{\kappa_3}}{\sqrt{\Gamma_3}} (\gamma_3 - \Gamma_3)$. Incidentally, the scale factor a in terms of ϕ_3 amounts to

$$a^{(\gamma_3 - \Gamma_3)} = \frac{1}{2\sqrt{\rho_{\phi_{30}}}} \left(e^{C_3(\phi_3 - \phi_{30})} - \rho_0 e^{-C_3(\phi_3 - \phi_{30})}\right). \tag{10.80}$$

By the way, the simplest solutions, expressible in terms of elementary functions, arise for $\gamma_3 = \Gamma_3$, and $\gamma_3 = 2\Gamma_3$, and amount respectively to

$$a^{\gamma_3} = \epsilon_a \gamma_3 \sqrt{\kappa_3 \rho_{30} + \rho_{\phi_{30}}} (t - t_0), \tag{10.81}$$

$$a^{\gamma_3} = z = \frac{1}{\rho_{\phi_{30}}} \left(\frac{\gamma_3^2}{4} \rho_{\phi_{30}}^2 \kappa_3 (t - t_0)^2 - \rho_{30}\right), \tag{10.82}$$

the first scale factor corresponds to the (2+1) power law solution, and the last scale factor gives rise to the (2+1) Barrow–Saich solution.

10.3.3 Barrow–Saich Solution; $\gamma = 2\Gamma$

From the general solutions given above, one can reproduce the Barrow and Saich (1993) solution arising for the particular branch $\gamma_4 = 2\Gamma_4$. Alternatively, one may use for the scale factor the expression (10.70). In the Barrow–Saich solution the matter and scalar field perfect fluids fulfill the equation of state: $p_4 + \rho_4 = \gamma_4 \rho_4$, $p_{\phi_4} + \rho_{\phi_4} = \frac{\gamma_4}{2} \rho_{\phi_4}$. The energy density conservation equations give $\rho_4 = A_4 a^{-2\gamma_4}$ and $\rho_{\phi_4} = A_{\phi_4} a^{-\gamma_4}$, where A_4, and A_{ϕ_4} are constants of integration. The expressions for the scale factor, scalar field and its potential are

$$a(t) = \left(\frac{\frac{3\kappa_4}{16}\gamma_4^2 A_{\phi 4}^2 (t-t_0)^2 - A_4}{A_{\phi 4}} \right)^{2/(3\gamma_4)}, \tag{10.83}$$

$$\phi_4(t) - \phi_{40} = \sqrt{\frac{2}{3}} \frac{2}{\sqrt{\gamma_4 \kappa_4}} \ln \left[\frac{3\gamma_4}{4} A_{\phi 4} \sqrt{\frac{\kappa_4}{3}} (t-t_0) + \sqrt{\frac{3\gamma_4^2}{16} \kappa_4 A_{\phi 4}^2 (t-t_0)^2 - A_4} \right], \tag{10.84}$$

$$V(\phi_4) = (4-\gamma_4) A_{\phi 4}^2 \frac{e^{\sqrt{3\kappa_4\gamma_4/2}(\phi_4 - \phi_{40})}}{\left[e^{\sqrt{3\kappa_4\gamma_4/2}(\phi_4 - \phi_{40})} - A_4 \right]^2}. \tag{10.85}$$

Using now the correspondence (10.19) and the relation (10.31), accompanied by the changes $\rho_4 \to \rho_3$, and $\rho_{\phi_{40}} \to \rho_{\phi_{30}}$, one obtains the following (2+1) spatially flat FRW Barrow–Saich counterpart:

$$a(t) = \left(\frac{\frac{\kappa_3}{4}\gamma_3^2 A_{\phi 3}^2 (t-t_0)^2 - A_3}{A_{\phi 3}} \right)^{1/\gamma_3}, \tag{10.86}$$

$$\phi_3(t) - \phi_{30} = \sqrt{\frac{2}{\gamma_3 \kappa_3}} \ln \left[\frac{\gamma_3}{2} A_{\phi 3} \sqrt{\kappa_3} (t-t_0) + \sqrt{\frac{\gamma_3^2}{4} \kappa_3 A_{\phi 3}^2 (t-t_0)^2 - A_3} \right], \tag{10.87}$$

$$V(\phi_3) = (4-\gamma_3) A_{\phi 3}^2 \frac{e^{\sqrt{2\kappa_3\gamma_3}(\phi_3 - \phi_{30})}}{\left[e^{\sqrt{2\kappa_3\gamma_3}(\phi_3 - \phi_{30})} - A_3 \right]^2}. \tag{10.88}$$

As far as the interpretation of these solutions is concerned, we direct readers to the original Barrow–Saich publication, where the inflationary behavior is clearly exhibited.

10.4 Single Scalar Field Spatially Flat FRW Solutions to $p_\phi + \rho_\phi = \Gamma\ \rho_\phi{}^\beta$

In this section we derive scalar field solutions to nonlinear state equations of the form $p_\phi + \rho_\phi = \Gamma\ \rho_\phi{}^\beta$, for any value of the parameter β except for $\beta = 1$ – the linear case. In particular, for $\beta = 1/2$ one recognizes the (2+1) Barrow–Burd–Lancaster, Barrow et al. (1986), and the (3+1) Madsen (1986) solutions. Cosmological constants are set equal to zero.

10.4.1 Spatially Flat (3+1) Solutions with $V(\phi) = A(\alpha\phi^{2/(1-\beta)} - \phi^{2\beta/(1-\beta)})$

For the derivation of solutions of this kind we consider the following state equation

$$p_{\phi_4} = \Gamma_4 \rho^\beta_{\phi_4} - \rho_{\phi_4}. \tag{10.89}$$

Consequently, (10.8) becomes

$$\frac{d}{da}\rho_{\phi_4} + \frac{3}{a}\Gamma_4\rho^\beta_{\phi_4} = 0, \tag{10.90}$$

which has, $\beta \neq 1$, the general integral

$$a(t) = a_0 \exp\left(-\frac{1}{3\Gamma_4(1-\beta)}\rho^{(1-\beta)}_{\phi_4}\right). \tag{10.91}$$

The equation for a, from (10.7) for $k = 0$, reads

$$\frac{\dot a}{a} = \epsilon_a\sqrt{\kappa_4/3}\rho^{1/2}_{\phi_4}, \tag{10.92}$$

therefore, substituting a from (10.91), one arrives at

$$-\frac{1}{3\Gamma_4\rho^\beta_{\phi_4}}\frac{d}{dt}\rho_{\phi_4} = \epsilon_a\sqrt{\kappa_4/3}\rho^{1/2}_{\phi_4}, \tag{10.93}$$

which possesses, for $\beta \neq 1/2$, the general solution

$$\rho_{\phi_4}(t) = \left(\rho_{\phi_{40}} - 3\epsilon_a\Gamma_4\sqrt{\kappa_4/3}(1/2-\beta)\,t\right)^{1/(1/2-\beta)}. \tag{10.94}$$

On the other hand, substituting the above expression into (10.91), one obtains $a(t)$, namely

$$a(t) = a_0 \exp\left[-\frac{1}{3\Gamma_4(1-\beta)}\left(\rho_{\phi_{40}} - 3\epsilon_a\Gamma_4\sqrt{\kappa_4/3}(1/2-\beta)\,t\right)^{2\frac{(1-\beta)}{(1-2\beta)}}\right]. \tag{10.95}$$

The equation to determine $\phi_4(t)$, $\dot\phi = \epsilon_\phi\sqrt{p_\phi + \rho_\phi}$, amounts to

$$\frac{d\phi_4}{dt} = \epsilon_\phi\sqrt{\Gamma_4}\rho_{\phi_4}{}^{\beta/2}, \tag{10.96}$$

which can be written as

$$d\phi_4 = -\epsilon_\phi \epsilon_a \frac{1}{\sqrt{3\Gamma_4 \kappa_4}} \rho_{\phi_4}{}^{-(1+\beta)/2} d\rho_{\phi_4}, \tag{10.97}$$

consequently

$$\phi_4 - \phi_{40} = -2\epsilon_\phi \epsilon_a \frac{1}{\sqrt{3\Gamma_4 \kappa_4}(1-\beta)} \rho_{\phi_4}{}^{(1-\beta)/2}. \tag{10.98}$$

Finally, the evaluation of $V(\phi_4)$ gives

$$V(\phi_4) = \rho_{\phi_4} - \frac{1}{2}\dot{\phi}_4{}^2 = \left[-\epsilon_a \epsilon_\phi \frac{1}{2}\sqrt{3\Gamma_4 \kappa_4}(1-\beta)(\phi_4 - \phi_{40})\right]^{2/(1-\beta)}$$
$$-\frac{\Gamma_4}{2}\left[-\epsilon_a \epsilon_\phi \frac{1}{2}\sqrt{3\Gamma_4 \kappa_4}(1-\beta)(\phi_4 - \phi_{40})\right]^{2\beta/(1-\beta)}. \tag{10.99}$$

Recall that these families of solutions arise for $\beta \neq 1/2$. The case $\beta = 1/2$ gives rise to the (3+1) Madsen cosmology.

10.4.2 Spatially Flat (2+1) Solutions with $V(\phi) = A(\alpha\phi^{2/(1-\beta)} - \phi^{2\beta/(1-\beta)})$

For the nonlinear state equation $p_{\phi_3} = \Gamma_3 \rho_{\phi_3}^\beta - \rho_{\phi_3}$, in the case of $\beta \neq 1$ and $\beta \neq 1/2$, the integrals of the dynamical equations are:
the scale factor can be given in the form

$$a(t) = a_0 \exp\left[-\frac{1}{2\Gamma_3(1-\beta)}\rho_{\phi_3}^{(1-\beta)}\right], \tag{10.100}$$

the scalar field energy density is

$$\rho_{\phi_3}(t) = \left[\rho_{\phi_{30}} - 2\epsilon_a \Gamma_3 \sqrt{\kappa_3}(1/2-\beta)\, t\right]^{1/(1/2-\beta)}, \tag{10.101}$$

thus, substituting this last expression into the scale factor a one has alternatively

$$a(t) = a_0 \exp\left[-\frac{1}{2\Gamma_3(1-\beta)}\left(\rho_{\phi_{30}} - 2\epsilon_a \Gamma_3 \sqrt{\kappa_3}(1/2-\beta)\, t\right)^{2\frac{(1-\beta)}{(1-2\beta)}}\right]. \tag{10.102}$$

The scalar field amounts to

$$\phi_3 - \phi_{30} = -\epsilon_\phi \epsilon_a \frac{1}{\sqrt{\Gamma_3 \kappa_3}(1-\beta)} \rho_{\phi_3}{}^{(1-\beta)/2}, \tag{10.103}$$

finally, the evaluation of $V(\phi_3)$ gives

$$V(\phi_3) = \rho_{\phi_3} - \frac{1}{2}\dot{\phi}_3{}^2 = \left[-\epsilon_a \epsilon_\phi \sqrt{\Gamma_3 \kappa_3}(1-\beta)(\phi_3 - \phi_{30})\right]^{2/(1-\beta)}$$
$$-\frac{\Gamma_3}{2}\left[-\epsilon_a \epsilon_\phi \sqrt{\Gamma_3 \kappa_3}(1-\beta)(\phi_3 - \phi_{30})\right]^{2\beta/(1-\beta)}. \tag{10.104}$$

It should be mentioned that all these solutions are determined under the condition $\beta \neq 1/2$ and $\beta \neq 1$. The case $\beta = 1/2$ yields the (2+1) Barrow–Burd–Lancaster solution, which is treated in detail below.

10.4.3 Barrow–Burd–Lancaster (2+1) and Madsen (3+1) Solutions

The derivation of the solutions in the case of $\beta = 1/2$ for flat FRW spacetimes subjected to the nonlinear state equations under consideration is described below.

Barrow–Burd–Lancaster (2+1) Solution

Barrow et al. (1986), (BBL), reported two exact solutions exhibiting the evolution of cosmological models containing self-interacting scalar fields with physically interesting potentials, in the zero-curvature FRW model. One of them is derived in what follows, starting from the nonlinear state equation

$$p_{\phi_3} = \Gamma_3 \rho_{\phi_3}^{1/2} - \rho_{\phi_3}, \tag{10.105}$$

the (10.17) for ρ_{ϕ_3} amounts to

$$\frac{d}{da}\rho_{\phi_3} + \frac{2}{a}\Gamma_3 \rho_{\phi_3}^{1/2} = 0. \tag{10.106}$$

Its general integral is given by

$$a = a_0 \exp\left(-\frac{1}{\Gamma_3}\rho_{\phi_3}^{1/2}\right). \tag{10.107}$$

From (10.16) for a, in general one has

$$\frac{\dot{a}}{a} = \epsilon_a \sqrt{\kappa_3}\rho_{\phi_3}^{1/2}. \tag{10.108}$$

Substituting this derivative of a into (10.106), one gets

$$\frac{1}{\Gamma_3}\frac{d}{dt}\rho_{\phi_3}{}^{1/2} + \epsilon_a \sqrt{\kappa_3}\rho_{\phi_3}{}^{1/2} = 0, \tag{10.109}$$

therefore, integrating one obtains

$$\rho_{\phi_3}(t) = \rho_{\phi_{30}} e^{-2\epsilon_a \Gamma_3 \sqrt{\kappa_3} t}. \tag{10.110}$$

Using this expression in (10.107) one arrives at $a(t)$, namely

$$a = a_0 \exp\left(-\frac{1}{\Gamma_3}\rho_{\phi_{30}}^{1/2} e^{-\epsilon_a \Gamma_3 \sqrt{\kappa_3} t}\right). \tag{10.111}$$

The equation for ϕ_3, $\dot{\phi} = \epsilon_\phi \sqrt{p_\phi + \rho_\phi}$, yields

$$\dot{\phi}_3 = \epsilon_\phi \sqrt{\Gamma_3}\rho_{\phi_3}^{1/4}, \tag{10.112}$$

therefore, using (10.110), one obtains

$$\phi_3 - \phi_{30} = -2\epsilon_\phi \epsilon_a \frac{1}{\sqrt{\kappa_3 \Gamma_3}} \rho_{\phi_{30}}^{1/4} \exp\left(-\frac{1}{2}\epsilon_a \Gamma_3 \sqrt{\kappa_3} t\right). \tag{10.113}$$

On the other hand, replacing

$$\rho_{\phi_3}^{1/4} = -\frac{1}{2}(\phi_3 - \phi_{30})\epsilon_a \epsilon_\phi \sqrt{\Gamma_3 \kappa_3}, \tag{10.114}$$

into $V_3 = \rho_{\phi_3} - \dot{\phi}_3^2/2$, one gets

$$V_3 = \frac{\Gamma_3^2 \kappa_3}{8}\left[\frac{1}{2}\kappa_3(\phi_3 - \phi_{30})^4 - (\phi_3 - \phi_{30})^2\right]. \tag{10.115}$$

To identify the derived solution with the original Barrow–Burd–Lancaster expressions – see Barrow et al. (1986) – one has to accomplish the following changes:

$$\Gamma_3 \longrightarrow \sqrt{8\mu/\kappa_3}, \rho_{\phi_{30}} \longrightarrow \frac{1}{2}\mu\kappa_3 A^4, -\epsilon_a \longrightarrow \epsilon_a = \pm 1, \tag{10.116}$$

in this manner one obtains:

$$\begin{aligned} a &= a_0 \exp\left(-\frac{\kappa_3}{4} A^2 e^{\epsilon_a \sqrt{8\mu} t}\right), \\ \rho_{\phi_3}(t) &= \frac{1}{2}\mu\kappa_3 A^4 e^{2\epsilon_a \sqrt{8\mu}\, t}, \\ \phi_3 - \phi_{30} &= \epsilon_\phi \epsilon_a A e^{\epsilon_a \sqrt{8\mu}\, t/2}, \\ V_3 &= \mu\left[\frac{1}{2}\kappa_3(\phi_3 - \phi_{30})^4 - (\phi_3 - \phi_{30})^2\right], \end{aligned} \tag{10.117}$$

where $\epsilon_a = \pm 1$, and ϕ_{30} is an integration constant.

Madsen (3+1) Solution

In this subsection we shall give a four-dimensional cosmological model fulfilling nonlinear state equation, $p_{\phi_4} = \Gamma_4 \rho_{\phi_4}^{1/2} - \rho_{\phi_4}$, which has been found by Madsen (1986). This inflationary solution admits symmetry breaking and is determined by:

the scale factor

$$a = a_0 \exp\left(-\frac{2}{3\Gamma_4}\rho_{\phi_{40}}^{1/2} e^{-3/2\epsilon_a \Gamma_4 \sqrt{\kappa_4/3}\, t}\right). \tag{10.118}$$

the scalar field energy density

$$\rho_{\phi_4}(t) = \rho_{\phi_{40}} e^{-3\epsilon_a \Gamma_4 \sqrt{\kappa_4/3} t}, \tag{10.119}$$

thus, a allows for the alternative representation

$$a = a_0 \exp\left(-\frac{2}{3\Gamma_4}\rho_{\phi_4}^{1/2}\right). \tag{10.120}$$

The scalar potential can be expressed as

$$\phi_4 - \phi_{40} = -\frac{4}{3}\epsilon_\phi \epsilon_a \frac{1}{\sqrt{\Gamma_4 \kappa_4/3}} \rho_{\phi_{40}}^{1/4} \exp\left(-\frac{3}{4}\epsilon_a \Gamma_4 \sqrt{\kappa_4/3}\, t\right), \tag{10.121}$$

while the potential can be brought to the form

$$V_4 = \frac{9}{32}\Gamma_4^2 \frac{\kappa_4}{3}\left[\frac{9}{8}\frac{\kappa_4}{3}(\phi_4 - \phi_{40})^4 - (\phi_4 - \phi_{40})^2\right]. \tag{10.122}$$

A more suitable representation of the Madsen's solution can be achieved by accomplishing the following changes

$$\Gamma_4 \longrightarrow \frac{2}{3}\frac{\sqrt{8\mu}}{\sqrt{\kappa_4/3}}, \rho_{\phi_{40}} \longrightarrow \frac{1}{6}\mu\kappa_4 A^4, -\epsilon_a \longrightarrow \epsilon_a = \pm 1; \tag{10.123}$$

in this way one obtains:

$$\begin{aligned} a &= a_0 \exp\left(-\frac{1}{4}\frac{\kappa_4}{3} A^2 e^{\epsilon_a \sqrt{8\mu} t}\right), \\ \rho_{\phi_4}(t) &= \frac{\mu}{2}\frac{\kappa_4}{3} A^4 e^{2\epsilon_a \sqrt{8\mu}\, t}, \\ \phi_4 - \phi_{40} &= \sqrt{\frac{2}{3}}\epsilon_\phi \epsilon_a A e^{\epsilon_a \sqrt{8\mu}\, t/2}, \\ V_4 &= \mu\left[\frac{9}{8}\frac{\kappa_4}{3}(\phi_4 - \phi_{40})^4 - (\phi_4 - \phi_{40})^2\right], \end{aligned} \tag{10.124}$$

where $\epsilon_a = \pm 1$, and ϕ_{40} is an integration constant.

It is really easy to establish the relationship existing between the BBL and Madsen solutions by means of the correspondence (10.19).

10.5 Scalar Field Solutions for a Given Scale Factor

In this section we present solutions for a scalar field alone, namely the second Barrow–Burd–Lancaster cosmology – see Barrow et al. (1986) – and its (3+1) generalization, such that a simple state equation of the form $p_\phi + \rho_\phi = \Gamma\, F(\rho_\phi)$ is difficult to establish. In order to reach our goal, we consider that the scale factor $a(t)$ is a known function of the argument t, and proceed to evaluate ρ_ϕ, next to integrate ϕ, and finally evaluate $V(\phi)$.

10.5.1 Second (2+1) BBL Solution

The approach we shall apply to derive the second BBL solution assumes the function $a(t)$ as a given one, namely

$$a(t) = t^2\sqrt{1 + \frac{A}{t^3}}. \tag{10.125}$$

From (10.16), one evaluates ρ_{ϕ_3},

$$\rho_{\phi_3} = \frac{1}{\kappa_3}\frac{\dot{a}^2}{a^2} = \frac{1}{4\kappa_3}\frac{(A+4t^3)^2}{t^2(A+t^3)^2}. \tag{10.126}$$

Further, since (10.18) reads $\dot{\rho}_{\phi_3} + 2(p_{\phi_3} + \rho_{\phi_3})\dot{a}/a = 0$, and $\dot{\phi}_3^{\,2} = p_{\phi_3} + \rho_{\phi_3}$, hence the equation for ϕ_3 to be integrated is

$$\dot{\rho_{\phi_3}} + 2\frac{\dot{a}}{a}\dot{\phi}_3^{\,2} = 0. \tag{10.127}$$

Substituting the derivative of ρ_{ϕ_3} from (10.126), one obtains

$$\dot{\phi}_3 = \frac{(2t^3 - A)}{\sqrt{2\kappa_3}\,t(A+t^3)}, \tag{10.128}$$

which has the following integral

$$\phi_3 = \frac{1}{\sqrt{2\kappa_3}}\ln\left(C_0\frac{A+t^3}{t}\right). \tag{10.129}$$

The evaluation of $V(\phi) = \rho_{\phi_3} - \dot{\phi}_3^{\,2}/2$ yields

$$V(\phi_3) = \frac{3}{\kappa_3}\frac{t}{A+t^3} = \frac{3}{\kappa_3}C_0\, e^{-\sqrt{2\kappa_3}\phi_3}. \tag{10.130}$$

Summarizing, the second (2+1) inflationary BBL solution is determined by

$$a(t) = t^2\sqrt{1+\frac{A}{t^3}}, \tag{10.131}$$

$$\phi_3 = \frac{1}{\sqrt{2\kappa_3}}\ln\left(C_0\frac{A+t^3}{t}\right), \tag{10.132}$$

$$V(\phi_3) = \frac{3}{\kappa_3}\frac{t}{A+t^3} = \frac{3}{\kappa_3}C_0\, e^{-\phi_3\sqrt{2\kappa_3}}, \tag{10.133}$$

where A and C_0 are constants.

To get an insight into the form of the conventional state equation, i.e., on the dependence of $F(\rho)$ on ρ, one expresses t in terms of $\rho := 4\kappa_3\rho_{\phi_3}$ by solving (10.126) with respect to t, which yields

$$t\sqrt{\rho} = 1 + \frac{1}{2}\sqrt{4+\rho^{1/2}A^{1/3}\Delta} + \frac{\epsilon_t}{2}\left(8 - \rho^{1/2}A^{1/3}\Delta + \frac{16 - 2\,\rho^{3/2}A}{\sqrt{4+\rho^{1/2}A^{1/3}\Delta}}\right)^{\frac{1}{2}}, \tag{10.134}$$

where $\Delta := \left(A\rho^{3/2} - 16\right)^{1/3}$. On the other hand, since

$$\dot{\phi}_3^{\,2} = p_{\phi_3} + \rho_{\phi_3} = \Gamma F(\rho_{\phi_3}) = \frac{1}{2\kappa_3}(2t^3 - A)^2/(t^2(A+t^3)^2),$$

substituting t from (10.134), one obtains a very involved function F on ρ.

10.5.2 (3+1) Generalization of the Second (2+1) BBL Solution

Using the relations (10.19) together with (10.31) we obtain the following (3+1) flat FRW cosmology:

$$a(t) = t^2\sqrt{1+\frac{A}{t^3}}, \tag{10.135}$$

$$\phi_4 = \frac{1}{\sqrt{\kappa_4}} \ln\left(C_0 \frac{A+t^3}{t}\right), \tag{10.136}$$

$$V(\phi_4) = \frac{1}{4\kappa_4}\frac{40t^6+32At^3+A^2}{t^2(t^3+A)^2}, \tag{10.137}$$

The expression of $V(\phi_4)$ in terms of ϕ_4, which is very involved, can be achieved by substituting the roots of t in terms of ϕ_4 from (10.136) into (10.137).

11

Einstein–Maxwell Solutions

The purpose of this chapter is to provide a new approach on the search of electromagnetic–gravitational solutions to the Einstein–Maxwell fields of the $(2+1)$ gravity in the presence of a cosmological constant, allowing for stationary and cyclic symmetries, establishing their relationship with known current solutions, and to point out the families allowing for black hole interpretation. The search and interpretation of this kind of solution has been the goal and realm of several authors' investigations, starting from quite different perspectives and using a variety of approaches, which have sometimes brought about duplication of results and efforts. Consequently, the completeness of the electromagnetic classes of stationary cyclic symmetric solutions under consideration will be demonstrated via straightforward integration of the field equations. A full characterization of the physical content of these solutions would require considerable work; for this reason, some short related comments are made close to those contained in the relevant references, if there are any, and also about newly discovered families with special emphasis on their black hole feature. Nevertheless, a full geometrical characterization based on the algebraic classification of the physical tensors has been produced.

From a general metric for stationary cyclic symmetric gravitational fields coupled to Maxwell electromagnetic fields within the $(2+1)$-dimensional gravity the uniqueness of wide families of exact solutions is established, including all uniform electromagnetic solutions possessing electromagnetic fields with vanishing covariant derivatives, all fields having constant electromagnetic invariants $F_{\mu\nu}\,F^{\mu\nu}$ and $T_{\mu\nu}\,T^{\mu\nu}$, the whole classes of hybrid electromagnetic solutions, and also wide classes of stationary solutions, derived for third-order nonlinear key equations. Certain of these families can be thought of as black hole solutions. For the most general set of Einstein–Maxwell equations, reducible to three nonlinear equations for the three unknown functions, two new classes of solutions – having anti-de Sitter spinning metric limits – are derived. The relationship of various families with those reported by different authors' solutions has been established. Among the classes of solutions with cosmological constant are: the electrostatic Peldan solution, the magnetostatic Peldan metric, the stationary uniform and spinning Clément classes, the constant electromagnetic invariant branches with the particular Kamata–Koikawa solution, the Ayón–Cataldo–García hybrid cyclic symmetric stationary black hole fields, and the no less important solutions generated via $SL(2,R)$ transformations where the Clément spinning charged solution, the Martínez–Teitelboim–Zanelli black hole solution, and Dias–Lemos metric merit mention.

11.1 Stationary Cyclic Symmetric Einstein–Maxwell Fields

11.1.1 Stationary Cyclic Symmetric Maxwell Fields

To begin with, we consider a stationary cyclic symmetric spacetime with signature $(-,+,+)$, i.e., a space endowed with stationary symmetry $\boldsymbol{k} = \boldsymbol{\partial_t}$, $\boldsymbol{k}\cdot\boldsymbol{k} < 0$, such that $\pounds_{\boldsymbol{k}}\boldsymbol{g} = 0$, and cyclic symmetry $\boldsymbol{m} = \boldsymbol{\partial_\phi}$, $\boldsymbol{m}\cdot\boldsymbol{m} > 0$, such that $\pounds_{\boldsymbol{m}}\boldsymbol{g} = 0$, with closed integral curves from 0 to 2π, which in turn commute $[\boldsymbol{k},\boldsymbol{m}] = 0$. Hence the Killing vector fields $\boldsymbol{k}$ and $\boldsymbol{m}$ generate the group $SO(2)\times R$. The electromagnetic field, described by the antisymmetric tensor field $\boldsymbol{F} = \frac{1}{2}F_{\mu\nu}\boldsymbol{dx}^\mu\wedge\boldsymbol{dx}^\nu$, is assumed to be stationary cyclic symmetric, i.e., $\pounds_{\boldsymbol{k}}\boldsymbol{F} = 0 = \pounds_{\boldsymbol{m}}\boldsymbol{F}$. It should be pointed out that, in contrast to the general (3+1) stationary cyclic symmetric spacetime, any $(2+1)$ stationary cyclic symmetric spacetime is necessarily circular, i.e., the circularity conditions

$$\boldsymbol{k}\wedge\boldsymbol{m}\wedge\boldsymbol{dk} = 0 = \boldsymbol{k}\wedge\boldsymbol{m}\wedge\boldsymbol{dm} \tag{11.1}$$

are identically fulfilled because of their 4-form character and hence there exists the discrete symmetry when simultaneously $t\to -t$ and $\phi\to-\phi$. One may find a coordinate system such that the metric tensor components $\boldsymbol{g}(\boldsymbol{k}\,\boldsymbol{dr}) = 0$ and $\boldsymbol{g}(\boldsymbol{m}\,\boldsymbol{dr}) = 0$, where the coordinate direction $\boldsymbol{dr}$ is orthogonal to the surface spanned by $\boldsymbol{k}\wedge\boldsymbol{m}$. Commonly one introduces the coordinate system $\{t,\phi,r\}$ in (2+1)-dimensional gravity .

The main goal of this section is to demonstrate of the following theorem:

Theorem 11.1 *The general form of stationary cyclic symmetric electromagnetic fields in (2+1) dimensions is given by*

$$*\,\boldsymbol{F} = a\boldsymbol{dt} + b\boldsymbol{d\phi} + c\frac{g_{rr}}{\sqrt{-g}}\boldsymbol{dr}, \tag{11.2}$$

where the constants a, b and c are subjected, by virtue of the Ricci circularity conditions, to the equations

$$a\,c = 0 = b\,c, \tag{11.3}$$

which gives rise to two disjoint branches

$$c \neq 0, *\boldsymbol{F} = c\frac{g_{rr}}{\sqrt{-g}}\boldsymbol{dr}, \tag{11.4}$$

and

$$c = 0, *\boldsymbol{F} = a\boldsymbol{dt} + b\boldsymbol{d\phi}, \tag{11.5}$$

with its own sub-classes $a = 0$ *or* $b = 0$.

Proof To establish that the field $*\boldsymbol{F}$ possesses the form given by (11.2) one uses the source-free Maxwell equations

$$\boldsymbol{dF} = 0 = \boldsymbol{d} * \boldsymbol{F}, \tag{11.6}$$

where $*$ denotes the Hodge star operation. Let us evaluate the exterior derivative of the t-component $*\boldsymbol{F}(\boldsymbol{k})$ of $*\boldsymbol{F}$,

$$\boldsymbol{d}*\boldsymbol{F}(\boldsymbol{k}) = \boldsymbol{d}\, i_{\boldsymbol{k}} * \boldsymbol{F} = \pounds_{\boldsymbol{k}} * \boldsymbol{F} - i_{\boldsymbol{k}}\boldsymbol{d} * \boldsymbol{F} = 0 - 0 \rightarrow *\boldsymbol{F}(\boldsymbol{k}) =: a = \text{const.}, \tag{11.7}$$

the first zero arises from the stationary character of the field $\boldsymbol{F}$, while the second one corresponds to the Maxwell equation. Similarly, for the ϕ–component $*\boldsymbol{F}(\boldsymbol{m})$ one has

$$\boldsymbol{d}*\boldsymbol{F}(\boldsymbol{m}) = \boldsymbol{d}\, i_{\boldsymbol{m}} * \boldsymbol{F} = \pounds_{\boldsymbol{m}} * \boldsymbol{F} - i_{\boldsymbol{m}}\boldsymbol{d} * \boldsymbol{F} = 0 \rightarrow *\boldsymbol{F}(\boldsymbol{m}) =: b = \text{const.} \tag{11.8}$$

In this manner we have established that the t and ϕ components of the dual field $*\boldsymbol{F}$ are constants given correspondingly by a and b. The component of $*\boldsymbol{F}$ along the vector direction $\boldsymbol{\partial_r}$ remains to be determined. For this purpose, consider the $t\phi$-component $\boldsymbol{F}(\boldsymbol{k},\boldsymbol{m})$ of the field $\boldsymbol{F}$, which can be expressed as $\boldsymbol{F}(\boldsymbol{k},\boldsymbol{m}) = i_{\boldsymbol{m}} i_{\boldsymbol{k}}\boldsymbol{F} = (-i_{\boldsymbol{m}} i_{\boldsymbol{k}} * *\boldsymbol{F} = i_{\boldsymbol{m}} * (\boldsymbol{k} \wedge *\boldsymbol{F}) = *(\boldsymbol{m} \wedge \boldsymbol{k} \wedge *\boldsymbol{F})) = - * \boldsymbol{F}(*(\boldsymbol{k} \wedge \boldsymbol{m}))$, thus its derivative yields

$$\begin{aligned} \boldsymbol{dF}(\boldsymbol{k},\boldsymbol{m}) &= \boldsymbol{d}(i_{\boldsymbol{m}} i_{\boldsymbol{k}}\boldsymbol{F}) = \boldsymbol{d}\, i_{\boldsymbol{m}}\,(i_{\boldsymbol{k}}\boldsymbol{F}) \\ &= (\pounds_{\boldsymbol{m}} - i_{\boldsymbol{m}}\boldsymbol{d})(i_{\boldsymbol{k}}\boldsymbol{F}) = i_{\boldsymbol{k}}\pounds_{\boldsymbol{m}}\boldsymbol{F} + i_{[\boldsymbol{k},\boldsymbol{m}]}\boldsymbol{F} - i_{\boldsymbol{k}}(\pounds_{\boldsymbol{m}} - i_{\boldsymbol{k}}\boldsymbol{d})\boldsymbol{F} \\ &= 0 \rightarrow \boldsymbol{F}(\boldsymbol{k},\boldsymbol{m}) =: c = \text{const.}. \end{aligned} \tag{11.9}$$

Since the constant c can be written as $c = - * \boldsymbol{F}(*(\boldsymbol{k} \wedge \boldsymbol{m}))$, to determine it, one evaluates $*(\boldsymbol{k} \wedge \boldsymbol{m})$. Identifying the Killing vectors accordingly with $\boldsymbol{k} = \boldsymbol{\partial_t}$ and $\boldsymbol{m} = \boldsymbol{\partial_\phi}$, then

$$* \,(\boldsymbol{k} \wedge \boldsymbol{m}) = -\sqrt{-g}\boldsymbol{dr} = -\sqrt{-g}\, g^{rr}\boldsymbol{\partial_r}, \tag{11.10}$$

thus

$$c = - * \boldsymbol{F}(-\sqrt{-g}\, g^{rr}\boldsymbol{\partial_r}) = \sqrt{-g}\, g^{rr} * \boldsymbol{F}(\boldsymbol{\partial_r}). \tag{11.11}$$

Conversely, from the above mentioned relation one determines the r component of the field $*\boldsymbol{F}$, namely $*\boldsymbol{F}(\boldsymbol{\partial_r}) = \frac{c}{\sqrt{-g}} g_{rr}$. In this manner, the structure of $\boldsymbol{F}$, explicitly given by (11.2), has been established.

The vanishing conditions (11.3) straightforwardly arise from the Ricci circularity conditions $\boldsymbol{m} \wedge \boldsymbol{k} \wedge \boldsymbol{R}(\boldsymbol{k}) = 0$ and $\boldsymbol{k} \wedge \boldsymbol{m} \wedge \boldsymbol{R}(\boldsymbol{m}) = 0$. Correspondingly, the vanishing conditions $a\,c = 0 = b\,c$ can be established immediately, as we shall see in the next section, from the Einstein equations $R_{\mu\nu} - \frac{1}{2} R\, g_{\mu\nu} = \kappa T_{\mu\nu} - \Lambda\, g_{\mu\nu}$, where the electromagnetic energy–momentum tensor components are defined through the electromagnetic field $F_{\mu\nu} = -F_{\nu\mu}$ as $4\pi\, T_{\mu\nu} = F_{\mu\sigma} F_{\nu}{}^{\sigma} - \frac{1}{4} g_{\mu\nu} F_{\alpha\sigma} F^{\alpha\sigma}$. □

A first formulation of this theorem with an outline of its demonstration has been reported by Ayón, Cataldo and García in Ayón–Beato et al. (2005).

11.1.2 General Stationary Metric and Einstein Equations

In general, in (2+1)-dimensional gravity any stationary cyclic symmetric metric can be given as

$$\boldsymbol{g} = g_{tt} dt^2 + 2 g_{t\phi} dt\, d\phi + g_{\phi\phi}\, d\phi^2 + g_{rr} dr^2. \tag{11.12}$$

When a Maxwell electromagnetic field is present, the field tensor, as we established previously, possesses the structure

$$\left(F^{\alpha\beta}\right) = \frac{1}{\sqrt{-g}} \begin{bmatrix} 0 & b & -\frac{c g_{rr}}{\sqrt{-g}} \\ -b & 0 & a \\ \frac{c g_{rr}}{\sqrt{-g}} & -a & 0 \end{bmatrix}, \tag{11.13}$$

where $g := \det(g_{\mu\nu})$, which makes apparent the fulfillment of the divergence equation

$$(\sqrt{-\det(g_{\mu\nu})} F^{\alpha\beta})_{;\beta} = 0$$

for constants a, b, and c. The Maxwell electromagnetic energy–momentum tensor is given as usual as

$$T_{\mu}{}^{\nu} = \frac{1}{4\pi} (F_{\mu\sigma}\, F^{\nu\sigma} - \frac{1}{4} \delta_{\mu}^{\nu} F_{\tau\sigma} F^{\tau\sigma}). \tag{11.14}$$

Without loss of generality, one can choose the coordinates for a stationary cyclic symmetric (2+1) metric, developed with respect to the cyclic symmetry $\boldsymbol{m} = \boldsymbol{\partial_\phi}$, in such a way that it becomes

$$\boldsymbol{g} = -\frac{F(r)}{H(r)} dt^2 + \frac{dr^2}{F(r)} + H(r) \left[d\phi + W(r) dt\right]^2. \tag{11.15}$$

On the other hand, if one chooses the stationary symmetry $\boldsymbol{k} = \boldsymbol{\partial_t}$ as the fundamental Killing field, the stationary cyclic symmetric $(2+1)$ metric can be written as

$$g = -\frac{F(r)}{h(r)}\left[dt - \omega(r)\, d\phi\right]^2 + h(r)\, d\phi^2 + \frac{dr^2}{F(r)},$$
$$F = F,\ H = h - \frac{F}{h}\omega^2,\ W\frac{H}{F} = \frac{\omega}{h},\ h = \frac{H\,F}{F - W^2\,H^2}. \tag{11.16}$$

Mostly we will use the metric (11.15) in the forthcoming developments, but occasionally the metric representation (11.16) will be used . When doing so, the derived expressions will be given in terms of the set $\{F(r), h(r), \omega(r)\}$ of structural functions. Omitting the dependence of the structural functions on the variable r, the Maxwell electromagnetic field contravariant tensor is given by

$$(F^{\mu\nu}) = \begin{bmatrix} 0 & b & -\frac{c}{F} \\ -b & 0 & a \\ \frac{c}{F} & -a & 0 \end{bmatrix}, \tag{11.17}$$

where a, b, and c are constants related with the character of the field. For instance, if only b is different from zero, while a and c vanish, the field is called (pure) electric field. When $a \neq 0$, $b = 0 = c$, one deals with a pure magnetic field; other possibilities do not receive a particular name. The covariant components $F_{\mu\nu}$ of the field tensor are given by:

$$F_{tr} = -b/H - W\,H(a - b\,W)/F,\ F_{t\phi} = c,\ F_{r\phi} = H(a - b\,W)/F. \tag{11.18}$$

The electromagnetic field quadratic invariant $FF := F_{\mu\nu}F^{\mu\nu}$ is given by

$$FF = -2\,\frac{c^2}{F} + 2\,\frac{H\,(a - W b)^2}{F} - 2\,\frac{b^2}{H}. \tag{11.19}$$

Notice that if one uses the vector-potential description of the electromagnetic field

$$\boldsymbol{F} := \frac{1}{2}F_{\mu\nu}\boldsymbol{dx}^{\mu} \wedge \boldsymbol{dx}^{\nu} = \boldsymbol{d}\left(A_{\mu}\boldsymbol{dx}^{\mu}\right) =: \boldsymbol{d}\,\boldsymbol{A} \tag{11.20}$$

one would have

$$\boldsymbol{F} = \boldsymbol{d}\int^{r}[(\frac{1}{H} - \frac{H}{F}W^2)b + \frac{H}{F}W\,a]dr \times \boldsymbol{dt} + \boldsymbol{d}\int^{r}[-\frac{H}{F}W\,b + \frac{H}{F}\,a]dr \times \boldsymbol{d\phi}$$
$$+\boldsymbol{d}\left[\frac{1}{2}c(t\boldsymbol{d\phi} - \phi\boldsymbol{dt})\right] = \boldsymbol{d}\,\boldsymbol{A}. \tag{11.21}$$

The energy–momentum tensor matrix (T_ν^μ) associated with the metric (11.15) is

$$\begin{bmatrix} \frac{b^2(H^2W^2-F)-a^2H^2-c^2H}{8\pi\,F\,H} & \frac{ac}{4\pi} & -\frac{a\left[b(H^2W^2-F)-aH^2W\right]}{4\pi\,F\,H} \\ \frac{cH(W\,b-a)}{4\,\pi F^2} & \frac{-b^2F+H^2(W\,b-a)^2+c^2H}{8\pi\,F\,H} & -\frac{c\left[b(H^2W^2-F)-aH^2W\right]}{4\pi\,F^2\,H} \\ \frac{bH(W\,b-a)}{4\,\pi F} & \frac{bc}{4\,\pi} & -\frac{b^2(H^2W^2-F)-a^2H^2+c^2H}{8\pi\,F\,H} \end{bmatrix} \tag{11.22}$$

and possesses the trace $T := T_\mu^\mu$ given by

$$T = -\frac{1}{8\,\pi}\frac{c^2}{F} + \frac{1}{8\,\pi}\frac{H\,(a - W\,b)^2}{F} - \frac{1}{8\,\pi}\frac{b^2}{H} = \frac{1}{16\pi}\,FF, \tag{11.23}$$

and the electromagnetic energy momentum quadratic invariant

$$TT = T_{\mu\nu}T^{\mu\nu} = \frac{3}{64\pi^2}\frac{\left[H^2(a - b\,W)^2 - b^2F - c^2H\right]^2}{F^2H^2} = \frac{3}{256\pi^2}\,FF^2. \tag{11.24}$$

The Einstein–Maxwell equations

$$E_{\mu\nu} := R_{\mu\nu} - \frac{R}{2}g_{\mu\nu} + \Lambda g_{\mu\nu} - 8\pi\,T_{\mu\nu} = 0 \tag{11.25}$$

for a negative cosmological constant $\Lambda = -1/l^2$ explicitly read:

$$\begin{aligned} E_t{}^t &= \frac{1}{2}\,F\frac{H_{,r,r}}{H} + \frac{1}{4}\,\frac{H_{,r}F_{,r}}{H} - \frac{1}{4}\frac{F}{H^2}H_{,r}{}^2 + \frac{1}{2}\,H^2W\,W_{,r,r} + HWW_{,r}H_{,r} \\ &\quad + \frac{1}{4}\,H^2W_{,r}{}^2 + b^2\frac{F - H^2W^2}{F\,H} + \frac{c^2}{F} + \frac{a^2H}{F} - \frac{1}{l^2}, \end{aligned} \tag{11.26a}$$

$$E_t{}^r = -2\,ca, \tag{11.26b}$$

$$\begin{aligned} E_t{}^\Phi &= \frac{1}{2}\,W\,F_{,r,r} - F\,W\,\frac{H_{,r,r}}{H} - W\,\frac{H_{,r}F_{,r}}{H} - 2\,a^2\,\frac{H\,W}{F} \\ &\quad - 2\,a\,b\,\frac{F - H^2W^2}{FH} - \frac{1}{2}\left(F + H^2W^2\right)\left(W_{,r,r} + 2W_{,r}\frac{H_{,r}}{H}\right) \\ &\quad + F\,W\,\frac{H_{,r}{}^2}{H^2} - H^2WW_{,r}{}^2, \end{aligned} \tag{11.26c}$$

$$E_r{}^t = -2c\,\frac{H}{F^2}\,(Wb - a)\,, \tag{11.26d}$$

$$\begin{aligned} E_r{}^r &= \frac{1}{4}\,\frac{H_{,r}F_{,r}}{H} - \frac{1}{4}\,F\frac{H_{,r}{}^2}{H^2} + \frac{1}{4}\,H^2W_{,r}{}^2 + \frac{b^2}{H} \\ &\quad - \frac{c^2}{F} - \frac{H}{F}(bW - a)^2 - \frac{1}{l^2}, \end{aligned} \tag{11.26e}$$

$$E_r{}^\Phi = -2\,c\,b\,\frac{F - H^2W^2}{F^2H} - 2\,c\,a\frac{H\,W}{F^2}, \tag{11.26f}$$

$$E_\Phi{}^t = \frac{1}{2}\,H^2W_{,r,r} + HW_{,r}H_{,r} - 2\,b\,\frac{H}{F}\,(Wb - a)\,, \tag{11.26g}$$

$$E_\Phi{}^r = -2\,bc, \tag{11.26h}$$

$$\begin{aligned} E_\Phi{}^\Phi = & -\frac{1}{l^2} + \frac{1}{2} F_{,r,r} - \frac{1}{2} F \frac{H_{,r,r}}{H} - \frac{3}{4}\frac{H_{,r} F_{,r}}{H} + \frac{3}{4} F \frac{H_{,r}{}^2}{H^2} - \frac{1}{2} H^2 W\, W_{,r,r} \\ & - H\, W\, W_{,r} H_{,r} - \frac{3}{4} H^2 W_{,r}{}^2 - b^2 \frac{F - H^2 W^2}{F\, H} + \frac{c^2}{F} - \frac{a^2 H}{F}. \end{aligned} \tag{11.26i}$$

The vanishing of $E_t{}^r$ and $E_\Phi{}^r$ yields respectively $ac = 0 = bc$. Therefore one can distinguish the branches:
$c = 0$ with a and b, not vanishing simultaneously; and
$c \neq 0$ with a and b vanishing simultaneously.
In the forthcoming sections we shall deal with the integration and characterization of each branch starting from the simplest static solutions.

11.1.3 Complex Extension and Real Cuts

It would be of some interest to add some lines about the complex extension of the metric under consideration. Accomplishing in the metric (11.15) the complex transformations

$$t \to i\,\Phi,\ \phi \to -i\,T, \tag{11.27}$$

one arrives at

$$g_c = \left(\frac{F}{H} - HW^2\right) d\Phi^2 + \frac{dr^2}{F} - H dT^2 + 2H\,W\,dT\,d\Phi, \tag{11.28}$$

which can be brought to the form

$$g_c = -\frac{\mathcal{F}}{\mathcal{H}} dT^2 + \frac{dr^2}{\mathcal{F}} + \mathcal{H}\,(d\Phi + \mathcal{W}\,dT)^2\,, \tag{11.29}$$

accompanied by the identification

$$\mathcal{F} = F,\, \mathcal{H} = \frac{F}{H} - HW^2,\, \mathcal{W} = \frac{H\,W}{\mathcal{H}}. \tag{11.30}$$

At the level of the field tensor $F^{\mu\nu}$ one has

$$(\mathcal{F}^{\mu\nu}) = \begin{bmatrix} 0 & \mathcal{B} & -\frac{\mathcal{C}}{\mathcal{F}} \\ -\mathcal{B} & 0 & \mathcal{A} \\ \frac{\mathcal{C}}{\mathcal{F}} & -\mathcal{A} & 0 \end{bmatrix} = \begin{bmatrix} 0 & -i\,a & \frac{c}{F} \\ i\,a & 0 & ib \\ -\frac{c}{F} & -ib & 0 \end{bmatrix}, \tag{11.31}$$

thus the following correspondence for the field constants arises

$$-i\,a \to \mathcal{B},\ i\,b \to \mathcal{A},\ -c \to \mathcal{C}. \tag{11.32}$$

Summarizing, one may say that the role of the Killingian coordinates has been interchanged: the timelike coordinate t becomes the spacelike Φ–coordinate, while the cyclic ϕ-coordinate becomes the new time-coordinate T. Correspondingly, one has to think of the tensor components of the participating quantities from this perspective. This procedure can be used to determine new classes of solutions from known ones. For instance, one can generate magnetic solutions from electric ones. The relations arising from this kind of complex transformations have been called "duality mapping" by Cataldo and Salgado (1996), although strictly there is no electric–magnetic duality in (2+1) dimensions.

11.1.4 Positive Λ Solutions

For completeness and to avoid duplication of works, it is worth noticing that solutions for positive cosmological constant are easily obtainable from the anti-de Sitter ($\Lambda = -1/l^2$) ones; first, notice that the Einstein equations for any sign of a cosmological constant Λ – positive or negative – are recovered from equations (11.26) simply by replacing $-1/l^2 \rightarrow \Lambda$. Next, having at disposal a concrete solution of the Einstein equations mentioned above (11.26), by replacing there l^2 by $-l^2$, one determines the corresponding metric structure for positive cosmological constant $\Lambda = 1/l^2$. This replacement is equivalent to accomplishing the complex change $l \rightarrow i\,l$ in the $\Lambda < 0$ solution, nevertheless one ought to take care of possible additional arrangements of constants, if any, and also of possible changes in the signature.

11.1.5 Characterizations of Einstein–Maxwell Solutions

The physical characterization is based on the determination of the local and global energy–momentum–mass quantities using the Brown–York approach (Brown and York Jr., 1993). As far as the algebraic–geometrical characterization is concerned, the eigenvalue problem for the electromagnetic field, energy–momentum and Cotton tensors is solved and their types are established.

The general form of electromagnetic fields for stationary cyclic symmetric (2+1) spacetimes is given by: $*\boldsymbol{F} = a\boldsymbol{dt} + b\boldsymbol{d\phi} + cg_{rr}/\sqrt{-g}\boldsymbol{dr}$, which splits into various sub-families: the electric $b \neq 0$ fields, the magnetic $a \neq 0$ fields, the uniform fields characterized by the vanishing of the covariant derivatives $F_{\alpha\beta;\gamma} = 0$, the class of stationary fields with constant invariant $F_{\mu\nu}F^{\mu\nu}$, and consequently, due to the structure of the electromagnetic fields with constant energy–momentum tensor invariants, the gravitational stationary cyclic solutions for the hybrid electromagnetic field $*\boldsymbol{F} = cg_{rr}/\sqrt{-g}\boldsymbol{dr}$; the explicit derivation of the solutions belonging to the quoted branches can be found in García (2009), and the evaluation of quasi-local energy–momentum quantities has been done in Garcia–Diaz (2013).

The families of Einstein–Maxwell solutions to be considered are: all uniform electromagnetic solutions possessing electromagnetic fields with vanishing covariant derivatives (stationary uniform and spinning Clément classes), all fields having constant electromagnetic field and energy–momentum tensors' invariants (Kamata–Koikawa solution), the whole classes of hybrid electromagnetic Ayon–Cataldo–García solutions, a new family of stationary electromagnetic solutions, the electrostatic and magnetostatic solutions with Peldan limit, the Clément spinning charged metric, the Martínez–Teitelboim–Zanelli black hole solution, and Dias–Lemos electromagnetic solution.

The application of the Hayward black hole dynamics formulation, Hayward (2008), and the Ashtekar isolated horizon approach, Ashtekar and Krishnan (2004), to the static and stationary black hole solutions reported here is straightforward.

Cotton Tensor Algebraic Classification

In $(n+1)$-dimensional spacetimes, for $n \geq 3$, the invariant decomposition of the Riemannian curvature tensor gives rise to the conformal Weyl tensor, the traceless Ricci tensor, and the scalar curvature; for the classification of gravity one classifies the Weyl tensor, and the classification of matter is achieved through the classification of the traceless Ricci tensor. For details, in $(3+1)$-dimensional spacetimes; see, for instance, the book by Stephani et al. (2003), and in the present book, Chapter 20.

In $(2+1)$-dimensional spacetimes there is no room for the conformal Weyl tensor, the Riemannian curvature tensor decomposes into the Ricci tensor, and the scalar curvature. The role of the conformal tensor in (2+1) gravity is played by the Cotton tensor – see Stephani et al. (2003) and Chapter 20 – which is defined by means of the Ricci tensor and the scalar curvature through their covariant derivatives

$$C^{\alpha\beta} = \epsilon^{\alpha\gamma\delta}(R^{\beta}{}_{\gamma} - \frac{1}{4}R\,\delta^{\beta}{}_{\gamma})_{;\delta},\; C^{\alpha}{}_{\alpha} = 0. \tag{11.33}$$

For the standard stationary (static) cyclic symmetric metric

$$ds^2 = -N^2dt^2 + L^{-2}\,dr^2 + K^2[d\phi + W dt]^2,$$

the traceless Cotton tensor, in the form $C^{\alpha}{}_{\beta}$, is

$$\mathcal{C} := (C^{\alpha}{}_{\beta}) = \begin{bmatrix} C^1{}_1 & 0 & C^1{}_3 \\ 0 & C^2{}_2 & 0 \\ C^3{}_1 & 0 & C^3{}_3 \end{bmatrix};\; C^1{}_1 + C^2{}_2 + C^3{}_3 = 0. \tag{11.34}$$

Determining the eigenvalues and eigenvectors of the Cotton matrix (11.34) one establishes the algebraic Cotton type of the spacetime one is dealing with. Accordingly, the characteristic equation for the eigenvalue λ amounts to

$$(C^2{}_2 - \lambda)\left((C^1{}_1 - \lambda)(C^3{}_3 - \lambda) - C^3{}_1\,C^1{}_3\right) = 0, \tag{11.35}$$

or, in terms of its solutions, as

$$(C^2{}_2 - \lambda)\left(\lambda + \frac{1}{2}C^2{}_2 + \frac{1}{2}\sqrt{(C^2{}_2 + 2\,C^1{}_1)^2 + 4\,C^3{}_1\,C^1{}_3}\right)$$
$$\times\left(\lambda + \frac{1}{2}C^2{}_2 - \frac{1}{2}\sqrt{(C^2{}_2 + 2\,C^1{}_1)^2 + 4\,C^3{}_1\,C^1{}_3}\right) = 0, \qquad (11.36)$$

while the eigenvector equations are

$$\begin{aligned} (C^1{}_1 - \lambda)V^1 + C^1{}_3 V^3 &= 0,\\ (C^2{}_2 - \lambda)V^2 &= 0,\\ C^3{}_1 V^1 + (C^3{}_3 - \lambda)V^3 &= 0. \end{aligned} \qquad (11.37)$$

For each eigenvalue the corresponding solution is:

$$\begin{aligned} \lambda_1 &= C^2{}_2, \ \mathbf{V1} = (0, V^2, 0),\\ \lambda_2 &= -\frac{1}{2}C^2{}_2 + \frac{1}{2}\sqrt{(C^2{}_2 + 2\,C^1{}_1)^2 + 4\,C^3{}_1\,C^1{}_3},\\ \mathbf{V2} &= (-\frac{C^1{}_3}{C^1{}_1 - \lambda_2}V^3, 0, V^3),\\ \lambda_3 &= -\frac{1}{2}C^2{}_2 - \frac{1}{2}\sqrt{(C^2{}_2 + 2\,C^1{}_1)^2 + 4\,C^3{}_1\,C^1{}_3},\\ \mathbf{V3} &= (-\frac{C^1{}_3}{C^1{}_1 - \lambda_3}V^3, 0, V^3). \end{aligned} \qquad (11.38)$$

Thus, the eigenvector $\mathbf{V1}$ – a real one – is oriented in the ρ-direction, the remaining two vectors $\mathbf{V2}$ and $\mathbf{V3}$ might be real vectors lying on the surface spanned by the t and ϕ coordinate directions or complex eigenvectors depending, correspondingly, upon whether the value of the radical $(C^2{}_2 + 2\,C^1{}_1)^2 + 4\,C^3{}_1\,C^1{}_3$ is positive or negative.

The nomenclature to be used for eigenvectors and algebraic types of tensors is borrowed from Plebański (1964); see also Plebański (1967), Chapter VI: timelike, spacelike, null, and complex vectors are denoted respectively by **T**, **S**, **N**, and **Z**. For algebraic types are used the symbols:

$$\{\lambda_1 T, \lambda_2 S_2, \lambda_3 S_3\} \equiv \{T, S, S,\}$$

meaning that the first real eigenvalue λ_1 gives raise to a timelike eigenvector $\mathbf{T}$, the second real eigenvalue λ_2 is associated with a spacelike eigenvector $\mathbf{S}_2$, finally the third real eigenvalue λ_3 is related to a spacelike eigenvector $\mathbf{S}_3$; for the sake of simplicity I use the typing $\{T, S, S\}$. It is clear that $\{N, N, S\}$ stands for the algebraic type allowing for two different real eigenvalues giving rise to two null eigenvectors while the third real root is associated with a spacelike eigenvector. When there is a single and a double real eigenvalues giving rise correspondingly to a timelike and spacelike eigenvector, the algebraic type is denoted by $\{T, 2\,S\}$; consequently, for a triple real eigenvalue, if that were the case, the types could be $\{3T\}$, $\{3N\}$, or $\{3S\}$. For a complex eigenvalue λ_Z, in general, the related eigenvectors occur to be complex and are denoted by $\mathbf{Z}$ and $\bar{\mathbf{Z}}$ its complex conjugated, the possible types are $\{T, Z, \bar{Z}\}$, $\{N, Z, \bar{Z}\}$, or $\{S, Z, \bar{Z}\}$.

In general, the spaces described by the stationary (static) cyclic symmetric metric above belong to the Cotton type I; if the three eigenvectors are real the type is I_R, otherwise the type is I_Z with eigenvectors $\mathbf{S}$, $\mathbf{N}$, $\mathbf{T}$, $\mathbf{Z}$, and $\bar{\mathbf{Z}}$. Following the notation above-proposed, the algebraic type for the Cotton tensor could be Type I: $\{S, S, S\}, \{S, Z, \bar{Z}\}$, and so on.

As far as the Jordan form is concerned, one achieves it by subjecting the matrix under consideration to (elementary) similarity transformations: adding to a row another row multiplied by a suitable constant or function in order to simplify the resulting row, or to add to a column another one multiplied by a constant or function. To bring $\mathcal{C}$, (11.34), the corresponding Jordan form, one has to eliminate the components $C^1{}_3$ and $C^3{}_1$, which is achieved by multiplying $\mathcal{C}$ from the right by the matrix $\mathcal{A}$, and from the left by the matrix $\mathcal{B}$, namely

$$\mathcal{A} = \begin{bmatrix} 1 & 0 & -C^1{}_3/C^1{}_1 \\ 0 & 1 & 0 \\ 0 & 0 & 1 \end{bmatrix}, \quad \mathcal{B} = \begin{bmatrix} 1 & 0 & 1 \\ 0 & \beta & 0 \\ -C^3{}_1/C^1{}_1 & 0 & 1 \end{bmatrix},$$

$$\mathcal{B}\,\mathcal{C}\,\mathcal{A} = \begin{bmatrix} C^1{}_1 & 0 & 0 \\ 0 & \beta C^2{}_2 & 0 \\ 0 & 0 & \gamma C^3{}_3 \end{bmatrix},$$

$$\gamma = 1 - \frac{C^1{}_3 C^3{}_1}{C^1{}_1 C^3{}_3}, \quad \beta := 1 + \frac{C^1{}_3 C^3{}_1}{C^1{}_1 C^2{}_2}, \tag{11.39}$$

and finally by multiplying the resulting matrix by $\mathcal{E} = \mathrm{diag}(1, 1/\beta, 1/\gamma)$, one arrives at

$$(\mathcal{B}\,\mathcal{C}\,\mathcal{A})\mathcal{E} \sim \mathcal{J}_I = \begin{bmatrix} \lambda_1 & 0 & 0 \\ 0 & \lambda_2 & 0 \\ 0 & 0 & -\lambda_1 - \lambda_2 \end{bmatrix}. \tag{11.40}$$

An alternative treatment of the Cotton tensor and conformal symmetries for (2+1)-dimensional spaces is given in Hall and Capocci (1999), and also in García et al. (2004), where the analysis on Cotton tensors in (n+1)-dimensions is also developed.

11.1.6 Static Cyclic Symmetric Equations for Maxwell Fields

In the forthcoming three sections, we derive all the static solutions of the Einstein–Maxwell equations (11.26); there are only three families within this class. For the static metric $W(r) = 0$, consequently the metric (11.15) becomes

$$g = -\frac{F(r)}{H(r)}dt^2 + \frac{dr^2}{F(r)} + H(r)d\phi^2, \tag{11.41}$$

and the Einstein–Maxwell equations simplify drastically:

$$\begin{aligned}
E_t{}^t &= \frac{1}{2}F\frac{H_{,r,r}}{H} + \frac{1}{4}\frac{H_{,r}F_{,r}}{H} - \frac{1}{4}F\frac{{H_{,r}}^2}{H^2} + \frac{b^2}{H} + \frac{c^2}{F} + \frac{a^2H}{F} - \frac{1}{l^2},\\
E_r{}^r &= \frac{1}{4}\frac{H_{,r}F_{,r}}{H} - \frac{1}{4}F\frac{{H_{,r}}^2}{H^2} + \frac{b^2}{H} - \frac{c^2}{F} - \frac{a^2H}{F} - \frac{1}{l^2},\\
E_\Phi{}^\Phi &= \frac{1}{2}F_{,r,r} - \frac{1}{2}F\frac{H_{,r,r}}{H} - \frac{3}{4}\frac{H_{,r}F_{,r}}{H} + \frac{3}{4}F\frac{{H_{,r}}^2}{H^2} - \frac{b^2}{H} + \frac{c^2}{F} - \frac{a^2H}{F} - \frac{1}{l^2},\\
E_t{}^r &= -2\,ca,\; E_t{}^\Phi = -2a\,b\frac{1}{H},\\
E_r{}^t &= 2a\,c\frac{H}{F^2},\; E_r{}^\Phi = -2b\,c\frac{1}{FH},\\
E_\Phi{}^t &= 2a\,b\frac{H}{F},\; E_\Phi{}^r = -2\,bc,
\end{aligned} \tag{11.42}$$

Each of these $E_\mu{}^\nu$ equations has to be equated to zero, therefore one can distinguish the following three families of static solutions:
the electric class: $b \neq 0, a = 0, c = 0$,
the magnetic class: $a \neq 0, b = 0, c = 0$,
the hybrid class: $c \neq 0, a = 0, b = 0$.
In the next sections we proceed to integrate each class separately.

11.2 Electrostatic Solutions; $b \neq 0, a = 0$

In this section, the electrostatic cyclic symmetric Einstein–Maxwell solutions are derived in the presence of a negative cosmological constant $\Lambda = -1/l^2$. It seems that static Einstein–Maxwell solutions with cosmological constant were first derived in Peldan (1993); in that publication Peldan mentioned his failure in finding any work done on explicit solutions to Einstein–Maxwell solutions with cosmological constant, although static and rotationally symmetric solutions with vanishing cosmological constant did exist, namely the Deser and Mazur (1985) and Melvin (1986) solutions. Moreover, in the BTZ publication, Bañados et al. (1992) also presented a charged rotating metric, which was assumed to be correct and to satisfy the Einstein–Maxwell equations. Nevertheless, according to the comments by Kamata and Koikawa (1995), and later by García (1999), it happened to be wrong; the azimuthal Maxwell equation yields $JQ/r^3 = 0$, hence one could distinguish the rotating BTZ solution and the charged static solution as two possible independent branches but not a single expression equipped with two non-vanishing simultaneously parameters as a "solution" as it was erroneously assumed in the BTZ publication. To my mind it is reasonable to name the electrostatic cyclic symmetric Einstein–Maxwell solution with cosmological constant "the charged static–Peldan solution." The subclass of solutions without Λ are explicitly given and identified, while the electrostatic Peldan solution with Λ is here analyzed in detail.

11.2.1 General Electrostatic Solutions

The substraction ${E_t}^t(a=0=c) - {E_r}^r(a=0=c)$ from (11.42) yields

$$\frac{d^2}{dr^2}H\,(r) = 0 \Rightarrow H(r) = C_0 + C_1\,r, \tag{11.43}$$

where C_0, and C_1 are constants of integration; C_1 at this stage is assumed to be different from zero; the zero case deserves special attention and will be treated separately. Substituting this structural function H into the equation ${E_t}^t(a = 0 = c)$ one arrives at a first-order differential equation for F

$$\frac{H_{,r}}{H}F_{,r} - \left(\frac{H_{,r}}{H}\right)^2 F + 4\frac{b^2}{H} - \frac{4}{l^2} = 0, \tag{11.44}$$

which by introducing an auxiliary function $f(r)$ through

$$F(r) = H(r)\,f(r) = (C_0 + C_1\,r)\,f(r),$$

reduces to the simple equation

$$\frac{df(r)}{dr} = \frac{4}{C_1\,l^2}\,\frac{C_1\,r + C_0 - b^2 l^2}{C_0 + C_1\,r} \tag{11.45}$$

with general integral

$$f = \frac{4}{C_1^2\,l^2}\left(K_0 + C_1\,r - b^2 l^2 \ln\left(C_0 + C_1\,r\right)\right), \tag{11.46}$$

where K_0 is a new integration constant, into which of course one has incorporated C_0.

Summarizing, one arrives at the metric

$$\begin{aligned} g &= -\frac{F(r)}{h(r)}dt^2 + \frac{dr^2}{F(r)} + h(r)d\phi^2,\\ F(r) &= \frac{4}{C_1^2\,l^2}\left[K_0 + h(r) - b^2 l^2 \ln h(r)\right] h(r),\\ h(r) &= C_1\,r + C_0. \end{aligned} \tag{11.47}$$

This solution is characterized by:
the vector field

$$\boldsymbol{A} = A_t\boldsymbol{dt} = \frac{b}{C_1}\ln h\,\boldsymbol{dt}, \tag{11.48}$$

the electromagnetic field tensors

$$F^{\mu\nu} = 2b\delta_{[t}{}^{\mu}\delta_{r]}{}^{\nu}, \; F_{\mu\nu} = -2\frac{b}{h(r)}\delta_{[\mu}{}^{t}\delta_{\nu]}{}^{r} \tag{11.49}$$

with field invariant

$$F_{\mu\nu}F^{\mu\nu} = -2\,\frac{b^2}{h}, \tag{11.50}$$

the energy momentum tensor

$$T_{\mu}{}^{\nu} = -\frac{b^2}{8\,\pi}\frac{1}{h}\left(\delta_{\mu}{}^{t}\delta_{t}{}^{\nu} + \delta_{\mu}{}^{r}\delta_{r}{}^{\nu} - \delta_{\mu}{}^{\phi}\delta_{\phi}{}^{\nu}\right), \tag{11.51}$$

with quadratic energy momentum invariant and trace

$$T_{\mu\nu}T^{\mu\nu} = \frac{3}{64\pi^2}\frac{b^4}{h^2}, \; T^{\mu}_{\mu} = -\frac{1}{8\pi}\frac{b^2}{h}. \tag{11.52}$$

A familiar representation of the above mentioned solution is achieved for the choice $C_0 = 0$, $C_1 = 2$, $K_0 = b^2\,l^2\,\ln 2r_0$, which yields

$$\begin{aligned} g &= -\left[\frac{2\,r}{l^2} - b^2\ln\frac{r}{r_0}\right]dt^2 + 2\,r d\phi^2 + \frac{dr^2}{2\,r\,\left[\frac{2\,r}{l^2} - b^2\ln\frac{r}{r_0}\right]}, \\ \boldsymbol{A} &= \frac{b}{2}\ln\frac{r}{r_0}\,\boldsymbol{dt}. \end{aligned} \tag{11.53}$$

This solution, endowed with mass, electric charge, a cosmological constant $-1/l^2$, and radial parameter, allows for a charged black hole interpretation. The mass may assume positive as well as negative values, whereas the charge is not upper-bounded.

11.2.2 Gott–Simon–Alpern, Deser–Mazur, and Melvin Electrostatic Solution

According to the existing references, Gott, Simon, and Alpern were the first authors to derive solutions within Maxwell theory in (2+1) gravity: Gott and Alpert (1984); Gott et al. (1986); they found, among other things, the electrostatic solution without cosmological constant given by Eq. (42) in Gott et al. (1986) in "Schwarzschild" coordinates.

Introducing in the above expressions, (11.47) and (11.48), the radial coordinate ρ through $C_0 + C_1\,r \to \rho^2$ together with $t \to C_1\,t/2, K_0 \to l^2\,k_0$, and by letting $1/l^2 \to 0$ one arrives at the electrostatic solution in the form

$$\begin{aligned} g &= -F dt^2 + \frac{d\rho^2}{F} + \rho^2 d\phi^2, \; F(\rho) = k_0 - 2b^2\ln\rho = \frac{\kappa}{2\pi}\,Q^2\ln\frac{\rho_c}{\rho}, \\ \boldsymbol{A} &= A_t\boldsymbol{dt} = b\,\ln\rho\,\boldsymbol{dt}. \end{aligned} \tag{11.54}$$

Some authors refer to the coordinate system $\{t, \rho, \phi\}$, in which the perimeter of the circle equates $2\pi\,\rho = \int_0^{2\pi}\rho\,\boldsymbol{d\phi}$, as to the "Schwarzschild" coordinates.

Two years later, Deser and Mazur (1985) published their version of the electrostatic solution for $\Lambda = 0$. Moreover, by then, the work by Melvin (1986) was published with the derivation of the electrostatic as well as the magnetostatic solutions for vanishing Λ.

Melvin (1986) introduced three kinds of coordinate systems, and defined K as the electric charge, while $\kappa = 8\pi\, G'$, M as the total mass, and the scale parameter $a = (1 - 4G'\,M)^2/(2\,K)$, among other quantities:
physical radial coordinates, in which the electrostatic solution amounts to

$$ds^2 = -N(\rho)^2\, dt^2 + d\rho^2 + S(\rho)^2 d\theta^2,\ d\rho = \frac{dS}{\sqrt{(1-4G'\,M)^2 - 2\,K\ln S}},$$
$$N = \sqrt{(1-4G'\,M)^2 - 2\,K\ln S}, \tag{11.55a}$$

Schwarzschild radial coordinates, in which the electrostatic solution is given by

$$ds^2 = -N(r)^2\, dt^2 + \psi(r)^2\, dr^2 + r^2 d\theta^2,$$
$$N(r) = \sqrt{(1-4G'\,M)^2 - 2\,K\ln r},\ \psi(r) = \frac{1}{N(r)}, \tag{11.55b}$$

conformal radial coordinates, in which the electrostatic solution becomes

$$ds^2 = -N(R)^2\, dt^2 + \Psi(R)^2\left(dR^2 + R^2\, d\theta^2\right),$$
$$N(R) = (1-4G'\,M) - \,K\ln R,$$
$$\Psi(R) = \exp\left[-4G'\,M\,\ln R - \frac{K}{2}(\ln R)^2\right]. \tag{11.55c}$$

Deser and Mazur (1985) reported the electrostatic solution, Eq. (12), in the conformal radial coordinate system (11.55c), with the following identifications

$$\{G', M, (1-4G'\,M), K, N, t\}_M \to \{G, m, \alpha, G\,e^2, \alpha\,N, t/\sqrt{\alpha}\}_{DM}.$$

There is a sign missprint in the function g_{rr} of Deser and Mazur (1985) , Eq. (12); in the factor in front of $(\ln r)^2$ should be $-G\,e^2\,(\ln r)^2$.

Kogan (1992) reported and analyzed the (electro- and magneto-) static solutions of the (2+1)-dimensional Einstein–Maxwell equations for both positive and negative signs of the gravitational constant κ; recall that in the three dimensions there is no restriction on its sign. The r-coordinate used there was such that $g_{rr} = 1$ for the signature used in the present report, consequently, the solutions are given in the (11.55a) representation.

11.2.3 Charged Static Peldan Solution with Λ

The Peldan electrostatic solution with cosmological constant in polar coordinates arises from the general expressions above, (11.47) and (11.48), by means of the coordinate and parameter changes $C_0 + C_1\, r \to \rho^2, t \to C_1\, t/2, K_0 \to -l^2\, m$. In this way one obtains

$$g = -F dt^2 + \frac{d\rho^2}{F} + \rho^2 d\phi^2,\ F(\rho) = \frac{\rho^2}{l^2} - m - 2b^2 \ln \rho,$$
$$\boldsymbol{A} = b \ln \rho\, \boldsymbol{dt}. \tag{11.56}$$

The corresponding field tensors are given by

$$T_\mu{}^\nu = -\frac{b^2}{8\pi}\frac{1}{\rho^2}\left(\delta_\mu{}^t\delta_t{}^\nu + \delta_\mu{}^\rho\delta_\rho{}^\nu - \delta_\mu{}^\phi\delta_\phi{}^\nu\right),\ F_{\mu\nu} = -2\frac{b}{\rho}\delta_{[\mu}{}^t\delta_{\nu]}{}^r, \tag{11.57}$$

To achieve the specific Peldan's writing, one has to accomplish the additional identifications $m \to -C_1, 1/l^2 \to -\lambda/2, b^2 \to q^2/4, t \to C_2 t, \rho \to r$.

The electrostatic Peldan (1993) solution, Eq. (71) – see Eq. (4.15) in García (2009) – in canonical representation is given by the metric

$$ds^2 = -N(\rho)^2 dt^2 + \frac{1}{L(\rho)^2} d\rho^2 + K(\rho)^2[d\phi + W(\rho)dt]^2,$$
$$L(\rho) = N(\rho) = \sqrt{\frac{\rho^2}{l^2} - 2b^2 \ln \rho - M},\ K(\rho) = \rho,\ W(\rho) = 0. \tag{11.58}$$

Mass and Energy

The surface energy density ϵ is

$$\epsilon(\rho, \epsilon_0) = -\frac{1}{\pi\rho}N(\rho) - \epsilon_0. \tag{11.59}$$

Consequently the global energy and mass are given by

$$E(\rho, \epsilon_0) = -2N(\rho) - 2\pi\,\rho\,\epsilon_0,$$
$$M(\rho, \epsilon_0) = -2\frac{\rho^2}{l^2} + 2m + 4b^2 \ln \rho - 2\pi\rho\, N(\rho)\epsilon_0. \tag{11.60}$$

Thus, for the natural choice of a vanishing reference energy density $\epsilon_0 = 0$, one has at the spatial infinity $\rho \to \infty$ that

$$\epsilon(\rho \to \infty, \epsilon_0 = 0) \approx -\frac{1}{\pi\, l} + \frac{l\, M}{2\pi\,\rho^2} + \frac{l\, b^2}{\pi\,\rho^2} \ln \rho,$$
$$E(\rho \to \infty, \epsilon_0 = 0) \approx -2\frac{\rho}{l} + \frac{M\, l}{\rho} + 2\frac{l\, b^2}{\rho} \ln \rho,$$
$$M(\rho \to \infty, \epsilon_0 = 0) \approx -2\frac{\rho^2}{l^2} + 2M + 4b^2 \ln \rho, \tag{11.61}$$

while if the reference energy corresponds to the anti-de Sitter spacetime $AdS(M_0)$, $\epsilon_0 = -\frac{1}{\pi\,\rho}\sqrt{\frac{\rho^2}{l^2} - M_0}$, $\epsilon_{0|\infty}(M_0) \approx -\frac{1}{\pi\, l} + \frac{l M_0}{2\pi\,\rho^2}$, then the energies and mass at spatial infinity are expressed as

$$\epsilon(\rho \to \infty, \epsilon_{0|\infty}(M_0)) \approx l\,\frac{M - M_0}{2\pi\,\rho^2} + \frac{l\, b^2}{\pi\,\rho^2} \ln \rho,$$
$$E(\rho \to \infty, \epsilon_{0|\infty}(M_0)) \approx l\frac{M - M_0}{\rho} + 2\frac{l b^2}{\rho} \ln \rho,$$
$$M(\rho \to \infty, \epsilon_{0|\infty}(M_0)) \approx M - M_0 + 2\, b^2 \ln \rho. \tag{11.62}$$

Comparing with the static BTZ one recognizes M as the BTZ *mass*. Notice that the energy and mass include an amount of energy due to the electric field through the logarithmical term; because of this dependence, these quantities diverge at infinity logarithmically.

Field, Energy–Momentum, and Cotton Tensors

The electromagnetic tensor field associated with the charged Peldan solution is given by

$$(F^{\alpha}{}_{\beta}) = \begin{bmatrix} 0 & \frac{b}{L^2\,\rho} & 0 \\ \frac{b\,L^2}{\rho} & 0 & 0 \\ 0 & 0 & 0 \end{bmatrix}, \tag{11.63}$$

Searching for its eigenvectors, one arrives at

$$\begin{aligned} &\lambda_1 = 0; \mathbf{V1} = [0,0,V^3], V_{\mu}V^{\mu} = \rho^2 {V^3}^2, \mathbf{V1} = \mathbf{S1}, \\ &\lambda_2 = \frac{b}{\rho}; \mathbf{V2} = [\frac{V^2}{L^2}, V^2, 0],\ V^{\mu}V_{\mu} = 0,\ \mathbf{V2} = \mathbf{N2}, \\ &\lambda_3 = -\frac{b}{\rho}; \mathbf{V3} = [-\frac{V^2}{L^2}, V^2, 0],\ V^{\mu}V_{\mu} = 0,\ \mathbf{V3} = \mathbf{N3}, \\ &\text{Type:}\{S, N, N\}. \end{aligned} \tag{11.64}$$

As far as the electromagnetic energy momentum tensor is concerned, its matrix amounts to

$$(T^{\alpha}{}_{\beta}) = \begin{bmatrix} -\frac{1}{8}\frac{b^2}{\pi\,\rho^2} & 0 & 0 \\ 0 & -\frac{1}{8}\frac{b^2}{\pi\,\rho^2} & 0 \\ 0 & 0 & \frac{1}{8}\frac{b^2}{\pi\,\rho^2} \end{bmatrix}, \tag{11.65}$$

with the following eigenvalues and their corresponding eigenvector

$$\begin{aligned} \lambda_1 &= -\frac{1}{8\pi}\frac{b^2}{\rho^2}; \mathbf{V1} = [V^1, V^2, 0], \\ V1^{\mu}V1_{\mu} &= (V^1)^2 g_{t\,t} + (V^2)^2 g_{\rho\rho},\ \mathbf{V1} = \{\mathbf{T1}, \mathbf{S1}, \mathbf{N1}\}, \\ \lambda_2 &= -\frac{1}{8\pi}\frac{b^2}{\rho^2}; \mathbf{V2} = [\tilde{V}^1, \tilde{V}^2, 0], \\ V2^{\mu}V2_{\mu} &= (\tilde{V}^1)^2 g_{t\,t} + (\tilde{V}^2)^2 g_{\rho\rho},\ \mathbf{V2} = \{\mathbf{T2}, \mathbf{S2}, \mathbf{N2}\}, \\ \lambda_3 &= \frac{1}{8\pi}\frac{b^2}{\rho^2}; \mathbf{V3} = [0,0,V^3], \\ V^{\mu}V_{\mu} &= (V^3)^2 g_{\phi\,\phi},\ \tilde{\mathbf{V}}3 = \mathbf{S3}. \end{aligned} \tag{11.66}$$

For $\mathbf{V1}$ and $\mathbf{V2}$, the character of these vectors depends on the sign of their magnitudes; for instance, choosing

$$\begin{aligned} &V^1 = s\,\sqrt{g_{\rho\,\rho}}/\sqrt{|g_{t\,t}|}\,V^2,\ s = \text{constant},\ \mathbf{V1}^{\mu}\mathbf{V1}_{\mu} = (1-s^2) g_{\rho\,\rho}\,(V^2)^2; \\ &s > 1 \rightarrow \mathbf{V1} = \mathbf{T},\ s = \pm 1 \rightarrow \mathbf{V1} = \mathbf{N},\ s < 1 \rightarrow \mathbf{V1} = \mathbf{S}. \end{aligned}$$

The spacelike vector **V3** is aligned along the circular Killing direction ∂_ϕ. Thus one may have the spacetime arrangement $\{\mathbf{T1}, \mathbf{S2}, \mathbf{S3}\}$, or $\{\mathbf{N1}, \mathbf{N2}, \mathbf{S3}\}$, and so on.

The Cotton tensor for electrostatic cyclic symmetric gravitational field is given by

$$(C^\alpha{}_\beta) = \begin{bmatrix} 0 & 0 & \frac{b^2}{2\rho^2} \\ 0 & 0 & 0 \\ -\frac{b^2}{2\rho^4}L^2 & 0 & 0 \end{bmatrix}. \tag{11.67}$$

The search for its eigenvectors yields

$$\begin{aligned} &\lambda_1 = 0; \mathbf{V1} = [0, V^2, 0],\ V1^\mu V1_\mu = (V^2)^2 g_{\rho\rho},\ \mathbf{V1} = \mathbf{S1}, \\ &\lambda_2 = \frac{i}{2}\frac{L\,b^2}{\rho^3}; \mathbf{V2} = [V^1, 0, \frac{i\,L}{\rho}V^1],\ \mathbf{V2} = \mathbf{Z}, \\ &\lambda_3 = -\frac{i}{2}\frac{L\,b^2}{\rho^3}; \mathbf{V3} = [V^1, 0, -\frac{i\,L}{\rho}V^1],\ \mathbf{V3} = \bar{\mathbf{Z}}, \end{aligned} \tag{11.68}$$

therefore the corresponding tensor type is

$$\text{Type I: } \{S, Z, \bar{Z}\}.$$

The eigenvectors **V2** and **V3** are complex conjugated, or, if one wishes, one may consider the component V^1 differently for each of the complex vectors. For the zero eigenvalue λ_1, the vector **V1** is a spacelike vector that it points along the ρ–coordinate direction.

It is worth pointing out that the field and Cotton tensors of the solutions generated via coordinate transformations, in particular $SL(2, R)$ transformations, applied onto this electrostatic cyclic symmetric Peldan solution will shear the eigenvalues λ_i of the corresponding field and Cotton tensors of the charged Peldan solution; recall that eigenvalues are invariant characteristics of tensors, although the components of the eigenvectors, in general, look different in different coordinate systems; this remark also applies to the (eigenvalues) eigenvectors of the seed and resulting solutions.

11.3 Magnetostatic Solutions; $a \neq 0, b = 0$

In this section, the magnetostatic cyclic symmetric Einstein–Maxwell solutions are derived in the presence of a negative cosmological constant $\Lambda = -1/l^2$. As was noted in the previous section, it seems that static Einstein–Maxwell solutions with a cosmological constant were first derived in Peldan (1993). For this class of solutions, Peldan left the Schwarzschild frame of coordinates and used those ones in which $g_{\phi\phi} = H(\rho)$.

The subclass of solutions without Λ are explicitly given and identified, while the magnetostatic Peldan and Hirschmann–Welch solution representation with Λ are analyzed in detail.

11.3.1 General Magnetostatic Solutions

To derive the magnetostatic solution, one starts from the addition ${E_t}^t + {E_r}^r$ from (11.42), which yields

$$\frac{d}{dr}\left[\frac{F}{H}\frac{d}{dr}H - 4\,r\,/l^2\right] = 0, \tag{11.69}$$

with integral

$$\frac{F}{H}\frac{d}{dr}H = 4\,\frac{r}{l^2} + C_1. \tag{11.70}$$

The substraction ${E_t}^t - {E_r}^r$ gives

$$F^2\frac{d^2}{dr^2}H + 4a^2H^2 = 0. \tag{11.71}$$

Substituting $F(r)$ from (11.70) into (11.71) one arrives at a first-order equation for $\frac{d}{dr}H$

$$\left(4\,r + C_1 l^2\right)^2\frac{d^2H}{dr^2} + 4\,a^2l^4\left(\frac{dH}{dr}\right)^2 = 0, \tag{11.72}$$

which is rewritten as

$$d(\frac{d}{dr}H)^{-1} = -a^2l^4 d\left(4\,r + C_1 l^2\right)^{-1} \tag{11.73}$$

with first integral

$$(\frac{d}{dr}H)^{-1} = \frac{C_2(4\,r + C_1\,l^2) - a^2l^4}{4r + C_1\,l^2}. \tag{11.74}$$

A subsequent integration gives

$$\begin{aligned} H(r) &= \frac{\left[R(r)l^2 + a^2l^4\ln\left(R(r)l^2\right) + C_3\right]}{4C_2^2}, \\ F(r) &= R(r)H(r), \\ R(r)\,l^2 &:= C_2\left(4\,r + C_1\,l^2\right) - a^2l^4, \end{aligned} \tag{11.75}$$

where (11.70) it has been used to evaluate $F(r)$. These structural functions completely determine the magnetostatic solution; without any loss of generality, by letting

$$C_2 \to C_1 l^2/4,\ C_1 \to 4\,(a^2\,l^2 + C_0)/(C_1\,l^2),\ C_3 \to K_0\,l^2 - a^2\,l^4\,\ln l^2,$$

the magnetostatic metric can be given as

$$
\begin{aligned}
g &= -h(r)dt^2 + \frac{dr^2}{H(r)h(r)} + H(r)d\phi^2,\\
H(r) &= \frac{4}{{C_1}^2\, l^2}\left[K_0 + h(r) + a^2 l^2 \ln h(r)\right],\\
F(r) &= H(r)\, h(r),\ h(r) := C_1\, r + C_0.
\end{aligned} \tag{11.76}
$$

This solution is characterized by:
the electromagnetic field vector

$$
\boldsymbol{A} = \frac{a}{C_1} \ln h \, \boldsymbol{d\phi}, \tag{11.77}
$$

the electromagnetic field tensors

$$
F^{\mu\nu} = -2a\delta_{[\phi}{}^{\mu}\delta_{r]}{}^{\nu},\ F_{\mu\nu} = -2\frac{a}{h(r)}\delta_{[\mu}{}^{\phi}\delta_{\nu]}{}^{r} \tag{11.78}
$$

with field invariant

$$
F_{\mu\nu}F^{\mu\nu} = 2\,\frac{a^2}{h(r)}, \tag{11.79}
$$

the energy–momentum tensor

$$
T_{\mu}{}^{\nu} = \frac{a^2}{8\,\pi}\frac{1}{h(r)}\left(-\delta_{\mu}{}^{t}\delta_{t}{}^{\nu} + \delta_{\mu}{}^{r}\delta_{r}{}^{\nu} + \delta_{\mu}{}^{\phi}\delta_{\phi}{}^{\nu}\right), \tag{11.80}
$$

with energy field invariants

$$
T_{\mu\nu}T^{\mu\nu} = \frac{3}{64}\frac{a^4}{\pi^2}\frac{1}{h(r)^2},\ T^{\mu}_{\mu} = a^2\frac{1}{8\pi}\frac{1}{h(r)}. \tag{11.81}
$$

This class of solutions allows for a hydrodynamics interpretation in terms of a perfect fluid energy–momentum tensor for a stiff fluid, $\mu = p$, where μ and p are, respectively, the fluid energy density and the fluid pressure. In fact, the energy momentum tensor for a perfect fluid is given by

$$
T_{\mu\nu} = (\mu + p)u_{\mu}u_{\nu} + p\, g_{\mu\nu}.
$$

Therefore choosing the fluid 4-velocity along the time direction $u^{\mu} = \delta^{\mu}_{t}/\sqrt{-g_{tt}}$ one establishes that $\mu = \frac{a^2}{8\pi}\frac{H}{F^2} = p$.

11.3.2 Melvin, and Barrow–Burd–Lancaster Magnetostatic Solution

Melvin (1986) derived the electric and the magnetic static solutions for vanishing cosmological constant $\Lambda = 0$. The corresponding solution can be obtained from the metric (11.76) introducing new coordinates according to $h = C_1 r + C_0 \rightarrow \rho^2, t \rightarrow t, \phi \rightarrow \phi\, C_1/2$, and setting $K_0 = 0$ arriving at

$$
g = -\rho^2 dt^2 + \frac{d\rho^2}{F(\rho)} + F(\rho)d\phi^2,\ F(\rho) = k_0 + 2a^2 \ln\rho, \tag{11.82}
$$

or by introducing $a^4\, r^2\, \mathrm{e}^{k_0/a^2} = k_0 + a^2 \ln \rho^2$, and scaling the variables t and ϕ one brings it to the form

$$g = e^{r^2}(-dt^2 + dr^2) + r^2 d\phi^2. \tag{11.83}$$

In the paragraph devoted to stiff perfect fluid, Barrow, Burd, and Lancaster – see Barrow et al. (1986) – pointed out that for a fluid aligned along the time-coordinate, "in $(2+1)$ dimensions the stiff fluid has an energy–momentum tensor identical to that of a static magnetic field," and they continued with a statement very close to the following: if one sets the electric field components $F_{0i} = 0$ and magnetic components $F_i^{\ j} = \epsilon_i^{\ j} \sqrt{2\mu}$ in the electromagnetic energy–momentum tensor $T_{\mu\nu} = F_{\mu\lambda}F_\nu^{\ \lambda} - g_{\mu\nu}F_{\alpha\lambda}F^{\lambda\alpha}/4$ reduces to the perfect fluid energy–momentum tensor $T_{\mu\nu} = (\mu + p)u_\mu u_\nu + p\, g_{\mu\nu}$, with energy density μ equalling the pressure p, $\mu = p$.

11.3.3 Peldan Magnetostatic Solution with Λ

Introducing in the metric (11.76) new coordinates according to $h = C_1 r + C_0 \to \rho^2, t \to t, \phi \to \phi\, C_1/2, K_0 \to k_0 l^2$ one gets

$$\begin{aligned} g &= -\rho^2 dt^2 + \frac{d\rho^2}{F(\rho)} + F(\rho) d\phi^2,\ F(\rho) = k_0 + \frac{\rho^2}{l^2} + 2a^2 \ln \rho, \\ \boldsymbol{A} &= a \ln \rho\, \boldsymbol{d\phi}, \end{aligned} \tag{11.84}$$

characterized by the field tensors

$$\begin{aligned} T_\mu^{\ \nu} &= \frac{a^2}{8\,\pi} \frac{1}{\rho^2} \left(-\delta_\mu^{\ t} \delta_t^{\ \nu} + \delta_\mu^{\ \rho} \delta_\rho^{\ \nu} + \delta_\mu^{\ \phi} \delta_\phi^{\ \nu} \right), \\ F_{\mu\nu} &= -2\frac{a}{\rho} \delta_{[\mu}^{\ \ \phi} \delta_{\nu]}^{\ \ \rho}. \end{aligned} \tag{11.85}$$

This solution has also been derived and analyzed in Peldan (1993), Eq. (86).

The Peldan (1993) magnetostatic solution – see also García (2009), Eq. (4.30) – with a negative cosmological constant is determined, in canonical representation, by

$$\begin{aligned} ds^2 &= -N(\rho)^2 dt^2 + \frac{1}{L(\rho)^2} d\rho^2 + K(\rho)^2 [d\phi + W(\rho) dt]^2, \\ L(\rho) &= K(\rho) = \sqrt{\frac{\rho^2}{l^2} + 2\, a^2 \ln \rho + m},\ N(\rho) = \rho,\ W(\rho) = 0. \end{aligned} \tag{11.86}$$

Notice that the metric functions L^2 and K^2 are positive functions for values of $\rho > \rho_{\mathrm{root}}$, where

$$\rho_{\mathrm{root}} = \exp\left[-\frac{m}{2\, a^2} - \frac{1}{2} \mathrm{LambertW}\, \left(\frac{1}{l^2\, a^2} \mathrm{e}^{-\frac{m}{a^2}} \right) \right],\ \mathrm{LW}\,(x) \exp(\mathrm{LW}\,(x)) = x,$$

where for short LW := LambertW , $\rho_{\rm root}$ is solution of the equation,

$$g^{\rho\rho}(\rho_{\rm root}) = L^2(\rho_{\rm root}) = 0,\ \rho_{\rm root}^2/l^2 + 2\,a^2 \ln \rho_{\rm root} + m = 0.$$

Therefore the coordinate ρ does not cover the expected range $0 \leq \rho \leq \infty$. For $\rho \leq \rho_{\rm root}$ the metric suffers an unacceptable signature change. This fact also points out on the non-existence of a horizon $\rho =$ const for the Peldan solution. Consequently, one has to modify the choice of the ρ-coordinate in order to be able to rich the origin of coordinates; with this purpose in mind a new coordinate system is chosen in the forthcoming paragraph (11.3.4).

Mass and Energy

The surface energy density ϵ is given by

$$\epsilon(\rho) = -\frac{1}{\pi K}(\frac{\rho}{l^2} + \frac{a^2}{\rho})\epsilon_0. \tag{11.87}$$

Consequently the global energy and mass are given by

$$\begin{aligned} E(\rho, \epsilon_0) &= -2\frac{\rho}{l^2} - 2\frac{a^2}{\rho} - 2\pi\, K\,\epsilon_0, \\ M(\rho, \epsilon_0) &= -2\frac{\rho^2}{l^2} - 2a^2 - 2\pi\,\rho\, K\,\epsilon_0. \end{aligned} \tag{11.88}$$

For the natural choice of a vanishing reference energy density $\epsilon_0 = 0$, one has at the spatial infinity $\rho \to \infty$ that

$$\begin{aligned} \epsilon(\rho \to \infty, \epsilon_0 = 0) &\approx -\frac{1}{\pi\, l} + l\,\frac{m - 2a^2}{2\pi\,\rho^2} + \frac{l\,a^2}{\pi\,\rho^2}\ln\rho, \\ E(\rho \to \infty, \epsilon_0 = 0) &= -2\frac{\rho}{l^2} - 2\frac{a^2}{\rho}, \\ M(\rho \to \infty, \epsilon_0 = 0) &= -2\frac{\rho^2}{l^2} - 2a^2, \end{aligned} \tag{11.89}$$

while if the reference energy is the one corresponding to the anti-de Sitter spacetime with M_0 parameter, $\epsilon_0 = -\frac{\rho}{\pi\, l^2}/\sqrt{\frac{\rho^2}{l^2} + M_0}$, $\epsilon_{0|\infty}(M_0) \approx -\frac{\sqrt{\Lambda}}{\pi} + \frac{M_0}{2\pi\sqrt{\Lambda}\,\rho^2}$, then the energies are expressed at spatial infinity as

$$\begin{aligned} \epsilon(\rho \to \infty, \epsilon_{0|\infty}(M_0)) &\approx l\,\frac{m - M_0 - 2a^2}{2\pi\,\rho^2} + \frac{a^2}{\pi\,\rho^2}\ln\rho, \\ E(\rho \to \infty, \epsilon_{0|\infty}(M_0)) &\approx \frac{m - M_0 - 2a^2}{\rho} + 2\frac{a^2}{\rho}\ln\rho, \\ M(\rho \to \infty, \epsilon_{0|\infty}(M_0)) &\approx m - M_0 - 2a^2 + 2\,a^2\ln\rho. \end{aligned} \tag{11.90}$$

Comparing these quantities with the corresponding ones of the static BTZ solution counterpart, one sees a complete correspondence for vanishing electromagnetic parameter a, thus one recognizes m as the mass parameter. Notice that the energy and mass include an amount of energy due to the magnetic field, in a way similar to the electric one, through logarithmical terms; because of this dependence, these quantities logarithmically diverge at infinity.

Field, Energy–Momentum and Cotton Tensors

The electromagnetic field tensor for this solution is given by

$$(F^{\alpha}{}_{\beta}) = \begin{bmatrix} 0 & 0 & 0 \\ 0 & 0 & \frac{L^2 a}{\rho} \\ 0 & -\frac{a}{L^2 \rho} & 0 \end{bmatrix}, \tag{11.91}$$

and is algebraically characterized by the following eigenvectors

$$\begin{aligned} &\lambda_1 = 0; \mathbf{V1} = [V^1, 0, 0],\ V^{\mu}V_{\mu} = -(V^1)^2(\rho^2),\ \mathbf{V1} = \mathbf{T1}, \\ &\lambda_2 = i\frac{a}{\rho}; \mathbf{V2} = [0, V^2, i\frac{1}{L^2}V^2], \mathbf{V2} = \mathbf{Z}, \\ &\lambda_3 = -i\frac{a}{\rho}; \mathbf{V3} = [0, V^2, -i\frac{1}{L^2}V^2], \mathbf{V3} = \bar{\mathbf{Z}}, \\ &\text{Type:}\{T, Z, \bar{Z}\}. \end{aligned} \tag{11.92}$$

As far as the electromagnetic energy momentum tensor is concerned, its matrix amounts to

$$(T^{\alpha}{}_{\beta}) \begin{bmatrix} -\frac{1}{8\pi}\frac{a^2}{\rho^2} & 0 & 0 \\ 0 & \frac{1}{8\pi}\frac{a^2}{\rho^2} & 0 \\ 0 & 0 & \frac{1}{8\pi}\frac{a^2}{\rho^2} \end{bmatrix} \tag{11.93}$$

with the following eigenvalues and their corresponding eigenvectors

$$\begin{aligned} &\lambda_1 = -\frac{1}{8\pi}\frac{a^2}{\rho^2}; \mathbf{T1} = [V^1, 0, 0],\ V^{\mu}V_{\mu} = -\rho^2\,(V^1)^2, \\ &\lambda_2 = \frac{1}{8\pi}\frac{a^2}{\rho^2}; \mathbf{S2} = [0, V^2, V^3],\ V^{\mu}V_{\mu} = (V^2)^2/\,L^2 + (V^3)^2\,L^2, \\ &\lambda_3 = \frac{1}{8\pi}\frac{a^2}{\rho^2}; \mathbf{S3} = [0, \tilde{V}^2, \tilde{V}^3],\ V^{\mu}V_{\mu} = (\tilde{V}^2)^2/\,L^2 + (\tilde{V}^3)^2\,L^2, \\ &\text{Type: }\{T, 2S\}. \end{aligned} \tag{11.94}$$

This tensor structure corresponds to that describing a perfect fluid energy momentum tensor, but this time for the state equation: *energy* = *pressure*. Again, the solutions generated from this metric by using coordinate transformations will possess this perfect fluid feature because the invariance of the eigenvalues.

The Cotton tensor for electrostatic cyclic symmetric gravitational field is given by

$$(C^{\alpha}{}_{\beta}) = \begin{bmatrix} 0 & 0 & \frac{a^2 L^2}{2\rho^4} \\ 0 & 0 & 0 \\ -\frac{a^2}{2\rho^2} & 0 & 0 \end{bmatrix}. \tag{11.95}$$

Searching for its eigenvectors, one arrives at

$$\begin{aligned} &\lambda_1 = 0; \mathbf{V1} = [0, V^2, 0],\ V^{\mu}V_{\mu} = (V^2)^2 g_{\rho\rho},\ \mathbf{V1} = \mathbf{S1}, \\ &\lambda_2 = \frac{i}{2}\frac{L a^2}{\rho^3}; \mathbf{V2} = [-\frac{i\,L}{\rho}V^3, 0, V^3],\ \mathbf{V2} = \mathbf{Z}, \\ &\lambda_3 = -\frac{i}{2}\frac{L\,a^2}{\rho^3}; \mathbf{V3} = [\frac{i\,L}{\rho}V^3, 0, V^3], \mathbf{V3} = \bar{\mathbf{Z}}, \\ &\text{Type I: } \{T, Z, \bar{Z}\}. \end{aligned} \tag{11.96}$$

The eigenvectors **V2** and **V3** are complex conjugated while the vector **V1**, associated to the zero eigenvalue, is spacelike – the only physically tractable ρ-direction vector in this case. It is worth pointing out that the solutions generated via coordinate transformations, in particular the $SL(2, R)$ transformations, applied onto this magneto-static cyclic symmetric metric will shear the eigenvalues λ_i of the Cotton tensor quoted above; recall that eigenvalues are invariant characteristics of tensors, although their components in different coordinate systems are different – this last also applies to the eigenvectors of the seed and the resulting solutions.

11.3.4 Hirschmann–Welch Solution with Λ

Accomplishing in the general magnetic static metric (11.76) the transformations

$$\begin{aligned} &C_1 r + C_0 \to (\rho^2 + r_+^2 - ml^2)/l^2 =: h(\rho), \\ &2\phi/(C_1\, l^2) \to \phi,\ a^2\, l^4 = \chi^2, K_0 = m, \end{aligned} \tag{11.97}$$

one ends with the Hirschmann and Welch (1996) representation of the magnetic solution

$$\begin{aligned} g = &-\frac{1}{l^2}(\rho^2 + r_+^2 - ml^2)dt^2 + [\rho^2 + r_+^2 + \chi^2 \ln(|h(\rho)|)]d\phi^2 \\ &+ \frac{l^2\,\rho^2\,d\rho^2}{(\rho^2 + r_+^2 - ml^2)[\rho^2 + r_+^2 + \chi^2 \ln(|h(\rho)|)]}, \\ h(\rho) = &\,(\rho^2 + r_+^2 - ml^2)/l^2, \end{aligned} \tag{11.98}$$

with vector potential

$$\boldsymbol{A} = \frac{1}{2}\chi \ln|(\rho^2 + r_+^2)/l^2 - m|\boldsymbol{d\Phi}, \tag{11.99}$$

For $\rho = 0$, one determines the constant r_+ fulfilling

$$r_+^2 + \chi^2 \ln |r_+^2/l^2 - m| = 0. \tag{11.100}$$

This solution is endowed with mass, magnetic charge, and radial parameters. The coordinate ρ ranges from zero to infinity. This magnetic solution does not allow the existence of an event horizon since timelike geodesics can reach the origin at finite proper time, while null geodesics approach the origin at finite affine parameter; hence it does not describe a magnetic black hole. Moreover the Ricci tensor, and consequently the curvature tensor, as well as the electromagnetic field, are well behaved in this spacetime.

Cataldo et al. (2004) commented on this static circular magnetic solution of the (2+1) Einstein–Maxwell equations, derived previously by other authors, and came to the conclusion that this solution, considered up to that moment as a two-parameter one, is in fact a one-parameter solution, which describes a distribution of a radial magnetic field in a (2+1) anti-de Sitter background spacetime, and that the mass parameter is just a pure gauge and can be rescaled to minus one.

Accomplishing in the original Peldan solution (11.86) the coordinate transformation

$$t \to t,\ \rho \to \sqrt{(\rho^2 + r_+^2 - m\, l^2)/l^2},\ \phi \to \phi\, l^2,\ \chi^2 := a^2\, l^2,$$

one obtains the Hirschmann and Welch (1996) representation of the magnetostatic solution, which is given by the metric functions

$$\begin{aligned}
ds^2 &= -N(\rho)^2 dt^2 + \frac{1}{L(\rho)^2}\, d\rho^2 + K(\rho)^2 [d\phi + W(\rho) dt]^2, \\
H(\rho) &= \frac{\rho^2 + r_+^2 - m\, l^2}{l^2},\ L(\rho) = \frac{\sqrt{H(\rho)}}{\rho} K(\rho), \\
K(\rho) &= \sqrt{\rho^2 + r_+^2 + \chi^2 \ln H(\rho)}, \\
N(\rho) &= \sqrt{H(\rho)},\ W(\rho) = 0.
\end{aligned} \tag{11.101}$$

In the original Hirschmann and Welch (1996) work there is a condition to be fulfilled by the parameter r_+, arising from the vanishing of K at $\rho = 0$, namely

$$r_+^2 + \chi^2 \ln(\frac{r_+^2}{l^2} - m) = 0 \to (\frac{r_+^2}{l^2} - m) \mathrm{e}^{(r_+^2/\chi^2)} = 1. \tag{11.102}$$

This equation has been used in the quoted publication to determine the conical angle deficit: as $\rho \to 0$ the behavior of

$$K^2/\rho^2 = 1 + \chi^2/\rho^2 \ln[(\rho^2 + r_+^2 - ml^2)/(r_+^2 - ml^2)]$$

is given by

$$(K^2/\rho^2)|_{\rho \to 0} \to \frac{r_+^2 - ml^2 + \chi^2}{r_+^2 - ml^2} = 1 + \frac{\chi^2}{l^2}\, \mathrm{e}^{(r_+^2/\chi^2)},$$

hence the spatial sector $(\frac{1}{L^2}\, d\rho^2 + K^2\, d\phi^2)_{|\rho\to 0}$ of the studied metric behaves as

$$\left[d\rho\,\frac{\mathrm{e}^{(r_+^2/2\chi^2)}}{\sqrt{1+\frac{\chi^2}{l^2}\,\mathrm{e}^{(r_+^2/\chi^2)}}}\right]^2 + \frac{\rho^2\,\mathrm{e}^{(r_+^2/\chi^2)}}{1+\frac{\chi^2}{l^2}\,\mathrm{e}^{(r_+^2/\chi^2)}}\left[d\phi\,\mathrm{e}^{(-r_+^2/2\chi^2)}(1+\frac{\chi^2}{l^2}\,\mathrm{e}^{(r_+^2/\chi^2)})\right]^2$$

$$=: d\tilde{\rho}^2 + \tilde{\rho}^2 d\tilde{\phi}^2,$$

hence the angles range

$$0 \le \tilde{\phi} \le 2\pi \to 0 \le \phi \le 2\pi\,\frac{\mathrm{e}^{(r_+^2/2\chi^2)}}{1+\frac{\chi^2}{l^2}\mathrm{e}^{(r_+^2/\chi^2)}} = 2\pi\,\frac{l\sqrt{r_+^2 - ml^2}}{r_+^2 - ml^2 + \chi^2},$$

thus, the conical singularity at $\rho = 0$, as pointed out in the HW paper, arises in ϕ with the period $T_\phi = 2\pi\nu := 2\pi\,\mathrm{e}^{(r_+^2/2\chi^2)}/(1+\frac{\chi^2}{l^2}\mathrm{e}^{(r_+^2/\chi^2)})$, consequently the angle deficit is $\delta T_\phi = 2\pi(1-\nu)$ as reported also in Dias and Lemos (2002).

Mass and Energy

For this electromagnetic field solution the surface energy density is given by

$$\epsilon(\rho,\epsilon_0) = -\frac{1}{\pi\, l}\,\frac{\rho^2 + r_+^2 - ml^2 + \chi^2}{\sqrt{\rho^2 + r_+^2 + \chi^2\ln H}\sqrt{\rho^2 + r_+^2 - l^2 m}} - \epsilon_0, \tag{11.103}$$

while the integral energy and mass amount to

$$E(\rho,\epsilon_0) = -\frac{2}{l}\,\frac{\rho^2 + r_+^2 - ml^2 + \chi^2}{\sqrt{\rho^2 + r_+^2 - l^2 m}} - 2\pi\, K\,\epsilon_0,$$

$$M(\rho,\epsilon_0) = -\frac{2}{l^2}(\rho^2 + r_+^2 - ml^2 + \chi^2) - 2\pi\, N\, K\epsilon_0. \tag{11.104}$$

The evaluation of the above functions independent of ϵ_0 behave at infinity according to

$$\epsilon(\rho\to\infty,\epsilon_0 = 0) \approx -\frac{1}{\pi\, l} + \frac{ml^2 - 2\chi^2}{2\pi\, l\,\rho^2} + \frac{\chi^2}{\pi\, l\,\rho^2}\ln\left(\frac{\rho}{l}\right),$$

$$E(\rho\to\infty,\epsilon_0 = 0) \approx -2\frac{\rho}{l} + \frac{m\, l^2 - r_+^2 - 2\chi^2}{l\,\rho},$$

$$M(\rho\to\infty,\epsilon_0 = 0) = -\frac{2}{l^2}(\rho^2 + r_+^2 - ml^2 + \chi^2). \tag{11.105}$$

Using in the expressions (11.104) the energy density for the anti-de Sitter solution counterpart, namely $\epsilon_0 = -\frac{1}{\pi\, l^2}\rho/\sqrt{M_0 + \frac{\rho^2}{l^2}}$, which at the spatial infinity behaves as $\epsilon_{0|\infty}(M_0) \approx -\frac{1}{\pi\, l} + \frac{l\, M_0}{2\pi\,\rho^2}$, the series expansions of the corresponding quantities at $\rho\to\infty$ result in

Table 11.3.1 *Electro(b)–Magneto(a)-Static Solutions*

References	§	Eqns.	elec.	mag.	Λ
Static E–M Eqns	11.1.6	(11.42)	√	√	√
Electrostatic	11.2.1	(11.47)	√	0	√
Gott and Alpert (1984)		(11.54)	√	0	0
Gott et al. (1986)					
Deser and Mazur (1985)		(11.55c)	√	0	0
Melvin (1986)		(11.55)	√	0	0
Kogan (1992)		(11.55a)	√	0	0
Peldan (1993)		(11.56)	√	0	√
Magnetostatic	11.3.1	11.76	0	√	√
Kogan (1992)		(11.55a)	0	√	0
Peldan (1993)		(11.86)	0	√	√
Hirschmann and Welch (1996)		(11.98)	0	√	√

$$
\begin{aligned}
\epsilon(\rho \to \infty, \epsilon_{0|\infty}(M_0)) &\approx -\frac{l\,M_0}{2\pi\,\rho^2} + \frac{ml^2 - 2\chi^2}{2\pi\,l\rho^2} + \frac{\chi^2}{\pi\,l\,\rho^2}\ln\left(\frac{\rho}{l}\right),\\
E(\rho \to \infty, \epsilon_{0|\infty}(M_0)) &\approx -\frac{l\,M_0}{\rho} + \frac{ml^2 - 2\chi^2}{l\,\rho} + 2\frac{\chi^2}{l\,\rho}\ln\left(\frac{\rho}{l}\right),\\
M(\rho \to \infty, \epsilon_{0|\infty}(M_0)) &\approx m - M_0 - 2\frac{\chi^2}{l^2} + 2\frac{\chi^2}{l^2}\ln\left(\frac{\rho}{l}\right).
\end{aligned}
\tag{11.106}
$$

Therefore, comparing with the energy characteristics of the BTZ solution, one concludes that the mass logarithmically diverges at spatial infinity, and that the role of mass is played by m.

Some data about the electro-magneto–static families of solutions can be represented schematically in the table above.

11.4 Cataldo Static Hybrid Solution

This class of electromagnetic static fields has no analog in $(3+1)$ Maxwell electromagnetism, where one finds fields $\boldsymbol{A} = F(r)\boldsymbol{dt}$ or $\boldsymbol{A} = F(r)\,\boldsymbol{d\phi}$ but not $\boldsymbol{A} = \frac{c}{2}(t\boldsymbol{d\phi} - \phi\boldsymbol{dt})$ which is the present case. The integration starts from the combination $E_t{}^t(a=0=b) + 2E_r{}^r(a=0=b) + E_\Phi{}^\Phi(a=0=b)$ from (11.42), which yields

$$
\frac{d^2}{dr^2}F - \frac{8}{l^2} = 0,
\tag{11.107}
$$

with integral

$$
F(r) = 4\,\frac{(r-r_1)\,(r-r_2)}{l^2}.
\tag{11.108}
$$

As the equation for $H(r)$ one may consider the first-order equation $E_r{}^r(a = 0, b = 0)$, which can be written as

$$
\left(\frac{H_{,r}}{2\,H} - \frac{F_{,r}}{4\,F}\right)^2 = \left(\frac{F_{,r}}{4\,F}\right)^2 - \frac{c^2}{F^2} - \frac{1}{l^2\,F}.
\tag{11.109}
$$

Evaluating the right–hand side of this equation, one arrives at

$$\left[\frac{d}{dr}\ln\left(\frac{H}{F^{1/2}}\right)\right]^2 = 4\frac{\left((r_2-r_1)^2-c^2l^4\right)}{l^4\,F^2}.$$

For definiteness we assume $r_2 > r_1$. Accomplishing the integration one obtains

$$\ln\left(\frac{H}{\sqrt{F}}\right) = \mp\frac{\sqrt{(r_2-r_1)^2-c^2l^4}}{2\,(r_2-r_1)} \times \ln\left(\frac{r-r_1}{r-r_2}\right).$$

Introducing the constant α through

$$\alpha = 1-\frac{l^4c^2}{(r_2-r_1)^2},\; c^2 = \frac{(r_2-r_1)^2\,(1-\alpha)}{l^4}, \tag{11.110}$$

one obtains $H(r)$ in the form

$$H\,(r) = K_0{}^2\,\sqrt{F\,(r)}\left(\frac{r-r_1}{r-r_2}\right)^{\pm\frac{\sqrt{\alpha}}{2}}. \tag{11.111}$$

Summarizing, this class of solutions is given by the metric

$$\begin{aligned}
g &= -\frac{F}{H}dt^2 + \frac{1}{F}dr^2 + H\,d\phi^2,\\
F &= \frac{4}{l^2}(r-r_1)(r-r_2),\\
H &= \frac{2\,K_0^2}{l}\,(r-r_1)^{(1\pm\sqrt{\alpha})/2}(r-r_2)^{(1\mp\sqrt{\alpha})/2},\\
\boldsymbol{A} &= \frac{c}{2}(t\boldsymbol{d\phi}-\phi\boldsymbol{dt}).
\end{aligned} \tag{11.112}$$

The field tensor characterization of this solution is given by

$$\begin{aligned}
F^{\mu\nu} &= -2\frac{c}{F}\delta^{[\mu}{}_t\delta^{\nu]}{}_\phi,\; F_{\mu\nu} = 2c\delta_{[\mu}{}^t\delta_{\nu]}{}^\phi,\\
T^\mu{}_\nu &= \frac{c^2}{8\,\pi\,F}\left(-\delta^\mu{}_t\delta^t{}_\nu + \delta^\mu{}_r\delta^r{}_\nu - \delta^\mu{}_\phi\delta^\phi{}_\nu\right),\\
FF &= -2\frac{c^2}{F},\; TT = \frac{3}{64}\frac{c^4}{\pi^2F^2},\\
T^\mu_\mu &= -\frac{1}{8}\frac{c^2}{\pi\,F}.
\end{aligned} \tag{11.113}$$

By scaling transformations of the Killingian coordinates ϕ and t, the arbitrary constant K_0 can be equated to 1.

Subjecting the metric (11.112) to a further coordinate transformation

$$t = \frac{1}{\sqrt{2}}K_0l^{\mp\sqrt{\alpha}/2}t',\, \phi = \frac{1}{\sqrt{2}K_0}l^{\pm\sqrt{\alpha}/2}\phi',\; r = \rho^2+r_1,\, M := \frac{1}{l^2}(r_2-r_1), \tag{11.114}$$

dropping primes, one brings the static hybrid metric to the form

$$g = -\rho^{1\mp\sqrt{\alpha}}\left(\frac{\rho^2}{l^2}-M\right)^{(1\pm\sqrt{\alpha})/2} dt^2 + \rho^{1\pm\sqrt{\alpha}}\left(\frac{\rho^2}{l^2}-M\right)^{(1\mp\sqrt{\alpha})/2} d\phi^2 + \left(\frac{\rho^2}{l^2}-M\right)^{-1} d\rho^2. \quad (11.115)$$

The electromagnetic field tensor under the above mentioned transformations becomes

$$F_{\mu\nu} = M\sqrt{1-\alpha}\,\delta_{[\mu}{}^{t}\delta_{\nu]}{}^{\phi},$$
$$T_{\nu}{}^{\mu} = \frac{M^2}{32\pi\,\rho^2}\frac{(1-\alpha)}{(\rho^2/l^2-M)}\times\left(-\delta_{\nu}{}^{t}\delta_{t}{}^{\mu}+\delta_{\nu}{}^{r}\delta_{r}{}^{\mu}-\delta_{\nu}{}^{\phi}\delta_{\phi}{}^{\mu}\right). \quad (11.116)$$

This solution corresponds to the static charged solution reported in Cataldo (2002), where the name of azimuthal static solution was coined.

11.4.1 Mass and Energy

The structural functions of the Cataldo static solution – see García (2009), Eq. (4.46), in its canonical representation – are given by

$$ds^2 = -N(\rho)^2 dt^2 + \frac{1}{L(\rho)^2}\,d\rho^2 + K(\rho)^2[d\phi + W(\rho)dt]^2,$$
$$N(\rho) = \rho^{(1/2-\sqrt{\alpha}/2)}(\rho^2/l^2-M)^{(1/4+\sqrt{\alpha}/4)},$$
$$L(\rho) = (\rho^2/l^2-M)^{(1/2)},$$
$$K(\rho) = \rho^{(1/2+\sqrt{\alpha}/2)}(\rho^2/l^2-M)^{(1/4-\sqrt{\alpha}/4)},\quad W(\rho)=0. \quad (11.117)$$

The corresponding surface densities are

$$\epsilon(\rho,\epsilon_0) = -\frac{1}{\pi l\rho}\frac{1}{\sqrt{\rho^2-Ml^2}}[\rho^2 - Ml^2\frac{(1+\sqrt{\alpha})}{2}] - \epsilon_0,$$
$$j_{\phi}(\rho) = 0 = J(\rho), \quad (11.118)$$

while the integral quantities amount to

$$E(\rho,\epsilon_0) = -2\pi\epsilon_0 l^{(\sqrt{\alpha}/2-1/2)}(\rho^2-Ml^2)^{(-\sqrt{\alpha}/4+1/4)}\rho^{(\sqrt{\alpha}/2+1/2)},$$
$$-l^{(\sqrt{\alpha}/2-3/2)}(\rho^2-Ml^2)^{(-\sqrt{\alpha}/4-1/4)}\rho^{(\sqrt{\alpha}/2-1/2)}[2\rho^2-(1+\sqrt{\alpha})Ml^2]$$
$$M(\rho,\epsilon_0) = -2\frac{\rho^2}{l^2} + (1+\sqrt{\alpha})M - \frac{2\pi}{l}\epsilon_0\rho\sqrt{\rho^2-Ml^2}. \quad (11.119)$$

The evaluation of energy and mass functions independent of ϵ_0 behave at infinity as

$$\epsilon(\rho\to\infty,\epsilon_0=0)\approx-\frac{1}{\pi\,l}+\frac{l\,\sqrt{\alpha}M}{2\pi\,\rho^2},$$
$$E(\rho\to\infty,\epsilon_0=0)\approx l^{\frac{1}{2}(1+\sqrt{\alpha})}[-2\frac{\rho}{l^2}+\frac{(1+\sqrt{\alpha})M}{2\rho}],$$
$$M(\rho\to\infty,\epsilon_0=0)\approx M(1+\sqrt{\alpha})-2\frac{\rho^2}{l^2}. \tag{11.120}$$

Using in the expressions (11.118) and (11.119) as the reference energy density the quantity $\epsilon_0=-\frac{1}{\pi\rho}\sqrt{-M_0+\frac{\rho^2}{l^2}}$, which at the spatial infinity behaves as $\epsilon_{0|\infty}(M_0)\approx-\frac{1}{\pi\,l}+\frac{M_0}{2\pi\,\rho^2}$ the series expansions of the corresponding quantities at $\rho\to$ infinity result in

$$\epsilon(\rho\to\infty,\epsilon_{0|\infty}(M_0))\approx\frac{l}{2\pi\,\rho^2}(-M_0+\sqrt{\alpha}M),$$
$$E(\rho\to\infty,\epsilon_{0|\infty}(M_0))\approx\frac{(-M_0+\sqrt{\alpha}M)}{\rho}\,l^{\frac{1}{2}(1+\sqrt{\alpha})},$$
$$M(\rho\to\infty,\epsilon_{0|\infty}(M_0))\approx-M_0+\sqrt{\alpha}M. \tag{11.121}$$

Therefore, comparing with the energy characteristics of the BTZ solution, one concludes that the mass parameter at spatial infinity is determined by the product $\sqrt{\alpha}M$, although the mass function diverges at infinity as fast as $1/\rho^2$, a similar behavior is exhibited by the energy density in that spatial region.

11.4.2 Field, Energy–Momentum, and Cotton Tensors

The electromagnetic field of this solution is given by

$$(F^{\alpha}{}_{\beta})=\frac{l\,M(1-\alpha)^{1/2}}{(\rho^2-Ml^2)^{1/2}\,\rho}\times\begin{bmatrix}0 & 0 & -\frac{1}{2}\frac{l^{\sqrt{\alpha}}}{\rho^{\sqrt{\alpha}}}\left(\rho^2-Ml^2\right)^{1/2\sqrt{\alpha}}\\ 0 & 0 & 0\\ -\frac{1}{2}\frac{\rho^{\sqrt{\alpha}}}{l^{\sqrt{\alpha}}}\left(\rho^2-Ml^2\right)^{-1/2\sqrt{\alpha}} & 0 & 0\end{bmatrix}, \tag{11.122}$$

and is characterized by the following eigenvalues and eigenvectors

$$\lambda_1=0;\mathbf{V1}=[0,V^2,0],\ V^{\mu}V_{\mu}=(V^2)^2g_{\rho\rho},\ \mathbf{V1}=\mathbf{S1},$$
$$\lambda_2=-1/2\,\frac{Ml\sqrt{1-\alpha}}{\sqrt{\rho^2-Ml^2}\rho};\mathbf{N2}=[l^{\sqrt{\alpha}}\left(\rho^2-Ml^2\right)^{-1/2\sqrt{\alpha}}\rho^{\sqrt{\alpha}},0,1],$$
$$V^{\mu}V_{\mu}=0,\ \mathbf{V2}=\mathbf{N2},$$
$$\lambda_3=1/2\,\frac{Ml\sqrt{1-\alpha}}{\sqrt{\rho^2-Ml^2}\rho};\mathbf{N3}=[-l^{\sqrt{\alpha}}\left(\rho^2-Ml^2\right)^{-1/2\sqrt{\alpha}}\rho^{\sqrt{\alpha}},0,1],$$

$$
\begin{aligned}
&V^{\mu}V_{\mu} = 0,\ \mathbf{V3} = \mathbf{N3},\\
&\text{Type}: \{S, N, N\}.
\end{aligned} \tag{11.123}
$$

On the other hand, the energy–momentum tensor, having the structure

$$
(T^{\alpha}{}_{\beta}) = -\frac{1}{32}\frac{M^2 l^2 (1-\alpha)}{\rho^2 (\rho^2 - M l^2)\,\pi}\begin{bmatrix} 1 & 0 & 0 \\ 0 & -1 & 0 \\ 0 & 0 & 1 \end{bmatrix}, \tag{11.124}
$$

allows for the eigenvalues $\lambda_1 = \frac{1}{32}\frac{M^2l^2(1-\alpha)}{\rho^2(\rho^2-Ml^2)\pi}$ and the other one, of multiplicity two, $\lambda_2 = \lambda_3 = -\frac{1}{32}\frac{M^2l^2(1-\alpha)}{\rho^2(\rho^2-Ml^2)\pi}$ with the corresponding eigenvectors

$$
\begin{aligned}
&\lambda_1 = \frac{1}{32}\frac{M^2 l^2 (1-\alpha)}{\rho^2 (\rho^2 - M l^2)\,\pi};\\
&\mathbf{V1} = [0, V^2, 0],\ V^{\mu}V_{\mu} = (V^2)^2 g_{\rho\rho},\ \mathbf{V1} = \mathbf{S1},\\
&\lambda_2 = -\frac{1}{32}\frac{M^2 l^2 (1-\alpha)}{\rho^2 (\rho^2 - M l^2)\,\pi};\\
&\mathbf{V2} = [V^1, 0, V^3],\ V^{\mu}V_{\mu} = (V^1)^2 g_{tt} + (V^3)^2 g_{\phi\,\phi},\\
&\mathbf{V2} = \mathbf{T2}, \mathbf{S2}, \mathbf{N2},\\
&\lambda_3 = -\frac{1}{32}\frac{M^2 l^2 (1-\alpha)}{\rho^2 (\rho^2 - M l^2)\,\pi};\\
&\mathbf{V3} = [\tilde{V}^1, 0, \tilde{V}^3],\ V^{\mu}V_{\mu} = (\tilde{V}^1)^2 g_{tt} + (\tilde{V}^3)^2 g_{\phi\,\phi},\\
&\mathbf{V3} = \mathbf{T3}, \mathbf{S3}, \mathbf{N3}.
\end{aligned} \tag{11.125}
$$

For $\mathbf{V2}$ and $\mathbf{V3}$, the character of the vector depends on the sign of its norm; for instance, by choosing

$$
\begin{aligned}
&V^1 = s\,\sqrt{g_{\phi\,\phi}}/\sqrt{|g_{tt}|}\,V^3,\ s = const.,\ \mathbf{V1}^{\mu}\mathbf{V1}_{\mu} = (1-s^2) g_{\phi\,\phi}\,(V^3)^2;\\
&s > 1 \rightarrow \mathbf{V1} = \mathbf{T},\ s = \pm 1 \ \rightarrow \mathbf{V1} = \mathbf{N},\ s < 1 \rightarrow \mathbf{V1} = \mathbf{S}.
\end{aligned}
$$

Recall that in (3+1) gravity the eigenvectors of the electromagnetic energy–momentum tensor (and at the same time of the electromagnetic field tensor) are null in pairs, i.e. they exhibit double coincidence. Hence in the (2+1) case under study one may think of the alignments $\{\mathbf{S}, 2\mathbf{N}\}$ or $\{\mathbf{N}, \mathbf{S}, \mathbf{N}\}$ as the corresponding reductions of electromagnetic field eigen-directions of (3+1) gravity.

To complete the characterization of this solution, it is reasonable to add some comments about the conformal Cotton tensor, which is given by

$$
(C^{\alpha}{}_{\beta}) = C_0\begin{bmatrix} 0 & 0 & \frac{\rho^{\sqrt{\alpha}}}{l\sqrt{\alpha}}\left(\rho^2 - M l^2\right)^{-\sqrt{\alpha}/2} \\ 0 & 0 & 0 \\ -\frac{l\sqrt{\alpha}}{\rho^{\sqrt{\alpha}}}\left(\rho^2 - M l^2\right)^{\sqrt{\alpha}/2} & 0 & 0 \end{bmatrix}, \tag{11.126}
$$

where $C_0=\frac{l^3\sqrt{\alpha}(\alpha-1)M^3}{8(\rho^2-Ml^2)^{3/2}\rho^3}$ with eigenvectors

$$
\begin{aligned}
&\lambda_1=0;\mathbf{V1}=[0,V^2,0],\ V^\mu V_\mu=(V^2)^2/L^2,\ \mathbf{V1}=\mathbf{S},\\
&\lambda_2=-\frac{i\sqrt{\alpha}l^3\,(1-\alpha)\,M^3}{8\,(\rho^2-Ml^2)^{3/2}\,\rho^3};\\
&\mathbf{V2}=[V^1=-i\rho^{\sqrt{\alpha}}\,(\rho^2-Ml^2)^{-\sqrt{\alpha}/2}\,l^{-\sqrt{\alpha}}V^3,0,V^3],\ \mathbf{V2}=\mathbf{Z},\\
&\lambda_3=\frac{i\sqrt{\alpha}l^3\,(1-\alpha)\,M^3}{8\,(\rho^2-Ml^2)^{3/2}\,\rho^3};\\
&\mathbf{V3}=[V^1=i\rho^{\sqrt{\alpha}}\,(\rho^2-Ml^2)^{-\sqrt{\alpha}/2\,l^{-\sqrt{\alpha}}}\,V^3,0,V^3],\ \mathbf{V3}=\bar{\mathbf{Z}},\\
&\text{Type I}:\{S,Z,\bar{Z}\}.
\end{aligned}
\tag{11.127}
$$

11.5 Uniform Electromagnetic Solutions $F_{\mu\nu\,;\sigma}=0$

To determine all uniform electromagnetic solutions, i.e., those possessing vanishing covariant derivatives of $F_{\mu\nu}$, $F_{\mu\nu\,;\sigma}=0$, one has to start the integration process from the differential relations arising from these conditions. The hybrid class $c\neq 0$ does not allow for such kind of solutions. The other families with $a\neq 0$ and (or) $b\neq 0$ give rise to nontrivial solutions.

11.5.1 General Uniform Electromagnetic Solution for $a\neq 0,\neq b$

A class of uniform electromagnetic stationary solutions, for $a\neq 0$, $b\neq 0$ and $c=0$, can be constructed by demanding the vanishing of the covariant derivatives of the electromagnetic tensor field, $F_{\mu\nu\,;\sigma}=0$, which yields two independent equations: $F_{t\phi;t}=0$ and $F_{tr;r}=0$. From the last one, one isolates dW/dr

$$
\frac{dW}{dr}=-b\frac{F}{H^3(a-b\,W)}\frac{dH}{dr},
\tag{11.128}
$$

which when used in the first equation $F_{t\phi;t}=0$ allows us to write

$$
\frac{dF}{dr}=\frac{F}{H}\frac{dH}{dr}-b^2\frac{F^2}{H^3(a-b\,W)^2}\frac{dH}{dr}.
\tag{11.129}
$$

As the next step, one substitutes recursively these first derivatives into the Einstein equations. One gets, among other relations, a simple expression for the equation ${E_\phi}^t$

$$
\frac{d^2H}{dr^2}=-4\frac{H^2}{F^2}(a-b\,W)^2,
\tag{11.130}
$$

which when substituted back into the Einstein equations reduces them to a single relation

$$
F(b^2l^2-H)=l^2H^2(a-b\,W)^2,
\tag{11.131}
$$

from which one has

$$W(r) = \frac{a}{b} \mp \frac{F}{lbH}\sqrt{b^2l^2 - H}. \tag{11.132}$$

Using the relation (11.131) in (11.130) one obtains

$$\frac{d^2H}{dr^2} = -\frac{4}{l^2F}(b^2l^2 - H). \tag{11.133}$$

On the other hand, substituting $W(r)$ from (11.132) into (11.129) one arrives at the relation

$$\frac{dF}{dr}(b^2l^2 - H) + F\frac{dH}{dr} = 0, \tag{11.134}$$

with integral

$$F(r) = \frac{b^2l^2 - H(r)}{l^2\,\beta^2}, \; \beta = \text{constant}. \tag{11.135}$$

The substitution of $F(r)$ from (11.135) into (11.133) yields

$$\frac{d^2H}{dr^2} = -4\beta^2, \tag{11.136}$$

hence

$$H(r) = -2\beta^2r^2 + c_1r + c_0. \tag{11.137}$$

Consequently the function $W(r)$ becomes

$$W(r) = \frac{a}{b} \mp \frac{1}{l^2\,b\,\beta}\frac{b^2l^2 - H}{H}. \tag{11.138}$$

No restriction arises from the remaining (11.128). Thus, we have determined the general uniform electromagnetic stationary cyclic symmetric solution given by the metric and the field vector

$$\begin{aligned} g &= -\frac{b^2\,l^2 - H(r)}{l^2\,\beta^2\,H(r)}dt^2 + \frac{l^2\,\beta^2\,dr^2}{b^2\,l^2 - H(r)} \\ &\quad + H(r)\left[d\phi + \left(\frac{a}{b} \mp \frac{1}{l^2\,b\,\beta}\frac{b^2l^2 - H}{H}\right)dt\right]^2, \\ H(r) &= -2\beta^2r^2 + C_1r + C_0, \\ \boldsymbol{A} &= -\beta\,r\,[\boldsymbol{d\phi} - \frac{1 \pm al^2\beta}{l^2\,b\,\beta}\boldsymbol{dt}], \end{aligned} \tag{11.139}$$

characterized by the uniform electromagnetic field tensors

$$\begin{aligned} F_{\mu\nu} &= -2\frac{1 \pm al^2\beta}{l^2b}\delta_{[\mu}{}^{t}\delta_{\nu]}{}^{r} \pm 2\beta\delta_{[\mu}{}^{r}\delta_{\nu]}{}^{\phi}, \\ 8\pi\,l^2T_\nu{}^\mu &= -(1 \pm 2al^2\beta)\left(\delta_\nu{}^t\delta_t{}^\mu - \delta_\nu{}^\phi\delta_\phi{}^\mu\right) \\ &\quad -\delta_\nu{}^r\delta_r{}^\mu \mp 2\,\beta\,bl^2\delta_\nu{}^\phi\delta_t{}^\mu + \frac{2a}{b}(1 \pm al^2\beta)\delta_\nu{}^t\delta_\phi{}^\mu. \end{aligned} \tag{11.140}$$

with FF invariant $FF = -2/l^2$.

Although the solution above has been derived for $\Lambda = -1/l^2$, the branch corresponding to $\Lambda = 1/l^2$ is achieved from the above expressions by changing $l^2 \to -l^2$.

11.5.2 Uniform "Stationary" Electromagnetic $A = r/(b\,l^2)(dt - \omega_0\,d\phi)$ Solutions

Consider now the case $a = 0 = c$ for the metric (11.16)

$$g = -\frac{F}{h}\left(dt - \omega\,d\phi\right)^2 + h\,d\phi^2 + \frac{dr^2}{F} = -\frac{F}{H}dt^2 + \frac{dr^2}{F} + H\left(d\phi + W\,dt\right)^2,$$
$$F = F,\, H = h - \frac{F}{h}\omega^2,\, W = \frac{\omega}{H}\frac{F}{h}. \tag{11.141}$$

The electromagnetic tensor amounts to $F^{\mu\nu} = 2\,b\,\delta^{\mu}{}_{[t}\delta^{\nu}{}_{r]}$, and

$$F_{\mu\nu} = -2\,\frac{b}{HF}(F - H^2\,W^2)\,\delta_{\mu}{}^{[t}\delta_{\nu}{}^{r]} + 2\,b\,\frac{HW}{F}\,\delta_{\mu}{}^{[\phi}\delta_{\nu}{}^{r]}$$
$$= -2\,\frac{b}{h(r)}\,\delta_{\mu}{}^{[t}\delta_{\nu}{}^{r]} + 2\,b\,\frac{\omega(r)}{h(r)}\,\delta_{\mu}{}^{[\phi}\delta_{\nu}{}^{r]}.$$

The covariant derivatives $F_{\phi r;r}$ and $F_{tr;r}$ of the field $F_{\mu\nu}$ are equal to zero if

$$\omega(r) = \omega_0,\; h(r) = h_0.$$

Therefore, the structural functions $\omega(r) = \omega_0$ and $h(r) = h_0$ are constants. The Einstein–Maxwell equations require the fulfillment of the equations

$$\frac{d^2}{d\,r^2}F(r) = \frac{4}{l^2} \to F(r) = \frac{2}{l^2}r^2 + c_1\,r + c_0,\; h_0 = b^2\,l^2. \tag{11.142}$$

Consequently the derived solution can be given as

$$g = -\frac{F}{h_0}(dt - \omega_0 d\phi)^2 + h_0\,d\phi^2 + \frac{dr^2}{F(r)},$$
$$F(r) = \frac{2r^2}{l^2} + c_1 r + c_0, h_0 = b^2\,l^2,$$
$$\boldsymbol{A} = \frac{r}{b\,l^2}(\boldsymbol{dt} - \omega_0\boldsymbol{d\phi}), \tag{11.143}$$

and hence by a shifting transformation of the t-coordinate the derived metric becomes a static one. The electromagnetic tensors characterizing this uniform "stationary" cyclic symmetric solution are given by

$$F_{\mu\,\nu} = -2\frac{1}{b\,l^2}\delta_{[\mu}{}^{t}\delta_{\nu]}{}^{r} + 2\frac{\omega_0}{b\,l^2}\delta_{[\mu}{}^{\phi}\delta_{\nu]}{}^{r},$$
$$T_{\mu}{}^{\nu} = \frac{1}{8\,\pi\,l^2}(-\delta_{\mu}{}^{t}\delta_{t}{}^{\nu} - \delta_{\mu}{}^{r}\delta_{r}{}^{\nu} + \delta_{\mu}{}^{\phi}\delta_{\phi}{}^{\nu}) + \frac{\omega_0}{4\,\pi\,l^2}\delta_{\mu}{}^{t}\delta_{\phi}{}^{\nu}. \tag{11.144}$$

This solution can be generated from the static solution, which is given in subsection 11.5.3 by the metric (11.147), via the transformation $t \to t - \omega_0\phi$, $\phi \to \phi$.

Clément Uniform "Stationary" Electromagnetic Solution

Clément (1993) reported the uniform "stationary" generalization of the electrostatic solution in the form of

$$
\begin{aligned}
g &= -\frac{F}{H_0}\left(dt - \omega_0 d\,\phi\right)^2 + \frac{dr^2}{F(r)} + H_0 d\phi^2,\\
F(r) &= \frac{2r^2}{l^2} + c_1 r + c_0,\\
\boldsymbol{A} &= \frac{1}{\sqrt{H_0}\,l}\, r\, \left(\boldsymbol{dt} - \omega_0 \boldsymbol{d}\,\boldsymbol{\phi}\right),
\end{aligned}
\tag{11.145}
$$

In Clément's parametrization one adopts $H_0 = \frac{\pi_0^2\, l^2}{4m}$, with $m = 1/(2\,\kappa)$.

No Uniform Generalization of the Electrostatic Solution for $\Lambda = 1/l^2$

On the other hand for $b \neq 0$ and positive cosmological constant $\Lambda = 1/l^2$ there is no a uniform electromagnetic stationary cyclic symmetric solution; the reason is hidden in the resulting erroneous signature. In the considered case, for the metric (11.16) with structural functions $F(r), h(r), \omega(r)$ the electromagnetic tensor amounts, for $a = 0$, to

$$
\begin{aligned}
F^{\mu\nu} &= 2\, b\, \delta^{\mu}{}_{[t}\delta^{\nu}{}_{r]},\\
F_{\mu\nu} &= -2\, b/h(r)\, \delta_{\mu}{}^{[t}\delta_{\nu}{}^{r]} + 2\, b\, \omega(r)/h(r)\, \delta_{\mu}{}^{[\phi}\delta_{\nu}{}^{r]}.
\end{aligned}
$$

Therefore $F_{\mu\nu;\lambda} = 0$ is achieved for $h(r) = h_0$, $\omega(r) = \omega_0$. The Einstein equations requires $E_r{}^r = b^2/h_0 + 1/l^2 = 0 \rightarrow h_0 = -b^2\, l^2$. The covariant tensor components g_{tt} and g_{rr} explicitly amount to $g_{tt} = F(r)/(b^2\, l^2) > 0$, $g_{rr} = 1/F(r) > 0$, which contradicts the adopted signature $\{-,+,+\}$, therefore this case does not represent a compatible solution.

11.5.3 Matyjasek–Zaslavskii Uniform Electrostatic $\boldsymbol{A} = r/(b\, l^2)\, dt$ Solution

The sub-branch $\{a = 0, b \neq 0, W(r) = 0\}$ of uniform electrostatic solutions arises for constant $H(r)$, $H(r) = H_0 = $ constant. The equation E_t^t implies that the constant H_0 has to be $H_0 = b^2\, l^2$. The remaining equation $E_t{}^{\phi}$ amounts to

$$
\frac{d^2\, F}{d\, r^2} - \frac{4}{l^2} = 0, \rightarrow F(r) = 2\frac{r^2}{l^2} + 4\, c_1 r + c_0;
\tag{11.146}
$$

consequently the metric and the field vector become

$$
g = -\frac{F(r)}{b^2\, l^2} dt^2 + \frac{dr^2}{F(r)} + b^2\, l^2 d\phi^2,\ \boldsymbol{A} = \frac{r}{b\, l^2}\, \boldsymbol{dt}.
\tag{11.147}
$$

The electromagnetic field tensors of this solution possess constant eigenvalues and also exhibit the uniform character; explicitly they are given by

$$F_{\mu\,\nu} = \begin{bmatrix} 0 & -\frac{1}{b\,l^2} & 0 \\ \frac{1}{b\,l^2} & 0 & 0 \\ 0 & 0 & 0 \end{bmatrix}, \; T_{\mu}{}^{\nu} = \begin{bmatrix} -\frac{1}{8\pi\,l^2} & 0 & 0 \\ 0 & -\frac{1}{8\pi\,l^2} & 0 \\ 0 & 0 & \frac{1}{8\pi\,l^2} \end{bmatrix}, \tag{11.148}$$

with constant field invariants given by

$$FF = -\frac{2}{l^2},\; TT = \frac{3}{64}\frac{1}{\pi^2\,l^4}.$$

Incorporating the constant $H_0 = b^2\,l^2$ in the new definitions of t and ϕ, $t/b\,l \to t$, $b\,l\phi \to \phi$, one can set $b\,l = 1$ in the metric (11.147).

For the sake of comparison with previous reports let us introduce hyperbolic functions:

$$F(r) = 2\frac{r^2}{l^2} + 4\,c_1 r + c_0 = |c_0 - 2\,l^2 c_1^2| \left(\frac{2(r + l^2 c_1)^2}{l^2|c_0 - 2\,l^2 c_1^2|} \pm 1\right),$$

$$\begin{aligned} +: r + l^2 c_1 &= \sqrt{l^2|c_0 - 2\,l^2 c_1^2|/2}\;\sinh(\alpha x),\; F(r) \to |c_0 - 2\,l^2 c_1^2|\cosh^2(\alpha x), \\ -: r + l^2 c_1 &= \sqrt{l^2|c_0 - 2\,l^2 c_1^2|/2}\;\cosh(\alpha x),\; F(r) \to |c_0 - 2\,l^2 c_1^2|\sinh^2(\alpha x), \\ c_0 = 2\,l^2 c_1^2 : r + l^2 c_1 &= \exp\sqrt{2}x/l,\; F(r) \to \frac{2}{l^2}\exp\left(2\sqrt{2}x/l\right). \end{aligned} \tag{11.149}$$

The above mentioned quantities, i.e., the metric (11.147) and the structural functions (11.149), determine the solution derived in Matyjasek and Zaslavskii (2004). Expressions (11.147) and (11.149) are equivalent to the Bertotti (1959) and Robinson (1959) uniform electromagnetic–gravitational field solution, for a constant slice of one of the spatial coordinates; the BR $(3+1)$ solution allows for a product of two surfaces of constant curvature as manifold. Moreover, the $2D$ BR metric sector ds^2_- reduces to $r^2_-\,d\phi^2$ in the $(2+1)$ case, therefore the (2+1)-dimensional uniform electrostatic field can be considered as a dimensional reduction of the $(3+1)$ Bertotti–Robinson solution.

With the aim of demonstrating the uniqueness of this class of uniform solutions with $H(r) = h_0 =$ constant, even in the framework of stationary fields, let us consider the general metric with $W(r)$:

In the case $\{a = 0 = c, b \neq 0, H(r) = h_0\}$, the combination of equations $E_t{}^t - E_r{}^r - WE_t{}^\phi = 2\,\frac{W(r)^2 h_0\,b^2}{F(r)}$ implies $W(r) = 0$ and consequently the gravitational field is static; further integration gives rise to the above mentioned uniform electrostatic fields.

Case $\{a \neq 0, b = 0 = c, H(r) = h_0\}$: from $E_t{}^t - E_r{}^r - WE_t{}^\phi = 2\,a^2 h_0/F(r)$ therefore there is no solution in this case.

Case $\{a = 0 = b, c \neq 0, H(r) = h_0\}$: the combination $E_t{}^t - E_r{}^r - WE_\phi{}^t = 2c^2/F(r)$, hence there is no solution.

Vanishing Mass, Energy and Momentum

The uniform electrostatic solution Matyjasek and Zaslavskii (2004) – see also García (2009), Eq. (5.20) – is given by the metric functions

$$ds^2 = -N(\rho)^2 dt^2 + \frac{1}{L(\rho)^2}\,d\rho^2 + K^2[d\phi + W dt]^2,$$
$$L(\rho) = N(\rho) = \sqrt{\frac{2}{l^2}\rho^2 + 4c_1\rho + c_0},\; K = 1,\; W = 0. \tag{11.150}$$

Since the surface energy density ϵ is proportional to ϵ_0, $\epsilon = -\epsilon_0$, consequently all the energy–mass quantities are given through it

$$\epsilon = -\epsilon_0,\; M(\rho,\epsilon_0) = -2\pi\, N(\rho)\epsilon_0,\; E(\rho,\epsilon_0) = -2\pi\epsilon_0. \tag{11.151}$$

Thus, for the natural choice of a vanishing reference energy density $\epsilon_0 = 0$ all the energy quantities vanish: $\epsilon = 0, M(\rho,0) = 0 = E(\rho,0)$. On the other hand, if the reference energy is the one corresponding to the anti-de Sitter spacetime, $\epsilon_0 = -\frac{1}{\pi\,\rho}\sqrt{\frac{\rho^2}{l^2} - M_0}$, the energies $M(\rho,\epsilon_0)$ and $E(\rho,\epsilon_0)$ will be again expressed through ϵ_0.

Field, Energy–Momentum, and Cotton Tensors

As far as the eigenvalue–vector properties of this solution go, one establishes straightforwardly that the Cotton tensor ought to vanish because of uniform character of the electromagnetic field, hence the (2+1) Matyjasek–Zaslavski gravitational field is conformally flat, $C^{\alpha}{}_{\beta} = 0$, Type O. On the other hand the electromagnetic field tensor

$$(F^{\alpha}{}_{\beta}) = \begin{bmatrix} 0 & \frac{1}{N^2 l} & 0 \\ \frac{N^2}{l} & 0 & 0 \\ 0 & 0 & 0 \end{bmatrix}, \tag{11.152}$$

allows for the eigenvectors

$$\lambda_1 = 0; \mathbf{V1} = \left[0, 0, V^3\right], V_\mu V^\mu = {V^3}^2,\; \mathbf{V1} = \mathbf{S1},$$
$$\lambda_2 = \frac{1}{l}; \mathbf{V2} = \left[V^1, N^2 V^1, 0\right],\; V^\mu V_\mu = 0,\; \mathbf{V2} = \mathbf{N2},$$
$$\lambda_3 = -\frac{1}{l}; \mathbf{V3} = \left[V^1, -N^2 V^1, 0\right],\; V^\mu V_\mu = 0,\; \mathbf{V3} = \mathbf{N3},$$
$$\text{Type I} : \{S, N, N\}. \tag{11.153}$$

For the electromagnetic energy–momentum tensor we have

$$(T^{\alpha}{}_{\beta}) = \begin{bmatrix} -\frac{1}{8l^2\pi} & 0 & 0 \\ 0 & -\frac{1}{8l^2\pi} & 0 \\ 0 & 0 & \frac{1}{8l^2\pi} \end{bmatrix}, \tag{11.154}$$

with eigenvectors

$$\begin{aligned}
\lambda_1 &= \frac{1}{8\,\pi\,l^2}; \mathbf{V1} = \left[0,0,V^3\right], V_\mu V^\mu = {V^1}^2, \mathbf{V1} = \mathbf{S1},\\
\lambda_{2,3} &= -\frac{1}{8\,\pi\,l^2}; \mathbf{V2,3} = \left[V^1,V^2,0\right],\\
V_\mu V^\mu &= -\frac{\left(N^2V^1 - V^2\right)\left(N^2V^1+V^2\right)}{N^2}\\
\mathbf{V2} &= \mathbf{T2}, \mathbf{S2}, \mathbf{N2},\ \ \mathbf{V3} = \mathbf{T3}, \mathbf{S3}, \mathbf{N3},
\end{aligned} \tag{11.155}$$

and therefore it allows for the types $\{S, 2T\}, \{S, 2N\}, \{S, 2S\}$.

11.5.4 Uniform "Stationary" Electromagnetic $A = r/(a\,l^2)(d\phi + W_0\,dt)$ Solutions

In the case of positive cosmological constant $\Lambda = 1/l^2$ there exists a uniform "stationary" magnetic solution with constant $W(r) = W_0$.

The electromagnetic tensor possesses the structure

$$F_{\mu\nu} = -2\,a\frac{H}{F}W\,\delta_\mu{}^{[t}\delta_\nu{}^{r]} - 2\,a\,\frac{H}{F}\,\delta_\mu{}^{[\phi}\delta_\nu{}^{r]},$$

and its covariant derivatives are zero if

$$\begin{aligned}
F_{rt;r} &= -\frac{a\,H^2}{2F}\frac{dW}{dr},\\
F_{r\phi;r} &= -\frac{a}{4\,F\,H}\left(-2H^2\frac{d}{dr}\frac{F}{H} + H^3\frac{dW}{dr}\right)
\end{aligned}$$

vanish. Hence

$$F(r) = \beta\,H(r),\ W(r) = W_0 = \text{constant}, \tag{11.156}$$

and consequently

$$F_{t\,r} = -W_0\frac{a}{\beta},\ F_{\phi\,r} = -\frac{a}{\beta}. \tag{11.157}$$

The Einstein equations reduce to

$$\begin{aligned}
E_r{}^r &= -\frac{a^2}{\beta} + \frac{1}{l^2} = 0 \rightarrow \beta = a^2\,l^2,\ 2\,E_t{}^t = \frac{d^2}{dr^2}F(r) + \frac{4}{l^2} = 0\\
&\rightarrow F(r) = -\frac{2}{l^2}r^2 + c_1 r + c_0.
\end{aligned} \tag{11.158}$$

The metric and fields for the derived solution can be expressed as

$$\begin{aligned}
g &= -a^2\,l^2 dt^2 + \frac{dr^2}{F(r)} + \frac{F(r)}{a^2 l^2}(d\phi + W_0 dt)^2,\\
F(r) &= -\frac{2r^2}{l^2} + c_1 r + c_0,\\
\boldsymbol{A} &= \frac{r}{a\,l^2}(\boldsymbol{d\phi} + W_0\boldsymbol{dt}),
\end{aligned} \tag{11.159}$$

with uniform electromagnetic field tensor is

$$F_{\mu\nu} = -2\frac{W_0}{a\,l^2}\delta^t_{[\mu}\delta^r_{\nu]} - 2\frac{1}{a\,l^2}\delta^\phi_{[\mu}\delta^r_{\nu]},$$

and energy–momentum tensor

$$8\pi\, T_\nu{}^\mu = \frac{1}{l^2}[-\delta^t_\nu\delta^\mu_t + \delta^r_\nu\delta^\mu_r + \delta^\phi_\nu\delta^\mu_\phi + 2\,W_0\delta^t_\nu\delta^\mu_\phi].$$

This solution is equivalent to the Clément's solution given by Eq. (26), $\Lambda = 1/l^2$ of Clément (1993).

11.5.5 No Uniform Stationary Magnetostatic Solution for $\Lambda = -1/l^2$

On the contrary, as far as to the stationary uniform electromagnetic branch with $a \neq 0$ and negative cosmological constant $\Lambda = -1/l^2$ is concerned, one establishes that there is no solution at all. Following a similar procedure as the one used in the previous case, where now $b = 0$, $F(r) = \beta\,H(r)$, $W(r) = W_0 =$ constant, the Einstein equation $E_r{}^r = -\frac{a^2}{\beta} - \frac{1}{l^2} = 0$ yields $\beta = -a^2\,l^2$. The covariant tensor components g_{rr} and $g_{\phi\,\phi}$ explicitly amount to $g_{rr} = 1/F(r) > 0$, $g_{\phi\,\phi} = -a^2\,l^2 F(r) < 0$, which yields a contradiction with the adopted signature $\{-,+,+\}$. Hence this case does not represent a solution compatible with the (2+1) metric signature.

11.6 Constant Electromagnetic Invariants' Solutions

This section is devoted to the derivation of the electromagnetic fields coupled to stationary (static) cyclic symmetric (2+1) gravitational fields such that their electromagnetic invariants FF, T and TT are constants; because of the proportionality of T and TT to FF, it is enough to establish under which conditions

$$F_{\mu\nu}F^{\mu\nu} = -2\,\frac{c^2}{F} + 2\,\frac{H\,(a - W b)^2}{F} - 2\,\frac{b^2}{H}$$

vanishes.

It should be pointed out that the hybrid $c \neq 0$ class of spacetimes does not allow for constant invariants' solution when a cosmological constant is present. Therefore it is sufficient to restrict oneself to the case $a \neq 0$ or $b \neq 0$.

In what follows we shall search for subclasses of solutions with constant electromagnetic invariant, namely those with $\{a \neq 0, b \neq 0, W(r) = W(r)\}$, $\{a = 0,\ b \neq 0, \omega(r) = \omega_0\}$, and $\{a \neq 0, b = 0\,, W(r) = W_0\}$ families of solutions.

At this stage it is worth pointing out that constant invariant electromagnetic fields contain, as subclasses, the covariantly constant electromagnetic field solutions, while the inverse statement does not hold.

11.6.1 General Constant Invariant $F_{\mu\nu}F^{\mu\nu} = 2\gamma$ for $a \neq 0 \neq b$

Restricting the present section to the study of the cases $a \neq 0$ and $b \neq 0$, the constancy of $FF = 2\gamma$ is guaranteed by

$$\frac{F}{H^2}(b^2 + \gamma\, H) = (a - b\, W)^2, \tag{11.160}$$

which yields

$$W(r) = \frac{a}{b} \pm \frac{\sqrt{F(r)}}{b\, H(r)}\sqrt{b^2 + \gamma\, H(r)}. \tag{11.161}$$

In general, the following combinations of the Einstein equations give

$$\begin{aligned} E_t{}^t + E_\phi{}^\phi + 2E_r{}^r &= \frac{d^2F}{dr^2} - \frac{8}{l^2} + 4\frac{b^2F - H^2(a - b\,W)^2}{F\,H} = 0, \\ 2HF(E_t{}^t - E_r{}^r - W\,E_\phi{}^t) &= F^2\frac{d^2H}{dr^2} + 4H^2(a - b\,W)^2 = 0. \end{aligned} \tag{11.162}$$

Using the relation (11.160) one brings (11.162) to the form

$$\frac{d^2F}{dr^2} = \frac{8}{l^2} + 4\gamma, \quad \frac{d^2H}{dr^2} + \frac{4\gamma}{F}H = -\frac{4\,b^2}{F}, \tag{11.163}$$

with solution for $F(r)$

$$F(r) = (\frac{4}{l^2} + 2\gamma)r^2 + C_1 r + C_0 = (\frac{4}{l^2} + 2\gamma)(r - r_1)(r - r_2), \tag{11.164}$$

and the function H, as solution of its second-order equation (11.163), is expressed in terms of hypergeometric functions. Nevertheless there exists a shortcut for the integration of the function $H(r)$ by noticing that $E_r{}^r$ contains only first derivatives of the structural functions: replacing W(r) from (11.161) in the quoted equation, after extracting square root, one arrives at

$$\gamma\, F\frac{dH}{dr} - (b^2 + \gamma\, H)\frac{dF}{dr} = \pm 4\frac{b}{l}\sqrt{1 + l^2\gamma}\,\sqrt{F}\,\sqrt{b^2 + \gamma\, H},$$

which, when introducing the auxiliary function $Q(r)^2 := b^2 + \gamma\, H(r)$, becomes

$$\frac{d}{dr}\frac{Q}{\sqrt{F}} \mp 2\frac{b}{l\,F}\sqrt{1 + l^2\gamma} = 0.$$

Integrating this equation, using $F(r)$ (11.164), one obtains

$$Q = \pm\, bl\frac{\sqrt{1 + l^2\gamma}}{2 + l^2\gamma}\frac{\ln(r - r_2) - \ln(r - r_1)}{r_2 - r_1} + \beta\sqrt{F(r)}. \tag{11.165}$$

Fulfilling a single constraint still remains; using the expressions of the function $W(r)$ and its derivatives from the $E_\mu{}^\nu$-equations as well as the derivatives of F(r), together with $H(r)$ in terms of $Q(r)$ and the derivative of the latter from (11.6.1), one gets a single equation

$$2(1 + l^2\gamma)\left(Q(r)^2 + b^2\right)\sqrt{F} + l\, b\, Q(r)\sqrt{1 + l^2\gamma}\frac{dF}{dr} = 0, \tag{11.166}$$

which is incompatible with $Q(r)$ determined in (11.165) except for $\gamma = -1/l^2$.

Hence, we conclude that there are no solutions for arbitrary constant electromagnetic invariant $FF = 2\,\gamma$.

11.6.2 Constant Electromagnetic Invariant $FF = \mp 2/l^2$ Solution

A class of constant electromagnetic invariants' stationary solutions with $a \neq 0$, $b \neq 0$ and $\gamma = -1/l^2$, $FF = -2/l^2$, arises for

$$W(r) = \frac{a}{b} \mp \frac{F}{b\,l\,H(r)}\sqrt{b^2\,l^2 - H}\,. \tag{11.167}$$

As in the previous case, $F(r)$ and $H(r)$ fulfill the (11.163) for $\gamma = -1/l^2$ and correspondingly their integrals are given by

$$F(r) = \frac{2}{l^2}r^2 + c_1 r + c_0,\ H(r) = b^2 l^2 - \beta^2 l^2\,F(r).$$

There are no further constraints from the field equations. Consequently the final result can be written as

$$\begin{aligned} g &= -\frac{b^2\,l^2 - H(r)}{l^2\,\beta^2\,H(r)}dt^2 + l^2\,\beta^2\,\frac{dr^2}{b^2\,l^2 - H(r)} \\ &\quad + H(r)\left[d\phi + \left(\frac{a}{b} \mp \frac{1}{l^2\,b\,\beta}\frac{b^2 l^2 - H}{H}\right)dt\right]^2, \\ H(r) &= -2\beta^2 r^2 + C_1 r + C_0, \\ \boldsymbol{A} &= -\beta\,r\,[\boldsymbol{d\phi} - \frac{1 \pm a l^2 \beta}{l^2\,b\,\beta}\boldsymbol{dt}], \end{aligned}$$

which coincides with the uniform solution (11.139). Therefore we have determined a class of uniform constant electromagnetic invariant stationary solutions for both non-vanishing constants $a \neq 0 \neq b$. Although the solution mentioned above has been derived for $\Lambda = -1/l^2$, the branch with positive $\Lambda = 1/l^2$ is achieved from the above-mentioned expressions by changing $l^2 \to -l^2$.

11.6.3 Constant Electromagnetic Invariant $FF = -2/l^2$ Solution for $b \neq 0$

The constant electromagnetic invariant solution with $b \neq 0$ can be determined as solution of the Einstein–Maxwell equations by considering the stationary metric in the form

$$\begin{aligned} g &= -\frac{F}{h}\left(dt - \omega\,d\phi\right)^2 + h\,d\phi^2 + \frac{dr^2}{F} = -\frac{F}{H}dt^2 + \frac{dr^2}{F} + H\left(d\phi + W\,dt\right)^2, \\ F &= F,\ H = h - \frac{F}{h}\omega^2,\ W = \frac{\omega}{H}\frac{F}{h}. \end{aligned}$$

Demanding the electromagnetic invariant FF in the case $b \neq 0 = a$ to be constant, one establishes

$$FF = -\frac{2b^2}{h(r)} \rightarrow h(r) = h_0 = \text{constant}. \tag{11.168}$$

Substituting $h(r) = h_0$ into the Einstein equations one obtains from ${E_\phi}^t$ that

$$\frac{d^2\omega}{dr^2} = -\frac{2}{F}\frac{dF}{dr}\frac{d\omega}{dr}, \tag{11.169}$$

which when used in ${E_t}^t - {E_r}^r$ yields

$$\frac{F}{h_0^2}(\frac{d\omega}{dr})^2 = 0 \rightarrow \omega(r) = \omega_0. \tag{11.170}$$

Replacing $\omega = \omega_0$ and $h = h_0$ in the remaining equations one establishes

$$\frac{d^2}{dr^2}F(r) = \frac{4}{l^2} \rightarrow F(r) = \frac{2}{l^2}r^2 + c_1\, r + c_0,\, h_0 = b^2\, l^2. \tag{11.171}$$

Therefore we arrive at a constant electromagnetic invariant solution in the form

$$\begin{aligned} g &= -\frac{F}{h_0}(dt - \omega_0 d\phi)^2 + h_0\, d\phi^2 + \frac{dr^2}{F(r)}, \\ F(r) &= \frac{2r^2}{l^2} + c_1 r + c_0, h_0 = b^2\, l^2, \\ \boldsymbol{A} &= \frac{r}{b\, l^2}(\boldsymbol{dt} - \omega_0 \boldsymbol{d\phi}), \end{aligned} \tag{11.172}$$

which in all respects is identical to the uniform electromagnetic solution (11.143) derived in the previous section. Notice that this solution exists only for negative cosmological constant, $\Lambda = -1/l^2$; there is no extension to $\Lambda = 1/l^2$. It is evident that this solution can be generated from the static one, (11.147), via the transformations $t \rightarrow t - \omega_0\phi$, $\phi \rightarrow \phi$.

For $\omega_0 = 0$, the above mentioned metric and field reduce to the Matyjasek–Zaslavskii solutions – see Section (11.5.3) – thus this class of constant electromagnetic invariants' static solutions is unique with the additional property of being a uniform static solution.

11.6.4 Constant Electromagnetic Invariant $FF = 2/l^2$ Stationary Solution for $a \neq 0$

In the case of positive cosmological constant $\Lambda = 1/l^2$ there exists a constant electromagnetic invariant stationary $a \neq 0$ solution. Requiring the constancy of the electromagnetic invariant $FF = 2a^2\frac{H}{F}$, one gets

$$FF = 2a^2\frac{H}{F} \rightarrow H(r) = \beta^2 F(r). \tag{11.173}$$

From ${E_\phi}^t$ one establishes

$$\frac{d^2}{dr^2}W = -\frac{2}{F}\frac{dF}{dr}\frac{dW}{dr}, \tag{11.174}$$

which when used in ${E_t}^t$–${E_r}^r$ yields

$$\frac{F}{\beta^4}(\frac{dW}{dr})^2 = 0 \rightarrow W(r) = W_0 \tag{11.175}$$

Using $W = W_0$ and $H(r) = \beta^2 F(r)$ in the remaining Einstein equations, one gets

$$\frac{d^2}{dr^2}F(r) + \frac{4}{l^2} = 0 \rightarrow F(r) = -\frac{2}{l^2}r^2 + c_1\, r + c_0. \tag{11.176}$$

Therefore we have established that there is a unique constant electromagnetic invariants' solution given by

$$\begin{aligned} g &= -a^2 l^2 dt^2 + \frac{dr^2}{F(r)} + \frac{F(r)}{a^2 l^2}(d\phi + W_0 dt)^2, \\ F(r) &= -\frac{2r^2}{l^2} + c_1 r + c_0, \\ \boldsymbol{A} &= \frac{r}{a\, l^2}(\boldsymbol{d\phi} + W_0 \boldsymbol{dt}), \end{aligned} \tag{11.177}$$

which is identical to the uniform electromagnetic solution (11.159) derived in the previous Section 11.5 dealing with uniform electromagnetic solutions. Notice that this solution exists only for positive cosmological constant, $\Lambda = 1/l^2$, there is no $\Lambda = -1/l^2$ solution within this class.

11.6.5 Vanishing Electromagnetic Invariant $FF = 0$ Solution

A particular family of stationary cyclic symmetric solutions arises by demanding the vanishing of the electromagnetic invariant $F_{\mu\nu}F^{\mu\nu}$,

$$F_{\mu\nu}F^{\mu\nu} = 2\,\frac{H\,(a - b\,W)^2}{F} - 2\,\frac{b^2}{H} = 0 \rightarrow W(r) = \frac{a}{b} \pm \frac{\sqrt{F}}{H}. \tag{11.178}$$

For the above mentioned $W(r)$, the equation ${E_r}^r$ gives

$$l^2\left(\frac{d\,F}{dr}\right)^2 - 16\,F = 0 \rightarrow F(r) = 4\,\frac{(r - C)^2}{l^2}. \tag{11.179}$$

After the substitution of $W(r)$ and F(r) into the Einstein equations, the remaining equation to be solved amounts to

$$(r - C)^2\,\frac{d^2}{dr^2}H + b^2 l^2 = 0 \rightarrow H(r) = C_0 + C_1\,r + b^2 l^2 \ln{(r - C)}\,.$$

The gravitational and electromagnetic fields of this solution can be given as

$$
\begin{aligned}
g &= -\frac{F}{H}dt^2 + \frac{dr^2}{F} + H(d\phi + W dt)^2, \\
H(r) &= C_0 + C_1 r + b^2 l^2 \ln(r - C), \; F(r) = 4\frac{(r-C)^2}{l^2}, \\
W(r) &= \frac{a}{b} \pm \frac{\sqrt{F}}{H}, \; \boldsymbol{A} = \mp\frac{l}{2}\ln(r-C)\,(a\boldsymbol{dt} + b\boldsymbol{d\phi}). \qquad (11.180)
\end{aligned}
$$

This solution is characterized by the field tensor

$$
\begin{aligned}
F_{\mu\nu} &= \pm\frac{l}{(r-C)}\left(a\delta_{[\mu}{}^{t}\delta_{\nu]}{}^{r} + b\delta_{[\mu}{}^{\phi}\delta_{\nu]}{}^{r}\right), \\
F^{\mu\nu} &= 2b\delta_t{}^{[\mu}\delta_r{}^{\nu]} - 2a\delta_\phi{}^{[\mu}\delta_r{}^{\nu]}, \qquad (11.181)
\end{aligned}
$$

with energy–momentum tensor

$$
T_\mu{}^\nu = \frac{l}{4\pi(r-C)}\left(-a\,b\delta_\mu{}^t\delta_t{}^\nu + a\,b\delta_\mu{}^\phi\delta_\phi{}^\nu + a^2\delta_\mu{}^t\delta_\phi{}^\nu - b^2\delta_\mu{}^\phi\delta_t{}^\nu\right). \qquad (11.182)
$$

Notice that the three invariants $F_{\mu\nu}F^{\mu\nu}$, T^μ_μ and $T_{\mu\nu}T^{\mu\nu}$ are equal to zero. Without any loss of generality one can always set $C = 0$.

This solution corresponds to a possible representation of the Kamata and Koikawa (1995) solution to be treated in detail in Section 11.6.5. It should be pointed out that this solution does not belong to the family of uniform solutions, i.e., the fields possessing vanishing covariant derivatives.

11.6.6 Kamata–Koikawa Solution

Kamata and Koikawa (1995) reported their electrically charged BTZ black hole with negative cosmological constant such that the Maxwell field is self (anti-self) dual, a condition which is imposed on the orthonormal basis components of the electric field and the magnetic field. This solution describes an electrically charged extreme black hole with mass M, angular momentum J, and electric charge Q. To achieve their representation one accomplishes in metric (11.180) the substitutions

$$
\begin{aligned}
&r = \rho^2,\, t \to t\sqrt{Q}/2,\, \phi \to \phi/\sqrt{Q},\, C_1 \to Q, l \to |\Lambda|^{-1/2}\, C_0 \to -b^2/\Lambda\,\ln\rho_0^2, \\
&C = \rho_0^2,\, H/Q \to K^2, W\,Q/2 \to \frac{a}{b}\frac{Q}{2} + N^\phi,\, F \to 4\,\rho^2\,L^2, \qquad (11.183)
\end{aligned}
$$

arriving at the solution

$$
\begin{aligned}
g &= -\rho^2\frac{L^2}{K^2}dt^2 + \frac{d\rho^2}{L^2} + K^2[d\phi + (\frac{a}{b}\frac{Q}{2} + N^\phi)dt]^2, \\
L^2 &= |\Lambda|(\rho - \rho_0^2/\rho)^2,\, K^2 = \rho^2 + \frac{b^2}{Q\,\Lambda}\ln\left(\frac{\rho^2-\rho_0^2}{\rho_0^2}\right),\, N^\phi = \pm\frac{\rho\,L}{K^2}, \\
\boldsymbol{A} &= \frac{b}{2\sqrt{Q\,|\Lambda|}}\ln(\frac{\rho^2-\rho_0^2}{\rho_0^2})\left(\boldsymbol{d\phi} + \frac{a}{b}\frac{Q}{2}\boldsymbol{dt}\right).
\end{aligned}
$$

The electromagnetic field tensors are

$$T_\mu{}^\nu = \frac{1}{8\pi}\frac{1}{\sqrt{|\Lambda|}(\rho^2-\rho_0^2)}\Big[a\,b(\delta_\mu{}^t\delta_t{}^\nu - \delta_\mu{}^\phi\delta_\phi{}^\nu)$$
$$-\frac{Q\,a^2}{2}\delta_\mu{}^\phi\delta_t{}^\nu - \frac{2b^2}{Q}\delta_\mu{}^t\delta_\phi{}^\nu\Big],$$
$$F_{\mu\nu} = \frac{\rho}{\sqrt{|\Lambda|}(\rho^2-\rho_0^2)}\left(a\sqrt{Q}\delta_{[\mu}{}^t\delta_{\nu]}{}^r + 2\frac{b}{\sqrt{Q}}\delta_{[\mu}{}^\phi\delta_{\nu]}{}^r\right). \tag{11.184}$$

Next, one restores the factor $\pi\,G$ in the above mentioned solution through the identifications of the physical parameters:

$$\rho_0^2 = 4\pi\,G\;Q^2/|\Lambda| = \frac{\epsilon}{2|\Lambda|^{1/2}}J,\; b = 2\sqrt{\pi\,G}Q^{3/2},\; a = \pm\,4\sqrt{\pi\,G}|\Lambda|^{1/2}Q^{1/2},$$

arriving at the metric (11.184) with structural functions

$$L^2 = |\Lambda|(\rho-\rho_0^2/\rho)^2,\;\; N^\phi = \pm\frac{\rho\,L}{K^2},\; K^2 = \rho^2+\rho_0^2\,\ln\left(\frac{\rho^2-\rho_0^2}{\rho_0^2}\right) \tag{11.185}$$

and electromagnetic field tensors

$$\boldsymbol{A} = Q\sqrt{\pi\,|\Lambda|}\ln\big(\frac{\rho^2-\rho_0^2}{\rho_0^2}\big)\times\left(\frac{1}{\sqrt{|\Lambda|}}\boldsymbol{d\phi}+\boldsymbol{dt}\right),$$
$$F_{\mu\nu} = -4Q\sqrt{\pi\,G}\frac{\rho}{\rho^2-\rho_0^2}\left(\delta_{[\mu}{}^t\delta_{\nu]}{}^r + \frac{1}{\sqrt{|\Lambda|}}\delta_{[\mu}{}^\phi\delta_{\nu]}{}^r\right),$$
$$T_\mu{}^\nu = \frac{Q^2\,G}{\rho^2-\rho_0^2}\left(-\delta_\mu{}^t\delta_t{}^\nu + \delta_\mu{}^\phi\delta_\phi{}^\nu - \frac{\delta_\mu{}^\phi\delta_t{}^\nu}{\sqrt{|\Lambda|}} + \sqrt{|\Lambda|}\delta_\mu{}^t\delta_\phi{}^\nu\right). \tag{11.186}$$

It should be pointed out that Clément (1993) also reported a metric expression and electromagnetic vector field describing a solution with vanishing electromagnetic invariants. Comments concerning the mass content of this solution can be found in Chan (1996). This solution is horizonless and consequently does not permit a black hole interpretation.

The Kamata and Koikawa (1995, 1997) solution, see also García (2009), Eq. (7.7), is defined by the metric and the structural functions

$$ds^2 = -N(\rho)^2dt^2 + \frac{1}{L(\rho)^2}\,d\rho^2 + K(\rho)^2[d\phi + W(\rho)dt]^2,$$
$$L(\rho) = \frac{\sqrt{\Lambda}}{\rho}(\rho^2-\rho_0^2),\; \sqrt{\Lambda} = 1/l,$$
$$K(\rho) = \sqrt{\rho^2 + \frac{Q^2}{\Lambda}\ln\big(\frac{\rho^2}{\rho_0^2}-1\big)},$$

$$N(\rho) = \rho L/K = \sqrt{\Lambda}(\rho^2 - \rho_0^2)/\sqrt{\rho^2 + \frac{Q^2}{\Lambda}\ln\left(\frac{\rho^2}{\rho_0^2} - 1\right)},$$

$$W(\rho) = \frac{(\rho^2 - \rho_0^2)\sqrt{\Lambda}}{[\rho^2 + \frac{Q^2}{\Lambda}\ln(\frac{\rho^2}{\rho_0^2} - 1)]} - \sqrt{\Lambda}, \tag{11.187}$$

which are used in the evaluation of the energy and mass.

Mass, Energy, and Momentum

The surface energy and momentum densities are respectively given by

$$\epsilon(\rho, \epsilon_0) = -\frac{1}{\pi\sqrt{\Lambda}}\frac{Q^2 - \Lambda\rho_0^2 + \Lambda\rho^2}{[\rho^2 + \frac{Q^2}{\Lambda}\ln(\frac{\rho^2}{\rho_0^2} - 1)]} - \epsilon_0,$$

$$j(\rho) = \frac{1}{\pi\sqrt{\Lambda}}\frac{\Lambda\rho_0^2 - Q^2 + Q^2\ln(\frac{\rho^2}{\rho_0^2} - 1)}{\sqrt{\rho^2 + \frac{Q^2}{\Lambda}\ln(\frac{\rho^2}{\rho_0^2} - 1)}}, \tag{11.188}$$

while the integral quantities amount to

$$J(\rho) = 2\frac{\Lambda\rho_0^2 - Q^2}{\sqrt{\Lambda}} + 2\frac{Q^2}{\sqrt{\Lambda}}\ln\left(\frac{\rho^2}{\rho_0^2} - 1\right),$$

$$E(\rho, \epsilon_0) = -\frac{2}{\sqrt{\Lambda}}\frac{Q^2 - \Lambda\rho_0^2 + \Lambda\rho^2}{\sqrt{\rho^2 + \frac{Q^2}{\Lambda}\ln(\frac{\rho^2}{\rho_0^2} - 1)}} - 2\pi K\epsilon_0,$$

$$M(\rho, \epsilon_0) = -2\Lambda\rho^2 + 2(2\Lambda\rho_0^2 - Q^2) + 2Q^2\ln\left(\frac{\rho^2}{\rho_0^2} - 1\right) - 2\pi\sqrt{\Lambda}(\rho^2 - \rho_0^2)\epsilon_0. \tag{11.189}$$

The evaluation of the functions above for vanishing ϵ_0, i.e. $\epsilon_0 = 0$, behave at $\rho \to \infty$ according to

$$\epsilon(\rho \to \infty, \epsilon_0 = 0) \approx -\frac{\sqrt{\Lambda}}{\pi} + \frac{\Lambda\rho_0^2 - Q^2}{\pi\sqrt{\Lambda}\rho^2} + 2\frac{Q^2}{\pi\sqrt{\Lambda}\rho^2}\ln\left(\frac{\rho}{\rho_0}\right),$$

$$j(\rho \to \infty) \approx \frac{\Lambda\rho_0^2 - Q^2}{\pi\sqrt{\Lambda}\rho} + 2\frac{Q^2}{\pi\sqrt{\Lambda}\rho}\ln\left(\frac{\rho}{\rho_0}\right),$$

$$J(\rho \to \infty) \approx 2\frac{\Lambda\rho_0^2 - Q^2}{\sqrt{\Lambda}} + 4\frac{Q^2}{\sqrt{\Lambda}}\ln\left(\frac{\rho}{\rho_0}\right),$$

$$E(\rho \to \infty, \epsilon_0 = 0) \approx -2\sqrt{\Lambda}\rho + 2\frac{\Lambda\rho_0^2 - Q^2}{\sqrt{\Lambda}\rho} + 2\frac{Q^2}{\sqrt{\Lambda}\rho}\ln\left(\frac{\rho}{\rho_0}\right),$$

$$M(\rho \to \infty, \epsilon_0 = 0) \approx -2\Lambda\rho^2 + 2(2\Lambda\rho_0^2 - Q^2) + 4Q^2\ln\left(\frac{\rho}{\rho_0}\right). \tag{11.190}$$

Using in the expressions (11.189) as reference energy density the quantity

$$\epsilon_0 = -\frac{1}{\pi\rho}\sqrt{-M_0 + \frac{\rho^2}{l^2}},$$

which at the spatial infinity behaves as $\epsilon_{0|\infty}(M_0) \approx -\frac{\sqrt{\Lambda}}{\pi} + \frac{M_0}{2\pi\sqrt{\Lambda}\,\rho^2}$, the series expansions of the corresponding quantities at $\rho \to \infty$ result in

$$\begin{aligned}
\epsilon(\rho \to \infty, \epsilon_{0|\infty}(M_0)) &\approx -\frac{M_0}{2\pi\sqrt{\Lambda}\,\rho^2} + \frac{\Lambda\rho_0^2 - Q^2}{\pi\sqrt{\Lambda}\,\rho^2} + 2\frac{Q^2}{\pi\sqrt{\Lambda}\,\rho^2}\ln\left(\frac{\rho}{\rho_0}\right),\\
E(\rho \to \infty, \epsilon_{0|\infty}(M_0)) &\approx -\frac{M_0}{\sqrt{\Lambda}\,\rho} + 2\frac{\Lambda\rho_0^2 - Q^2}{\sqrt{\Lambda}\,\rho} + 4\frac{Q^2}{\sqrt{\Lambda}\,\rho}\ln\left(\frac{\rho}{\rho_0}\right),\\
M(\rho \to \infty, \epsilon_{0|\infty}(M_0)) &\approx -M_0 + 2(\Lambda\rho_0^2 - Q^2) + 4Q^2\ln\left(\frac{\rho}{\rho_0}\right). \qquad (11.191)
\end{aligned}$$

Therefore, comparing with the energy characteristics of the BTZ solution, one arrives at the conclusion that there is no mass parameter of the kind M present in the BTZ solution. All characteristic functions logarithmically diverge at spatial infinity.

Field, Energy–Momentum and Cotton Tensors

The electromagnetic field tensor

$$(F^{\alpha}{}_{\beta}) = \begin{bmatrix} 0 & -\frac{Q\rho\, q}{\Lambda(\rho^2-\rho_0{}^2)^2} & 0 \\ \frac{qQ\Lambda(\rho^2-\rho_0{}^2)}{\rho} & 0 & -\frac{qQ\sqrt{\Lambda}(\rho^2-\rho_0{}^2)}{\rho} \\ 0 & -\frac{Q\rho\, q}{\sqrt{\Lambda}(\rho^2-\rho_0{}^2)^2} & 0 \end{bmatrix}, \qquad (11.192)$$

allows for a triple zero eigenvalue and the following set of eigenvectors

$$\begin{aligned}
&\lambda_{1,2,3} = 0;\ \mathbf{V} = [V^1, V^2, \sqrt{\Lambda}\,V^1],\ V^{\mu}\,V_{\mu} = 0,\ \mathbf{V} = \mathbf{N},\\
&\text{Type N} : \{3N\}. \qquad (11.193)
\end{aligned}$$

The electromagnetic energy momentum tensor with vanishing invariants is

$$(T^{\alpha}{}_{\beta}) = \frac{1}{4\pi}\frac{Q^2}{\sqrt{\Lambda}\,(\rho^2 - \rho_0{}^2)}\begin{bmatrix} -\sqrt{\Lambda} & 0 & -\Lambda \\ 0 & 0 & 0 \\ 1 & 0 & \sqrt{\Lambda} \end{bmatrix}, \qquad (11.194)$$

while the Cotton tensor for this electromagnetic–gravitational stationary cyclic symmetric field is given by

$$(C^{\alpha}{}_{\beta}) = \frac{Q^2}{(\rho^2 - \rho_0{}^2)}\begin{bmatrix} -\sqrt{\Lambda} & 0 & 1 \\ 0 & 0 & 0 \\ -\Lambda & 0 & \sqrt{\Lambda} \end{bmatrix}. \qquad (11.195)$$

It is clear that both the Cotton and Maxwell tensors possess the same eigenvalues, namely the triple zero eigenvalue $\lambda = 0$. Searching for the eigenvectors of these tensors, one arrives at

$$\lambda_{1,2,3} = 0;\ \mathbf{V} = [V^1, V^2, \sqrt{\Lambda}\, V^1],\ V^\mu\, V_\mu = \frac{(V^2)^2 \rho^2}{\Lambda\,(\rho - \rho_0)^2},\ \mathbf{V} = \mathbf{S},$$
$$\mathbf{V}(V^2 = 0) = \mathbf{N},$$
$$\text{Type} : \{S, 2N\}. \tag{11.196}$$

The eigenvectors are spacelike or null vectors depending on the non-vanishing or vanishing value of the component V^2 . One may consider them different, one to another, having different V^1 and V^2 components. The most degenerate cases are $\{3S\}$ and $\{3N\}$.

11.6.7 Proper Kamata–Koikawa Solution, $\rho_0 = \pm Q/\sqrt{\Lambda}$

The proper Kamata–Koikawa solution is defined by the metric and structural functions of (11.187) for $\Lambda\rho_0^2 - Q^2 = 0$, i.e., $\rho_0 = \pm Q/\sqrt{\Lambda}$, namely

$$ds^2 = -N(\rho)^2 dt^2 + \frac{1}{L(\rho)^2}\, d\rho^2 + K(\rho)^2 [d\phi + W(\rho) dt]^2,$$
$$L(\rho) = \frac{\sqrt{\Lambda}}{\rho}(\rho^2 - \rho_0^2),$$
$$K(\rho) = \sqrt{\rho^2 + \frac{Q^2}{\Lambda}\ln\left(\frac{\rho^2}{\rho_0^2} - 1\right)},$$
$$N(\rho) = \rho\, L/\, K = \sqrt{\Lambda}(\rho^2 - \rho_0^2)/\sqrt{\rho^2 + \frac{Q^2}{\Lambda}\ln\left(\frac{\rho^2}{\rho_0^2} - 1\right)},$$
$$W(\rho) = \frac{(\rho^2 - \rho_0^2)\sqrt{\Lambda}}{[\rho^2 + \frac{Q^2}{\Lambda}\ln\left(\frac{\rho^2}{\rho_0^2} - 1\right)]} - \sqrt{\Lambda}, \tag{11.197}$$

The surface energy and momentum densities are respectively given by

$$\epsilon(\rho, \epsilon_0) = -\frac{\sqrt{\Lambda}}{\pi}\frac{\rho^2}{[\rho^2 + \frac{Q^2}{\Lambda}\ln\left(\frac{\rho^2}{\rho_0^2} - 1\right)]},$$
$$j(\rho) = \frac{1}{\pi\sqrt{\Lambda}}\frac{Q^2 \ln\left(\frac{\rho^2}{\rho_0^2} - 1\right)}{\sqrt{\rho^2 + \frac{Q^2}{\Lambda}\ln\left(\frac{\rho^2}{\rho_0^2} - 1\right)}}, \tag{11.198}$$

while the integral quantities amount to

$$J(\rho) = 2\,\frac{Q^2}{\sqrt{\Lambda}}\ln\left(\frac{\rho^2}{\rho_0^2} - 1\right),$$
$$E(\rho, \epsilon_0) = -2\sqrt{\Lambda}\,\frac{\rho^2}{\sqrt{\rho^2 + \frac{Q^2}{\Lambda}\ln\left(\frac{\rho^2}{\rho_0^2} - 1\right)}} - 2\pi\, K\epsilon_0,$$
$$M(\rho, \epsilon_0) = -2\Lambda\,\rho^2 + 2\Lambda\rho_0^2 + 2Q^2 \ln\left(\frac{\rho^2}{\rho_0^2} - 1\right) - 2\pi\,\sqrt{\Lambda}\,(\rho_0^2 - \rho_0^2)\epsilon_0. \tag{11.199}$$

The evaluation of the functions above for vanishing ϵ_0, i.e. $\epsilon_0 = 0$, behaves at spatial infinity according to

$$\begin{aligned}
\epsilon(\rho \to \infty, \epsilon_0 = 0) &\approx -\frac{\sqrt{\Lambda}}{\pi} + 2\frac{Q^2}{\pi\sqrt{\Lambda}\,\rho^2}\ln\left(\frac{\rho}{\rho_0}\right),\\
j(\rho \to \infty) &\approx 2\frac{Q^2}{\pi\sqrt{\Lambda}\,\rho}\ln\left(\frac{\rho}{\rho_0}\right),\\
J(\rho \to \infty) &\approx 4\frac{Q^2}{\sqrt{\Lambda}}\ln\left(\frac{\rho}{\rho_0}\right),\\
E(\rho \to \infty, \epsilon_0 = 0) &\approx -2\sqrt{\Lambda}\rho + 2\frac{Q^2}{\sqrt{\Lambda}\,\rho}\ln\left(\frac{\rho}{\rho_0}\right),\\
M(\rho \to \infty, \epsilon_0 = 0) &\approx -2\Lambda\,\rho^2 + 2\Lambda\rho_0^2 + 4Q^2\ln\left(\frac{\rho}{\rho_0}\right).
\end{aligned} \tag{11.200}$$

Using in the expressions (11.199) as reference energy density the quantity

$$\epsilon_{0|\infty}(M_0) \approx -\frac{\sqrt{\Lambda}}{\pi} + \frac{M_0}{2\pi\sqrt{\Lambda}\,\rho^2},$$

the series expansions of the corresponding quantities at $\rho =$ infinity result in

$$\begin{aligned}
\epsilon(\rho \to \infty, \epsilon_{0|\infty}(M_0)) &\approx -\frac{M_0}{2\pi\sqrt{\Lambda}\,\rho^2} + 2\frac{Q^2}{\pi\sqrt{\Lambda}\,\rho^2}\ln\left(\frac{\rho}{\rho_0}\right),\\
E(\rho \to \infty, \epsilon_{0|\infty}(M_0)) &\approx -\frac{M_0}{\sqrt{\Lambda}\,\rho} + 4\frac{Q^2}{\sqrt{\Lambda}\,\rho}\ln\left(\frac{\rho}{\rho_0}\right),\\
M(\rho \to \infty, \epsilon_{0|\infty}(M_0)) &\approx -M_0 + 4Q^2\ln\left(\frac{\rho}{\rho_0}\right).
\end{aligned} \tag{11.201}$$

In the work by Chan (1996) there are some comments addressed to the evaluation of the global momentum, energy and mass of the proper Kamata–Koikawa solution: the exact and the approximated expressions of the momentum J coincide with the corresponding ones given in (Chan, 1996, Eq. 9) and (Chan, 1996, Eq. 7). Moreover, the mass M at spatial infinity, (11.200), coincides with the M (Chan, 1996, Eq. 10.) for a zero background energy density with the correct extra term $-2\Lambda\,\rho^2$. From my point of view, it is to be recommended to accomplish series expansions of the quantities under consideration to determine how fast they approach zero or diverge at spatial infinity. From this perspective, the evaluation of the energy density ϵ and the global energy E yield to quantities different from zero at spatial infinity, although they both approach faster to zero as $\rho \to \infty$ than the momentum and mass. Comparing with the energy characteristics of the BTZ solution, one concludes that the mass, energy and momentum functions logarithmically diverge at spatial infinity.

11.7 Ayón–Cataldo–Garcia Stationary Hybrid Solution

The main goal of this section is to derive the stationary cyclic symmetric spacetime corresponding to the case $c \neq 0$, i.e., for the vector potential

$$\boldsymbol{A} = \frac{c}{2}(t\boldsymbol{d\phi} - \phi\boldsymbol{dt}) \rightarrow *\boldsymbol{F} = c\frac{g_{rr}}{\sqrt{-g}}\boldsymbol{dr}. \tag{11.202}$$

It is worthwhile to point out that this case has no analog in stationary axial symmetric spacetimes of the standard (3+1) Einstein–Maxwell theory. The set of field equations is given by: $\{E_t{}^t, E_t{}^\phi, E_r{}^r, E_\phi{}^t, E_\phi{}^\phi\}$. In the forthcoming subsections two main families of solutions exhibiting the hybrid feature of the vector potential are derived.

The starting point in the integration process of the system of field equations is $E_\phi{}^t(a = 0 = b) = 0$, (11.26g), which possesses a first integral of the form

$$W_{,r} = \frac{J}{H^2}, \tag{11.203}$$

where J is an integration constant. The combination $E_t{}^t + 2E_r{}^r + E_\phi{}^\phi$, for $a = 0 = b$, yields

$$F_{,r,r} - \frac{8}{l^2} = 0, \tag{11.204}$$

which possesses the general solution

$$F = \frac{4}{l^2}(r - r_1)(r - r_2), \tag{11.205}$$

where r_1 and r_2 are constant of integration. Next, using $W_{,r}$ from (11.203) in $E_r{}^r(a = 0 = b)$, (11.26e), one arrives at

$$\frac{1}{4}\left(\frac{H_{,r}}{H} - \frac{1}{2}\frac{F_{,r}}{F}\right)^2 - \frac{J^2}{4H^2 F} = \frac{F_{,r}{}^2}{16F^2} - \frac{c^2}{F^2} - \frac{1}{l^2 F}. \tag{11.206}$$

The evaluation the right-hand side of this equation gives the same result as in the static case, thus one gets

$$\left[\frac{d}{dr}\ln\left(\frac{H}{F^{1/2}}\right)\right]^2 - \frac{J^2}{H^2 F} = 4\frac{(r_2 - r_1)^2\alpha}{l^4 F^2}, \tag{11.207}$$

where α is defined through

$$c^2 = \frac{(r_2 - r_1)^2}{l^4}(1 - \alpha).$$

From the above equation it becomes apparent that H can be sought in the form of

$$H(r) = h(r)\sqrt{F(r)}. \tag{11.208}$$

Replacing $H(r)$ from above into (11.207) one obtains an equation for $h(r)$ which can be given as

$$\frac{dh}{\sqrt{\alpha_0{}^2 h^2 + J^2}} = \mp\frac{dr}{F}, \; \alpha_0 := 2(r_2 - r_1)\sqrt{\alpha}/l^2, \tag{11.209}$$

with integral

$$\ln\left(\alpha_0\, h+\sqrt{\alpha_0{}^2\, h^2+J^2}\right)=\ln\left[k_1\left(\frac{r-r_1}{r-r_2}\right)^{\pm\sqrt{\alpha}/2}\right], \qquad (11.210)$$

where k_1 is an integration constant. Therefore $h(r)$ can be expressed as

$$h(r)=\frac{l^2\, k_1}{4(r_2-r_1)\sqrt{\alpha}}\left[\left(\frac{r-r_1}{r-r_2}\right)^{\pm\sqrt{\alpha}/2}-\frac{J^2}{k_1^2}\left(\frac{r-r_1}{r-r_2}\right)^{\mp\sqrt{\alpha}/2}\right]. \qquad (11.211)$$

The integration of the (11.203) for W does not present problem.

Summarizing the derived above results, one has that this family of solutions can be given by

$$\begin{aligned}
g &= -\frac{F}{H}dt^2+\frac{dr^2}{F}+H(d\phi+W dt)^2,\\
F &= \frac{4}{l^2}(r-r_1)(r-r_2),\\
H(r) &= l\frac{\sqrt{(r-r_1)(r-r_2)}}{2K_1(r_2-r_1)\sqrt{\alpha}}\times\left[\left(\frac{r-r_1}{r-r_2}\right)^{\pm\frac{\sqrt{\alpha}}{2}}-K_1^2\, J^2\left(\frac{r-r_1}{r-r_2}\right)^{\mp\frac{\sqrt{\alpha}}{2}}\right],\\
W(r) &= W_0\pm\frac{4}{l^2}J\, K_1^2\sqrt{\alpha}(r_2-r_1)\times\left[\left(\frac{r-r_1}{r-r_2}\right)^{\pm\sqrt{\alpha}}-K_1^2\, J^2\right]^{-1},\\
\boldsymbol{A} &= \frac{c}{2}(t\boldsymbol{d\phi}-\phi\boldsymbol{dt}),
\end{aligned} \qquad (11.212)$$

where the constant K_1 stands for $1/k_1$, $K_1=1/k_1$, and W_0 is an integration constant. Recall that the parameter α is related to c, r_1 and r_2 through $c^2=\frac{(r_2-r_1)^2}{l^4}(1-\alpha)$.

Correspondingly, the electromagnetic field tensors are

$$\begin{aligned}
F_{\mu\nu} &= 2c\delta_{[\mu}{}^{t}\delta_{\nu]}{}^{\phi},\\
T_{\mu}{}^{\nu} &= \frac{c^2}{8\pi\, F}(-\delta_{\mu}{}^{t}\delta_{t}{}^{\nu}+\delta_{\mu}{}^{r}\delta_{r}{}^{\nu}-\delta_{\mu}{}^{\phi}\delta_{\phi}{}^{\nu}),
\end{aligned} \qquad (11.213)$$

with invariants

$$F_{\mu\nu}F^{\mu\nu}=-2\frac{c^2}{F},\ T_{\mu\nu}T^{\mu\nu}=\frac{3}{64}\frac{c^4}{\pi^2F^2},\ T^{\mu}_{\mu}=-\frac{1}{8}\frac{c^2}{\pi\, F}.$$

This solution has been reported, for the first time to our knowledge, in Ayón–Beato et al. (2005). The static hybrid solution (11.112) arises from the stationary one above by setting $J=0=W_0$ and identifying $4(r_2-r_1)\sqrt{\alpha}\, K_0^2=l^2/K_1$.

11.7.1 ACG Hybrid Solution Allowing for BTZ Limit

To achieve a representation of this hybrid solution in terms of the radial coordinate ρ, such that at the limit of vanishing electromagnetic parameter

$c = 0 \to \alpha = 1$, the stationary BTZ solution would arise, one has to accomplish the coordinate transformations

$$t = \frac{l}{4K_1}\frac{1-K_1^2J^2}{r_2-r_1}T, \tag{11.214a}$$

$$\phi = \Phi - \left(W_0\frac{l}{4K_1}\frac{1-K_1^2J^2}{r_2-r_1} + \frac{K_1}{l}J\right)T, \tag{11.214b}$$

$$r = \frac{1}{1-K_1^2J^2}\left(r_1 - r_2K_1^2J^2 - 2\frac{K_1}{l}(r_2-r_1)\rho^2\right), \tag{11.214c}$$

where with $\{T,\rho,\Phi\}$ are denoted the corresponding BTZ coordinates, which ought to be accompanied with the identification

$$J^2K_1 = -R_{(-)} := -\left(M\,l - \sqrt{M^2l^2-J^2}\right),\; M = -\frac{1+K_1^2J^2}{2\,l\,K_1}. \tag{11.215}$$

In this way this solution can be given in the standard representation as

$$\begin{aligned}
g &= -\frac{\rho^2 f(\rho)}{H(\rho)}dT^2 + \frac{d\rho^2}{f(\rho)} + H(\rho)\left[d\Phi + W(\rho)dT\right]^2,\\
f(\rho) &= \frac{\rho^2}{l^2} - M + \frac{J^2}{4\rho^2},\; R_\pm := M\,l \pm \sqrt{M^2l^2-J^2},\\
H(\rho) &= \frac{\sqrt{2\rho^2-lR_-}\sqrt{2\rho^2-lR_+}}{4\sqrt{\alpha}K_1\sqrt{M^2l^2-J^2}}\left[-(2\rho^2-lR_-)^{\sqrt{\alpha}/2}(2\rho^2-lR_+)^{-\sqrt{\alpha}/2}\right.\\
&\quad\left.+J^2K_1^2(2\rho^2-lR_-)^{-\sqrt{\alpha}/2}(2\rho^2-lR_+)^{\sqrt{\alpha}/2}\right],\\
W(\rho) &= \frac{R_-}{Jl}\left[(2\rho^2-lR_+)^{\sqrt{\alpha}}(2\sqrt{\alpha}\sqrt{M^2l^2-J^2}+R_-)R_-\right.\\
&\quad\left.-(2\rho^2-lR_-)^{\sqrt{\alpha}}J^2\right]\times\left[(2\rho^2-lR_+)^{\sqrt{\alpha}}R_-^2 - (2\rho^2-lR_-)^{\sqrt{\alpha}}J^2\right]^{-1}.
\end{aligned} \tag{11.216}$$

When the electromagnetic field is turned off, $c = 0 \to \alpha = 1$, the above metric components reduce to

$$g_{TT} = M - \frac{\rho^2}{l^2},\, g_{T\Phi} = \frac{J}{2},\, g_{\Phi\Phi} = \rho^2,\, g_{\rho\rho} = \left(\frac{\rho^2}{l^2} - M + \frac{J^2}{4\rho^2}\right)^{-1}, \tag{11.217}$$

which correspond to the BTZ ones.

This solution possesses mass M, angular momentum J, electromagnetic parameter α, and negative cosmological constant, and describes a black hole.

The metric defining this kind of stationary electromagnetic solution, see Ayón-Beato et al. (2005) and García (2009), can be given in the standard form as

$$g = -N^2 dt^2 + \frac{1}{L^2} d\rho^2 + K^2[d\phi + W dt]^2 = -\frac{\rho^2 f}{H} dt^2 + \frac{d\rho^2}{f}$$
$$+H(d\phi + W dt)^2, \; f(\rho) = \frac{\rho^2}{l^2} - M + \frac{J^2}{4\rho^2},$$
$$H(\rho) = \frac{\sqrt{2\rho^2 - lR_-}\sqrt{2\rho^2 - lR_+}}{4K_1\sqrt{\alpha}S_M}\left[-(2\rho^2 - lR_-)^{\sqrt{\alpha}/2}(2\rho^2 - lR_+)^{-\sqrt{\alpha}/2}\right.$$
$$\left.+J^2 K_1^2 (2\rho^2 - lR_-)^{-\sqrt{\alpha}/2}(2\rho^2 - lR_+)^{\sqrt{\alpha}/2}\right],$$
$$W(\rho) = -\frac{R_-}{J\,l}\left[(2\rho^2 - lR_+)^{\sqrt{\alpha}}(2\sqrt{\alpha}S_M + R_-)R_-\right.$$
$$\left.-(2\rho^2 - lR_-)^{\sqrt{\alpha}} J^2\right] \times \left[(2\rho^2 - lR_+)^{\sqrt{\alpha}} R_-^2 - (2\rho^2 - lR_-)^{\sqrt{\alpha}} J^2\right]^{-1},$$
$$K_1 := -\frac{R_-}{J^2}, \; S_M := \sqrt{M^2\, l^2 - J^2}. \tag{11.218}$$

The structural functions appearing in the definitions of the energy and momentum quantities are expressed as

$$N(\rho) = \sqrt{\frac{\rho^2 f(\rho)}{H(\rho)}}, \; L(\rho) = \sqrt{f(\rho)}, \; K(\rho) = \sqrt{H(\rho)}, \; W(\rho) = W(\rho). \tag{11.219}$$

The corresponding electromagnetic tensors are given as

$$F_{\mu\,\nu} = -\frac{\sqrt{1-\alpha}}{l}\sqrt{M^2\,l^2 - J^2}\delta_{[\mu}{}^{t}\delta_{\nu]}{}^{\phi},$$
$$8\,\pi\,T_\mu{}^\nu = \frac{\alpha - 1}{4\,l^2}\frac{M^2\,l^2 - J^2}{\rho^2\, f(\rho)}[\delta_\mu{}^T\delta_T{}^\nu - \delta_\mu{}^\rho\delta_\rho{}^\nu + \delta_\mu{}^\Phi\delta_\Phi{}^\nu]. \tag{11.220}$$

When the electromagnetic field is turned off, $\alpha = 1$, the above metric components reduce to

$$g_{TT} = M - \frac{\rho^2}{l^2}, g_{T\Phi} = -\frac{J}{2}, g_{\Phi\Phi} = \rho^2, g_{\rho\rho} = \left(\frac{\rho^2}{l^2} - M + \frac{J^2}{4\rho^2}\right)^{-1},$$

which correspond to the BTZ ones.

11.7.2 Mass, Energy, and Momentum

In terms of the structural metric functions the momentum quantities they allow for very simple expressions

$$\jmath(\rho) = \frac{J}{2\pi}\frac{1}{\sqrt{H(\rho)}}, \; J(\rho) = J, \tag{11.221}$$

while the energy and mass characteristics become

$$\epsilon(\rho, \epsilon_0) = -\frac{1}{2\pi}\frac{\sqrt{f(\rho)}}{H(\rho)}\frac{d}{d\rho}H(\rho) - \epsilon_0, \tag{11.222}$$

$$E(\rho, \epsilon_0) = -\frac{\sqrt{f(\rho)}}{\sqrt{H(\rho)}} \frac{d}{d\rho} H(\rho) - 2\pi\epsilon_0 \sqrt{H(\rho)}, \tag{11.223}$$

$$M(\rho, \epsilon_0) = -\rho \frac{f(\rho)}{H(\rho)} \frac{d}{d\rho} H(\rho) - J\, W(\rho) - 2\rho\pi\epsilon_0 \sqrt{f(\rho)}, \tag{11.224}$$

Because of the involved dependence of the metric functions upon the ρ coordinate, the evaluation of the energy quantities will be done in the approximation of the spatial infinity. The momentum density at infinity becomes

$$j(\rho \to \infty) \approx \frac{J}{2\pi} \frac{\alpha^{1/4}}{\rho}, \tag{11.225}$$

while the global momentum remains constant in the whole space

$$J(\rho) = J. \tag{11.226}$$

It becomes apparent then that the role of the momentum parameter is played and coincides with J.

The approximated at $\rho \to \infty$ surface energy density, global energy and mass, for zero base energy density ϵ_0 are given by

$$\begin{aligned}
\epsilon(\rho \to \infty, \epsilon_0 = 0) &\approx -\frac{1}{\pi\, l} + \frac{l\, M\sqrt{\alpha}}{2\pi\, \rho^2}, \\
E(\rho \to \infty, \epsilon_0 = 0) &\approx -2\frac{\rho}{l\, \alpha^{1/4}} + \frac{l}{2\, \rho\alpha^{1/4}}(1 + \sqrt{\alpha})M, \\
M(\rho \to \infty, \epsilon_0 = 0) &\approx -2\frac{\rho^2}{l^2} + 2M + \frac{\sqrt{\alpha} - 1}{l}\sqrt{M^2\, l^2 - J^2}.
\end{aligned} \tag{11.227}$$

Using in the expressions (11.222)–(11.224) as reference energy density the energy corresponding to the anti-de Sitter metric, $\epsilon_0 = -\frac{1}{\pi\rho}\sqrt{\frac{\rho^2}{l^2} - M_0}$, $\epsilon_{0|\infty}(M_0) \approx -\frac{1}{\pi\, l} + \frac{l\, M_0}{2\pi\, \rho^2}$, the series expansions of the global energy and mass quantities at $\rho \to \infty$ result in

$$\begin{aligned}
\epsilon(\rho \to \infty, \epsilon_{0|\infty}(M_0)) &\approx \frac{l}{2\pi\, \rho^2}(\sqrt{\alpha}M - M_0), \\
E(\rho \to \infty, \epsilon_{0|\infty}(M_0)) &\approx l\, \frac{(\sqrt{\alpha}M - M_0)}{\rho\alpha^{1/4}}, \\
M(\rho \to \infty, \epsilon_{0|\infty}(M_0)) &\approx M - M_0 + \frac{\sqrt{\alpha} - 1}{l}\sqrt{M^2\, l^2 - J^2}.
\end{aligned} \tag{11.228}$$

Comparing with the energy characteristics of the BTZ solution, the mass parameter is an involved quantity depending on M, the momentum J, and the charge α, namely $M + \frac{\sqrt{\alpha}-1}{l}\sqrt{M^2\, l^2 - J^2}$, although the mass function is finite at spatial infinity. On the other hand, if $M\, l >> J$ then $\sqrt{\alpha}M$ becomes the mass parameter. The energy density and global energy are proportional at infinity to $1/\rho^2$ and $1/\rho$ correspondingly.

Field, Energy–Momentum, and Cotton Tensors

The coordinate system $\{t, r, \phi\}$ is more adequate in the derivation of the eigenvalue–vector characteristics of the considered solution of García (2009), Eq.(8.11). The metric in $\{t, r, \phi\}$-coordinates is given by

$$g = \begin{bmatrix} -F/H + H\,W^2 & 0 & HW \\ 0 & 1/F & 0 \\ HW & 0 & H \end{bmatrix}, \tag{11.229}$$

with structural functions

$$\begin{aligned} F(r) &= 4\,\frac{(r-r1)\,(r-r2)}{l^2}, \\ H(r) &= \frac{l}{2}\,\frac{\sqrt{(r-r1)\,(r-r2)}}{K1\,\sqrt{\alpha}\,(r2-r1)}\left[\left(\frac{r-r1}{r-r2}\right)^{1/2\,\sqrt{\alpha}}\right. \\ &\quad \left. -J^2\,K1^2\left(\frac{r-r1}{r-r2}\right)^{-1/2\,\sqrt{\alpha}}\right], \\ W(r) &= W0 - 2\,K1\,J\left(\frac{r-r1}{r-r2}\right)^{-1/2\,\sqrt{\alpha}}\frac{\sqrt{(r-r1)\,(r-r2)}}{l\,H(r)}. \end{aligned} \tag{11.230}$$

The electromagnetic field tensor is given by

$$(F^{\alpha}{}_{\beta}) = \begin{bmatrix} -c\frac{H\,W}{F} & 0 & -c\frac{H}{F} \\ 0 & 0 & 0 \\ -c\frac{(F-H^2W^2)}{HF} & 0 & c\frac{HW}{F} \end{bmatrix},\ c = \frac{(r2-r1)\,\sqrt{1-\alpha}}{l^2}. \tag{11.231}$$

In the search of its eigenvectors, one arrives at

$$\begin{aligned} &\lambda_1 = 0; \mathbf{V1} = [0, V^2, 0], V_\mu V^\mu = \frac{{V^2}^2}{F},\ \mathbf{V1} = \mathbf{S1}, \\ &\lambda_2 = \frac{c}{\sqrt{F}}; \\ &\mathbf{V2} = [V^1, 0, -\frac{\left(HW + \sqrt{F}\right)V^1}{H}],\ V^\mu V_\mu = 0,\ \mathbf{V2} = \mathbf{N2}, \\ &\lambda_3 = -\frac{c}{\sqrt{F}}; \\ &\mathbf{V3} = [V^1, 0, \frac{\left(-HW + \sqrt{F}\right)V^1}{H}],\ V^\mu V_\mu = 0,\ \mathbf{V3} = \mathbf{N3}, \end{aligned} \tag{11.232}$$

hence this tensor is of the type $\{S, N, N\}$.

For the energy–momentum tensor

$$(T^{\alpha}{}_{\beta}) = \begin{bmatrix} -\frac{1}{8\pi}\frac{c^2}{F} & 0 & 0 \\ 0 & \frac{1}{8\pi}\frac{c^2}{F} & 0 \\ 0 & 0 & -\frac{1}{8\pi}\frac{c^2}{F} \end{bmatrix}, \tag{11.233}$$

the eigenvectors are

$$\begin{aligned} \lambda_1 &= \frac{1}{8\pi}\frac{c^2}{F}; \mathbf{V1} = [0, V^2, 0], V_\mu V^\mu = \frac{{V^2}^2}{F}, \mathbf{V1} = \mathbf{S1}, \\ \lambda_{2,3} &= -\frac{1}{8\pi}\frac{c^2}{F}; \mathbf{V2,3} = [V^1, 0, V^3], \\ V_\mu V^\mu &= -\frac{{V^1}^2 F}{H} + \left(V^1 W + V^3\right)^2 H = \frac{{V^1}^2 F\left(Z^2 - 1\right)}{H}, \\ \mathbf{V2} &= \{\mathbf{T2}, \mathbf{N2}, \mathbf{S2}\}, \ \ \mathbf{V3} = \{\mathbf{T3}, \mathbf{N3}, \mathbf{S3}\}, \end{aligned} \tag{11.234}$$

hence it allows for the types: $\{S, 2T\}, \{S, 2N\}, \{S, 2S\}$.

The Cotton tensor is determined through

$$(C^{\alpha}{}_{\beta}) = \begin{bmatrix} C^1{}_1 & 0 & C^1{}_3 \\ 0 & 0 & 0 \\ C^3{}_1 & 0 & -C^1{}_1 \end{bmatrix},$$

$$\begin{aligned} C^1{}_1 &= -C^3{}_3 = -\frac{c^2}{32\pi F^2}\left(W\,\mathcal{Q} + 2\,\frac{FJ}{H}\right), \\ C^1{}_3 &= -\frac{c^2}{32\pi F^2}\,\mathcal{Q}, \\ C^3{}_1 &= -\frac{c^2}{32\pi F^2 H^2}\left[-WH^2\left(W\mathcal{Q} + 4\,\frac{FJ}{H}\right) - F\mathcal{Q}\right], \\ \mathcal{Q} &:= 2\,F\frac{d}{dr}\,H - H\frac{d}{dr}\,F; \\ \mathcal{Q} &= -\frac{2}{lK1}\sqrt{(r - r1)\,(r - r2)}\left[\left(\frac{r - r1}{r - r2}\right)^{1/2\,\sqrt{\alpha}}\right. \\ &\left. + J^2 K1^2\left(\frac{r - r1}{r - r2}\right)^{-1/2\,\sqrt{\alpha}}\right], \end{aligned} \tag{11.235}$$

possesses the following set of eigenvectors

$$\begin{aligned} &\lambda_1 = 0;.\mathbf{V1} = [0, V^2, 0], \ V_\mu V^\mu = {V^2}^2/F, \ \mathbf{V1} = \mathbf{S1}, \\ &\lambda_2 = \frac{1}{4}i\,\frac{\sqrt{\alpha}c^2\,(r2 - r1)}{\pi\,F^{3/2}l^2}; \\ &\mathbf{Z} = [-c^2V^3\,\mathcal{Q}\left(c^2W\,\mathcal{Q} + 2\,\frac{c^2J\,F}{H} + 32\,\lambda_2\,F^2\pi\right)^{-1}, 0, V^3], \end{aligned}$$

$$\lambda_3 = -\frac{1}{4} i\, \frac{\sqrt{\alpha} c^2\,(r2 - r1)}{\pi\, F^{3/2} l^2};$$
$$\bar{\mathbf{Z}} = [-c^2 V^3\, \mathcal{Q}\left(c^2 W\, Q + 2\,\frac{c^2 J\, F}{H} + 32\,\lambda_3\, F^2 \pi\right)^{-1}, 0, V^3]. \quad (11.236)$$

therefore its type is Type I: $\{S, Z, \bar{Z}\}$.

11.7.3 Constant Electromagnetic Invariants' Hybrid Solution for $\Lambda = 0$

This section is devoted to the studied of the hybrid electromagnetic stationary solution with constant electromagnetic invariant FF and by virtue of the field structure, constant T and TT. The constant character of $FF = -\frac{2c^2}{F(r)}$ is achieved by requiring $F(r) = F_0$, and consequently all electromagnetic invariants equal to constants

$$FF = -\frac{2c^2}{F_0},\; TT = \frac{3}{64}\frac{c^4}{\pi^2 F_0^2},\; T_\mu{}^\mu = -\frac{c^2}{8\pi\, F_0}. \quad (11.237)$$

Again the integration of the Einstein equations start from $E_\phi{}^t(a = 0 = b) = 0$, which gives the relation

$$\frac{d}{dr} W(r) = \frac{J}{H(r)^2} \quad (11.238)$$

for the integration of the function $H(r)$ the substitution of $\frac{d}{dr}W(r)$ and $F(r) = F_0$ into the remaining Einstein equations requires the cosmological constant to vanish, $\Lambda = 0$. Under such conditions, the equation for $H(r)$ becomes

$$F_0 \left(\frac{d\, H(r)}{dr}\right)^2 - J^2 + 4\, c^2\, F_0 H(r)^2 = 0$$

with solution

$$H(r) = \epsilon_H\, \frac{J}{2\,c}\, \sqrt{F_0}\, \sin\frac{2\,c}{F_0}(r - C_0),\; \epsilon_H = \pm 1, \quad (11.239)$$

which, used in (11.238), after integration yields

$$W(r) = W_0 + \epsilon_W\, \frac{2\,c}{J}\, \cot\frac{2\,c}{F_0}(r - C_0),\; \epsilon_W = \pm 1, \quad (11.240)$$

where ϵ_H and ϵ_W assume their signs independently; one has to take care on the ranges of the variable r to guarantee a correct signature. Moreover, notice that the integration constant C_0 can be always equated to zero. Therefore the corresponding metric and electromagnetic field vector amount to

$$g = -\frac{F_0}{H} dt^2 + \frac{dr^2}{F_0} + H(d\phi + W dt)^2,$$
$$H = \epsilon_H\, \frac{J}{2\,c}\, \sqrt{F_0}\, \sin\frac{2\,c}{F_0}\, r,\; W = W_0 + \epsilon_W\, \frac{2\,c}{J}\, \cot\frac{2\,c}{F_0}\, r,$$
$$\mathbf{A} = \frac{c}{2}(\boldsymbol{t d\phi} - \boldsymbol{\phi dt}). \quad (11.241)$$

The electromagnetic field tensors are

$$
\begin{aligned}
F_{\mu\nu} &= 2c\delta_{\mu}{}^{[t}\delta_{\nu}{}^{\phi]},\\
T_{\mu}{}^{\nu} &= \frac{c^2}{8\pi\, F_0}[-\delta_{\mu}{}^{t}\delta_{t}{}^{\nu} + \delta_{\mu}{}^{r}\delta_{r}{}^{\nu} - \delta_{\mu}{}^{\phi}\delta_{\phi}{}^{\nu}].
\end{aligned} \tag{11.242}
$$

By means of scaling transformations F_0 can always be set equal to unit, $F_0 = 1$, hence this solution is endowed with two effective parameters c and J.

11.8 Stationary Solutions for $a \neq 0$ or $b \neq 0$

This section deals with the search of stationary solutions for the branches where one of the electromagnetic constants is zero, $a \neq 0 = b$ or $b \neq 0 = a$. For these families, the integration problem reduces to finding the solution of a master fourth-order (reducible to a third-order) nonlinear equation for $F(r)$, and to fitting a differential constraint on the found structural functions $F(r)$ and $H(r)$ ($\mathcal{H}(r)$). The integration of $W(r)$ ($\mathcal{W}(r)$) is trivial.

11.8.1 Stationary Magneto-Electric Solution for $a \neq 0 = b$

If the structural function $W(r)$ is different from a constant (the constant case will be treated at the end of this paragraph) then $E_{\phi}{}^{t}$ reads

$$
E_t{}^{\phi} = \frac{d}{dr}\left(H^2\frac{d}{dr}W\right) = 0, \tag{11.243}
$$

which yields

$$
\frac{d}{dr}W = \frac{J}{H^2}. \tag{11.244}
$$

The remaining independent Einstein–Maxwell equations arise respectively from combinations $(4\,E_r{}^{r} + 2E_t{}^{t} + 2E_{\phi}{}^{\phi})$, $(-2H(E_r{}^{r} - E_t{}^{t} + W\,E_t{}^{\phi})/F)$, and $E_r{}^{r}$:

$$
\begin{aligned}
EQ_F &= \frac{d^2\,F}{d\,r^2} - 4\,a^2\frac{H}{F} - 8\frac{1}{l^2} = 0,\\
EQ_H &= \frac{d^2}{dr^2}H + 4a^2\frac{H^2}{F^2} = 0,\\
E_r{}^{r} &= \frac{1}{4H}\frac{d\,H}{d\,r}\frac{d\,F}{d\,r} - \frac{F}{4H^2}\left(\frac{d\,H}{d\,r}\right)^2 + \frac{J^2}{4\,H^2} - a^2\frac{H}{F} - \frac{1}{l^2} = 0.
\end{aligned} \tag{11.245}
$$

The equation $E_r{}^{r}$ can be written in the form

$$
\left(\frac{1}{2H}\frac{dH}{dr} - \frac{1}{4F}\frac{dF}{dr}\right)^2 - \frac{1}{16\,F^2}\left(\frac{dF}{dr}\right)^2 + \frac{a^2H}{F^2} - \frac{1}{4}\frac{J^2}{FH^2} + \frac{1}{l^2F} = 0. \tag{11.246}
$$

On the other hand, using EQ_F one expresses H in terms of F and its derivative

$$H(r) = \frac{1}{4\,a^2}\left(\frac{d^2F}{dr^2} - \frac{8}{l^2}\right) F. \tag{11.247}$$

Substituting the above $H(r)$ into EQ_H (11.245) one gets

$$F\,\frac{d^4F}{dr^4} + 2\,\frac{d^3F}{dr^3}\,\frac{dF}{dr} + 2\left(\frac{d^2F}{dr^2}\right)^2 - \frac{24}{l^2}\,\frac{d^2F}{dr^2} + \frac{64}{l^4} = 0. \tag{11.248}$$

Therefore, integrating, if possible, (11.248) for $F(r)$, substituting the solution $F(r)$ into (11.247) one determines $H(r)$. The resulting functions $F(r)$ and $H(r)$ ought to fulfil the (11.246) or ${E_r}^r$ equation from (11.245). By integrating the linear first-order (11.244) one determines $W(r)$.

The contravariant components of electromagnetic tensor are

$$F^{\mu\nu} = -2\,a\,\delta^{\mu}{}_{[\phi}\delta^{\nu}{}_{r]}. \tag{11.249}$$

The equation (11.248) for $F(r)$ can be reduced to a third-order nonlinear equation. The problem for deriving solutions in this branch actually resides in this equation.

Another possibility arises with the introduction of the auxiliary function $h(r)$ by means of

$$H(r) = F(r)^{1/2}\,h(r), \tag{11.250}$$

the (11.246) acquires the form

$$\begin{aligned} EQ_h = {} & -l^2\,h^2\left(\frac{dF}{dr}\right)^2 + 4\,l^2 F^2\left(\frac{dh}{dr}\right)^2 \\ & +16\,l^2\,a^2\,h^3\sqrt{F} - 4\,l^2\,J^2 + 16\,h^2\,F = 0 \end{aligned} \tag{11.251}$$

and one could try to determine solutions for this variant.

"Stationary" Magneto-Electric $\boldsymbol{A} = A(r)(\boldsymbol{d\phi} - J_0\boldsymbol{dt})$ Solution

A particular solution to (11.248) is given by $F(r)$ from (11.76), namely

$$F(r) = \frac{4\,h(r)}{{C_1}^2\,l^2}\left[K_0 + h(r) + a^2 l^2 \ln h(r)\right], \quad h(r) := C_1\,r + C_0 \tag{11.252}$$

which, being substituted into (11.247), leads to

$$H(r) = \frac{4}{{C_1}^2\,l^2}\left[K_0 + h(r) + a^2 l^2 \ln h(r)\right]. \tag{11.253}$$

Entering with these particular solutions $F(r)$ and $H(r)$ in the constraint (11.246) one arrives at

$$W(r) = -J_0 = \text{constant}. \tag{11.254}$$

Summarizing, this solution is given by the same structural functions (11.76) of the magnetostatic solution except that in the present case the function $W(r)$ is a constant. The corresponding metric line element and field vector can be written as

$$g = -h(r)dt^2 + \frac{dr^2}{H(r)\,h(r)} + H(r)(d\phi - J_0 d\,t)^2,$$
$$\boldsymbol{A} = \frac{a}{C_1}\ln h(r)(\boldsymbol{d\phi} - J_0\boldsymbol{dt}). \tag{11.255}$$

The electromagnetic field tensors and their invariants are given by

$$F^{\mu\nu} = 2\,a\,\delta^{\mu}{}_{[r}\delta^{\nu}{}_{\phi]},\; FF = 2\frac{a^2}{h}, \tag{11.256}$$

and

$$T_{\nu}{}^{\mu} = \frac{a^2}{8\pi\,h}\left[-\delta^t_{\nu}\delta^{\mu}_t + \delta^r_{\nu}\delta^{\mu}_r + \delta^{\phi}_{\nu}\delta^{\mu}_{\phi} - 2\,J_0\delta^t_{\nu}\delta^{\mu}_{\phi}\right],\; T = \frac{3}{64\pi^2}\frac{a^4}{h^2}. \tag{11.257}$$

Because of the structure of the energy–momentum tensor above, this solution can be interpreted as a rigidly rotating perfect fluid

$$T_{\mu}\nu = (\rho + p)u_{\mu}u_{\nu} + p\,g_{\mu\nu},\; u^{\mu} = \frac{1}{\sqrt{F/H}}(\delta^{\mu}{}_t + J_0\delta^{\mu}{}_{\phi}), \tag{11.258}$$

with energy density ρ and pressure p given by

$$\rho = \frac{1}{8\pi}\frac{a^2}{h} = p.$$

This solution can be generated via transformations $t \to t$, $\phi \to \phi - J_0 t$ from the magnetostatic solution (11.76).

Clément "Rotating" Electromagnetic $\boldsymbol{A} = A(r)(\boldsymbol{d\phi} + \omega_0\boldsymbol{dt})$ Solution

Clément (1993) published the dual family of electromagnetic "stationary" cyclic symmetric solutions, Eq. (24), changing signature and $V_{Cl} \to -V$, given by

$$g = V(d\phi + \omega_0 dt)^2 + \frac{1}{\xi_0^2}\frac{d\rho^2}{2\rho V} - 2\rho dt^2,$$
$$V = -2\Lambda\rho + \frac{\pi_1^2}{4m}\ln(\frac{\rho}{\rho_0}),$$
$$\boldsymbol{A} = -\frac{\pi_1}{2}\ln(\frac{\rho}{\rho_0})(\boldsymbol{d\phi} + \omega_0\boldsymbol{dt}), \tag{11.259}$$

where m, π_1, ξ_0 and ρ_0 are constants, $\Lambda = \pm 1/l^2$ stands for the cosmological constant of both signs; for anti-de Sitter $\Lambda = -1/l^2$. The parameter ω_0 is related to the angular momentum constant.

It is worth noticing that the Clément expressions (11.259) satisfy the (2+1) Einstein–Maxwell equations if $\xi_0^2 = 1$ and for $2\,m = 1/\kappa$; for the adopted in the Clément's convention, $\kappa \neq 1$, $G_{\mu\nu} + \Lambda\,g\mu\nu = 4\pi\kappa\,T_{\mu\nu}$, the evaluation of the right-hand side of the Einstein equations for the structural functions (11.259), for $\xi_0^2 = 1$, yields

$$G_\nu{}^\mu = \frac{\pi_1^2}{8m\rho}[-\delta^t_\nu\delta^\mu_t + \delta^r_\nu\delta^\mu_r + \delta^\phi_\nu\delta^\mu_\phi + 2\,\omega_0\delta^t_\nu\delta^\mu_\phi],$$

while the right-hand side amounts to

$$4\pi\,\kappa\,T_\nu{}^\mu = \frac{\kappa\,\pi_1^2}{4\rho}[-\delta^t_\nu\delta^\mu_t + \delta^r_\nu\delta^\mu_r + \delta^\phi_\nu\delta^\mu_\phi + 2\,\omega_0\delta^t_\nu\delta^\mu_\phi]$$

hence $2m = 1/\kappa$.

If one were adopting $\kappa = 1$, then modifying the electromagnetic vector $\boldsymbol{A}$ to be $\boldsymbol{A}_{mod} = -\frac{\pi_1}{2\sqrt{2}\,m}\ln(\frac{\rho}{\rho_0})(\boldsymbol{d\phi} + \omega_0\boldsymbol{dt})$, one would arrive at the solution in our convention.

It is apparent that these Clément's solutions correspond to a variant of the solution derived in the previous Section (11.8.1), with the identification $r \to \rho$ followed by minor scaling transformations of t and ϕ. Notice also that the above generalization with $W(r) = \omega_0 \neq 0$ of the magnetostatic solution (11.76) can be determined applying to it $SL(2,R)$ transformations of the form $\phi \to \phi + \omega_0\,t$, $t \to t$.

11.8.2 Stationary Electromagnetic Solution for $b \neq 0 = a$

A straightforward way to derive the equations and solutions of this class of fields is just by using the complex extension of the stationary magnetic field we derived in the previous subsection taking into account the specific structure of the functions (11.30) of the extended metric (11.29) and the metric components from (11.255) of the magnetic solution together with the formal change $a^2 \to -b^2$.

Another close possibility is to accomplish the substitution

$$F = \mathcal{F},\, H = \frac{\mathcal{F}}{\mathcal{H}} - \mathcal{H}\mathcal{W}^2, W = \frac{\mathcal{H}\,\mathcal{W}}{H}, \tag{11.260}$$

in the corresponding Einstein equations for this case $b \neq 0 = a$, arriving at the following set of independent field equations

$$\begin{aligned}
EQ_{\mathcal{F}} &= \frac{d^2\,\mathcal{F}}{d\,r^2} + 4\,b^2\frac{\mathcal{H}}{\mathcal{F}} - 8\frac{1}{l^2} = 0,\\
EQ_{\mathcal{H}} &= \frac{d^2}{dr^2}\mathcal{H}^2 - 4b^2\frac{\mathcal{H}^4}{\mathcal{F}^2} = 0,\\
E_r^r &= \frac{1}{4\mathcal{H}}\frac{d\,\mathcal{H}}{d\,r}\frac{d\,\mathcal{F}}{d\,r} - \frac{\mathcal{F}}{4\mathcal{H}^2}(\frac{d\,\mathcal{H}}{d\,r})^2 + \frac{1}{4\,\mathcal{H}^2}\,J^2 + b^2\frac{\mathcal{H}}{\mathcal{F}} - \frac{1}{l^2} = 0,\\
\frac{d}{dr}\mathcal{W} &= \frac{J}{\mathcal{H}^2}.
\end{aligned} \tag{11.261}$$

Continuing with the parallelism, isolating $\mathcal{H}$ from $EQ_{\mathcal{F}}$ and replacing it into $EQ_{\mathcal{H}}$ one obtains

$$\mathcal{F}\frac{d^4\mathcal{F}}{dr^4}+2\frac{d^3\mathcal{F}}{dr^3}\frac{d\mathcal{F}}{dr}+2\left(\frac{d^2\mathcal{F}}{dr^2}\right)^2-\frac{24}{l^2}\frac{d^2\mathcal{F}}{dr^2}+\frac{64}{l^4}=0. \tag{11.262}$$

Thus, as before, the first step in the integration of the problem depends upon the (11.262) for $\mathcal{F}(r)$, structurally identical to (11.248). Substituting the solution $\mathcal{F}(r)$ into $EQ_{\mathcal{H}}$ from (11.261) one determines $\mathcal{H}$. The resulting functions $\mathcal{F}(r)$ and $\mathcal{H}$ have to fulfill $E_r{}^r$ from (11.261). By integrating the linear first-order equation for $\mathcal{W}$ one determines $\mathcal{W}(r)$.

"Stationary" Electromagnetic $\boldsymbol{A}=A(r)(\boldsymbol{dt}+J_0\boldsymbol{d\phi})$ Solution

Hitherto, the only known particular solution for $\mathcal{F}(r)$ of (11.262) and its corresponding solutions for $\mathcal{H}$ and $\mathcal{W}$ is

$$\mathcal{F}=\frac{4}{{C_1}^2\,l^2}\left[K_0+h(r)-b^2l^2\ln h(r)\right]\,h(r),\ h(r):=C_1\,r+C_0,$$
$$\mathcal{H}=\frac{\mathcal{F}}{h},\ \mathcal{W}=-J_0=\text{constant}. \tag{11.263}$$

Substituting these expressions into (11.260), one gets

$$F=\mathcal{F}=\mathcal{H}h,\ H=h-\mathcal{H}{J_0}^2,\ W=-J_0\frac{\mathcal{H}}{h-\mathcal{H}{J_0}^2},$$
$$\mathcal{H}:=\frac{4}{{C_1}^2\,l^2}\left[K_0+h(r)-b^2l^2\ln h(r)\right], \tag{11.264}$$

therefore, the corresponding metric and field vector can be written as

$$g=-\mathcal{H}(dt+J_0d\phi)^2+\frac{dr^2}{\mathcal{H}\,h(r)}+h(r)d\phi^2,$$
$$\boldsymbol{A}=\frac{b}{C_1}\ln h(r)(\boldsymbol{dt}+J_0\boldsymbol{d\phi}). \tag{11.265}$$

The electromagnetic field tensors and their invariants are given by

$$F^{\mu\nu}=2\,b\,\delta^{\mu}{}_{[t}\delta^{\nu}{}_{r]},\ F_{\mu\nu}=-2\,\frac{b}{h(r)}\,\delta_{\mu}{}^{[t}\delta_{\nu}{}^{r]}-2\,b\,\frac{J_0}{h(r)}\,\delta_{\mu}{}^{[\phi}\delta_{\nu}{}^{r]},\ FF=-2\frac{b^2}{h},$$
$$T_{\nu}{}^{\mu}=\frac{b^2}{8\pi\,h}[-\delta^t_{\nu}\delta^{\mu}_t-\delta^r_{\nu}\delta^{\mu}_r+\delta^{\phi}_{\nu}\delta^{\mu}_{\phi}-2\,J_0\delta^t_{\nu}\delta^{\mu}_{\phi}],\ TT=\frac{3}{64\pi^2}\frac{b^4}{h^2}.$$

As we shall see in the forthcoming section, this stationary electromagnetic solution can be generated via transformations $t\to t+J_0\phi$, $\phi\to\phi$ from the electrostatic solution (11.47).

Clément "Rotating" Electromagnetic $\boldsymbol{A}=A(r)(\boldsymbol{dt}-\omega_0\boldsymbol{d\phi})$ Solution

Clément (1993) also published a class of electromagnetic "stationary" cyclic symmetric metrics, Eq. (23), changing signature, given by

$$\boldsymbol{g} = -U(\boldsymbol{dt} - \omega_0 \boldsymbol{d\phi})^2 + \frac{1}{\xi_0^2}\frac{\boldsymbol{d\rho}^2}{2\rho U} + 2\rho \boldsymbol{d\phi}^2,$$
$$U = -2\Lambda\rho - \frac{\pi_0^2}{4m}\ln(\frac{\rho}{\rho_0}),$$
$$\boldsymbol{A} = \frac{\pi_0}{2}\ln(\frac{\rho}{\rho_0})(\boldsymbol{dt} - \omega_0 \boldsymbol{d\phi}), \tag{11.266}$$

where m, π_0, ξ_0 and ρ_0 are constant parameters, $\Lambda = \pm 1/l^2$ stands for the cosmological constant of both signs; for anti-de Sitter $\Lambda = -1/l^2$. The parameter ω_0 is a constant related to the angular momentum. The evaluation of the right-hand side of the Einstein equations for the structural functions (11.266), for $\xi_0^2 = 1$, yields

$$G_\nu{}^\mu = -\frac{\pi_0^2}{8mr}[\delta^t_\nu \delta^\mu_t + \delta^r_\nu \delta^\mu_r - \delta^\phi_\nu \delta^\mu_\phi - 2\,\omega_0 \delta^\phi_\nu \delta^\mu_t],$$

while the energy–momentum tensor in the left-hand side, for the vector $\boldsymbol{A}$, amounts to

$$4\pi\, T_\nu{}^\mu = -\frac{\pi_0^2}{4r}[\delta^t_\nu \delta^\mu_t + \delta^r_\nu \delta^\mu_r - \delta^\phi_\nu \delta^\mu_\phi - 2\,\omega_0 \delta^\phi_\nu \delta^\mu_t],$$

therefore Einstein–Maxwell equations are fulfilled if $2m = 1/\kappa$ or, for $\kappa = 1$, modifying the electromagnetic vector $\boldsymbol{A}$ to be $\boldsymbol{A}_{mod} = \frac{\pi_0}{2\sqrt{2\,m}}\ln(\frac{\rho}{\rho_0})(\boldsymbol{dt} - \omega_0 \boldsymbol{d\phi})$. Recall that additionally one has to set $\xi_0^2 = 1$.

It is clear that this solution is equivalent to the one treated in Section 11.8.2 for the identification $C_1\, r + C_0 \to \rho$ accompanied with minor scaling transformations of t and ϕ.

Notice that this branch of rotating solutions with $W(r) = \omega_0$ can be determined from the static electric field solution, i.e., metric (11.47) and vector $\boldsymbol{A}$ (11.48), via $SL(2,R)$ transformations: $t \to t - \omega_0\phi$, $\phi \to \phi$.

Constant W Electric Solution

In the case $W(r) = -J =$ constant the equation $E_\phi{}^t$, from (11.26), reduces to $b^2\, H\, J/F = 0$, then $J = 0 \to W = 0$. Hence, the set of equations reduces to the corresponding one of the static case.

11.9 García Stationary Solutions for $a \neq 0$ and $b \neq 0$

It is clear that the derivation of a general solution to the whole system of Einstein–Maxwell equations (11.26) is far from being an easy task. Nevertheless, some simplifications of the system of equations can be achieved by a useful change of the structural functions and combinations of the Einstein equations; the integration problem on the whole for the three structural functions is constrained to three differential equations without any further restrictions. Although we could not find sufficiently general classes of solutions, we were able to determine new families of solutions within particular combinations of elementary functions.

11.9.1 Alternative Representation of the Einstein Equations

Having in mind the derivation of other possible families of Einstein–Maxwell solutions with a and b different from zero, it is desirable to have at hand the most simple set of equations. For this purpose, introducing $W(r) = \Omega(r)/H(r)$, the independent Einstein equations can be written as;

$$EQ_{H2} = F^2 \frac{d^2}{dr^2} H + 4(a\,H - b\Omega)^2 = 0, \tag{11.267}$$

$$EQ_{\Omega 2} = H\,F^2 \frac{d^2\Omega}{dr^2} + 4\,\Omega(a\,H - b\Omega)^2 + 4b\,F(a\,H - b\Omega) = 0, \tag{11.268}$$

$$EQ_{F1} = F\frac{dF}{dr}\,H\frac{dH}{dr} - F^2\left(\frac{dH}{dr}\right)^2 - 4\,H\,(a\,H - b\Omega)^2$$
$$+F\left(H\frac{d\Omega}{dr} - \Omega\frac{dH}{dr}\right)^2 + 4F\,H(b^2 - H/l^2) = 0, \tag{11.269}$$

$$EQ_{F2} = \Omega\,H^2\,\frac{d^2F}{dr^2} - 2\,\Omega\,H\frac{dH}{dr}\frac{dF}{dr} + 2\Omega\left(\frac{dH}{dr}\right)^2 F$$
$$-2\,\Omega\left(H\frac{d\Omega}{dr} - \Omega\frac{dH}{dr}\right)^2 = 0. \tag{11.270}$$

It is worth pointing out that the equation EQ_{F2} can be considered as an integrability condition of the system of equations; differentiating the EQ_{F1} one obtains the second derivative $\frac{d^2F}{dr^2}$ together with second derivatives of H and Ω, which can be replaced through EQ_{H2} and $EQ_{\Omega 2}$; next substituting $\frac{d^2F}{dr^2}$ into EQ_{F2} one arrives at an equation of the form $\left(H\,\frac{dF}{dr} + F\,\frac{dH}{dr}\right) \times EQ_{F1}$, equal to zero by virtue of the same EQ_{F1}. Although one can adopt a different point of view; the EQ_{F2} equation arises as the differentiation of EQ_{F1} together with the use of EQ_{H2} and $EQ_{\Omega 2}$, and therefore it is not an independent equation.

Using the experience gathered until now, we shall search for particular solutions of the form

$$\begin{aligned} F(r) &= P(r) + Q(r)\ln(r), \\ H(r) &= A(r) + B(r)\ln(r), \\ W(r) &= \Omega(r)/H(r), \\ \Omega(r) &= V(r) + Z(r)\ln(r), \end{aligned} \tag{11.271}$$

where it is assumed the explicit dependence on $\ln(r)$. Substituting these guessed functions into the quoted system of equations and equating to zero the coefficients of different powers of $\ln(r)$, one arrives at a very large nonlinear system of equations; since there are five nontrivial Einstein equations, then one may expect 40 secondary equations. For instance, from equations arising from the coefficients of $\ln(r)$ to the seventh power, one has

$$E^t_{t\,ln^7} = B^2\,Q\,l^2(Z\frac{dB}{dr} - B\frac{dZ}{dr})^2 - 2\,B^3\,Z\,Q\,l^2(Z\frac{d^2B}{dr^2} - B\frac{d^2Z}{dr^2}),$$

$$E^\phi_{\phi\,ln^7} + 3E^r_{r\,ln^7} = 2\,l^2B^3Q(Z\frac{d^2B}{dr^2} - B\frac{d^2Z}{dr^2}),$$

hence

$$Z\frac{d^2B}{dr^2} - B\frac{d^2Z}{dr^2} = 0,\ Z\frac{dB}{dr} - B\frac{dZ}{dr} = 0,$$

therefore

$$Z(r) = c_1 B(r).$$

After a very lengthy and time-consuming integration process we succeeded in getting two branches of stationary electromagnetic solutions of the Einstein–Maxwell equations. The structural functions H and W possess a multiplicative factor a/b which can be absorbed by re–scaling of the Killingian coordinates according to: $\sqrt{|a/b|}t \to t$ and $\sqrt{|b/a|}\phi \to \phi$, $\sqrt{|a|\,|b|} = \pm\alpha$.

11.9.2 García Stationary Electromagnetic Solution with BTZ limit

This class of solution, depending on three parameters, is given by

$$\begin{aligned}
g &= -\frac{F(r)}{H(r)}d\,t^2 + \frac{dr^2}{F(r)} + H(r)\,[d\phi + W(r)dt]^2\,,\\
F(r) &= 4\frac{r^2}{l^2} + 2\frac{r}{l}\left(l\,w_1 + \sqrt{l^2w_1^2 - 4}\right)[w_0 + W_0\ln(r)]\,,\\
H(r) &= \frac{r}{l}\sqrt{l^2w_1^2 - 4} - [w_0 + W_0\ln(r)]\,,\\
W(r) &= \Omega(r)/H(r),\ \Omega(r) := w_0 + w_1\,r + W_0\ln(r),\\
W_0 &:= -\frac{1}{2}\,l^2\alpha^2\left(l^2w_1^2 - 2 - l\,w_1\sqrt{l^2w_1^2 - 4}\right),\\
\boldsymbol{A} &= \frac{1}{4}\alpha\left(l\,w_1 - \sqrt{l^2w_1^2 - 4}\right)(\boldsymbol{dt} - \boldsymbol{d\phi})\ln(r).
\end{aligned} \qquad (11.272)$$

and electromagnetic tensors

$$F^{\mu\nu} = 2\,\alpha\,\left(\delta^\mu{}_{[t}\delta^\nu{}_{r]} - \delta^\mu{}_{[\phi}\delta^\nu{}_{r]}\right), \qquad (11.273)$$

$$(T^\mu{}_\nu) = \begin{bmatrix} -\frac{[F+H^2(1-W^2)]}{C} & 0 & \frac{2[F+H^2W(1-W)]}{C} \\ 0 & -\frac{[F-H^2(1-W)^2]}{C} & 0 \\ \frac{-2H^2(1-W)}{C} & 0 & \frac{[F+H^2(1-W^2)]}{C} \end{bmatrix}, \qquad (11.274)$$

where $C = \frac{8\pi\,F\,H}{\alpha^2}$. The electromagnetic invariants are

$$FF = -2\frac{\alpha^2}{H} + 2\frac{\alpha^2}{F}\,H\,(1 - W)^2,$$

and

$$TT = \frac{3}{64}\frac{\alpha^4}{\pi^2}\frac{1}{F^2 H^2}[-F + H^2 (1-W)^2]^2.$$

Since this solution uses the BTZ solution as a limit for $\alpha = 0$, it is natural to search for new coordinates in which the BTZ standard structure will become apparent. First one determines the radial transformation $r = \beta_0(\rho^2 + \gamma_0)$; since $g_{rr} \to g_{\rho\rho}$, then

$$F(r) = 4\frac{r^2}{l^2} + 2\frac{r}{l}\left(l\,w_1 + \sqrt{l^2 w_1^2 - 4}\right) w_0 \to F(\rho) = \frac{\rho^4}{l^2} - M\rho^2 + \frac{J^2}{4},$$

hence

$$\gamma_0^2/l^2 + \gamma_0 M + J^2/4 = 0 \to \gamma_0/l = -\frac{Ml}{2} \mp \frac{1}{2}\sqrt{l^2M^2 - J^2},$$

$$w_0\left(l\,w_1 + \sqrt{l^2 w_1^2 - 4}\right) = \pm 2\,\beta\sqrt{l^2M^2 - J^2}.$$

Next, the structure of the Killingian transformations is of the form

$$t = \alpha_t\, T + \beta_t\, \Phi,\ \ \phi = \delta_t\, \Phi.$$

Substituting these relations into the metric and comparing the metric components with the corresponding ones of the BTZ-metric one establishes that

$$w_0 = \frac{1}{J}\sqrt{l^2M^2 - J^2}\left(-l\,M + \sqrt{l^2M^2 - J^2}\right),\ \beta = -1,\ w_1 = 2\frac{M}{J}. \tag{11.275}$$

Therefore, the coordinate transformations to be used in the electromagnetic solution in order to get the proper BTZ limit when the electromagnetic α–parameter is switched off is given by

$$\begin{aligned}
r &= -\rho^2 + \frac{M\,l^2}{2} + \frac{l}{2}\sqrt{l^2M^2 - J^2},\\
t &= \frac{1}{\sqrt{2}}\left(l^2M^2 - J^2\right)^{-1/4}\left(\sqrt{\frac{J}{l}}\,T - l\,M\sqrt{\frac{l}{J}}\,\Phi\right),\\
\phi &= \frac{1}{\sqrt{2}}\sqrt{\frac{l}{J}}\left(l^2M^2 - J^2\right)^{1/4}\,\Phi,\\
\alpha &= Q\sqrt{J}.
\end{aligned} \tag{11.276}$$

Under these transformations the metric, in its canonical representation, becomes

$$\begin{aligned}
ds^2 &= -N(\rho)^2 dT^2 + \frac{1}{L(\rho)^2}\,d\rho^2 + K(\rho)^2[d\Phi + W(\rho)dT]^2,\\
H(\rho) &:= \frac{H_n}{H_d};\\
H_n &= 4\,\rho^2\,(\rho^2 - M\,l^2)(M^2\,l^2 - J^2) - J^2 Q^4\, l^6\, R_-^2\,(\ln|\,Z(\rho)\,|)^2\\
&\quad -2Q^2\, l^3\sqrt{M^2\,l^2 - J^2}\,[M\,J^2\,l^2 - 2\,\rho^2 R_-\sqrt{M^2\,l^2 - J^2}]\,\ln|\,Z(\rho)\,|,\\
H_d &= 4\,(M^2\,l^2 - J^2)(\rho^2 - M\,l^2) - 2\,l^3 Q^2\, R_-^2\sqrt{M^2\,l^2 - J^2}\,\ln|\,Z(\rho)\,|,
\end{aligned}$$

$$\begin{aligned}
L(\rho)^2 &= \frac{\rho^2}{l^2} - M + \frac{J^2}{4\,\rho^2} + \frac{l\,Q^2}{2\,\rho^2}(2\rho^2\,R_- - l\,J^2)\ln|\,Z(\rho)\,|,\\
K(\rho)^2 &= H(\rho),\\
N(\rho)^2 &= \rho^2\frac{L(\rho)^2}{H(\rho)},\\
W(\rho)\,H_n &= J\,Q^4\,l^5\,R_-^3\,(\ln|\,Z(\rho)\,|)^2 - 2\,J\,(M^2\,l^2 - J^2)(\rho^2 - M\,l^2),\\
&\quad + Q^2\,l^2\,J\,\sqrt{M^2\,l^2 - J^2}\,[J^2\,l + 2\,l\,R_-^2 - 2\rho^2\,R_-]\ln|\,Z(\rho)\,|\\
Z(\rho) &:= \rho^2 - \frac{l}{2}(M\,l - \sqrt{M^2\,l^2 - J^2}) = \rho^2 - \frac{l\,R_-}{2},\\
R_\pm &:= M\,l \pm \sqrt{M^2\,l^2 - J^2},
\end{aligned} \tag{11.277}$$

with electromagnetic vector potential

$$\boldsymbol{A} := -\frac{\sqrt{l}\,Q}{\sqrt{2}}\frac{\ln|\,\rho\,|}{\sqrt[4]{l^2M^2 - J^2}}\left(R_-\boldsymbol{dt} - \frac{J}{l}\boldsymbol{d\phi}\right), \tag{11.278}$$

therefore the non-vanishing covariant components of the electromagnetic field tensor $F_{\mu\nu}$ are

$$F_{t\rho} = \frac{\sqrt{l}\,Q}{\sqrt{2}}\frac{R_-}{\sqrt[4]{l^2M^2 - J^2}}\frac{1}{\rho},\ F_{\phi\rho} = \frac{Q}{\sqrt{2}}\frac{J}{\sqrt{l}}\frac{1}{\sqrt[4]{l^2M^2 - J^2}}\frac{1}{\rho}. \tag{11.279}$$

Notice that the above gravitational–electromagnetic field, as was pointed out previously, when the electromagnetic field is switched off, $Q = 0$, becomes the rotating BTZ solution, while for vanishing rotation, $J = 0$, the corresponding solution is represented by the static BTZ metric, i.e., the AdS metric with M-parameter.

Notice that $(g_{tt}g_{\phi\phi} - g_{t\phi}^2) = -\rho^2\,F(\rho)$. The correspondence of this function representation of this electromagnetic solution with the BTZ solution in the limit of vanishing electromagnetic parameter α becomes apparent:

$$F(\rho) = \frac{\rho^2}{l^2} - M + \frac{J^2}{4\rho^2},\ H(\rho) = \rho^2,\ W(\rho) = -\frac{J}{2\,\rho^2}.$$

Thus, this anti-de Sitter solution has three parameters: mass M, angular momentum J, and electromagnetic parameter α. Because of its close similarity to the BTZ solution, it could represent a black hole.

Mass, Energy and Momentum

The evaluation of the surface energy density at spatial infinity $\rho \to \infty$ for $\epsilon_0 = 0$ yields

$$\begin{aligned}
\epsilon(\rho \to \infty, \epsilon_0 = 0) \approx &-\frac{1}{\pi\,l} + \frac{l\,M}{2\pi\rho^2} - \frac{l^2\,J^2\,Q^2}{2\pi\rho^2\,\sqrt{M^2l^2 - J^2}}\\
&+ \frac{l^3\,Q^2\,M}{\pi\,\rho^2\,\sqrt{M^2l^2 - J^2}}\,R_-\,\ln(\rho),
\end{aligned} \tag{11.280}$$

while the momentum quantities amount to

$$j(\rho \to \infty) \approx \frac{J}{2\pi\,\rho} - \frac{l^2\,JQ^2}{2\pi\rho\,\sqrt{M^2l^2 - J^2}}\,R_- + \frac{l^2 JQ^2\,R_-}{\pi\,\rho\,\sqrt{M^2l^2 - J^2}}\,\ln(\rho)$$

$$J(\rho \to \infty) \approx J - \frac{l^2\,JQ^2}{\sqrt{M^2l^2 - J^2}}\,R_- + \frac{2\,l^2 JQ^2\,R_-}{\sqrt{M^2l^2 - J^2}}\,\ln(\rho). \tag{11.281}$$

The integral energy and mass characteristic at spatial infinity can be evaluated from the generic expressions $E(\rho) = 2\,\pi\,K\,\epsilon(\rho)$, $M(\rho) = N\,E(\rho) - W\,J(\rho)$.

The behavior of the corresponding functions with $\epsilon_0 = 0$ as $\rho \to \infty$ is

$$\begin{aligned} E(\rho \to \infty, \epsilon_0 = 0) &\approx -\frac{2\rho}{l} + \frac{l\,M}{\rho} - \frac{l^2\,J^2\,Q^2}{\rho\sqrt{M^2l^2 - J^2}} \\ &\quad + \frac{l^2\,Q^2}{\rho\,\sqrt{M^2l^2 - J^2}}\,R_-^2\,\ln(\rho), \\ M(\rho \to \infty, \epsilon_0 = 0) &\approx -2\frac{\rho^2}{l^2} + 2\,M - \frac{l\,J^2\,Q^2}{\sqrt{M^2l^2 - J^2}} \\ &\quad + \frac{2\,l\,Q^2}{\sqrt{M^2l^2 - J^2}}\,R_-^2\,\ln(\rho). \end{aligned} \tag{11.282}$$

The series expansions of the expressions of the energy and mass evaluated for the reference energy density $\epsilon_0 = -\frac{1}{\pi\rho}\sqrt{-M_0 + \frac{\rho^2}{l^2}}$, which at the spatial infinity behaves as $\epsilon_{0|\infty} \approx -\frac{1}{\pi\,l} + \frac{l\,M_0}{2\pi\,\rho^2}$, at $\rho \to \infty$ occur to be

$$\begin{aligned} \epsilon(\rho \to \infty, \epsilon_{0|\infty}) &\approx \frac{l}{2\pi\rho^2}(M - M_0) - \frac{l^2\,J^2\,Q^2}{2\pi\rho^2\,\sqrt{M^2l^2 - J^2}} \\ &\quad + \frac{l^3 Q^2\,M\,R_-}{\pi\,\rho^2\,\sqrt{M^2l^2 - J^2}}\,\ln(\rho), \\ E(\rho \to \infty, \epsilon_{0|\infty}) &\approx \frac{l(M - M_0)}{\rho} - \frac{l^2\,J^2\,Q^2}{\rho\sqrt{M^2l^2 - J^2}} \\ &\quad + \frac{2\,l^3 Q^2\,M}{\rho\,\sqrt{M^2l^2 - J^2}}\,R_-\,\ln(\rho), \\ M(\rho \to \infty, \epsilon_{0|\infty}) &\approx M - M_0 - \frac{l\,J^2\,Q^2}{\sqrt{M^2l^2 - J^2}} \\ &\quad + \frac{2\,l^2\,M\,Q^2}{\sqrt{M^2l^2 - J^2}}\,R_-\,\ln(\rho). \end{aligned} \tag{11.283}$$

For vanishing electromagnetic field charge Q, which gives rise to the rotating BTZ black hole, the mass, and the energy–momentum quantities become just the mass–energy–momentum characteristics of the BTZ solution; hence one concludes that the parameters M and J are related to the mass and momentum respectively. Moreover, in the electromagnetic case, the momentum, mass, energy functions logarithmically diverges at spatial infinity.

Field, Energy–Momentum, and Cotton Tensors

To determine the eigenvector structure of the García solution it is more convenient to use another of its representation in the coordinates $\{\tau, r, \sigma\}$, namely

$$g = \begin{bmatrix} -F/H + H^2 & 0 & HW \\ 0 & 1/F & 0 \\ HW & 0 & H \end{bmatrix}, \qquad (11.284)$$

where the metric functions are

$$\begin{aligned} F(r) &= 4\frac{r^2}{l^2} + 2\frac{r}{l}\left(l w_1 + \sqrt{l^2 {w_1}^2 - 4}\right)(w_0 + W_0 \ln(r)), \\ H(r) &= -w_0 - W_0 \ln(r) + \frac{r}{l}\sqrt{l^2 {w_1}^2 - 4}, \\ W(r) &= \frac{(w_0 + W_0 \ln(r) + w_1 r)}{\left(-w_0 - W_0 \ln(r) + \frac{r}{l}\sqrt{l^2 {w_1}^2 - 4}\right)}, \\ W_0 &= -\frac{l^2\alpha^2}{4}\left(l w_1 - \sqrt{l^2 {w_1}^2 - 4}\right)^2. \end{aligned} \qquad (11.285)$$

To achieve the metric structure studied in the previous paragraph, one subjects the above metric to the coordinate transformation

$$\begin{aligned} \tau &= \frac{1}{\sqrt{2\,J l}}\frac{(J\,t - l^2 M\,\phi)}{\sqrt[4]{l^2M^2 - J^2}}, r = -\rho^2 + \frac{l^2 M}{2} + \frac{l}{2}\sqrt{l^2M^2 - J^2}, \\ \sigma &= \frac{\sqrt{l}}{\sqrt{2}\sqrt{J}}\sqrt[4]{l^2M^2 - J^2}\,\phi, \end{aligned}$$

together with

$$\begin{aligned} w_0 &= -\frac{\sqrt{l^2M^2 - J^2}}{J}R_-,\ w_1 = 2\frac{M}{J},\ W_0 = -\frac{l^2\alpha^2}{J^2}R_-^2, \\ R_- &= \left(lM - \sqrt{l^2M^2 - J^2}\right). \end{aligned}$$

followed by the change of the charge $\alpha \to J^{1/2}Q$.

In these coordinates, the Maxwell electromagnetic field tensor is given by

$$(F^{\alpha}{}_{\beta}) = \begin{bmatrix} 0 & \alpha/F & 0 \\ \alpha\left[F - H^2W(W-1)\right]/H & 0 & -H\alpha\,(W-1) \\ 0 & -\alpha/F & 0 \end{bmatrix}, \qquad (11.286)$$

and is characterized by the following set of eigenvectors

$$\begin{aligned} \lambda_1 &= 0; \mathbf{V1} = [\frac{H^2\,(W-1)}{F - H^2W(W-1)}V^3, 0, V^3], \\ V_\mu V^\mu &= \frac{HF\left(F - H^2\,(W-1)^2\right)}{\left[F - H^2W\,(W-1)\right]^2}{V^3}^2,\ \mathbf{V1} = \mathbf{T1},\ \mathbf{S1}, \\ \lambda_2 &= \frac{\sqrt{F - H^2(W-1)^2}}{\sqrt{HF}}\alpha; \end{aligned}$$

$$\mathbf{V2} = [\frac{\alpha\, V^2}{\lambda_2\, F\,(r)}, V^2, -\frac{\alpha\, V^2}{\lambda_2\, F}],\ V^\mu V_\mu = 0,\ \mathbf{V2} = \mathbf{N2},\ \ \mathbf{Z},$$
$$\lambda_3 = -\frac{\sqrt{F - H^2(W-1)^2}}{\sqrt{HF}}\alpha;$$
$$\mathbf{V3} = [\frac{\alpha\, V^2}{\lambda_3\, F}, V^2, -\frac{\alpha\, V^2}{\lambda_3\, F}],\ V^\mu V_\mu = 0,\ \mathbf{V3} = \mathbf{N3}, \bar{\mathbf{Z}}. \qquad (11.287)$$

This kind of tensor exhibits an enormous variety of types.

For the Maxwell energy–momentum tensor matrix $(T^\alpha{}_\beta)$ is given as

$$\begin{bmatrix} -\frac{1}{8\pi}\frac{\alpha^2\left[F-H^2(W^2-1)\right]}{F\,H} & 0 & \frac{1}{4\pi}\frac{H\alpha^2(W-1)}{F} \\ 0 & -\frac{1}{8\pi}\frac{\alpha^2\left[F-H^2(W-1)^2\right]}{F\,H} & 0 \\ \frac{1}{4\pi}\frac{\alpha^2\left[F-H^2W(W-1)\right]}{F\,H} & 0 & \frac{1}{8\pi}\frac{\alpha^2\left[F-H^2(W-1)(W+1)\right]}{F\,H} \end{bmatrix}, \qquad (11.288)$$

one has the following eigenvalues and eigenvectors

$$\lambda_{1,2} = -\frac{1}{8\pi}\frac{\alpha^2\left[F - H^2\,(W-1)^2\right]}{F\,H}; \mathbf{V1,2} = [-V^3, V^2, V^3],$$
$$V_\mu V^\mu = \frac{V^{2\,2}}{F} - \frac{V^{3\,2}[F - H^2\,(W-1)^2]}{H},$$
$$\mathbf{V1} = \mathbf{T1},\ \mathbf{S1}, \mathbf{V2} = \mathbf{T2},\ \mathbf{S2},$$
$$\lambda_3 = \frac{1}{8\pi}\frac{\alpha^2\left[F - H^2\,(W-1)^2\right]}{F\,H};$$
$$\mathbf{V3} = [\frac{H^2 V^3\ (W-1)}{F - H^2 W(W-1)}, 0, V^3],$$
$$V_\mu V^\mu = -\frac{V^{3\,2} H F\left[F - H^2\,(W-1)^2\right]}{\left[F - H^2 W\,(W-1)\right]^2},$$
$$\mathbf{V3} = \mathbf{T3},\ \mathbf{S3}. \qquad (11.289)$$

Therefore one has a quite big choice of algebraic types of this tensor.

Finally, the Cotton tensor can be given as

$$(C^\alpha{}_\beta) = \begin{bmatrix} C^1{}_1 & 0 & C^1{}_3 \\ 0 & C^2{}_2 & 0 \\ C^3{}_1 & 0 & -C^1{}_1 - C^2{}_2 \end{bmatrix}, \qquad (11.290)$$

$$C^1{}_1 = -\frac{1}{32\,\pi}\frac{HW\alpha^2\,(W-1)^2\,(HF_{,r} - FH_{,r})}{F^2} + \frac{1}{32\,\pi}\frac{\alpha^2\,(W-1)\,F_{,r}}{F}$$
$$-\frac{1}{32\,\pi}\frac{\alpha^2\,(3\,W-2)\,H_{,r}}{H} - \frac{1}{16\,\pi}\,\alpha^2 W_{,r}, \qquad (11.291)$$

with components

$$C^1{}_3 = -\frac{1}{32\,\pi}\,\frac{\alpha^2 H\,(W-1)^2\,(H\,F_{,r} - F\,H_{,r})}{F^2} - \frac{1}{32\,\pi}\,\frac{\alpha^2 H_{,r}}{H}, \tag{11.292}$$

$$\begin{aligned} C^2{}_2 = &-\frac{1}{16\,\pi}\,\frac{\alpha^2\,(W-1)\,(F_{,r}H - 2\,FH_{,r})}{F\,H} \\ &+\frac{1}{16\,\pi}\,\frac{\alpha^2\left(H^2\,(W-1)^2 + F\right)W_{,r}}{F}, \end{aligned} \tag{11.293}$$

$$\begin{aligned} C^3{}_1 = &\frac{1}{32\,\pi}\,\frac{\alpha^2\,(W-1)\,(-F\,(1+W) + W^2H^2\,(W-1))\,F_{,r}}{F^2} \\ &-\frac{1}{32\,\pi}\,\frac{(F + H^2W^2)\,\alpha^2\left(-F + H^2\,(W-1)^2\right)H_{,r}}{H^3F}, \\ &-\frac{1}{16\,\pi}\,\frac{W\alpha^2\left(-F + H^2\,(W-1)^2\right)W_{,r}}{F}, \end{aligned} \tag{11.294}$$

$$\begin{aligned} C^3{}_3 = &\frac{1}{32\,\pi}\,\frac{H^2W\alpha^2\,(W-1)^2\,F_{,r}}{F^2} - \frac{1}{16\,\pi}\,\frac{\alpha^2H^2\,(W-1)^2\,W_{,r}}{F}, \\ &-\frac{1}{32\,\pi}\,\frac{\alpha^2\left(H^2W\,(W-1)^2 + F\,(W-2)\right)H_{,r}}{F\,H} \end{aligned} \tag{11.295}$$

possesses, in general, three different eigenvalues, with the possibility of complex conjugated roots, namely

$$\begin{aligned} \lambda_1 &= C^2{}_2, \\ \lambda_2 &= -1/2\,C^2{}_2 + 1/2\,\sqrt{(C^1{}_1 + C^2{}_2)^2 + 4\,C^1{}_3\,C^3{}_1}, \\ \lambda_3 &= -1/2\,C^2{}_2 - 1/2\,\sqrt{(C^1{}_1 + C^2{}_2)^2 + 4\,C^1{}_3\,C^3{}_1}]. \end{aligned} \tag{11.296}$$

The set of eigenvector equations reduces to

$$\begin{aligned} V^1\,(C^1{}_1 - \lambda) + C^1{}_3\,V^3 &= 0, \\ \left(C^2{}_2 - \lambda\right)V^2 &= 0, \\ C^3{}_1\,V^1 - V^3(C^1{}_1 + C^2{}_2 + \lambda) &= 0 \end{aligned} \tag{11.297}$$

with solutions

$$\begin{aligned} &\lambda_1 = C^2{}_2;\, \mathbf{V1} = [0, V^2, 0],\ V_\mu V^\mu = {V^2}^2/F,\ \mathbf{V1} = \mathbf{S}1, \\ &\lambda_2 = -1/2\,C^2{}_2 + 1/2\,\sqrt{(C^1{}_1 + C^2{}_2)^2 + 4\,C^1{}_3\,C^3{}_1}; \\ &\mathbf{V2} = [-\frac{C^1{}_3\,V^3}{C^1{}_1 - \lambda_2}, 0, V^3],\ V^\mu V_\mu = \{0, > 0, < 0\}, \\ &\mathbf{V2} = \mathbf{S2}, \mathbf{N2},\ \mathbf{Z}, \\ &\lambda_3 = -1/2\,C^2{}_2 - 1/2\,\sqrt{(C^1{}_1 + C^2{}_2)^2 + 4\,C^1{}_3\,C^3{}_1}; \\ &\mathbf{V3} = [-\frac{C^1{}_3\,V^3}{C^1{}_1 - \lambda_3}, 0, V^3],\ V^\mu V_\mu = \{0, > 0, < 0\}, \\ &\mathbf{V3} = \mathbf{S3}, \mathbf{N3}, \bar{\mathbf{Z}}. \end{aligned} \tag{11.298}$$

In general, one is dealing with a Type I Cotton tensor; there is a big choice of eigenvectors.

11.9.3 García Stationary Solution with BTZ-Counterpart Limit

The second possible solution in the studied class is given by

$$\begin{aligned}
g &= -\frac{\mathcal{F}}{\mathcal{H}}dT^2 + \frac{dr^2}{\mathcal{F}} + \mathcal{H}\,(d\Phi + \mathcal{W}\,dT)^2\,,\\
\mathcal{F}(r) &= 4\frac{r^2}{l^2} + 2\frac{r}{l}(l\,w_1 + \sqrt{l^2 w_1^2 - 4})\,(w_0 + W_0\ln(r))\,,\\
\mathcal{H}(r) &= \frac{\mathcal{H}_n}{\mathcal{H}_d},\\
\mathcal{H}_n &= \mathcal{F}(r) - \Omega(r)^2,\\
\mathcal{H}_d &= \frac{r}{l}\sqrt{l^2\,w_1^2 - 4} - (w_0 + W_0\ln(r))\,,\\
\mathcal{W}(r) &= \frac{\Omega(r)}{\mathcal{H}(r)},\ \Omega(r) = w_0 + W_0\,\ln(r) + w_1\,r,\\
W_0 &:= \frac{1}{2}\,l^2\alpha^2\left(l^2 w_1^2 - 2 - l\,w_1\sqrt{l^2 w_1^2 - 4}\right),
\end{aligned} \tag{11.299}$$

with electromagnetic tensors

$$F^{\mu\nu} = 2\,\alpha\,\left(\delta^{\mu}{}_{[t}\delta^{\nu}{}_{r]} + \delta^{\mu}{}_{[\phi}\delta^{\nu}{}_{r]}\right), \tag{11.300}$$

$$(T^{\mu}{}_{\nu}) = \begin{bmatrix} -\frac{[\mathcal{F}+\mathcal{H}^2(1-\mathcal{W}^2)]}{\mathcal{C}} & 0 & -\frac{2[\mathcal{F}-\mathcal{H}^2\mathcal{W}(1+\mathcal{W})]}{\mathcal{C}} \\ 0 & -\frac{[\mathcal{F}-\mathcal{H}^2(1+\mathcal{W})^2]}{\mathcal{C}} & 0 \\ \frac{2\mathcal{H}^2(1+\mathcal{W})}{\mathcal{C}} & 0 & \frac{[\mathcal{F}+\mathcal{H}^2(1-\mathcal{W}^2)]}{\mathcal{C}} \end{bmatrix}, \tag{11.301}$$

where $\mathcal{C} = \frac{8\pi\,\mathcal{F}\,\mathcal{H}}{\alpha^2}$, and w_0 and w_1 are parameters related to mass and angular momentum, while α is an electromagnetic parameter; the electromagnetic invariants are

$$\begin{aligned}
FF &= 2\frac{\alpha^2}{\mathcal{F}\mathcal{H}}\left[-\mathcal{F} + \mathcal{H}^2\,(1+\mathcal{W})^2\right],\\
TT &= \frac{3}{64}\frac{\alpha^4}{\pi^2}\frac{1}{\mathcal{F}^2\,\mathcal{H}^2}[-\mathcal{F} + \mathcal{H}^2\,(1+\mathcal{W})^2]^2.
\end{aligned} \tag{11.302}$$

The calligraphic capital letters have been used above to make their relationship evident to those structural functions arising as real cuts of the complex extensions of the studied class of metric, see (11.29). This solution of the Einstein–Maxwell equations can be considered also as a real cut of the complex version of the stationary electromagnetic solution with BTZ-limit given in the previous paragraph; the structural functions $\mathcal{F}$, $\mathcal{H}$, and $\mathcal{W}$ can be constructed according to (11.30) with F, H, and W from (11.271) accompanied by the replacement of the sign in front of α^2, ${\alpha^2}_{el} \to -{\alpha^2}_{mg}$. If one searches for the anti-de Sitter limit of this solution, one would arrive at an alternative real cut of the BTZ solution, namely to the "BTZ solution counterpart," or "BTZ counterpart" for short.

$$g_c = -\rho^2 \frac{\mathcal{F}}{\mathcal{H}} dT^2 + \frac{d\rho^2}{\mathcal{F}} + \mathcal{H}\,(d\Phi + \mathcal{W}\,dT)^2\,,$$
$$\mathcal{F} = \frac{\rho^2}{l^2} - M + \frac{J^2}{4\,\rho^2},\ \mathcal{H} = \frac{\rho^2}{l^2} - M,\ \mathcal{W} = \frac{J}{2\mathcal{H}}. \tag{11.303}$$

Recall that in the above mentioned metric one can again introduce the radial coordinate by changing

$$\rho^2 \to R^2 + M\,l^2,\ \mathcal{H} \to R^2, \mathcal{W} \to \frac{J}{2R^2},$$
$$\mathcal{F}(\rho) \to F(R) = \frac{R^2}{l^2} + M + \frac{J^2}{4\,R^2}. \tag{11.304}$$

Since this solution possesses the BTZ counterpart as a limit for $\alpha = 0$, it is pertinent to search for new coordinates in which the BTZ solution counterpart structure (11.303) will become apparent.

Transformation to BTZ-Counterpart Coordinates

The constants and the coordinate transformations to be used in this case are given by

$$w_0 = -\frac{1}{J}\sqrt{l^2M^2 - J^2}\left(l\,M - \sqrt{l^2M^2 - J^2}\right),\ w_1 = 2\frac{M}{J},$$
$$r = -\rho^2 + \frac{M\,l^2}{2} + \frac{l}{2}\sqrt{l^2M^2 - J^2},$$
$$\phi = \frac{1}{\sqrt{2}}\left(l^2M^2 - J^2\right)^{-1/4}\left(M\,l\sqrt{\frac{l}{J}}\,T + \sqrt{\frac{J}{l}}\,\Phi\right),$$
$$t = \frac{1}{\sqrt{2}}\sqrt{\frac{l}{J}}\left(l^2M^2 - J^2\right)^{1/4}\,T. \tag{11.305}$$

Under these transformations the solution amounts to

$$g = -\rho^2\,\frac{F(\rho)}{H(\rho)}dT^2 + \frac{d\rho^2}{F(\rho)} + H(\rho)(d\Phi + W(\rho)dT)^2,$$

$$F = \frac{\rho^2}{l^2} - M + \frac{J^2}{4\rho^2} + \frac{l\,\alpha^2}{2\,J\,\rho^2}\left(J^2\,l - 2\,R_{(-)}\,\rho^2\right)\ln|r|,$$

$$H(\rho) = \frac{H_n}{H_d},$$
$$H_n := 4\,J^3l^3\left(l^2M - \rho^2\right)\sqrt{l^2M^2 - J^2}R_{(+)}^2\,\alpha^2\ln|r|$$
$$-4\left(l^2M - \rho^2\right)^2\left(l^2M^2 - J^2\right)R_{(+)}^4 - l^6J^6\alpha^4\left(\ln|r|\right)^2$$

$$H_d := -2l^5 J^3 \sqrt{l^2M^2-J^2}R_{(+)}^2\alpha^2 \ln|r| \\ +4\,l^2\left(l^2M-\rho^2\right)(l^2M^2-J^2)R_{(+)}^4,$$

$$\begin{aligned} W(\rho) &= \frac{l^2}{J}\frac{\Omega(\rho)}{H_n}, \\ \Omega(\rho) &= -l^5 J^6 R_{(+)}\alpha^4 \left(\ln|r|\right)^2 \\ &\quad +l^2\,J^3\left[l\left(R_{(+)}^2+2J^2\right)\sqrt{l^2M^2-J^2}+\left(J^2-R_{(+)}^2\right)\rho^2\right]R_{(+)}^2\alpha^2\ln|r| \\ &\quad +2\,J^2\left(l^2\,M^2-J^2\right)\left(\rho^2-l^2M\right)R_{(+)}^4, \\ r &:= -\rho^2+\frac{M\,l^2}{2}+\frac{l}{2}\sqrt{l^2M^2-J^2},\ R_{(\pm)} := M\,l\pm\sqrt{l^2M^2-J^2}. \end{aligned} \tag{11.306}$$

The correspondence of this representation with the BTZ solution counterpart in the limit of vanishing electromagnetic parameter α is evident.

Because of the complexity of the system of equations, we have found very hard to determine other branches, if any, of exact solutions in the general case.

11.10 Generating Solutions via $SL(2,R)$–Transformations

This section deals with $SL(2,R)$ transformations applied on static solutions to construct stationary cyclic symmetric classes of solutions, namely the electric and magnetic stationary families.

Let us consider the general metric

$$g = g_{tt}dt^2+2g_{t\phi}dtd\phi+g_{\phi\phi}\,d\phi^2+g_{rr}dr^2,$$

and accomplish here a $SL(2,R)$ transformations of the Killingian coordinates t and ϕ

$$t=\alpha\tilde{t}+\beta\tilde{\phi},\ \phi=\gamma\tilde{t}+\delta\tilde{\phi},\ \Delta:=\alpha\delta-\beta\gamma\neq 0. \tag{11.307}$$

The transformed metric components are given by

$$\begin{aligned} g_{\tilde{t}\tilde{t}} &= \alpha^2\,g_{tt}+2\alpha\gamma\,g_{t\phi}+\gamma^2\,g_{\phi\phi},\ g_{\tilde{t}\tilde{\phi}}=\alpha\beta\,g_{tt}+(\alpha\delta+\beta\gamma)\,g_{t\phi}+\gamma\delta\,g_{\phi\phi}, \\ g_{\tilde{\phi}\tilde{\phi}} &= \beta^2\,g_{tt}+2\beta\delta\,g_{t\phi}+\delta^2\,g_{\phi\phi},\ g_{rr}=g_{rr}, \end{aligned} \tag{11.308}$$

while under the considered transformations the electromagnetic field tensor (11.13) becomes

$$\begin{aligned} (F^{\tilde{\alpha}\tilde{\beta}}) &= \frac{1}{\sqrt{-\tilde{g}}}\begin{bmatrix} 0 & \tilde{b} & -\frac{\tilde{c}g_{rr}}{\sqrt{-\tilde{g}}} \\ -\tilde{b} & 0 & \tilde{a} \\ \frac{\tilde{c}g_{rr}}{\sqrt{-\tilde{g}}} & -\tilde{a} & 0 \end{bmatrix}, \\ \tilde{g} &= \det(g_{\tilde{\mu}\tilde{\nu}}), \end{aligned} \tag{11.309}$$

where the new constant are given in terms of the original ones through

$$\begin{aligned}\tilde{a} &= \frac{\alpha\, a + \gamma\, b}{\alpha\delta - \beta\gamma}, \quad a = \delta\tilde{a} - \gamma\tilde{b},\\ \tilde{b} &= \frac{\beta\, a + \delta\, b}{\alpha\delta - \beta\gamma}, \quad b = -\beta\tilde{a} + \alpha\tilde{b},\\ \tilde{c} &= \frac{c}{\alpha\delta - \beta\gamma}.\end{aligned} \tag{11.310}$$

Notice that

$$g_{\tilde{t}\tilde{t}}g_{\tilde{\phi}\tilde{\phi}} - g^2_{\tilde{t}\tilde{\phi}} = (g_{tt}g_{\phi\phi} - g^2_{t\phi})(\alpha\delta - \beta\gamma)^2 = -F\Delta^2,$$

therefore, in concrete applications it is more useful to use normalized transformations with $\Delta = \alpha\,\delta - \beta\,\gamma = 1$.

The electromagnetic tensor is form-invariant under the above-mentioned $SL(2,R)$ transformations if the field constants a, b, and c are identified by means of Eq. (11.310). This property, on its turn, yields to the form-invariance of the electromagnetic energy–momentum tensor $T_\mu{}^\nu = 1/(4\pi)(F_{\mu\sigma}\,F^{\nu\sigma} - 1/4\delta^\nu_\mu F_{\tau\sigma}F^{\tau\sigma})$, and consequently to the form-invariance of the Einstein–Maxwell equations.

Therefore, starting with an electromagnetic solution in which a single electric ($b \neq 0$) or magnetic ($a \neq 0$) field is present, by accomplishing the above-mentioned $SL(2,R)$ transformations, one can generate solutions with both electric and magnetic fields $\tilde{b} \neq 0, \tilde{a} \neq 0$ present. Conversely, if one originally has had a solution endowed with both constant parameters a and b then, via transformations, one could achieve a branch of solutions with one single parameter. At this level, one may argue that one deals with one specific solution in its different coordinate representations. But there exists a second point of view in $(2+1)$ gravity: to end with a new solution one has to change the variety, i.e., the topology, requiring the ranges of change of the new variable to be, for instance, the same as the ranges of the original variables. This procedure can be considered as a generating solution technique and it has been used to construct stationary solutions starting from static solutions as we shall show in the forthcoming sections.

For the metric (11.15), subjected to the above mentioned $SL(2,R)$ transformations, one gets

$$\begin{aligned}g_{\tilde{t}\tilde{t}} &= -\alpha^2\,\frac{F}{H} + H\,(\alpha\,W + \gamma)^2,\\ g_{\tilde{t}\tilde{\phi}} &= -\alpha\,\beta\,\frac{F}{H} + H\,(\delta + \beta\,W)\,(\gamma + \alpha\,W),\\ g_{\tilde{\phi}\tilde{\phi}} &= -\beta^2\,\frac{F}{H} + H\,(\beta\,W + \delta)^2,\\ g_{rr} &= \frac{1}{F},\end{aligned} \tag{11.311}$$

hence, the expressions of the new structural functions are given in the form

$$\tilde{H} = -\beta^2 \frac{F}{H} + H(\delta + \beta W)^2,$$
$$\tilde{W}\tilde{H} = -\alpha\beta \frac{F}{H} + H(\delta + \beta W)(\gamma + \alpha W),\ \tilde{F} = F. \tag{11.312}$$

The transformed electromagnetic field tensor $F^{\tilde{\mu}\tilde{\nu}}$, as it should be, exhibits its form-invariant property

$$\left(F^{\tilde{\mu}\tilde{\nu}}\right) = \begin{bmatrix} 0 & \tilde{b} & -\frac{\tilde{c}}{F} \\ -\tilde{b} & 0 & \tilde{a} \\ \frac{\tilde{c}}{F} & -\tilde{a} & 0 \end{bmatrix}, \tag{11.313}$$

where as before the new field constant parameters are related to the old ones according to (11.310).

Although the metrics generated via $SL(2,R)$ transformations should be given as subsections in this section, because of the important place that various of these solutions occupy in the field, I have decided to report them in separate sections together with their properties.

11.11 Transformed Electrostatic $b \neq 0$ Solutions

Starting with the general electrostatic Maxwell solution (11.47) with metric

$$g = -\frac{F}{H}dt^2 + \frac{1}{F}dr^2 + H d\phi^2,$$
$$F(r) = 4\frac{H(r)}{C_1^2\, l^2}\left[K_0 + H(r) - b^2 l^2 \ln H(r)\right],$$
$$H(r) = C_1\, r + C_0, \tag{11.314}$$

under normalized $SL(2.R)$ transformations

$$t = \frac{\alpha}{\sqrt{\Delta}}\tilde{t} + \frac{\beta}{\sqrt{\Delta}}\tilde{\phi},\ \phi = \frac{\gamma}{\sqrt{\Delta}}\tilde{t} + \frac{\delta}{\sqrt{\Delta}}\tilde{\phi},\ \Delta = \alpha\delta - \beta\gamma \neq 0 \tag{11.315}$$

(in the general (non-normalized) case the same expressions hold except for the absence of Δ, set simply $\Delta = 1$), the new metric, the rotated one, acquires the form

$$\left(g_{\tilde{\mu}\tilde{\nu}}\right) = \begin{bmatrix} -\frac{\alpha^2}{\Delta}\frac{F}{H} + \frac{\gamma^2}{\Delta}H & 0 & -\frac{\alpha\beta}{\Delta}\frac{F}{H} + \frac{\delta\gamma}{\Delta}H \\ 0 & \frac{1}{F} & 0 \\ -\frac{\alpha\beta}{\Delta}\frac{F}{H} + \frac{\delta\gamma}{\Delta}H & 0 & -\frac{\beta^2}{\Delta}\frac{F}{H} + \frac{\delta^2}{\Delta}H \end{bmatrix}, \tag{11.316}$$

the electromagnetic field tensor becomes

$$\left(F^{\tilde{\mu}\tilde{\nu}}\right) = \begin{bmatrix} 0 & \frac{\delta\, b}{\sqrt{\Delta}} & 0 \\ -\frac{\delta\, b}{\sqrt{\Delta}} & 0 & \frac{\gamma\, b}{\sqrt{\Delta}} \\ 0 & -\frac{\gamma\, b}{\sqrt{\Delta}} & 0 \end{bmatrix}, \tag{11.317}$$

while the electromagnetic energy–momentum tensor amounts to

$$\left(T_{\tilde{\mu}}{}^{\tilde{\nu}}\right) = \begin{bmatrix} -\frac{1}{8\pi}\frac{(\alpha\,\delta+\beta\,\gamma)b^2}{H\,\Delta} & 0 & \frac{1}{4\pi}\frac{\gamma\,\alpha\,b^2}{H\,\Delta} \\ 0 & -\frac{1}{8\pi}\frac{b^2}{H} & 0 \\ -\frac{1}{4\pi}\frac{\delta\,\beta\,b^2}{H\,\Delta} & 0 & \frac{1}{8\pi}\frac{(\alpha\,\delta+\beta\,\gamma)b^2}{H\,\Delta} \end{bmatrix}. \tag{11.318}$$

Explicitly, the new metric is given by the nonzero components

$$\begin{aligned} g_{\tilde{t}\tilde{t}} &= -\frac{\alpha^2}{\Delta}\frac{1}{H(r)\,g_{rr}} + \frac{\gamma^2}{\Delta}H(r),\ g_{\tilde{t}\tilde{\phi}} = -\frac{\alpha\beta}{\Delta}\frac{1}{H(r)\,g_{rr}} + \frac{\delta\gamma}{\Delta}H(r), \\ g_{\tilde{\phi}\tilde{\phi}} &= -\frac{\beta^2}{\Delta}\frac{1}{H(r)\,g_{rr}} + \frac{\delta^2}{\Delta}H(r), \\ g_{rr} &= \frac{1}{4\,H(r)}\frac{C_1^2\,l^2}{\left[K_0 + H(r) - b^2l^2\ln H(r)\right]}, \\ H(r) &= C_1 r + C_0. \end{aligned} \tag{11.319}$$

For general $SL(2,R)$ transformations, with non-vanishing entries, the electromagnetic field tensor $F^{\mu\nu}$ allows for the presence of both electric and magnetic fields, corresponding to new b and a different from zero.

If one accomplishes the transformation of the dependent variable r to the radial (polar) coordinate ρ, arc $= \rho\, d\phi$, one chooses

$$H(r) = C_1\, r + C_0 = \rho^2,\ C_1 = 2. \tag{11.320}$$

11.11.1 Stationary Electromagnetic Solution

In particular, for the $SL(2,R)$ transformation

$$t = \tilde{t} - \omega\tilde{\phi},\ \phi = \tilde{\phi},\ \alpha = 1,\ \beta = -\omega,\ \gamma = 0,\ \delta = 1,\ \ \Delta = 1, \tag{11.321}$$

one obtains a new solution, the rotated one, with metric components

$$\begin{aligned} g_{\tilde{t}\tilde{t}} &= -\frac{4}{C_1^2\,l^2}\left[K_0 + H(r) - b^2l^2\ln H(r)\right], \\ g_{\tilde{t}\tilde{\phi}} &= \omega\,\frac{4}{C_1^2\,l^2}\left[K_0 + H(r) - b^2l^2\ln H(r)\right],\ g_{\tilde{\phi}\tilde{\phi}} = H(r) - \omega^2\frac{1}{H(r)\,g_{rr}}, \\ g_{rr} &= \frac{1}{4\,H(r)}\frac{C_1^2\,l^2}{\left[K_0 + H(r) - b^2l^2\ln H(r)\right]}, \\ H(r) &= C_1 r + C_0. \end{aligned} \tag{11.322}$$

The electromagnetic field tensor is given by

$$(F^{\tilde{\mu}\tilde{\nu}}) = \begin{bmatrix} 0 & b & 0 \\ -b & 0 & 0 \\ 0 & 0 & 0 \end{bmatrix}, (T_{\tilde{\mu}}{}^{\tilde{\nu}}) = \begin{bmatrix} -\frac{1}{8\pi}\frac{b^2}{H} & 0 & 0 \\ 0 & -\frac{1}{8\pi}\frac{b^2}{H} & 0 \\ \frac{1}{4\pi}\frac{\omega b^2}{H} & 0 & \frac{1}{8\pi}\frac{b^2}{H} \end{bmatrix}.$$

Therefore, by means of a $SL(2, R)$ transformation applied to the static electric cyclic symmetric $(2+1)$ Einstein–Maxwell solution one can generate a unique electromagnetic stationary cyclic symmetric solution in the sense of the structure of the field tensor $F^{\mu\nu}$, which is equal in all respects to the electromagnetic solution determined by the metric (11.265). It is worth mentioning that Clément (1993) reported a solution, Eq. (24), belonging to this class.

11.11.2 Clément Spinning Solution

The so-called Clément's spinning charged BTZ solution, derived in Clément (1996), deserves special attention. It arises as a result of a $SL(2, R)$ transformation of the electrostatic solution given in terms of the radial coordinate $\rho \to r$. Here the main Clément results are reproduced in a way quite close to the cited work. Setting $C_1 = 2$, which is equivalent to $t \to t\,C_1/2$, accomplishing the coordinate transformation $H(r) = C_1 r + C_0 \to r^2$, and introducing the definitions $r_0 = \exp(K_0/(2b^2 l^2))$, and $b^2 = 4\pi\, G\, Q^2$, the metric (11.314) becomes

$$\begin{aligned} g &= -F(r)dt^2 + \frac{dr^2}{F(r)} + r^2 d\phi^2, \\ F(r) &= \frac{K_0}{l^2} + \frac{r^2}{l^2} - b^2 \ln r^2 = \frac{r^2}{l^2} - 8\pi G Q^2 \ln \frac{r}{r_0}, \\ \boldsymbol{A} &= 2\,Q\,\sqrt{\pi\, G}\,\ln \frac{r}{r_0}\,\boldsymbol{dt}. \end{aligned} \tag{11.323}$$

To establish the range of values of r_0 allowing the existence of a black hole, let us consider $F(r)$ in the form

$$F(r) = \frac{r^2}{l^2}(1 - \frac{k}{r^2}\ln\frac{r^2}{r_0^2}),\ k = 4\pi G Q^2 l^2, \tag{11.324}$$

the factor $(1 - \frac{k}{r^2}\ln\frac{r^2}{r_0^2})$ vanishes in the set of points r_h determined through the LambertW function, LambertW(x) exp(LambertW(x)) $= x$, namely

$$r_h^2 = -k\,\mathrm{LambertW}(-r_0^2/k), \tag{11.325}$$

which is positive for $r_0^2 = k\,\exp(-1)\epsilon$, $0 < \epsilon \leq 1$, or explicitly

$$r_0^2 \leq 4\pi G Q^2 l^2/\mathrm{e}. \tag{11.326}$$

Subjecting the metric (11.323) and the vector potential $\boldsymbol{A}$ to the transformation at uniform angular velocity

$$t \to t - \omega\phi,\ \phi \to \phi - \frac{\omega}{l^2}\,t,\ \alpha = 1,\ \beta = -\omega,\ \gamma = -\frac{\omega}{l^2},\ \delta = 1, \tag{11.327}$$

one arrives at the metric

$$\begin{aligned} g &= -(F - \frac{\omega^2}{l^4} r^2)dt^2 + 2\omega(F - \frac{r^2}{l^2})dt d\phi + (r^2 - \omega^2 F)d\phi^2 + \frac{dr^2}{F}, \\ F &= \frac{r^2}{l^2} - 4\pi G Q^2 \ln\frac{r^2}{r_0^2}, \\ \boldsymbol{A} &= Q\sqrt{\pi\,G}\ln\frac{r^2}{r_0^2}\ (\boldsymbol{dt} - \omega\boldsymbol{d\phi}). \end{aligned} \tag{11.328}$$

One could arrive at this result by using the metric components (11.319) with transformation coefficients from (11.327) and setting $C_1 = 2$, $\Delta = 1$.

By choosing the axial symmetry as fundamental, the metric (11.328) can be brought to the form

$$\begin{aligned} g &= -r^2\,(1 - \omega^2/l^2)^2\,\frac{\mathcal{F}(r)}{\mathcal{H}(r)}dt^2 + \frac{dr^2}{\mathcal{F}(r)^2} + \mathcal{H}(r)\,(d\phi + \mathcal{W}(r)dt)^2\,, \\ \mathcal{F} &= F = \frac{r^2}{l^2} - 4\pi G Q^2 \ln\frac{r^2}{r_0^2},\ \mathcal{W} = \omega\,\frac{F^2 - r^2/l^2}{\mathcal{H}} = -4\pi G Q^2 \frac{\omega}{\mathcal{H}}\ln\frac{r^2}{r_0^2}, \\ \mathcal{H} &= r^2 - \omega^2\,F = r^2(1 - \frac{\omega^2}{l^2}) + \omega^2\,4\pi G Q^2 \ln\frac{r^2}{r_0^2}. \end{aligned} \tag{11.329}$$

The Clément spinning charged BTZ solution is endowed with three parameters Q, r_0, and ω. It allows for a black hole interpretation.

Alternatively, introducing the scaling transformation $r = l/\bar{l}\times\bar{r}$, the definitions $\bar{l}^2 = l^2 - \omega^2$, $|\omega| < l$, and $\bar{r}_0 = \bar{l}/l \times r_0$, the proper Clément (1996) solution, dropping the bar from $r \to \rho$, is given in its canonical representation as

$$\begin{aligned} ds^2 &= -N(\rho)^2 dt^2 + \frac{1}{L(\rho)^2}\,d\rho^2 + K(\rho)^2[d\phi + W(\rho)dt]^2, \\ K(\rho)^2 &= H(\rho) := \rho^2 + 4\pi\,G\,\omega^2\,Q^2 \ln\,(\frac{\rho^2}{\rho_0^2}), \\ L(\rho)^2 &= F(\rho) := \frac{\rho^2}{l^2} - \frac{4\pi\,G\,(l^2 - \omega^2)\,Q^2}{l^2}\ln\,(\frac{\rho^2}{\rho_0^2}), \\ L(\rho) &= \sqrt{F(\rho)}, K(\rho) = \sqrt{H(\rho)}, N(\rho) = \rho\sqrt{\frac{F(\rho)}{H(\rho)}}, \\ W(\rho) &= -\omega\frac{4\pi\,G\,Q^2}{H(\rho)}\ln\,(\frac{\rho^2}{\rho_0^2}), \\ \boldsymbol{A} &= 2\,Q\,\sqrt{\pi\,G}\ln\frac{\rho}{\rho_0}\ (\boldsymbol{dt} - \omega\boldsymbol{d\phi}); \end{aligned} \tag{11.330}$$

see also García (2009), Eq. (11.24). The corresponding electromagnetic fields are given by

$$F_{\mu\nu} = -\frac{4Q}{\rho}\sqrt{\pi G}(\delta_{[\mu}{}^{t}\delta_{\nu]}{}^{r} - \omega\delta_{[\mu}{}^{\phi}\delta_{\nu]}{}^{\rho}),$$

$$T_{\mu}{}^{\nu} = -\frac{G Q^2}{2l^2}\frac{l^2+\omega^2}{\rho^2}\delta_{\mu}{}^{t}\delta_{t}{}^{\nu} - \frac{GQ^2}{l^2}\frac{\omega}{\rho^2}\delta_{\mu}{}^{t}\delta_{\phi}{}^{\nu} + \frac{GQ^2}{2l^2}\frac{l^2+\omega^2}{\rho^2}\delta_{\mu}{}^{\phi}\delta_{\phi}{}^{\nu}$$
$$+\frac{GQ^2\omega}{\rho^2}\delta_{\mu}{}^{\phi}\delta_{t}{}^{\nu} - \frac{GQ^2}{2l^2}\frac{l^2-\omega^2}{\rho^2}\delta_{\mu}{}^{r}\delta_{r}{}^{\nu}. \tag{11.331}$$

This solution is a black hole if the condition of the form (11.326),

$$\bar{\rho}_0^2 \leq 4\pi G Q^2 \bar{l}^2/\mathrm{e}, \tag{11.332}$$

is fulfilled. It possesses two horizons, at which $F(\rho)$ vanishes, which are roots of the relation

$$\rho^2 - \bar{l}^2 8\pi G Q^2 \ln\frac{\rho}{\bar{\rho}_0} = 0, \tag{11.333}$$

which are given by the LambertW function, see (11.325). The largest root determines the event horizon at $\rho = \rho_+ = \rho_h$, while the inner one is a Cauchy horizon at $\rho = \rho_-$, with $\rho_+ > \rho_- > \bar{\rho}_0$. Since the metric function H changes sign for a certain value $\rho = \rho_c < \bar{\rho}_0$, similarly as the rotating BTZ solution, thus there are closed timelike curves in the region inside the radius ρ_c. It is apparent that the metric and the electromagnetic field are singular at $\rho = 0$.

The length rescaling was chosen in such a manner that $H \to \rho^2$ at spatial infinity and $g(11.330) \to g_{\mathrm{BTZ}}$. According to Clément, one may formally define mass and angular momentum parameters $M(\rho_1)$ and $J(\rho_1)$ by identifying, at a given scale $\rho = \rho_1$, the values of the structural functions with the corresponding BTZ values. Nevertheless, the mass and angular momentum defined in this way occur to be ρ_1-dependent and diverge logarithmical as $\rho_1 \to \infty$.

Mass, Energy and Momentum

The evaluation of the surface energy and momentum densities yields

$$\epsilon(\rho,\epsilon_0) = -\frac{1}{\pi\rho}\frac{L(\rho)}{K(\rho)^2}\left(\rho^2 + 4\pi G\omega^2 Q^2\right) - \epsilon_0,$$
$$j(\rho) = -4G\omega Q^2\rho[1 - \ln(\frac{\rho^2}{\rho_0^2})]\frac{1}{K(\rho)}, \tag{11.334}$$

while the integral quantities can be evaluated from the generic expressions

$$J(\rho) = 2\pi K(\rho)\, j(\rho),\ E(\rho) = 2\pi K(\rho)\,\epsilon(\rho),\ M(\rho) = N(\rho)\,E(\rho) - W(\rho)J.$$

In this manner one arrives at

$$J(\rho) = -8\pi G\omega Q^2(1 - \ln(\frac{\rho^2}{\rho_0^2})),$$
$$E(\rho,\epsilon_0) = -2\frac{L}{\rho K}\left(\rho^2 + 4\pi G\omega^2 Q^2\right) - 2\pi K(\rho)\,\epsilon_0,$$

$$M(\rho, \epsilon_0) = -2\frac{\rho^2}{l^2} - 8\frac{\pi\, G\omega^2\, Q^2}{l^2} + 8\pi\, G\, Q^2 \ln(\frac{\rho^2}{\rho_0^2})$$
$$-2\,\pi\, K(\rho)\, N(\rho)\, \epsilon_0. \tag{11.335}$$

The evaluation of the corresponding functions for the base energy $\epsilon_0 = 0$ yield at spatial infinity $\rho \to \infty$

$$j(\rho \to \infty) \approx -4\frac{G\,\omega\, Q^2}{\rho} + 8\frac{G\,\omega Q^2}{\rho} \ln(\frac{\rho}{\rho_0}),$$
$$J(\rho \to \infty) \approx -8\pi\, G\,\omega\, Q^2(1 - 2\ln(\frac{\rho}{\rho_0})),$$
$$\epsilon(\rho \to \infty, \epsilon_0 = 0) \approx -\frac{1}{\pi\, l} - 4\frac{G\,\omega^2\, Q^2}{l\rho^2} + 4\frac{G\, Q^2}{l\rho^2}(l^2 + \omega^2)\ln(\frac{\rho}{\rho_0}),$$
$$E(\rho \to \infty, \epsilon_0 = 0) \approx -\frac{2\rho}{l} - 8\frac{\pi\, G\,\omega^2\, Q^2}{l\rho} + 8\frac{\pi\, G\, Q^2 l}{\rho} \ln(\frac{\rho}{\rho_0}),$$
$$M(\rho \to \infty, \epsilon_0 = 0) \approx -2\frac{\rho^2}{l^2} - 8\frac{\pi\, G\omega^2\, Q^2}{l^2} + 16\pi\, G\, Q^2 \ln(\frac{\rho}{\rho_0}). \tag{11.336}$$

Using in the expressions (11.334) and (11.335) as the reference energy density the quantity $\epsilon_0 = -\frac{1}{\pi\rho}\sqrt{-M_0 + \frac{\rho^2}{l^2}}$, which at the spatial infinity behaves as $\epsilon_{0|\infty}(M_0) \approx -\frac{1}{\pi\, l} + \frac{l M_0}{2\pi\, \rho^2}$, the series expansions of the corresponding quantities at $\rho \to \infty$ result in

$$\epsilon(\rho \to \infty, \epsilon_{0|\infty}(M_0)) \approx -\frac{l\, M_0}{2\pi\, \rho^2} - 4\frac{G\,\omega^2\, Q^2}{l\rho^2}$$
$$+4\frac{G\, Q^2}{l\rho^2}(l^2 + \omega^2)\ln(\frac{\rho}{\rho_0}),$$
$$E(\rho \to \infty, \epsilon_{0|\infty}(M_0)) \approx -\frac{l\, M_0}{\rho} - 8\frac{\pi\, G\,\omega^2\, Q^2}{l\rho}$$
$$+8\frac{\pi\, G\, Q^2}{l\rho}(l^2 + \omega^2)\ln(\frac{\rho}{\rho_0}),$$
$$M(\rho \to \infty, \epsilon_{0|\infty}(M_0)) \approx -M_0 - 8\frac{\pi\, G\omega^2\, Q^2}{l^2}$$
$$+8\pi\, G\, Q^2 \frac{l^2 + \omega^2}{l^2} \ln(\frac{\rho}{\rho_0}). \tag{11.337}$$

Comparing with the energy characteristics of the BTZ solution, one concludes that a mass parameter M similar to the BTZ mass is absent; instead, a term in the mass function due to the product of the rotation ω and the charge Q is present. Notice that $E(\rho)$ and $M(\rho)$ logarithmically diverge at spatial infinity. The momentum parameter is due to the product of $\omega\, Q$, and hence is not a free parameter.

Cotton Tensor

The Cotton characterization of this solution is given by

$$\left(C^{\alpha}_{\beta}\right) = \begin{bmatrix} \frac{1}{8}\frac{\omega\left(F(\rho)l^2+\rho^2\right)}{\rho\left(l^2-\omega^2\right)}\frac{d^3}{d\rho^3}F\left(\rho\right) & 0 & -\frac{1}{8}\frac{\left(F(\rho)\omega^2+\rho^2\right)l^2}{\rho\left(l^2-\omega^2\right)}\frac{d^3}{d\rho^3}F\left(\rho\right) \\ 0 & 0 & 0 \\ \frac{1}{8}\frac{\left(F(\rho)l^4+\omega^2\rho^2\right)}{l^2\rho\left(l^2-\omega^2\right)}\frac{d^3}{d\rho^3}F\left(\rho\right) & 0 & -\frac{1}{8}\frac{\omega\left(F(\rho)l^2+\rho^2\right)}{\rho\left(l^2-\omega^2\right)}\frac{d^3}{d\rho^3}F\left(\rho\right) \end{bmatrix}, \tag{11.338}$$

where

$$F(\rho) = \frac{\rho^2}{l^2} + 4\,Q^2\pi\,G\ln\left(\rho_0^2\right) - 4\,Q^2\pi\,G\ln\left(\rho^2\right),$$
$$\frac{d^3}{d\rho^3}F\left(\rho\right) = -16\,\frac{Q^2\pi\,G}{\rho^3} \tag{11.339}$$

The eigenvalue problem yields

$$\begin{aligned} &\lambda_1 = 0;\\ &\mathbf{V1} = [0, V^2, 0],\ V_\mu V^\mu = V^{2^2},\ \mathbf{V1} = \mathbf{S1},\\ &\lambda_2 = 2\,\frac{\sqrt{-F\left(\rho\right)}Q^2\pi\,G}{\rho^3};\\ &\mathbf{V2} = [V^1, 0, \frac{V^1\left(\omega\,\rho + \sqrt{-F\left(\rho\right)}l^2\right)}{l^2\left(\rho + \sqrt{-F\left(\rho\right)}\omega\right)}] = \ \mathbf{Z},\\ &\lambda_3 = -2\,\frac{\sqrt{-F\left(\rho\right)}Q^2\pi\,G}{\rho^3};\\ &\mathbf{V3} = [V^1, 0, \frac{V^1\left(\omega\,\rho - \sqrt{-F\left(\rho\right)}l^2\right)}{l^2\left(\rho - \sqrt{-F\left(\rho\right)}\omega\right)}] = \bar{\mathbf{Z}}. \end{aligned} \tag{11.340}$$

It is apparent that the Cotton tensor is of Type I: $\{S, Z, \bar{Z}\}$.

Kamata–Koikawa Limit

It should be pointed out that Clément (1993) also reported the so-called self-dual solution, published later in Kamata and Koikawa (1995, 1997). By accomplishing the limiting transition $\omega \to \pm l \Rightarrow \bar{l} \to 0$, of the metric structural functions (11.330), while the other parameters Q and $\bar{r}_0$ remain fixed, one arrives then at the metric (11.330) with structural functions and vector field

$$\begin{aligned} &F = \frac{r^2}{l^2},\ W = \mp\frac{l}{H}\,8\pi G Q^2\ln\frac{r}{\bar{r}_0},\ H = r^2 + l^2\,8\pi G Q^2\ln\frac{r}{\bar{r}_0},\\ &\boldsymbol{A} = Q\ln\frac{r}{\bar{r}_0}\ (\boldsymbol{dt} \mp l\,\boldsymbol{d\phi}). \end{aligned} \tag{11.341}$$

Notice that this solution does not possess an horizon; in the limiting transition $\bar{l} \to 0$, the horizon does not survive since it disappears below $\bar{l} = (4\pi G Q^2)^{1/2}$, as quoted by Clément.

The proper KK representation of this one-parameter solution is achieved by accomplishing the radial transformation and scaling of parameters

$$r^2 = r_{KK}^2 - r_{0KK}^2, \, r_{0KK} = (4\pi G Q^2 l^2)^{1/2}, \, r_0 = r_{0KK}/\sqrt{\mathrm{e}}, \quad (11.342)$$

and the subscripts are self-explanatory. It is worth also noticing that a derivation and analysis of the KK solution has been accomplished in Cataldo and Salgado (1996), too.

11.11.3 Martínez–Teitelboim–Zanelli Solution

Martínez, Teitelboim and Zanelli – see Martínez et al. (2000) – reported a generalization of the BTZ black hole spacetime equipped with an electric charge Q, the mass M and the angular momentum J. The main features of this charged black hole, among others, following the quoted paper, are: the total M, J and Q, which are boundary terms at infinity; the extreme black hole can be thought of as a particle moving with the speed of light; and the inner horizon of the rotating uncharged black hole is unstable under the perturbation of a small electric charge. According to the quoted reference, this electrically charged black hole is pathological in the sense it exists for arbitrary values of the mass and that there is no upper bound on the electric charge.

The starting point is the electrostatic metric (11.314) given in terms of the polar coordinate r,

$$H(r) = C_1 r + C_0 \to r^2, C_1 = 2, \, K_0/l^2 \to -\tilde{M}, b^2 \to \frac{1}{4}\tilde{Q}^2, \quad (11.343)$$

therefore $F(r) = \frac{r^2}{l^2} - \tilde{M} - \frac{1}{4}\tilde{Q}^2 \ln r^2$. Using the metric components (11.319) with transformation coefficients from the "rotation boost" transformation

$$t \to \frac{1}{\sqrt{1-\omega^2/l^2}}\left(t - \omega\phi\right), \, \phi \to \frac{1}{\sqrt{1-\omega^2/l^2}}\left(\phi - \frac{\omega}{l^2}\,t\right), \quad (11.344)$$

one arrives at the metric

$$g = -\left[\frac{r^2}{l^2} - \frac{1}{1-\omega^2/l^2}(\tilde{M} + \frac{\tilde{Q}^2}{4}\ln r^2)\right] dt^2 - 2\frac{\omega\, dt d\phi}{1-\omega^2/l^2}(\tilde{M} + \frac{\tilde{Q}^2}{4}\ln r^2)$$
$$+ \left[r^2 + \frac{\omega^2}{1-\omega^2/l^2}(\tilde{M} + \frac{\tilde{Q}^2}{4}\ln r^2)\right] d\phi^2 + \frac{dr^2}{r^2/l^2 - \tilde{M} - \frac{1}{4}\tilde{Q}^2 \ln r^2}. \quad (11.345)$$

The electromagnetic field tensor is given by

$$F_{\mu\nu} = \frac{\tilde{Q}}{r\sqrt{1-\omega^2/l^2}}\left(\delta_{[\mu}{}^t\delta_{\nu]}{}^r + \omega\, l^2 \delta_{[\mu}{}^t\delta_{\nu]}{}^\phi\right). \quad (11.346)$$

The angular momentum, charge, and mass can be evaluated via quasi-local definitions, see below; the presence of logarithmic terms in the structural metric functions yields to divergences at infinity of the energy–momentum quantities. As pointed out by the authors, the divergence in the mass can be handled by enclosing the system in a large circle of radius r_0 in which will be bound $M(r_0)$ – the energy within r_0 – and the electrostatic energy outside r_0 given by $-Q^2 \ln r_0/2$, thus the total mass (independent of r_0 and finite) is given by $\tilde{M} = M(r_0) - Q^2 \ln r_0/2$.

Dropping tilde and replacing $r \to \rho$, this metric can be brought to the canonical form

$$
\begin{aligned}
ds^2 &= -N(\rho)^2 dt^2 + \frac{1}{L(\rho)^2}\, d\rho^2 + K(\rho)^2 [d\phi + W(\rho)dt]^2,\\
K(\rho)^2 &= H(\rho) = \rho^2 + \frac{l^2\,\omega^2}{l^2-\omega^2}(M + \frac{Q^2}{4}\ln\rho^2),\\
L(\rho)^2 &= F(\rho) = \frac{\rho^2}{l^2} - M - \frac{Q^2}{4}\ln\rho^2,\\
N(\rho) &= \rho\frac{L(\rho)}{\sqrt{H(\rho)}} = \rho\frac{L(\rho)}{K(\rho)},\\
W(\rho) &= -\frac{\omega\, l^2}{l^2-\omega^2}\frac{1}{H(\rho)}(M + \frac{Q^2}{4}\ln\rho^2).
\end{aligned}
\tag{11.347}
$$

Mass, Energy and Momentum

The evaluation of the surface energy and momentum densities yields

$$
\begin{aligned}
\epsilon(\rho,\epsilon_0) &= -\frac{1}{2\pi}\frac{L}{K^2}\left(2\rho + \frac{l^2\,\omega^2}{l^2-\omega^2}\frac{Q^2}{2\rho}\right) - \epsilon_0,\\
j(\rho) &= \frac{\rho}{\pi}\frac{l^2\omega}{l^2-\omega^2}\frac{L}{N\,K^2}\left(M + \frac{1}{2}Q^2\ln\rho - \frac{Q^2}{4}\right),
\end{aligned}
\tag{11.348}
$$

while the integral quantities can be evaluated from the generic expressions for $\{J(\rho), E(\rho), M(\rho)\}$. The evaluation of the corresponding functions with $\epsilon_0 = 0$ behave at infinity according to

$$
\begin{aligned}
\epsilon(\rho\to\infty,\epsilon_0=0) &\approx -\frac{1}{\pi\, l} + \frac{[2\,M\,(l^2+\omega^2) - \omega^2\,Q^2]}{4\,\pi(l^2-\omega^2)\rho^2} + \frac{l\,Q^2}{4\,\pi\rho^2}\frac{l^2+\omega^2}{l^2-\omega^2}\ln\rho,\\
j(\rho\to\infty) &\approx \frac{\omega\, l^2}{4\pi\rho}\frac{4\,M - Q^2}{l^2-\omega^2} + \frac{\omega\,Q^2\,l^2}{2\pi\rho(l^2-\omega^2)}\ln\rho,\\
J(\rho\to\infty) &\approx \frac{\omega\,l^2}{2}\frac{4\,M-Q^2}{l^2-\omega^2} + \frac{\omega\,Q^2\,l^2}{l^2-\omega^2}\ln\rho\\
&= J_{(\mathrm{MTZ\ Eq.82})} + \frac{\omega\,Q^2\,l^2}{l^2-\omega^2}\ln\rho,\\
E(\rho\to\infty,\epsilon_0=0) &\approx -\frac{2\rho}{l} + \frac{l}{2\rho}\frac{2Ml^2-\omega^2\,Q^2}{l^2-\omega^2} + \frac{l^3\,Q^2}{2\rho(l^2-\omega^2)}\ln\rho,\\
M(\rho\to\infty,\epsilon_0=0) &\approx -\frac{2\rho^2}{l^2} + \frac{4\,Ml^2-\omega^2\,Q^2}{2(l^2-\omega^2)} + \frac{l^2\,Q^2}{l^2-\omega^2}\ln\rho.
\end{aligned}
\tag{11.349}
$$

Using in the expressions (11.348) and $\epsilon_0 = -\frac{1}{\pi\rho}\sqrt{-M_0+\frac{\rho^2}{l^2}}$, as reference energy density, which at the spatial infinity behaves as $\epsilon_{0|\infty} \approx -\frac{1}{\pi l} + \frac{l M_0}{2\pi\,\rho^2}$, the series expansions of the corresponding quantities at $\rho =$ infinity result in

$$\begin{aligned}
\epsilon(\rho\to\infty,\epsilon_{0|\infty}) &\approx \frac{l}{2\pi\,\rho^2}\,(M-M_0) + \frac{l\,\omega^2}{4\,\pi(l^2-\omega^2)\rho^2}(4\,M-Q^2)\\
&+\frac{l\,Q^2}{4\,\pi\rho^2}\frac{l^2+\omega^2}{l^2-\omega^2}\ln\rho,\\
E(\rho\to\infty,\epsilon_{0|\infty}) &\approx \frac{l}{\rho}\,(M-M_0) + \frac{l\omega^2}{2\rho\,(l^2-\omega^2)}(4\,M-Q^2)\\
&+\frac{l\,Q^2}{2\,\rho}\frac{l^2+\omega^2}{l^2-\omega^2}\ln\rho,\\
M(\rho\to\infty,\epsilon_{0|\infty}) &\approx M-M_0 + \frac{\omega^2}{2(l^2-\omega^2)}(4\,M-Q^2) + \frac{Q^2}{2}\frac{l^2+\omega^2}{l^2-\omega^2}\ln\rho\\
&= -M_0 + M_{(\mathrm{MTZEq.81})}(\omega\to\omega/l) + \frac{Q^2}{2}\frac{l^2+\omega^2}{l^2-\omega^2}\ln\rho.
\end{aligned} \tag{11.350}$$

Notice that the charges Q used above differs from $Q_{(\mathrm{MTZ,Eq.83})}$,

$$Q_{(\mathrm{MTZEq.83})} = \frac{l}{\sqrt{l^2-\omega^2}}Q \tag{11.351}$$

Therefore, comparing with the energy characteristics of the BTZ solution, one concludes that the parameter M can be considered as the BTZ mass, and the energy and mass functions logarithmically diverges at spatial infinity.

Field, Energy–Momentum, and Cotton Tensors

The Maxwell field tensor for this MTZ solution is given by

$$(F^{\alpha}{}_{\beta}) = \begin{bmatrix} 0 & \frac{1}{2}\frac{Ql}{\rho\sqrt{l^2-\omega^2}\,L^2} & 0\\ \frac{1}{2}\frac{QlL^2}{\rho\sqrt{l^2-\omega^2}} & 0 & -\frac{1}{2}\frac{\omega\,lL^2\,Q}{\rho\sqrt{l^2-\omega^2}}\\ 0 & \frac{1}{2}\frac{\omega\,Q}{l\rho\sqrt{l^2-\omega^2}\,L^2} & 0\end{bmatrix},$$

$$L^2 = \frac{\rho^2}{l^2} - m - \frac{1}{4}\,Q^2\ln\left(\rho^2\right), \tag{11.352}$$

while its eigenvalues and the corresponding eigenvectors amount to

$$\begin{aligned}
\lambda_1 &= 0; \mathbf{V1} = [V^3\,\omega, 0, V^3],\\
V_\mu V^\mu &= \frac{l^2-\omega^2}{l^2}\rho^2{V^3}^2,\ \mathbf{V1} = \mathbf{S1},\\
\lambda_2 &= -\frac{1}{2}\frac{Q}{\rho}; \mathbf{V2} = [V^1, \frac{V^1\sqrt{l^2-\omega^2}}{l}L^2, V^1\frac{\omega}{l^2}],\\
V_\mu V^\mu &= 0, \mathbf{V2} = \mathbf{N2}
\end{aligned}$$

$$\lambda_3 = \frac{1}{2}\frac{Q}{\rho}; \mathbf{V3} = [V^1, -\frac{V^1\sqrt{l^2-\omega^2}}{l}L^2, V^1\frac{\omega}{l^2}],$$
$$V_\mu V^\mu = 0, \mathbf{V3} = \mathbf{N3},$$
$$\text{Type: } \{S, N, N\}. \tag{11.353}$$

The energy–momentum tensor

$$(T^\alpha{}_\beta) = \begin{bmatrix} -\frac{1}{32\pi}\frac{Q^2(l^2+\omega^2)}{\rho^2(l^2-\omega^2)} & 0 & \frac{1}{16\pi}\frac{l^2\omega Q^2}{\rho^2(l^2-\omega^2)} \\ 0 & -\frac{1}{32\pi}\frac{Q^2}{\rho^2} & 0 \\ -\frac{1}{16\pi}\frac{\omega Q^2}{\rho^2(l^2-\omega^2)} & 0 & \frac{1}{32\pi}\frac{Q^2(l^2+\omega^2)}{\rho^2(l^2-\omega^2)} \end{bmatrix}, \tag{11.354}$$

possesses the following eigenvalues and eigenvectors

$$\lambda_1 = \frac{1}{32\pi}\frac{Q^2}{\rho^2}; \mathbf{V1} = [V^3\omega, 0, V^3],$$
$$V_\mu V^\mu = \frac{{V^3}^2\rho^2(l^2-\omega^2)}{l^2}, \mathbf{V1} = \mathbf{S1},$$
$$\lambda_{2,3} = -\frac{1}{32\pi}\frac{Q^2}{\rho^2}; \mathbf{V2} = [\frac{l^2V^3}{\omega}, V^2, V^3],$$
$$V_\mu V^\mu = -\frac{l^2(l^2-\omega^2){V^3}^2}{\omega^2}L^2 + \frac{{V^2}^2}{L^2},$$
$$\mathbf{V2} = \{\mathbf{T2}, \mathbf{N2}, \mathbf{S2}\}, \mathbf{V3} = \{\mathbf{T3}, \mathbf{N3}, \mathbf{S3}\},$$
$$\text{Type: } \{S, 2N\}, \text{and so on.} \tag{11.355}$$

For the Cotton tensor

$$(C^\alpha{}_\beta) = \begin{bmatrix} -\frac{1}{8}\frac{(\rho^2+l^2L^2)\omega Q^2}{\rho^4(l^2-\omega^2)} & 0 & \frac{1}{8}\frac{(\rho^2+\omega^2L^2)l^2Q^2}{\rho^4(l^2-\omega^2)} \\ 0 & 0 & 0 \\ -\frac{1}{8}\frac{(\omega^2\rho^2+l^4L^2)Q^2}{\rho^4l^2(l^2-\omega^2)} & 0 & \frac{1}{8}\frac{(\rho^2+l^2L^2)\omega Q^2}{\rho^4(l^2-\omega^2)} \end{bmatrix}, \tag{11.356}$$

the eigenvalues and the corresponding eigenvectors are

$$\lambda_1 = 0; \mathbf{V1} = [0, V^2, 0], V_\mu V^\mu = \frac{{V^2}^2}{L^2}\mathbf{V1} = \mathbf{S1},$$
$$\lambda_2 = \frac{1}{8}\frac{Q^2\sqrt{-L^2}}{\rho^3}; \mathbf{V2} = [V^1, 0, \frac{V^1\left(\omega\rho + \sqrt{-L^2}l^2\right)}{l^2\left(\rho+\sqrt{-L^2}\omega\right)}] = \mathbf{Z},$$
$$\lambda_3 = -\frac{1}{8}\frac{Q^2\sqrt{-L^2}}{\rho^3}; \mathbf{V3} = [V^1, 0, \frac{V^1\left(\omega\rho - \sqrt{-L^2}l^2\right)}{l^2\left(\rho-\sqrt{-L^2}\omega\right)}] = \bar{\mathbf{Z}},$$
$$\text{Type I: } \{S, Z, \bar{Z}\}. \tag{11.357}$$

Transformation of the Clément Stationary Solution to the MTZ Metric
Since these two metric representations arise from $SL(2,R)$ transformations applied on the same static charged metric, there ought to be a link between them; in fact, time and angular coordinates are related by the scaling transformations

$$\{t,\,\phi\}_C = \frac{1}{\sqrt{1-\omega^2/l^2}}\{t,\,\phi\}_{MTZ}, \tag{11.358}$$

for the same radial coordinate r and angular parameter ω, while the constant parameters are related according to

$$\{16\pi\, G\, Q^2,\, 4\pi\, G\, Q^2 \ln {r_0}^2\}_C = \{\tilde{Q}^2,\, -\tilde{M}\}_{MTZ}. \tag{11.359}$$

In this way, the Clément (1996) solution (11.328) is transformed into the MTZ, Martínez et al. (2000) solution (11.345).

11.12 Transformed Magnetostatic $a \neq 0$ Solutions

The stationary rotating generalization of the magnetostatic metric (11.76)

$$g = -\frac{F}{H}dt^2 + \frac{1}{F}dr^2 + H d\phi^2,\; F(r) = H(r)\, h(r),$$
$$H(r) = \frac{4}{C_1^2\, l^2}\left[K_0 + h(r) + a^2 l^2 \ln h(r)\right],\; h(r) := C_1\, r + C_0,$$

is determined by means of the $SL(2,R)$ transformations

$$t = \frac{\alpha}{\sqrt{\Delta}}\tilde{t} + \frac{\beta}{\sqrt{\Delta}}\tilde{\phi},\; \phi = \frac{\gamma}{\sqrt{\Delta}}\tilde{t} + \frac{\delta}{\sqrt{\Delta}}\tilde{\phi},\; \Delta = \alpha\delta - \gamma\beta,$$

which lead to the new metric in the form

$$(g_{\tilde{\mu}\tilde{\nu}}) = \begin{bmatrix} -\frac{\alpha^2}{\Delta}\frac{F}{H} + \frac{\gamma^2}{\Delta}H & 0 & -\frac{\alpha\beta}{\Delta}\frac{F}{H} + \frac{\gamma\delta}{\Delta}H \\ 0 & \frac{1}{F} & 0 \\ -\frac{\alpha\beta}{\Delta}\frac{F}{H} + \frac{\gamma\delta}{\Delta}H & 0 & -\frac{\beta^2}{\Delta}\frac{F}{H} + \frac{\delta^2}{\Delta}H \end{bmatrix},$$

which is accompanied with the electromagnetic field tensor

$$(F^{\tau\sigma}) = \begin{bmatrix} 0 & a\frac{\beta}{\sqrt{\Delta}} & 0 \\ -a\frac{\beta}{\sqrt{\Delta}} & 0 & a\frac{\alpha}{\sqrt{\Delta}} \\ 0 & -a\frac{\alpha}{\sqrt{\Delta}} & 0 \end{bmatrix}. \tag{11.360}$$

The corresponding Maxwell energy–momentum tensor becomes

$$(T_{\tilde{\mu}}{}^{\tilde{\nu}}) = \begin{bmatrix} -\frac{(\alpha\,\delta+\beta\,\gamma)a^2}{8\pi\Delta}\frac{H}{F} & 0 & \frac{\gamma\,\alpha\,a^2}{4\pi\Delta}\frac{H}{F} \\ 0 & \frac{a^2 H}{8\pi\, F} & 0 \\ -\frac{\delta\,\beta\,a^2}{4\pi\Delta}\frac{H}{F} & 0 & \frac{(\alpha\,\delta+\beta\,\gamma)a^2}{8\pi\,\Delta}\frac{H}{F} \end{bmatrix}. \tag{11.361}$$

Explicitly, the nonzero metric components are

$$g_{\tilde{t}\tilde{t}} = -\frac{\alpha^2}{\Delta} h(r) + \frac{\gamma^2}{\Delta} \frac{1}{h(r)\, g_{rr}}, \; g_{\tilde{t}\tilde{\phi}} = -\frac{\alpha\beta}{\Delta} h(r) + \frac{\delta\gamma}{\Delta} \frac{1}{h(r)\, g_{rr}},$$
$$g_{\tilde{\phi}\tilde{\phi}} = -\frac{\beta^2}{\Delta} h(r) + \frac{\delta^2}{\Delta} \frac{1}{h(r)\, g_{rr}},$$
$$g_{rr} = \frac{1}{4\,h(r)} \frac{C_1^2\, l^2}{\left[K_0 + h(r) + a^2 l^2 \ln h(r)\right]}, \; h(r) = C_1 r + C_0. \tag{11.362}$$

11.12.1 Stationary Magneto-Electric Solution

In particular, for the $SL(2,R)$ transformation

$$t = \tilde{t}, \; \phi = -\frac{\omega}{l^2}\tilde{t} + \tilde{\phi}, \; \alpha = 1, \; \beta = 0, \; \gamma = -\frac{\omega}{l^2}, \; \delta = 1, \tag{11.363}$$

one obtains a new solution with metric components

$$g_{\tilde{t}\tilde{t}} = -h(r) + \frac{\omega^2}{l^4} \frac{1}{h(r)\, g_{rr}}, \; g_{\tilde{t}\tilde{\phi}} = -\frac{\omega}{l^2} \frac{1}{h(r)\, g_{rr}},$$
$$g_{\tilde{\phi}\tilde{\phi}} = \frac{1}{h(r)\, g_{rr}}, \; g_{rr} = \frac{1}{4\,h(r)} \frac{C_1^2\, l^2}{\left[K_0 + h(r) + a^2 l^2 \ln h(r)\right]},$$
$$h(r) = C_1 r + C_0, \tag{11.364}$$

which is equal to the constant $W = \omega$ stationary magneto-electric solution (11.255). Therefore, by means of a $SL(2,R)$ transformation applied to the magnetostatic cyclic symmetric $(2+1)$ Einstein–Maxwell solution, one can generate a unique electromagnetic stationary cyclic symmetric solution in the sense of the structure of the field tensor $F^{\mu\nu}$. Clément (1993) reported a field belonging to this class of solutions; see Eq. (Cl.23).

11.12.2 Dias–Lemos Magnetic BTZ–Solution Counterpart

Dias and Lemos (2002) published a rotating magnetic solution in (2+1) gravity: the magnetic counterpart of the spinning charged BTZ solution, i.e., a point source generating a magnetic field. Also, it was established that both the static and rotating magnetic solutions possess negative mass and that there is an upper bound for the intensity of the magnetic field source and for the value of the angular momentum.

A simple representation of this solution can be achieved from our transformed magnetic metric (11.362) by setting

$$t = \tilde{t}, \; \phi = -\frac{\omega}{l^2}\tilde{t} + \tilde{\phi}, \; \alpha = 1, \; \beta = -\omega, \; \gamma = -\frac{\omega}{l^2}, \; \delta = 1, \; \Delta = 1,$$
$$C_1 = 2, \; h(r) = C_1 r + C_0 \to r^2,$$

obtaining

$$g = -(r^2 - \frac{\omega^2}{l^4}F)dt^2 + (F - \omega^2 r^2)d\phi^2 - 2\omega(\frac{F}{l^2} - r^2)dtd\phi + \frac{dr^2}{F(r)},$$
$$F = \frac{K_0}{l^2} + \frac{r^2}{l^2} + a^2 \ln r^2. \tag{11.365}$$

The proper Dias–Lemos representation uses a more involved definition of the transformed r-coordinate and parameterizations of the $SL(2,R)$ transformations, namely

$$h(r) = C_1 r + C_0 \to (\rho^2 + r_+^2 - ml^2)/l^2,\ C_1 = 2/l^2,\ \chi^2 = a^2 l^4,$$
$$t = \sqrt{1+\omega^2}\tilde{t} - l\,\omega\tilde{\phi}\ \phi = -\frac{\omega}{l}\tilde{t} + \sqrt{1+\omega^2}\,\tilde{\phi};$$

dropping tildes, one has

$$g = -\left[h - \frac{\omega^2}{l^2}\left(m\,l^2 + \chi^2 \ln|h|\right)\right] dt^2$$
$$-2\frac{\omega}{l}\sqrt{\omega^2+1}\left[m\,l^2 + \chi^2 \ln|h|\right] dt\, d\phi$$
$$+[h\,l^2 + (\omega^2+1)\left(m\,l^2 + \chi^2 \ln|h|\right)]d\phi^2$$
$$+\frac{l^2\,\rho^2\,d\rho^2}{(\rho^2 + r_+^2 - ml^2)[\rho^2 + r_+^2 + \chi^2 \ln|h|]},$$
$$h = (\rho^2 + r_+^2 - ml^2)/l^2, \tag{11.366}$$

with vector potential

$$\boldsymbol{A} = \frac{1}{2}\chi \ln|h(\rho)|[-\frac{\omega}{l}\boldsymbol{dt} + \sqrt{1+\omega^2}\,\boldsymbol{d\Phi}]. \tag{11.367}$$

Notice that in the above mentioned representation, the equation

$$r_+^2 + \chi^2 \ln|(\rho^2 + r_+^2)/l^2 - m| = 0, \tag{11.368}$$

used in the Eq. (3.2) of Dias and Lemos (2002) was not used here. According to these authors, this rotating magnetic spacetime is null and timelike geodesically complete, and as such horizonless.

It it noteworthy to point out that by subjecting the static Hirschman–Welch metric (11.101) to the $SL(2,R)$ transformation

$$t \to \sqrt{1+\omega^2}t - \omega\,l\phi,\ \rho \to \rho, \phi \to -\frac{\omega}{l}t + \sqrt{1+\omega^2}\phi,$$

one arrives at the Dias and Lemos (2002) solution too, see also García (2009), Eq. (11.41), given in the canonical representation by metric

$$ds^2 = -N(\rho)^2 dt^2 + \frac{1}{L(\rho)^2}d\rho^2 + K(\rho)^2[d\phi + W(\rho)dt]^2,$$
$$H(\rho) := (\rho^2 + r_+^2 - ml^2)/l^2,$$

$$L(\rho) = \frac{\sqrt{H(\rho)}}{\rho}\sqrt{\rho^2 + r_+^2 + \chi^2 \ln H(\rho)},$$
$$K(\rho) = \sqrt{\rho^2 + r_+^2 + \omega^2 l^2 m + (1+\omega^2)\chi^2 \ln H(\rho)},$$
$$N(\rho) = \rho\frac{\sqrt{L(\rho)}}{\sqrt{K(\rho)}},\ W(\rho) = -\frac{\omega\sqrt{1+\omega^2}}{l}\frac{[ml^2 + \chi^2 \ln H(\rho)]}{K(\rho)^2}. \tag{11.369}$$

Notice that this metric, in the case of vanishing charge $\chi = 0$, yields to an alternative coordinate representation of the rotating BTZ solution, with parameter ω, namely

$$ds^2 = -\left(\frac{\rho^2 + r_+^2}{l^2} - (1+\omega^2)\, m\, l^2\right) dt^2 + \frac{\rho^2 l^2}{(\rho^2 + {r_+}^2)(\rho^2 + r_+^2 - m\, l^2)}\, d\rho^2$$
$$-2\,\omega\, m\, l\,\sqrt{1+\omega^2} d\phi\, dt + (\rho^2 + {r_+}^2 + l^2\omega^2) d\phi^2, \tag{11.370}$$

which differs from the standard BTZ solution representations.
Accomplishing in the above metric the transformations

$$t \to l^2\, t,\ \rho \to \sqrt{\rho^2/l^2 - ml^2\omega^2 - {r_+}^2 + M}$$

and identifying the parameters according to

$$m = \frac{M}{1+2\,\omega^2},\ M^2\frac{(1+\omega^2)\omega^2\, l^2}{(2\omega^2+1)^2} = J^2/4 \to$$
$$\omega^2 = \frac{1}{2}\frac{Ml \pm \sqrt{M^2l^2 - J^2}}{\sqrt{M^2l^2 - J^2}},\ m = \frac{\sqrt{M^2l^2 - J^2}}{l^3},$$

one arrives at the BTZ solution counterpart representation

$$ds^2 = -\rho^2 dt^2 + \left(\frac{\rho^2}{l^2} + M + \frac{J^2}{4\rho^2}\right)^{-1} d\rho^2 - J\, d\phi\, dt + \left(\frac{\rho^2}{l^2} + M\right) d\phi^2.$$

On the other hand, by replacing $\rho \to \sqrt{\rho^2 - m\, l^2\omega^2 - {r_+}^2}$ one arrives at the middle of the road metric

$$ds^2 = -\left(\frac{\rho^2}{l^2} - (1+2\,\omega^2)\, m\right) dt^2 - 2\,\omega\, m\, l\,\sqrt{1+\omega^2} d\phi\, dt + \rho^2 d\phi^2$$
$$+\left(\frac{\rho^2}{l^2} - (1+2\,\omega^2) m + \frac{l^2 m^2\omega^2(1+\omega^2)}{\rho^2}\right)^{-1} d\rho^2, \tag{11.371}$$

which, identifying

$$m = \frac{M}{1+2\,\omega^2},\ l^2 m^2\omega^2(1+\omega^2) = J^2/4 \to$$
$$\omega^2 = \frac{1}{2}\frac{Ml \pm \sqrt{M^2l^2 - J^2}}{\sqrt{M^2l^2 - J^2}},\ m = \frac{\sqrt{M^2l^2 - J^2}}{l},\ 2\sqrt{1+\omega^2}\omega\, ml \to J,$$

gives rise to the standard description of the stationary BTZ black hole metric. Therefore, as the vacuum limit of the DL metric (11.369) one may consider the rotating BTZ solution counterpart, and consequently one may think of it as the reference vacuum solution in the evaluation of the quasi-local energy, momentum and mass.

Mass, Energy and Momentum

The surface energy and momentum densities are given by

$$\epsilon(\rho,\epsilon_0) = -\rho\,\frac{L}{\pi\, l^2\, K^2\, H}\,(\rho^2 + r_+^2 - ml^2 + (1+\omega^2)\chi^2) - \epsilon_0,$$
$$j(\rho,\epsilon_0) = \omega\sqrt{1+\omega^2}\,\frac{\rho}{\pi\, l}\,\frac{L}{N\,K^2}[m\,l^2 - \chi^2 + \chi^2 \ln H], \tag{11.372}$$

while the integral quantities amount to

$$\begin{aligned}
J(\rho,\epsilon_0) &= 2\omega\sqrt{1+\omega^2}\,\frac{\rho}{l}\,\frac{L}{N\,K}[m\,l^2 - \chi^2 + \chi^2 \ln H] \\
&= \frac{2}{l}\omega\sqrt{1+\omega^2}\,[m\,l^2 - \chi^2 + \chi^2 \ln H], \\
E(\rho,\epsilon_0) &= -2\frac{\rho}{l^2}\,\frac{L}{K\,H}\,P(\rho) - 2\pi\, K\epsilon_0, \\
M(\rho,\epsilon_0) &= -2\frac{\rho}{l^2}\,\frac{NL}{HK}\,P(\rho) - W\,J - 2\pi\, N\, K\epsilon_0 = 2\,m \\
&\quad -\frac{2}{l^2}(\rho^2 + \chi^2 + r_+^2) + 2\frac{\omega^2}{l^2}[ml^2 - \chi^2 + \chi^2 \ln H] - 2\pi\, N\, K\epsilon_0, \\
P(\rho) &:= \rho^2 + r_+^2 - ml^2 + (1+\omega^2)\chi^2.
\end{aligned} \tag{11.373}$$

The evaluation of the main parts of above functions, i.e., the corresponding functions independent of ϵ_0 behave at infinity according to

$$\begin{aligned}
j(\rho\to\infty) &\approx \frac{\omega}{l\pi\rho}\sqrt{1+\omega^2}[m\,l^2 - \chi^2 + 2\chi^2 \ln(\frac{\rho}{l})], \\
J(\rho\to\infty) &\approx 2\frac{\omega}{l}\sqrt{1+\omega^2}[m\,l^2 - \chi^2 + 2\chi^2 \ln(\frac{\rho}{l})], \\
\epsilon(\rho\to\infty,\epsilon_0=0) &\approx -\frac{1}{\pi\, l} + \frac{ml^2 - 2\chi^2}{2\pi\, l\,\rho^2} + \frac{\chi^2}{\pi\, l\,\rho^2}\ln(\frac{\rho}{l}) \\
&\quad + \omega^2[\frac{m\,l^2 - \chi^2}{\pi\, l\,\rho^2} + 2\frac{\chi^2}{\pi\, l\,\rho^2}\ln(\frac{\rho}{l})], \\
E(\rho\to\infty,\epsilon_0=0) &\approx -\frac{2\rho}{l} + \frac{ml^2 - r_+^2 - 2\chi^2}{l\,\rho} \\
&\quad + \frac{\omega^2}{l\,\rho}[m\,l^2 - 2\chi^2 + 2\chi^2 \ln(\frac{\rho}{l})], \\
M(\rho\to\infty,\epsilon_0=0) &\approx 2m - 2\frac{1}{l^2}(\rho^2 + r_+^2 + \chi^2) \\
&\quad + 2\frac{\omega^2}{l^2}[m\,l^2 - \chi^2 + 2\chi^2 \ln(\frac{\rho}{l})].
\end{aligned} \tag{11.374}$$

Using in the expressions (11.373) as reference energy density the quantity $\epsilon_0 = -\frac{1}{\pi\rho}\sqrt{\frac{\rho^2}{l^2} - M_0}$, which at the spatial infinity behaves as $\epsilon_{0|\infty} \approx -\frac{1}{\pi l} + \frac{l M_0}{2\pi\,\rho^2}$, the series expansions of the corresponding quantities at $\rho =$ infinity result in

$$\begin{aligned}
\epsilon(\rho \to \infty, \epsilon_{0|\infty}) \approx\ & \frac{l}{2\pi\,\rho^2}(m - M_0) - \frac{1}{2\pi\,l\,\rho^2}(\chi^2 - \chi^2 \ln(\frac{\rho}{l})) \\
& + \frac{\omega^2}{\pi\,l\,\rho^2}[ml^2 - \chi^2 + 2\chi^2 \ln(\frac{\rho}{l})], \\
E(\rho \to \infty, \epsilon_{0|\infty}) \approx\ & \frac{l}{\rho}(m - M_0) - 2\frac{\chi^2}{l\,\rho} + 2\frac{\chi^2}{l\,\rho}\ln(\frac{\rho}{l}) \\
& + 2\frac{\omega^2}{l\,\rho}[m\,l^2 - \chi^2 + 2\chi^2 \ln(\frac{\rho}{l})], \\
M(\rho \to \infty, \epsilon_{0|\infty}) \approx\ & m - M_0 - 2\frac{\chi^2}{l^2} + 2\frac{\chi^2}{l^2}\ln(\frac{\rho}{l}) \\
& + 2\frac{\omega^2}{l^2}[m\,l^2 - \chi^2 X + 2\chi^2 \ln(\frac{\rho}{l})].
\end{aligned} \tag{11.375}$$

Therefore, comparing with the energy characteristics of the BTZ solution, one concludes that the mass logarithmically diverges at spatial infinity. For vanishing rotation parameter ω one recovers the static solution in the representation of Hirschman–Welch and certainly the corresponding energy quantities. The parameter m can be considered as the BTZ mass.

Field, Energy and Cotton Tensors

To determine the algebraic types of the electromagnetic field, energy–momentum, and Cotton tensors it is more convenient to work with the DL metric in the form

$$g = \begin{bmatrix} -h - \omega^2\left(h - L^2\right) & 0 & l\sqrt{1+\omega^2}\omega\left(h - L^2\right) \\ 0 & \frac{\rho^2}{h\ L^2 l^2} & 0 \\ l\sqrt{1+\omega^2}\omega\left(h - L^2\right) & 0 & l^2\left(-\omega^2 h + L^2 + L^2\omega^2\right) \end{bmatrix}, \tag{11.376}$$

where

$$L^2 = \frac{\chi^2 \ln(h) + \rho^2}{l^2},\ h = \frac{\rho^2 + Mg}{l^2},\ Mg = r_+^2 - ml^2. \tag{11.377}$$

In this representation, the electromagnetic field tensor becomes

$$(F^\alpha{}_\beta) = \begin{bmatrix} 0 & -\frac{\rho\,\chi\,\omega}{l^3 L^2 h} & 0 \\ -\frac{\omega\,\chi\,L^2}{l\rho} & 0 & \frac{\chi\sqrt{1+\omega^2}L^2}{\rho} \\ 0 & -\frac{\rho\,\chi\sqrt{1+\omega^2}}{L^2\,l^4 h} & 0 \end{bmatrix}, \tag{11.378}$$

with the following eigenvalues and their corresponding eigenvectors

$$\begin{aligned}
&\lambda_1 = 0; \mathbf{V1} = [V^1, 0, \frac{\omega}{l\sqrt{1+\omega^2}} V^1], V_\mu V^\mu = -\frac{h}{1+\omega^2} {V^1}^2, \mathbf{V1} = \mathbf{T1},\\
&\lambda_2 = -i\frac{\chi}{\sqrt{h}l^2}; \mathbf{V2} = [i\frac{\omega\rho}{lL^2\sqrt{h}} V^2, V^2, i\frac{\sqrt{1+\omega^2}\rho}{L^2 l^2\sqrt{h}} V^2], \mathbf{V2} = \mathbf{Z},\\
&\lambda_3 = i\frac{\chi}{\sqrt{h}l^2}; \mathbf{V3} = [-i\frac{\omega\rho}{lL^2\sqrt{h}} V^2, V^2, -i\frac{\sqrt{1+\omega^2}\rho}{L^2 l^2\sqrt{h}} V^2], \mathbf{V3} = \bar{\mathbf{Z}},\\
&\text{Type}: \{T, Z, \bar{Z}\}.
\end{aligned} \tag{11.379}$$

As far as to the electromagnetic energy momentum tensor is concerned, its matrix is given by

$$(T^\alpha{}_\beta) = \begin{bmatrix} -\frac{1}{8\pi}\frac{\chi^2(1+2\omega^2)}{l^2(\rho^2+Mg)} & 0 & \frac{1}{4\pi}\frac{\omega\chi^2\sqrt{1+\omega^2}}{l(\rho^2+Mg)} \\ 0 & \frac{1}{8\pi}\frac{\chi^2}{l^2(\rho^2+Mg)} & 0 \\ -\frac{1}{8\pi}\frac{\omega\chi^2\sqrt{1+\omega^2}}{l^3(\rho^2+Mg)} & 0 & \frac{1}{8\pi}\frac{\chi^2(1+2\omega^2)}{l^2(\rho^2+Mg)} \end{bmatrix}, \tag{11.380}$$

with the following eigenvalues and their corresponding eigenvectors

$$\begin{aligned}
\lambda_1 &= -\frac{1}{8\pi}\frac{\chi^2}{l^2(\rho^2+Mg)}; \mathbf{V1} = [V^1, 0, \frac{\omega}{l\sqrt{1+\omega^2}} V^1],\\
V_\mu V^\mu &= -\frac{h}{1+\omega^2}{V^1}^2, \mathbf{V1} = \mathbf{T1},\\
\lambda_2 &= \frac{1}{8\pi}\frac{\chi^2}{l^2(\rho^2+Mg)}; \mathbf{V2} = [\frac{\omega l}{\sqrt{1+\omega^2}} V^3, V^2, V^3],\\
V^\mu V_\mu &= \frac{l^2L^2}{1+\omega^2}{V^3}^2 + \frac{\rho^2}{hL^2l^2}{V^2}^2, \mathbf{V2} = \mathbf{S2},\\
\lambda_3 &= \frac{1}{8\pi}\frac{\chi^2}{l^2(\rho^2+Mg)}; \mathbf{V3} = [\frac{\omega l}{\sqrt{1+\omega^2}}\tilde{V}^3, \tilde{V}^2, \tilde{V}^3],\\
V^\mu V_\mu &= \frac{l^2L^2}{1+\omega^2}(\tilde{V}^3)^2 + \frac{\rho^2}{hL^2l^2}(\tilde{V}^2)^2, \mathbf{V3} = \mathbf{S3},\\
&\text{Type}: \{T, 2\,S\}.
\end{aligned} \tag{11.381}$$

This tensor structure corresponds to that of a perfect fluid energy momentum tensor, but this time for the state equation: *energy* = *pressure*. Again, the solutions generated from this metric by using coordinate transformations possesses this perfect fluid feature because of the invariance of the eigenvalues.

The Cotton tensor for stationary cyclic symmetric gravitational field is given by

$$(C^\alpha{}_\beta) = \begin{bmatrix} -\frac{\chi^2}{2}\frac{\omega\sqrt{1+\omega^2}(h+L^2)}{h^2l^5} & 0 & \frac{\chi^2}{2}\frac{(\omega^2h+L^2+L^2\omega^2)}{h^2l^4} \\ 0 & 0 & 0 \\ -\frac{\chi^2}{2}\frac{(h+\omega^2h+L^2\omega^2)}{h^2l^6} & 0 & \frac{\chi^2}{2}\frac{\omega\sqrt{1+\omega^2}(h+L^2)}{h^2l^5} \end{bmatrix}. \tag{11.382}$$

Searching for its eigenvectors, one arrives at

$$\begin{aligned}
&\lambda_1 = 0; \mathbf{V1} = [0, V^2, 0], V_\mu V^\mu = \frac{\rho^2}{h\ L^2 l^2} V^{1^2}, \mathbf{V1} = \mathbf{S},\\
&\lambda_2 = \frac{i}{2} \frac{\chi^2 L}{h^{3/2} l^5};\\
&\mathbf{V2} = [\frac{V^3 l \left(-i L\sqrt{h} + \omega\sqrt{1+\omega^2}h + \omega\sqrt{1+\omega^2}L^2\right)}{h + \omega^2 h + L^2\omega^2}, 0, V^3], \mathbf{V2} = \mathbf{Z},\\
&\lambda_3 = -\frac{i}{2} \frac{\chi^2 L}{h^{3/2} l^5};\\
&\mathbf{V3} = [\frac{V^3 l \left(i L\sqrt{h} + \omega\sqrt{1+\omega^2}h + \omega\sqrt{1+\omega^2}L^2\right)}{h + \omega^2 h + L^2\omega^2}, 0, V^3], \mathbf{V3} = \bar{\mathbf{Z}},\\
&\text{Type I: } \{T, Z, \bar{Z}\}.
\end{aligned} \tag{11.383}$$

The eigenvectors **V2** and **V3** are complex conjugated while the vector **V1**, associated to the zero eigenvalue, is the only physically meaningful spacelike direction in this case.

The data of the stationary rotationally symmetric solutions generated by means of $SL(2,R)$ transformations can be represented schematically by the table:

Table 11.12.1 *Stationary electromagnetic solutions*

Integration and $SL(2,R)$ transformations			
References	via	Eqns.	$\Lambda = \pm 1/l^2$
electrostatic seed	SL–T	§ 11.11	$-1/l^2$
Clément (1996)	SL–T	§ 11.11.2	$\pm 1/l^2$
Martínez et al. (2000)	SL–T	§ 11.11.3	$-1/l^2$
magnetostatic seed	SL–T	§ 11.12	$-1/l^2$
Dias and Lemos (2002)	SL–T	§ 11.12.2	$-1/l^2$
electromagnetic		§ 11.1.2	$-1/l^2$
Kamata and Koikawa (1995)	Int	§11.6.7	$\pm 1/l^2$
Ayón–Beato et al. (2005)	Int	§ 11.7	$\pm 1/l^2$
García (2009) and	Int	§ 11.9	$\pm 1/l^2$
Garcia–Diaz (2013)	Int	(11.277), (11.9.3)	$\pm 1/l^2$

11.13 Transformed Cataldo Hybrid Static Solution

Applying the $SL(2,R)$ transformation,

$$t = \frac{\alpha}{\sqrt{\Delta}}\tilde{t} + \frac{\beta}{\sqrt{\Delta}}\tilde{\phi}, \phi = \frac{\gamma}{\sqrt{\Delta}}\tilde{t} + \frac{\delta}{\sqrt{\Delta}}\tilde{\phi},\ \Delta = \alpha\delta - \gamma\beta,$$

to the Cataldo azimuthal electrostatic solution (11.115) one arrives at a new stationary solution having the static BTZ solution as a limit, namely

$$g = -(\frac{\alpha^2}{\Delta} H_{(-)} - \frac{\gamma^2}{\Delta} H_{(+)})dt^2 + 2(-\frac{\alpha\beta}{\Delta} H_{(-)}$$
$$+\frac{\gamma\delta}{\Delta} H_{(+)})dtd\phi + (-\frac{\beta^2}{\Delta} H_{(-)} + \frac{\delta^2}{\Delta} H_{(+)})d\phi^2 + \left(\frac{\rho^2}{l^2} - M\right)^{-1} dr^2,$$
$$H_{(+)} := \rho^{1+\sqrt{\alpha_0}} \left(\frac{\rho^2}{l^2} - M\right)^{(1-\sqrt{\alpha_0})/2}, \; H_{(-)} := \rho^{1-\sqrt{\alpha_0}} \left(\frac{\rho^2}{l^2} - M\right)^{(1+\sqrt{\alpha_0})/2}. \tag{11.384}$$

The electromagnetic tensors are given by

$$F_{\mu\nu} = \frac{M\sqrt{1-\alpha_0}}{2} \, \delta_{[\mu}{}^{t} \delta_{\nu]}{}^{\phi},$$
$$T_{\mu}{}^{\nu} = \frac{M^2}{32\pi\rho^2} \frac{(1-\alpha_0)}{\rho^2/l^2 - M} (-\delta_{\mu}{}^{t}\delta_{t}{}^{\nu} + \delta_{\mu}{}^{\rho}\delta_{\rho}{}^{\nu} - \delta_{\mu}{}^{\phi}\delta_{\phi}{}^{\nu}). \tag{11.385}$$

In the seed hybrid static metric α_0 has been used instead of the original α to avoid confusion with the parameters appearing in the $SL(2,R)$ transformations.

Therefore the transformed electromagnetic tensors stay unchanged. Hence, by means of an $SL(2,R)$ transformation applied to the hybrid static cyclic symmetric (2+1) Einstein–Maxwell solution, one generates a family of hybrid, stationary cyclic symmetric solutions with structurally unique field tensors $F_{\mu\nu}$ and $T_{\mu}{}^{\nu}$.

In particular, one could choose the rotation boost transformation

$$t \to \frac{1}{\sqrt{1-\omega^2/l^2}} (t - \omega\phi) \, , \; \phi \to \frac{1}{\sqrt{1-\omega^2/l^2}} \left(\phi - \frac{\omega}{l^2} t\right),$$

where the parameter ω can be related to the angular momentum constant.

The generalization of the Cataldo static solution, via $SL(2,R)$ transformations, is given by

$$ds^2 = -N(\rho)^2 dt^2 + \frac{1}{L(\rho)^2} d\rho^2 + K(\rho)^2 [d\phi + W(\rho)dt]^2,$$
$$K(\rho)^2 = \frac{\rho(\rho^2 - Ml^2)^{1/2}}{l\,\Delta} \left[\delta^2 \, l^{\sqrt{\alpha_0}} \rho^{\sqrt{\alpha_0}} (\rho^2 - Ml^2)^{-\sqrt{\alpha_0}/2} \right.$$
$$\left. -\beta^2 \, l^{-\sqrt{\alpha_0}} \rho^{-\sqrt{\alpha_0}} (\rho^2 - Ml^2)^{\sqrt{\alpha_0}/2} \right],$$
$$N(\rho)^2 = \frac{\rho^2}{l^2} \frac{\rho^2 - Ml^2}{K(\rho)^2}, \quad \Delta := (\alpha\delta - \beta\gamma) \neq 0,$$
$$W(\rho) = -\frac{1}{\Delta} \frac{\rho}{l} \frac{\sqrt{\rho^2 - Ml^2}}{K(\rho)^2} \left[\alpha\beta \, l^{-\sqrt{\alpha_0}} \rho^{-\sqrt{\alpha_0}} (\rho^2 - Ml^2)^{\sqrt{\alpha_0}/2} \right.$$
$$\left. - \gamma\delta \, l^{\sqrt{\alpha_0}} \rho^{\sqrt{\alpha_0}} (\rho^2 - Ml^2)^{-\sqrt{\alpha_0}/2} \right],$$
$$L(\rho)^2 = \frac{\rho^2}{l^2} - M. \tag{11.386}$$

11.13.1 Mass, Energy and Momentum

The corresponding surface densities are

$$\epsilon(\rho,\epsilon_0) = -\frac{1}{2\pi\, l\sqrt{1-Ml^2/\rho^2}}\left[\beta^2(1-\frac{Ml^2}{\rho^2})^{\frac{\sqrt{\alpha_0}}{2}}(2-(1-\sqrt{\alpha_0})\frac{Ml^2}{\rho^2}\right.$$
$$\left. -\,\delta^2 l^{2\sqrt{\alpha_0}}(1-\frac{Ml^2}{\rho^2})^{-\frac{\sqrt{\alpha_0}}{2}}(2-(1+\sqrt{\alpha_0})\frac{Ml^2}{\rho^2})\right]$$
$$\times\left[\beta^2(1-\frac{Ml^2}{\rho^2})^{\frac{\sqrt{\alpha_0}}{2}} - \delta^2\, l^{2\sqrt{\alpha_0}}(1-\frac{Ml^2}{\rho^2})^{-\frac{\sqrt{\alpha_0}}{2}}\right]^{-1} - \epsilon_0, \quad (11.387)$$

while the local momentum is

$$j(\rho) = \beta\delta\frac{\sqrt{\alpha_0}M}{\pi\Delta\, K(\rho)}, \quad (11.388)$$

therefore the product $\beta\delta$ is related with the rotation properties of the considered solution. The integral quantities amount to

$$\begin{aligned} J(\rho) &= 2\,\pi\, K(\rho)\, j(\rho) = 2\beta\delta\,\frac{\sqrt{\alpha_0}M}{\Delta},\\ E(\rho) &= 2\,\pi\, K(\rho)\,\epsilon(\rho),\\ M(\rho) &= N(\rho)\, E(\rho) - W(\rho)\, J(\rho). \end{aligned} \quad (11.389)$$

The evaluation of energy and mass functions independent of ϵ_0 behave at infinity as

$$\epsilon(\rho\to\infty,\epsilon_0=0) \approx -\frac{1}{\pi\, l} + \frac{\sqrt{\alpha_0}l\, M}{2\pi\rho^2}\frac{\delta^2 l^{2\sqrt{\alpha_0}}+\beta^2}{\delta^2\, l^{2\sqrt{\alpha_0}}-\beta^2},$$

$$E(\rho\to\infty,\epsilon_0=0) \approx -2\rho\, l^{-3/2-\sqrt{\alpha_0}/2}\frac{\sqrt{\delta^2\, l^{2\sqrt{\alpha_0}}-\beta^2}}{\sqrt{\Delta}}$$
$$-\,l^{1/2-\sqrt{\alpha_0}/2}\frac{M}{2\rho\sqrt{\Delta}}\frac{[(\sqrt{\alpha_0}-1)\beta^4+2\beta^2\delta^2\, l^{2\sqrt{\alpha_0}}-(\sqrt{\alpha_0}+1)\delta^4\, l^{4\sqrt{\alpha_0}}]}{(\delta^2\, l^{2\sqrt{\alpha_0}}-\beta^2)^{3/2}},$$

$$M(\rho\to\infty,\epsilon_0=0) = M\; - 2\frac{\rho^2}{l^2}$$
$$+M\left[\sqrt{\alpha_0}\frac{\delta^2\, l^{2\sqrt{\alpha_0}}+\beta^2}{\delta^2\, l^{2\sqrt{\alpha_0}}-\beta^2} + 2\sqrt{\alpha_0}\frac{\beta\delta}{\Delta}\frac{\alpha\beta-\gamma\delta\, l^{2\sqrt{\alpha_0}}}{\delta^2\, l^{2\sqrt{\alpha_0}}-\beta^2}\right]. \quad (11.390)$$

Using in expressions (11.387)–(11.389) as reference energy density at the spatial infinity the quantity which behaves as $\epsilon_{0|\infty} \approx -\frac{1}{\pi\, l}+\frac{l\, M_0}{2\pi\,\rho^2}$, the series expansions of the corresponding quantities at $\rho\to$ infinity result in

$$\epsilon(\rho\to\infty,\epsilon_{0|\infty}) \approx \frac{l}{2\pi\,\rho^2}(\sqrt{\alpha_0}M - M_0) - \frac{l\,\sqrt{\alpha_0}M}{\pi\,\rho^2}\frac{\beta^2}{\delta^2 l^{2\sqrt{\alpha_0}}-\beta^2},$$

$$
\begin{aligned}
E(\rho\to\infty,\epsilon_{0|\infty}) &\approx l^{1/2-\sqrt{\alpha_0}/2}\frac{\sqrt{\delta^2\, l^{2\sqrt{\alpha_0}}-\beta^2}}{\rho\sqrt{\Delta}}(\sqrt{\alpha_0}M-M_0)\\
&\quad+2l^{1/2-\sqrt{\alpha_0}/2}\frac{\beta^2\sqrt{\alpha_0}M}{\rho\sqrt{\delta^2 l^{2\sqrt{\alpha_0}}-\beta^2}},\\
M(\rho\to\infty,\epsilon_{0|\infty}) &\approx \sqrt{\alpha_0}M-M_0\\
&\quad-\frac{\sqrt{\alpha_0}M\beta}{\Delta}\frac{\gamma\delta^2\, l^{2\sqrt{\alpha_0}}-2\alpha\beta\delta+\beta^2\gamma}{\delta^2 l^{2\sqrt{\alpha_0}}-\beta^2}.
\end{aligned}
\tag{11.391}
$$

Therefore, comparing with the energy characteristics of the BTZ solution, one concludes that role of the mass parameter is played by the product $\sqrt{\alpha_0}M$. At spatial infinity the mass function is finite, and the energy density and global energy approach infinity as fast as $1/\rho^2$ and $1/\rho$ correspondingly.

11.14 Summary on Electromagnetic Maxwell Solutions

In the framework of the (2+1)-dimensional Einstein–Maxwell theory with cosmological constant, different families of exact solutions for cyclic symmetric stationary (static) metrics have been derived. For the static classes and also hybrid, static and stationary families, their uniqueness is proven by the integration procedure used. The completeness and relationship of all uniform electromagnetic $F_{\mu\mu;\alpha}=0$, and constant invariant $F_{\mu\nu}F^{\mu\nu}=2\gamma$ solutions is achieved. The uniqueness of the stationary families of solutions has been partially established; various specific branches of solutions in the general case are determined via a straightforward integration. In this systematic approach all known electromagnetic stationary cyclic symmetric solutions are properly identified. It seems to be a rule that electrically charged solutions allow for a black hole interpretation, while for the magnetic classes such a black hole feature seems to be absent.

Their energy–momentum densities and global energy–momentum–mass quantities have been evaluated using the Brown–York approach. As reference characteristics those corresponding to the stationary or static BTZ–AdS with parameter M_0 solutions have been used. The electric and magnetic solutions, and their generalizations through $SL(2,R)$ transformations exhibit at the spatial infinity $\rho\to\infty$ the following generic behavior

$$
\begin{aligned}
J(\rho\to\infty) &\approx \alpha_J\, J+\beta_J\ln\rho,\\
\epsilon(\rho\to\infty,\epsilon_{0|\infty}(M_0)) &\approx \frac{l}{2\pi\,\rho^2}(\alpha_M M-\alpha_{M_0}M_0)+\alpha_Q\frac{Q^2}{2\pi\,\rho^2}\\
&\quad+\frac{a}{2\pi\,\rho}J+\frac{A}{\pi\,\rho^2}\ln\rho,\\
E(\rho\to\infty,\epsilon_{0|\infty}(M_0)) &\approx \frac{l}{\rho}(\beta_M M-\beta_{M_0}M_0)+\beta_2\frac{Q^2}{\rho}+\frac{b}{\rho}J+2\frac{B}{\rho}\ln\rho,\\
M(\rho\to\infty,\epsilon_{0|\infty}(M_0)) &\approx \gamma_M M-\gamma_{M_0}M_0+\gamma_Q Q^2+c\,J+C\ln\rho,
\end{aligned}
$$

where α_J, β_M,..., γ_Q are constant numerical factors related to the physical parameters: J momentum, M mass, ..., Q electromagnetic charge.

The momentum, energy and mass of the hybrid solutions behaves at spatial infinity $\rho \to \infty$ as follows

$$\begin{aligned}
J(\rho \to \infty) &\approx \alpha_J\, J + \beta_J, \\
\epsilon(\rho \to \infty, \epsilon_{0|\infty}(M_0)) &\approx \frac{l}{2\pi\, \rho^2}(\alpha_M M - \alpha_{M_0} M_0) + \alpha_Q \frac{Q^2}{2\pi\, \rho^2} + \frac{a}{2\pi\, \rho^2} J, \\
E(\rho \to \infty, \epsilon_{0|\infty}(M_0)) &\approx \frac{l}{\rho}(\beta_M M - \beta_{M_0} M_0) + \beta_Q \frac{Q^2}{\rho} + \frac{b}{\rho} J, \\
M(\rho \to \infty, \epsilon_{0|\infty}(M_0)) &\approx \gamma_M M - \gamma_{M_0} M_0 + \gamma_Q Q^2 + c\, J,
\end{aligned}$$

where the charge Q is related to the electromagnetic parameter α.

Moreover, the eigenvectors for their electromagnetic field, Cotton and energy–momentum tensors have been explicitly determined; the static and stationary Peldan electric classes, the Martínez–Teitelboim–Zanelli and the Clément solutions exhibit the following algebraic types:

$$\text{Field} : \{S, N, N\},\ \text{Energy} : \{S, 2T\},\ \{S, 2N\}, \{S, 2S\},$$
$$\text{Cotton} : \{S, Z, \bar{Z}\},$$

while the static and stationary Peldan magnetic families, the Hirschmann–Welch and the Dias–Lemos solutions exhibit the following algebraic types:

$$\text{Field} : \{T, Z, \bar{Z}\},\ \text{Energy} : \{T, 2S\};\ \text{Cotton} : \{S, Z, \bar{Z}\}.$$

The García solution allows for the set of types:

$$\text{Field} : \{T, Z, \bar{Z}\},\ \{S, Z, \bar{Z}\},\ \{T, 2N\},\ \{S, 2N\},\ \{3N\},$$

$$\text{Energy} : \{T, 2T\}, \{T, 2S\}, \{T, 2N\}, \cdots, \{S, 2N\}, \{3N\},$$

$$\text{Cotton} : \{S, Z, \bar{Z}\}.$$

The hybrid Cataldo and Ayón–Cataldo–García solutions fall into the types:

$$\text{Field} : \{S, N, N\}, \text{Energy} : \{S, 2T\}, \{S, 2S\}, \{S, 2N\},$$

$$\text{Cotton} : \{S, Z, \bar{Z}\}.$$

The Kamata–Koikawa belongs to types:

$$\text{Field} : \{3S\}, \text{Energy} : \{3S\}, \{3N\},\ \text{Cotton} : \{3S\}, \{3N\}.$$

Finally, the Matyjasek–Zaslavski solution exhibits the types:

$$\text{Field} : \{S, N, N\}, \text{Energy} : \{S, 2T\}, \{S, 2S\},\ \{S, 2N\},$$

$$\text{Cotton} : \{0\}.$$

Recall that algebraic structures $\{T, 2S\}$ are thought of as perfect fluids.

12

Black Holes Coupled To Nonlinear Electrodynamics

In (3+1) gravity it is well known that the vacuum plus a cosmological constant Λ, i.e., the (anti) de Sitter–Kottler (1918) solution, is a regular non-asymptotically flat solution (the scalar curvature is equal to 4Λ and all the invariants of the conformal Weyl tensor are zero.) On the other hand, Einstein–Maxwell electrovacuum asymptotically flat metrics endowed with timelike and spacelike symmetries do not allow for the existence of regular black hole solutions. In order to be able to derive regular (black hole) solutions one has to enlarge the class of electrodynamics to nonlinear ones; as an example, the regular solution by Ayón–Beato and García (1998), which is a solution to gravitational–nonlinear electromagnetic fields; the first examples belonging to this class are the Borde (1997) model and the Ayón–García regular charged static black hole.

In (2+1) gravity, in the vacuum case, all solutions are locally Minkowski (the Riemann tensor is zero); the extension to the vacuum plus cosmological constant allows for the existence of the static and the rotating anti-de Sitter regular black holes; see Bañados et al. (1992). The static (2+1)-charged black hole with cosmological constant is singular (when radial coordinate goes to zero the curvature and the Ricci square invariants blow up). Similarly, as in (3+1) gravity, one may search for regular solutions in (2+1) gravity incorporating nonlinear electromagnetic fields to which one imposes the weak energy conditions in order to have physically plausible matter-field distributions. One may look for regular solutions with nonlinear electromagnetic fields of the Born–Infeld type (Born and Infeld, 1934); Salazar et al. (1987); Salazar et. al (1984); Gibbons and Rasheed (1995); Fradkin and Tseytlin (1985); Deser and Gibbons (1998), and/or electrodynamics of wider spectra (Cataldo and García, 1999, 2000; Cataldo et al., 2000).

The Born–Infeld electrodynamics (Born and Infeld, 1934) is free from certain singularities appearing in the classical Maxwell electromagnetic field theory. Nonlinear electromagnetic Lagrangians, in particular the Born–Infeld Lagrangian, arise in open string theory (the low-energy effective action for an electromagnetic field is precisely the Born–Infeld action; see Fradkin and Tseytlin, 1985); string theory has emerged as the most promising candidate for the consistent quantization of gravity. In particular, the open string theory has Born–Infeld coupled vector fields, but it is not clear that this remains the case after compactification to a $3D$ space with negative cosmological constant Λ. It is worth mentioning that there is renewed interest in Born–Infeld theory in various contexts (Frolov et al., 1996 and Salazar et al., 1987).

Using our experience in the determination of exact solutions in the standard Einstein gravity with nonlinear electromagnetic fields, in this chapter we deal with $(2+1)$ solutions, mostly static, having nonlinear electrodynamics sources depending on a single electromagnetic invariant. Stationary generalizations of the derived static solutions can be constructed using $SL(2, R)$ transformations.

12.1 Nonlinear Electrodynamics in (2 + 1) Dimensions

In this section, we are using electromagnetic Lagrangian $L(F)$ depending upon a single invariant $F = 1/4F^{ab}F_{ab}$, which we demand in the weak field limit to be equal to the Maxwell Lagrangian

$$L(F) \longrightarrow -F/4\pi,$$

the corresponding energy–momentum tensor has to fulfill the weak energy conditions: for any timelike vector u^a, $u^a u_a = -1$ (we are using signature $- + +$) one requires

$$T_{ab}u^a u^b \geq 0, \;\; q_a q^a \leq 0,$$

where $q^a = T^a_b u^b$. This invariant F can be expressed in terms of the electric (vector) and magnetic (scalar) fields: in a Lorentzian frame, for an observer moving with the 3-velocity v^a, the electric and the magnetic fields are correspondingly defined as

$$E_a = -F_{ab}v^b, \;\; B = -\frac{1}{2}\epsilon_{abc}F_{bc}v^a, \tag{12.1}$$

where Latin indices, to keep the notation used in the original publications, run the values $0, 1, 2$ and ϵ_{abc} is the totally anti-symmetric Levi–Civita symbol with $\epsilon_{012} = 1$, usually the v^a is oriented along the time coordinate, i.e., $v^a = \delta^a_t$, with such a choice

$$E_a = F_{0a}, \;\; B = -F_{12}. \tag{12.2}$$

Thus the invariant can be expressed by $F \equiv \frac{1}{4}F^{ab}F_{ab} = \frac{1}{2}(B^2 - E^2)$.

In general one can construct a (2+1)-Einstein theory coupled with nonlinear electrodynamics starting from the action

$$S = \int \sqrt{-g}\left(\frac{1}{16\pi}(R - 2\Lambda) + L(F)\right) d^3x, \tag{12.3}$$

with the electromagnetic Lagrangian $L(F)$ unspecified explicitly at this stage. We are using units in which $c = G = 1$. The ambiguity in the definition of the gravitational constant (there is no Newtonian gravitational limit in (2+1) dimensions) allows us to maintain the factor $1/16\pi$ in the action to keep the parallelism with (3+1) gravity. Varying this action with respect to gravitational field gives the Einstein equations

$$G_{ab} + \Lambda g_{ab} = 8\pi T_{ab},$$

$$T_{ab} = g_{ab}L(F) - F_{ac}F_b{}^c L_{,F}, \tag{12.4}$$

while the variation with respect to the electromagnetic potential A_a entering in $F_{ab} = A_{b,a} - A_{a,b}$, yields the electromagnetic field equations

$$\nabla_a \left(F^{ab}L_{,F}\right) = 0, \tag{12.5}$$

where $L_{,F}$ stands for the derivative of $L(F)$ with respect to F.

12.2 General Nonlinear Electrostatic Solution

In this section a (2+1) static family of (black hole) metrics coupled to nonlinear electric fields is derived; the source to the Einstein equations is a nonlinear electrodynamics, satisfying the weak energy conditions, which in the weak field limit becomes the (2+1)-Maxwell field.

Concrete solutions to the dynamical equations above we present for the static metric

$$ds^2 = -f(r)dt^2 + \frac{dr^2}{f(r)} + r^2 d\Omega^2, \tag{12.6}$$

where $f(r)$ is an unknown function of the variable r.

Assume that the electromagnetic field $F_{\mu\nu}$ possesses non-vanishing components $E := F_{tr}$ and $B := F_{\Omega r}$. It occurs that the scalar magnetic field B, vanishes as a consequence of the Einstein equations; the Ricci tensor components, evaluated for the metric (12.6), yield the following relation

$$A := R_{tt} + f^2 R_{rr} = 0, \tag{12.7}$$

while the evaluation of the same relation using the electromagnetic energy-momentum gives

$$A = -8\pi L_{,F}(\frac{f}{r}B)^2. \tag{12.8}$$

Therefore, the scalar magnetic field B should be equated to zero, $B = 0$, thus the only case to be treated is just the one with the electric field E,

$$F_{ab} = E(r)\left(\delta^t_a \delta^r_b - \delta^r_a \delta^t_b\right). \tag{12.9}$$

The Maxwell equations reduce to

$$\frac{d}{dr}\left(rEL_{,F}\right)=0, \quad \left(\frac{d}{dr}\left(\frac{f}{r}BL_{,F}\right)=0\right). \tag{12.10}$$

Therefore, Maxwell equations allow for the integral

$$E(r)L_{,F}=-\frac{q}{4\pi r}. \tag{12.11}$$

where q is an integration constant chosen in that way in order to obtain the Maxwell limit.

The invariant F, which reduces now to

$$2F=-E^2(r), \tag{12.12}$$

permits to express the electric field E in term of F. Thus, using (12.12), the derivative, L_F can be expressed as a function of r as follows

$$L_{,r}=\frac{q}{4\pi r}E_{,r}. \tag{12.13}$$

The Einstein's equations equivalently can be written as

$$R_{ab}=8\pi\left(T_{ab}-Tg_{ab}\right)+2\Lambda g_{ab}. \tag{12.14}$$

From (12.4) using (12.9) and (12.12) the trace becomes

$$T=3L(F)+2E^2(r)L_{,F}. \tag{12.15}$$

As was pointed out above, the Lagrangian $L(F)$ must satisfy: (i) correspondence to Maxwell theory, i.e. $L(F)\longrightarrow -L/4\pi$, and (ii) the weak energy conditions: $T_{ab}u^a u^b\geq 0$ and $q_a q^a\leq 0$, where $q^a=T^a_b u^b$ for any timelike vector u^a; in our case the first inequality requires

$$-\left(L+E^2L_{,F}\right)\geq 0, \tag{12.16}$$

which can be stated equivalently as

$$L\leq EL_{,E}\longrightarrow L\leq\frac{q}{4\pi r}E. \tag{12.17}$$

The norm of the energy flux q_a is always less than or equal to zero; for u^a along the time coordinate, $u^a=\delta^a_t/\sqrt{-g_{tt}}$, one has the inequality $q_a q^a=-(L+L_{,F}E^2)^2\leq 0$.

As far as the Einstein equations are concerned, the $R_{tt}(=-f^2R_{_{rr}})$ and $R_{\Omega\Omega}$ components yield respectively the equations

$$f_{,rr}+\frac{f_{,r}}{r}=-4\Lambda+16\pi\left(2L(F)+E^2L_{,F}\right), \tag{12.18}$$

$$f_{,r}=-2\Lambda r+16\pi r\left(L(F)+E^2L_{,F}\right). \tag{12.19}$$

If one replaces $f_{,r}$ from (12.19) and its derivative $f_{,rr}$ into (12.18) one arrives, taking into account the equation (12.13), at an identity. Therefore one can forget the equation (12.18) and integrate the relevant Einstein equation (12.19):

$$f(r) = -M - \Lambda r^2 + 16\pi \int r \left[L(F(r)) + E^2 L_{,F}\right] dr. \tag{12.20}$$

Summarizing, we have obtained a wide class of solutions, depending on a Lagrangian $L(E)$, given by:
the metric

$$ds^2 = -f(r)dt^2 + \frac{dr^2}{f(r)} + r^2 d\Omega^2, \tag{12.21}$$

the structural function

$$f(r) = -M - (\Lambda - 2C)r^2 + 4q \int \left[r \int \frac{E_{,r}}{r} dr - E \right] dr, \tag{12.22}$$

which is obtained from (12.20) by using (12.13) and (12.12), where C is a constant of integration, and the Lagrangian $L(E)$ is constrained to

$$L_{,r} = \frac{q}{4\pi r} E_{,r}. \tag{12.23}$$

We recall that the Lagrangian and the energy momentum tensor have to fulfill the conditions (12.17) quoted above.

In the forthcoming sections, various particular classes are presented for different choices of the Lagrangian functions.

12.2.1 Static Charged Peldan Solution

As a particular example belonging to the linear Maxwell theory, one has the static charged Peldan (1993) solution characterized by the function

$$-g_{tt} = f = -M + \frac{r^2}{l^2} - 2q^2 \ln r, \tag{12.24}$$

the Lagrangian and the electric field

$$L(E) = -\frac{1}{4\pi} F = \frac{1}{8\pi} E^2 = \frac{1}{8\pi} \frac{q^2}{r^2}, \quad E(r) = \frac{q}{r}, \tag{12.25}$$

where C in (12.22) has been equated to zero and $\Lambda = -1/l^2$. It is worth pointing out that the static charged Peldan black hole is singular at $r = 0$.

12.3 Cataldo–García Nonlinear EBI Charged Black Hole

A nonlinear charged version of the $(2+1)$ AdS black hole solution is derived. The source to the Einstein equations is a Born–Infeld electromagnetic field, which in the weak field limit becomes the (2+1)-Maxwell field. The obtained

Einstein–Born–Infeld solution for certain range of the parameters (mass, charge, cosmological and the Born–Infeld constants) represents the static circularly symmetric Cataldo and García (1999) black hole.

The Born–Infeld nonlinear electrodynamics Lagrangian is given by

$$L(F) = -\frac{b^2}{4\pi}\left(\sqrt{1+2\frac{F}{b^2}}-1\right), \tag{12.26}$$

where the constant b is the Born–Infeld parameter. Notice that this Lagrangian reduces to the Maxwell one in the limit when $b^2 \longrightarrow \infty$, $L(F) = -F/4\pi$. The field equations of the Einstein–Born–Infeld (EBI) theory, amount to

$$G_{ab} + \Lambda g_{ab} = 2\left(\frac{F_{ac}F_b{}^c}{\sqrt{1+2F/b^2}} - b^2 g_{ab}(\sqrt{1+2\frac{F}{b^2}}-1)\right), \tag{12.27}$$

together with the electromagnetic field equations

$$\nabla_a\left(\frac{F^{ab}}{\sqrt{1+2F/b^2}}\right) = 0. \tag{12.28}$$

As a concrete solution of the Einstein–Born–Infeld dynamical equations we present a static self-consistent solution with metric (12.6). The invariant F is given by $2F = -E^2(r)$. The electromagnetic field equations (12.5) yield

$$\partial_r\left(\frac{rE(r)}{\sqrt{1-E^2/b^2}}\right) = 0, \tag{12.29}$$

which integrates as

$$E(r) = \frac{q}{\sqrt{r^2+q^2/b^2}}, \tag{12.30}$$

where q is an integration constant having the meaning of the charge, as one could expect. One also arrives at this result by integrating the master equation (12.23). In the Maxwell limit, we obtain from the last expression the right $E = q/r$ in (2+1) dimensions. The Born–Infeld field is characterized by a charge density distribution ρ_e, which can be evaluated from the Maxwell equations $\mathrm{div}\vec{E} = \nabla\vec{E} = 2\pi\rho_e$, which in the considered case amount to

$$\mathrm{div}\vec{E} = \frac{1}{r}\frac{d}{dr}(rE(r)) = 2\pi\rho_e, \tag{12.31}$$

substituting here $E(r)$ from (12.30) one obtains

$$\rho_e = \frac{qr_0^2}{2\pi r(r^2+r_0^2)^{3/2}}, \tag{12.32}$$

where $r_0 = q/b$. It is easy to verify that the surface integral of ρ_e is equal to q, in fact

$$\int_0^\infty \rho_e dA = qr_0^2\int_0^\infty \frac{dr}{(r^2+r_0^2)^{3/2}} = q. \tag{12.33}$$

It is worth pointing out the regular behavior of the vector $\vec{E}$ of the electric field and the surface charge distribution ρ_e; the same regular behavior one encounters for the static spherically symmetric electric field in (3+1) Born–Infeld theory.

12.3.1 Static Cyclic Symmetric EBI Solution

As far as the Einstein equation are concerned, the R_{tt} and $R_{\Omega\Omega}$ components yield respectively the equations

$$f_{,rr} + \frac{f_{,r}}{r} = -4\Lambda - \frac{4q^2}{r\sqrt{r^2+q^2/b^2}} - 8b^2\left(\frac{r}{\sqrt{r^2+q^2/b^2}} - 1\right), \quad (12.34)$$

$$f_{,r} = -2\Lambda r - \frac{4q^2}{\sqrt{r^2+q^2/b^2}} - 4b^2 r\left(\frac{r}{\sqrt{r^2+q^2/b^2}} - 1\right). \quad (12.35)$$

The general integral of the $R_{\Omega\Omega}$–equation is given by

$$f(r) = -M - (\Lambda - 2b^2)r^2 - 2b^2 r\sqrt{r^2+q^2/b^2} - 2q^2\ln(r + \sqrt{r^2+q^2/b^2}). \quad (12.36)$$

From this last expression one sees that there is a contribution of the Born–Infeld field to the term with the cosmological constant. The Lagrangian and the electric field are given by

$$L(F) = -\frac{b^2}{4\pi}\left(\sqrt{1+2\frac{F}{b^2}} - 1\right) = -\frac{b^2}{4\pi}\left(\frac{r}{\sqrt{r^2+q^2/b^2}} - 1\right),$$
$$E(r) = \frac{q}{\sqrt{r^2+q^2/b^2}}. \quad (12.37)$$

Invariants

Having the explicit metric, one can easily calculate the curvature tensor components:

$$R_{0110} = 2b^2 - \Lambda - \frac{2b^2 r}{\sqrt{r^2+q^2/b^2}}, \quad (12.38)$$

$$R_{0202} = f(r)\left(\Lambda r^2 - 2b^2r^2 + 2b^2 r\sqrt{r^2+\frac{q^2}{b^2}}\right), \quad (12.39)$$

and

$$R_{1221} = f(r)^{-1}\left(\Lambda r^2 - 2b^2r^2 + 2b^2 r\sqrt{r^2+\frac{q^2}{b^2}}\right). \quad (12.40)$$

It is surprising that the covariant metric and curvature components do not exhibit a singular behavior in the neighborhood of the origin at $r = 0$. Nevertheless, this solution is singular at $r = 0$ in the sense that its invariant curvature characteristics – Ricci and Kretschmann scalars – such as the Ricci scalar and the Ricci square blow up at $r = 0$, (the Riemann square does not need to be evaluated since in (2+1) dimensions the Riemann tensor is given in terms of the Ricci tensor, curvature scalar and the metric tensor). Additionally, in (2+1) dimensions one considers the behavior of the invariant $\det(R_{ab})/\det(g_{ab})$, thus one has to evaluate the invariants (Weinberg, 1972)

$$R, \ R_{ab}R^{ab}, \ \frac{\det(R_{ab})}{\det(g_{ab})}$$

at critical points. In our case these three invariants are given as

$$R = 6\Lambda - 12b^2 + \frac{4(2q^4 + 5q^2b^2r^2 + 3b^4r^4)}{b^2r\,(r^2 + q^2/b^2)^{3/2}}, \tag{12.41}$$

$$\begin{aligned} R_{ab}R^{ab} &= 3(4b^2 - 2\Lambda)^2 - \frac{8b^2(4b^2 - 2\Lambda)}{r}\left[\frac{3r^2 + 2q^2/b^2}{\sqrt{r^2 + q^2/b^2}}\right] \\ &\quad + 8b^4\left[2 + \frac{r^2}{r^2 + q^2/b^2} + \frac{3(r^2 + q^2/b^2)^2}{r^2}\right], \end{aligned} \tag{12.42}$$

and

$$\frac{\det R_{ab}}{\det g_{ab}} = \left(\frac{2\Lambda - 4b^2}{r^2 + \frac{4b^2}{r^3}\sqrt{r^2 + q^2/b^2}}\right)\left(-2\Lambda + 4b^2 - \frac{2b^2(2r^2 + q^2/b^2)}{r\sqrt{r^2 + q^2/b^2}}\right)^2. \tag{12.43}$$

Since these scalars go to infinity at $r \to 0$, we conclude that they are singular at this point.

12.3.2 Cataldo–García Black Hole to EBI

This solution is a black hole. To establish this assertion one has to demonstrate the existence of horizons, which require the vanishing of the g_{tt} component, i.e., $f(r) = 0$. The roots of this equation give the location of the horizons (inner and outer in our case). Since this equation is a transcendent one, we are not able, as is usual for charged (2+1) black holes, to express the roots analytically, even for the charged static Peldan solution; the roots are expressed in terms of the Lambert $W(x)$ function. To overcome this difficulty we study the extreme case, in which the derivative of $\partial_r(f(r)) = 0$ gives

$$r_{extr} = \frac{2qb}{\sqrt{\Lambda^2 - 4b^2\Lambda}} > 0 \tag{12.44}$$

for $\Lambda < 0$. Now entering r_{extr} into $f(r) = 0$ one obtains a relation between mass, charge, cosmological constant and the Born–Infeld parameter, which can be solved explicitly for the mass (the extreme one):

$$M_{extr} = -2q^2 \ln\left(\frac{q}{b}\sqrt{\frac{\Lambda - 4b^2}{\Lambda}}\right). \tag{12.45}$$

We have an extreme black hole if $\Lambda < 0$, $M_{extr} > 0$ and $q^2 < b^2$; this last constraint arises from $\Lambda < 4b^2q^2/(q^2 - b^2)$ when one demands $M_{extr} > 0$. Fixing the values of the M_{extr} for given values of q, b and Λ, one has a black hole solution with inner and outer horizons when $M > M_{extr}$. For $M < M_{extr}$ one has a soliton solution, i.e., there are no horizons at all and we have a naked singularity. It may occur that for certain values of the parameters there is only one positive root of the equation $f(r) = 0$: the horizon $r_h > 0$. In such a case one has also a black hole solution. A similar analysis can be carried out for other r_{extr} which arises for $\Lambda > 0$. At infinity, for weak electromagnetic field this BI solution asymptotically behaves as the static charged Peldan solution.

Moreover, for $\Lambda = -1/l^2$ this BI solution at infinity behaves as anti-de Sitter spacetime. At the origin $r = 0$, this BI solution is singular. If one requires additionally the vanishing of the cosmological constant, one arrives at a solution reported in Gott et al. (1986).

As regards the analytical extension of our solution, one may follow step by step the procedure presented in standard textbooks (for instance, Wald, 1984) to determine the Kruskal–Szekeres coordinates. First one has to integrate for the tortoise r_* coordinate: $r_* = \int 1/f(r)dr$, which in our case has no expression in terms of elementary functions; next one defines the null coordinate u and v by $u = t - r_*$, $v = t + r_*$; in these coordinates the studied metric acquires the form

$$ds^2 = -f(r)dudv + r^2 d\Omega^2, \tag{12.46}$$

where r has to be interpreted as function of u and v, $r_* = r_*(u, v)$. Further, one introduces null Kruskal–Szekeres coordinates $U = -e^{-\alpha u}$ and $V = e^{\beta V}$ where α and β are to be chosen appropriately, finally one introduces the Kruskal–Szekeres coordinates $T = (U + V)/2$, X= (V-U)/2, arriving at the Kruskal extension.

If one were interested in the thermodynamics of the obtained solution one would need to evaluate the temperature of the black hole, which is given in terms of its surface gravity by Visser (1992); Brown et al. (1994):

$$k_{_B} T_{_H} = \frac{\hbar}{2\pi} k. \tag{12.47}$$

For a spherically symmetric (and for circularly symmetric in (2+1) dimensions) system the surface gravity can be computed via (for our signature)

$$k = -\lim_{r \to r_+} \left[\frac{1}{2}\frac{\partial_r g_{tt}}{\sqrt{-g_{tt}g_{rr}}}\right], \tag{12.48}$$

where r_+ is the outermost horizon. For our solution we have from (12.22), (12.47 and (12.48) that

$$k_B T = \frac{\hbar}{4\pi}\left(-2(\Lambda - 2b^2)r_+ - 4b^2\sqrt{r_+^2 + \frac{q^2}{b^2}}\right). \tag{12.49}$$

Since in our case there is no analytical expression of r_+ in terms of elementary functions, one cannot give a parameter-dependent expression of (12.49). It is easy to check that when $q = 0$, T in (12.49) reduces to the BTZ temperature. In the extremal case (12.44), the temperature vanishes in (12.49). The entropy can be trivially obtained using the entropy formula $S = 4\pi r_+$. Other thermodynamic quantities such as heat capacity and chemical potential can be computed as in Brown et al. (1994). We recall that most of these quantities in the literature are evaluated for metrics given in terms of polynomial functions.

Notice that the 4*D* Einstein–Born–Infeld counterpart – the Kottler–Born–Infeld black hole; Salazar et. al (1984); Gibbons and Rasheed (1995) – can be given by the metric (12.6) with

$$\begin{aligned} f(r) = 1 - 2M/r - (\Lambda/3 - 2b^2/3)r^2 - \frac{2}{3}b^2\sqrt{r^4 + q^2/b^2} \\ -2\frac{q^2}{3r}\int^r \frac{dr}{\sqrt{r^4 + q^2/b^2}}, \end{aligned} \tag{12.50}$$

where now $d\Omega^2 = d\theta^2 + \sin^2\theta d\phi^2$. As in the (2+1) case, there is also a contribution to the cosmological constant term of the nonlinear field. The corresponding electric field is given by

$$E(r) = \frac{q}{\sqrt{r^4 + q^2/b^2}}. \tag{12.51}$$

Notice that the electric field in this case is regular everywhere. This gravitational field asymptotically behaves as the Kottler charged solution, with the structural function and electromagnetic field of the form

$$\begin{aligned} f(r) &= 1 - \frac{2M}{r} - \frac{\Lambda}{3}r^2 + \frac{q^2}{r^2} + O\left(\frac{1}{r^6}\right), \\ E(r) &= \frac{q}{r^2} + O\left(\frac{1}{r^3}\right). \end{aligned}$$

By canceling Λ one obtains an asymptotically flat solution.

12.4 Regular Black Hole Solution

Following Ayón–Beato and García (1998), a static class of regular everywhere solution can be constructed; it is given by the metric (12.6) with structural function of the form

$$f(r) = -M - \Lambda r^2 - q^2\ln(r^2 + a^2) \tag{12.52}$$

where M, a, q and Λ are free parameters. The corresponding Lagrangian and the electric field are given by

$$L(r) = \frac{q^2}{8\pi}\frac{(r^2 - a^2)}{(r^2 + a^2)^2}, \quad E(r) = q\frac{r^3}{(r^2 + a^2)^2}. \tag{12.53}$$

$L(r)$ and the electric field satisfy the weak energy conditions (12.17). To express the Lagrangian in terms of F or equivalently E, one has to write r in terms of E by solving the quartic equation for $r(E)$, this will give rise an explicit r containing radicals of E, which introduced in $L(r)$, finally will bring L as function of E. The expression $L(E)$ is not quite illuminating, thus we omit it here.

12.4.1 Regularity

To establish that this solution is regular one has to evaluate the curvature Ricci and Kretschmann scalars. The non-vanishing curvature components, which are regular at $r = 0$, are given by:

$$R_{0110} = \frac{q^2(a^2 - r^2)}{(r^2 + a^2)^2} + \Lambda, \tag{12.54}$$

$$R_{0202} = -f(r)\left(\frac{q^2r^2}{r^2 + a^2} + \Lambda r^2\right), \tag{12.55}$$

$$R_{1212} = f(r)^{-1}\left(\frac{q^2r^2}{r^2 + a^2} + \Lambda r^2\right), \tag{12.56}$$

where 0,1,2 stand respectively for t, r and Ω.

Evaluating the invariants R, and $R_{ab}R^{ab}$ one has

$$R = \frac{2q^2(r^2 + 3a^2)}{(r^2 + a^2)^2} + 6\Lambda \tag{12.57}$$

$$R_{ab}R^{ab} = 12\Lambda^2 + 4q^4\frac{r^4 + 2r^2a^2 + 3a^4}{(r^2 + a^2)^4} + \frac{8\Lambda q^2(3a^2 + r^2)}{(r^2 + a^2)^2}. \tag{12.58}$$

Since the metric, the electric field and these invariants behave regularly for all values of r, we conclude that this solution is curvature regular everywhere. Nevertheless, for solutions without any horizon or black hole solutions with an inner and outer horizons, at $r = 0$ a conical singularity may arise. At $r = 0$ the function $f(r)$ becomes $f(0) = -M - q^2\ln(a^2)$. Thus for M positive, $M > 0$, and a in the range $0 < a < 1$, the value of $f(0)$ will be $f(0) = -M + q^2\ln(1/a)^2$, which will be positive, say $f(0) := \beta^2$, if $\ln(1/a)^2 > M/q^2$. In such a case, for $0 < \beta < 1$ the solutions will show angular deficit since the angular variable Ω, which originally runs $0 \leq \Omega < 2\pi$ will now run $0 \leq \Omega < 2\beta\pi$; the parameter a can be expressed in terms of β, q and M as $a^2 = \exp[-(\beta^2 + M)/q^2]$. For $\beta = 1$, there will be no angular deficit, the ratio of the perimeter of a small circle around $r = 0$ to its radius, as this last tends to zero, will be 2π. If one allows M to be negative, $M < 0$, and a to take values in the interval $0 < a < 1$, then $f(0)$ will be always positive; in this case one can adopt the following parametrization: $-M = \beta^2 \cos^2 \alpha$, $q^2 \ln(1/a)^2 = \beta^2 \sin^2 \alpha$, therefore $f(0) = \beta^2$. One will have angular deficit if $0 < \beta < 1$, and for $\beta = 1$ the resulting (2+1) spacetime will be free of singularities. Another possibility with positive $f(0) = \beta^2$ arises for $M < 0$, and $a > 1$; $f(0)$ can be parameterized as $-M = \beta^2 \cosh^2 \alpha$, $q^2 \ln(1/a)^2 = \beta^2 \sinh^2 \alpha$. Again the values taken by β will govern the existence of angular deficit, for $\beta = 1$ the solutions will be regular. If $f(0)$ is negative, $f(0) =: -\beta^2$, the character of the coordinates t and r changes, the coordinate t becomes spacelike, while r is now timelike and one could think of the singularities, if any, as causal structure singularities because they could arise at the "time" $r = 0$. In what follows we shall treat the parameter a as a free one, having in mind the above restrictions to have solutions free of conical singularities.

12.4.2 Horizons

To establish that this solution represents a black hole, one has to demonstrate the existence of horizons, which require the vanishing of the g_{tt} component, i.e., $f(r) = 0$. The roots of this equation give the location of the horizons (inner and outer in our case). The roots – at most four – of the equation $f(r) = 0$ can be expressed in terms of the Lambert $W(r)$ function

$$r_{1,2,3,4} = \pm[\exp(\frac{\Lambda a^2 - M}{q^2} - LW[\frac{\Lambda}{q^2} \exp\left(\frac{\Lambda a^2 - M}{q^2}\right)]) - a^2]^{\frac{1}{2}}.$$

There arise various cases which depend upon the values of the parameters: four real roots (two positive and two negative roots: the negative roots have to be ignored), two complex and two real roots, two complex and one real positive root (the extreme case), and four complex roots (no black hole solutions). This analytical expression for the Lambert function can be used in all calculations, recall that Lambert function fulfills the following equation

$$\ln(LW(x)) + LW(x) = \ln(x).$$

Analytically one can completely treat the extreme black hole case; for it, the derivative of $f(r)$ has to be zero, $\partial_r(f(r)) = 0$, at the r_{extr}, this gives

$$r_{extr} = \sqrt{-a^2 - \frac{q^2}{\Lambda}} > 0 \tag{12.59}$$

for $\Lambda < 0$. From this expression one concludes that the following inequality holds: $a^2 < -q^2/\Lambda$. Entering now r_{extr} into $f(r) = 0$ one obtains a relation between the parameters involved, which can be solved explicitly for the mass (the extreme one):

$$M_{extr} = a^2\Lambda + q^2\left(1 + \ln\left[\frac{-\Lambda}{q^2}\right]\right), \tag{12.60}$$

this M_{extr} varies its values depending on the values given to the parameters a, q and Λ. We have an extreme black hole characterized by negative cosmological constant, $\Lambda < 0$, and positive extreme mass, $M_{extr} > 0$, if the parameter a is restricted by the inequality $a^2 < -(q^2(1 + \ln(-\Lambda/q^2)))/\Lambda$.

For other values of the mass M, one distinguishes the following branches: if $M > M_{extr}$ one has a black hole solution, and if $M < M_{extr}$ there are no horizons.

12.4.3 Thermodynamics

Similarly, as this question was treated in the previous section, here one can also evaluate the temperature (12.47), of the black hole through its surface gravity (12.48). For our present solution we have from (12.22), (12.47) and (12.48) that

$$k_B T = \frac{\hbar}{2\pi}\left(-\Lambda r_+ - \frac{q^2 r_+}{r_+^2 + a^2}\right). \tag{12.61}$$

The entropy can be trivially obtained using the entropy formula $S = 4\pi r_+$.

To achieve the maximal extension of our regular black solutions one has to proceed step by step through the procedure presented above, determining first the Kruskal–Szekeres coordinates.

12.5 Coulomb-Like Black Hole Solution

The electromagnetic tensor for a Coulomb potential in $(3 + 1)$ Maxwell electrodynamics is trace free, $T = T_{ab}g^{ab} = 0$, Thus, it is of some interest to establish which solution arises if one requires the vanishing of the trace T in $(2 + 1)$ electrodynamics. Demanding the vanishing of

$$T = T_{ab}g^{ab} = 3L(F) - 4FL_{,F}, \tag{12.62}$$

one obtains

$$L = C\,|F|^{3/4}, \tag{12.63}$$

where C is a constant of integration, and bars denote moduli. Taking into account (12.12) one gets that

$$L = C\,E^{3/2}. \tag{12.64}$$

Entering this L into (12.13) one gets that $E = (q^2/6\pi C)^2 1/r^2$. By choosing now $C = \sqrt{|q|}/6\pi$ we arrive at the electric field

$$E(r) = \frac{q}{r^2}, \tag{12.65}$$

which coincides with the standard Coulomb field for a point charge of the Maxwell theory in the Minkowski space.

The Lagrangian in this case is given by

$$L = \frac{q^2}{6\pi r^3} = \frac{\sqrt{|q|}}{6\pi} E^{3/2}. \tag{12.66}$$

It is easy to check that this Lagrangian satisfies the weak energy conditions: $q_a q^a \leq 0$, where $q^a = T^a_b u^b$ for any timelike vector u^a and on the other hand

$$-\left(L + E^2 L_{,F}\right) = \frac{q^2}{12\pi r^3} \geq 0. \tag{12.67}$$

We rewrite the Einstein's equations equivalently as

$$R_{ab} = 8\pi T_{ab} + 2\Lambda g_{ab}, \tag{12.68}$$

where it has been considered that the trace of T_{ab} is equal to zero, $T = 0$. The Einstein equations for $R_{tt}(= -f^2 R_{,rr})$ and $R_{\Omega\Omega}$ components yield respectively the equations

$$f_{,rr} + \frac{f_{,r}}{r} = -2\Lambda + \frac{2q^2}{3r^3}, \tag{12.69}$$

$$f_{,r} = -2\Lambda r - \frac{4q^2}{3r^2}. \tag{12.70}$$

It is easy to show that equation (12.69), by virtue equation the Maxwell equations is just an identity. Therefore the only Einstein equation to be integrated is (12.70), which gives

$$f(r) = C_O - \Lambda r^2 + \frac{4q^2}{3r}, \tag{12.71}$$

where C_O is a constant of integration. We will see now that the constant C_O can be expressed in terms of asymptotic values of the mass. For the circularly symmetric metric (12.6) the quasilocal energy $E(r)$ and the quasilocal mass $M(r)$ at a radial boundary r can be shown to be respectively

$$E = 2(\sqrt{f_0(r)} - \sqrt{f(r)}), \tag{12.72}$$

$$M(r) = \sqrt{f(r)E(r)}, \tag{12.73}$$

where $f_0(r) = g_0^{rr}(r)$ is a background metric function which determines the zero of the energy. The function $f_0(r)$ can be obtained simply by setting constants of integration of our solution (12.71) to some special values that then specify the reference spacetime. We set in (12.71) $q = C_O = 0$ as the background and as a consequence it is the vacuum anti-de Sitter spacetime. The same background function was used in Brown et al. (1994); Chan (1996) for analogous calculations. Now for $\Lambda = -1/l^2 < 0$ we have that $\sqrt{f_0(r)} = r/l$ (for a de Sitter spacetime $\Lambda > 0$) and then the quasi-local energy and the quasi-local mass are given respectively by

$$E(r) = \frac{r}{l} - \sqrt{C_O + \frac{r^2}{l^2} + \frac{4}{3}\frac{q^2}{r}}, \tag{12.74}$$

$$M(r) = 2\frac{r}{l}\sqrt{C_O + \frac{r^2}{l^2} + \frac{4}{3}\frac{q^2}{r}} - 2\left(C_O + \frac{r^2}{l^2} + \frac{4}{3}\frac{q^2}{r}\right). \tag{12.75}$$

As $r \longrightarrow \infty$, the analogous ADM mass is defined to be $M =: M(\infty)$. In our case we see from (12.74) and (12.75) that $E(\infty)$ vanishes and $M(\infty) =: M = -C_O$ respectively. Then we get that the constant C_O has the sense of the asymptotic observable mass $M > 0$ and we can write

$$f(r) = -M - \Lambda r^2 + \frac{4q^2}{3r}. \tag{12.76}$$

In the following we analyze the obtained solutions. The case $\Lambda > 0$ corresponds to an asymptotically de Sitter spacetime. For $\Lambda < 0$, one is dealing with an asymptotically anti-de Sitter spacetime. For vanishing cosmological constant, $\Lambda = 0$, one has an asymptotically flat solution coupled with a Coulomb-like field.

12.5.1 Horizons for the Coulomb-Like Solution

To establish the existence of horizons, one has to require the vanishing of the g_{tt} component, i.e., $f(r) = 0$. The roots of this equation are

$$r_1 = \frac{h}{3\Lambda} - \frac{M}{h}, \tag{12.77}$$

$$r_2 = -\frac{h}{6\Lambda} + \frac{M}{2h} + \frac{i\sqrt{3}}{2}\left(\frac{h}{3\Lambda} + \frac{M}{h}\right), \tag{12.78}$$

$$r_3 = -\frac{h}{6\Lambda} + \frac{M}{2h} - \frac{i\sqrt{3}}{2}\left(\frac{h}{3\Lambda} + \frac{M}{h}\right), \tag{12.79}$$

where

$$h = \left[\left(18q^2 + 3\sqrt{3\left(\frac{M^3 + 12q^4\Lambda}{\Lambda}\right)}\right)\Lambda^2\right]^{1/3}. \tag{12.80}$$

These equations give the location of the horizons (if there are any). Since the coordinate r ranges over positive values from 0 to infinity, we exclude the roots which are negative. Because the positive character of $M > 0$, the complex or real character of these roots depends crucially on Λ values. For $\Lambda > 0$ there is only one real root. For $\Lambda < 0$ there are two possibilities: two complex roots and one real; or three real roots. To obtain only real roots we must cancel the term $h/3\Lambda + M/h$. Then we have that $h = \pm\sqrt{-3\Lambda M}$. If $M^3/\Lambda + 12q^4 = 0$ then one has an extreme black hole, in such case $M = M_{extr} = -(12q^4\Lambda)^{1/3}$ and the roots become

$$r_1 = 2\left(\frac{2q^2}{3\Lambda}\right)^{1/3}, \quad r_2 = r_3 = r_{extr} = -\left(\frac{2q^2}{3\Lambda}\right)^{1/3}. \tag{12.81}$$

From these expressions we see that for $\Lambda > 0$ there is not a positive r_{extr}, although the horizon $r_1 > 0$ the behavior of the f function is quite peculiar, for $r > r_1$, $f(r) < 0$, while for $r < r_1$, $f(r) > 0$. For $\Lambda < 0$, we have a positive extreme horizon (since $r_1 < 0$ it does no represent an horizon).

For $\Lambda > 0$ (and $M > 0$) we have that $\alpha > 0$ and then we always have one real root $r_1 > 0$ and two complex roots r_2 and r_3. From (12.71), hence $f(r)$ is a decreasing function for $r > 0$.

For $\Lambda < 0$ the roots r_2 and r_3 can be real or complex roots (in this case always $r_1 < 0$). For $M^3 + 12q^4\Lambda > 0$ or $q^2 > \sqrt{-M/3\Lambda}/2$ we have that these roots are complex. For $M^3 + 12q^4\Lambda < 0$ or $0 \leq q^2 < \sqrt{-M/3\Lambda}/2$ we can write the real roots as

$$r_\pm = -2\sqrt{\frac{-M}{3\Lambda}}\cos\left(\frac{1}{3}\arccos\left[\frac{2q^2}{\sqrt{-\frac{M^3}{3\Lambda}}}\right] \pm \frac{2\pi}{3}\right). \tag{12.82}$$

In this case these roots represent the horizons of a black hole; the outer horizon (the event horizon) is r_+, while the inner horizon is r_-. When $q = 0$ we have that $r_- = 0$ and $r_+ = \sqrt{-M/\Lambda}$, which is the horizon of the static BTZ metric. Using the negative cosmological constant as $\Lambda = -1/l^2$ one gets

$$k_B T = \frac{\hbar}{4\pi}\left(-\sqrt{\frac{16M}{3l^2}}\cos\left(\frac{1}{3}\arccos\left[\frac{2q^2}{\sqrt{\frac{M^3l^2}{3}}}\right] + \frac{2\pi}{3}\right)\right)$$
$$-\frac{q^2}{Ml^2}\cos^{-2}\left(\frac{1}{3}\arccos\left[\frac{2q^2}{\sqrt{\frac{M^3l^2}{3}}}\right] + \frac{2\pi}{3}\right). \tag{12.83}$$

It is easy to check that when $q = 0$, T in (12.83) reduces to the BTZ temperature. In the extremal case (12.81), the temperature vanishes in (12.83). The maximal extension of black solutions can be achieved by using the Kruskal–Szekeres approach.

Another interesting case arises when $\Lambda = 0$, yielding a charged asymptotically flat (2+1)-dimensional black hole with a cosmological horizon at

$$r_H = \frac{4q^2}{3M}, \quad T = -\frac{3\hbar}{16\pi}\frac{M^2}{q^4} < 0. \tag{12.84}$$

12.6 Stationary Nonlinear Electrodynamics Black Holes

The standard way to search for solutions within the class of stationary cyclic symmetric metrics:

$$ds^2 = -N(\rho)^2 dt^2 + \frac{1}{L(\rho)^2} d\rho^2 + K(\rho)^2 [d\phi + W(\rho)dt]^2,$$

for nonlinear electrodynamics determined by a Lagrangian $L(F)$ depending upon a single invariant $F = 1/4F^{ab}F_{ab}$, to which one demands in the weak field limit to become the Maxwell Lagrangian $L(F) \longrightarrow -F/4\pi$, consists in solving the Einstein equations

$$G_{ab} + \Lambda g_{ab} = 8\pi T_{ab},$$

with

$$T_{ab} = g_{ab} L(F) - F_{ac} F_b{}^c L_{,F},$$

and the electromagnetic field equations

$$\nabla_a \left(F^{ab} L_{,F} \right) = 0,$$

where $L_{,F}$ stands for the derivative of $L(F)$ with respect to F. The field $F_{ab} = A_{b,a} - A_{a,b}$, where A_a is the electromagnetic potential A_a. Although this task is solvable for particular cases, in general is time consuming.

Thus, as in the search for Maxwell stationary (2+1) solutions, it is easier to generate solutions via $SL(2,R)$ transformations applied to solutions given by (diagonal) static metrics. Consider the static metric

$$g = g_{tt} dt^2 + g_{\phi\phi}\, d\phi^2 + g_{\rho\rho} d\rho^2,$$

and accomplish here a $SL(2,R)$ – transformations of the Killingian coordinates t and ϕ

$$t = \alpha\tilde{t} + \beta\tilde{\phi},\ \phi = \gamma\tilde{t} + \delta\tilde{\phi},\ \Delta := \alpha\delta - \beta\gamma \neq 0,$$

then the transformed metric components will be

$$g_{\tilde{t}\tilde{t}} = \alpha^2\, g_{tt} + \gamma^2\, g_{\phi\phi},\ g_{\tilde{t}\tilde{\phi}} = \alpha\beta\, g_{tt} + \gamma\delta\, g_{\phi\phi},$$
$$g_{\tilde{\phi}\tilde{\phi}} = \beta^2\, g_{tt} + \delta^2\, g_{\phi\phi},\ g_{\rho\rho} = g_{\rho\rho}.$$

Requiring the new coordinate $\tilde{\phi}$ to be cyclic, $0 \leq \tilde{\phi} \leq 2\pi$, one is generating stationary solutions from static ones. In this manner one generates the stationary nonlinear solution with arbitrary $L(F(\rho))$, the stationary Cataldo–García black hole to EBI, the stationary generalization of the regular and of the Coulomb-like solutions. Explicit expressions are easily derivable.

13

Dilaton Field Minimally Coupled to $(2+1)$ Gravity

Using the Schwarzschild coordinate frame for a static cyclic symmetric metric in $(2+1)$ gravity coupled minimally to a dilaton logarithmically depending on the radial coordinate in the presence of an exponential potential together with a Maxwell electric field, by solving first-order linear Einstein equations, the general solution is derived and identified with the Chan–Mann dilaton solution, and, in the charged case, with the Chan–Mann charged dilaton solution. In these coordinates, a new stationary dilaton solution is obtained; it does not allow for a de Sitter–Anti-de Sitter limit at spatial infinity, where its structural functions increase indefinitely. On the other hand, it is horizonless and allows for a naked singularity at the origin of coordinates; moreover, one can identify at a large radial coordinate a (quasi-local) mass parameter and in the whole space a constant angular momentum.

Via a general $SL(2,R)$ transformation, applied on the static cyclic symmetric metric, a family of stationary (charged) dilaton solutions has been generated. A particular $SL(2,R)$ transformation is identified, which gives rise to the rotating Chan–Mann (charged) dilaton solution. All the exhibited solutions have been characterized by their quasi-local energy, mass, and momentum through their series expansions at spatial infinity. The algebraic structure of the Ricci–energy-momentum, and Cotton tensors is given explicitly. A summary of results is presented in section 13.9.

13.1 Scalar Field Minimally Coupled to Einstein Gravity

The literature on stationary rotating scalar field solutions is rather scarce among them Chan (1997), Chan and Mann (1996); most of the known solutions of this class are static. The static cyclic symmetric metric can always be described by three structural functions appearing in the metric

$$ds^2 = -A(r)^2dt^2 + B(r)^2dr^2 + C(r)^2dt^2,$$

leaving still a freedom in the choice of the r–coordinate, which can be used to fix the metric structure, for instance:

the Schwarzschild coordinate frame: $ds^2 = -A(r)^2dt^2 + B(r)^2dr^2 + r^2dt^2$, or the $g_{tt} = -1/g_{rr}$ coordinate frame: $ds^2 = -F(r)^2dt^2 + dr^2/F(r)^2 + H(r)^2dt^2$. In one of the well-known works on dilaton Chan and Mann (1994), this last metric was used to argue that its use simplifies the calculations; this is partially true when it is also assumed that $H(r) = r^{N/2}$; nevertheless, the field equations to be integrated are of the second order. The r-gauge freedom allows one to fix only one structural function and leave two structural metric functions undetermined in the class of static circularly symmetric metric in (2+1) gravity; the Einstein field equations yield further constraints fixing the dependence on r of the remaining structural functions.

In the Schwarzschild coordinate frame, $g_{\theta\theta} = r^2$, the equations to be integrated, for a dilaton logarithmically depending on r and associated to an exponential potential, are first-order equations for each of the (different) structural functions. Thus, one can claim on its generality and uniqueness under the restrictions imposed on the dilaton field. It should be pointed out that in the Chan–Mann papers the solutions were derived under the ansatz $g_{\theta\theta} = \gamma^2 r^N$. In this section, the integration of the field equations is done for the static cyclic symmetric metric in the r-gauge fixed according to $g_{tt} = -1/g_{rr} = F(r)$, and $g_{\theta\theta} = H(r)$; it happens that the resulting two possible solutions can be identified, via parameter choice, with the static Chan and Mann (1994) solution.

On the other hand, dealing with stationary cyclic symmetric metrics for logarithmically dependent dilaton field, there exists a single solution for the Schwarzschild coordinate $g_{\theta\theta} = r^2$–gauge.

The action to be considered in this work dealing with (2+1)-dimensional gravity is given by

$$\mathcal{S} = \int d^3x\sqrt{-g}\left[R - 4\,\nabla_\mu\Psi\,\nabla^\mu\Psi + 2\,\mathrm{e}^{b\Psi}\Lambda\right], \tag{13.1}$$

where Ψ is the massless minimally coupled scalar field, and R is the scalar curvature. The variations of this action yield the dynamical equations

$$\begin{aligned} EQ_{\mu\nu} &:= R_{\mu\nu} = 4\,\nabla_\mu\Psi\,\nabla_\nu\Psi - 2g_{\mu\nu}\mathrm{e}^{b\,\Psi}\Lambda, \\ &\nabla^\mu\nabla_\mu\Psi + \frac{b}{4}\,\Lambda\,\mathrm{e}^{b\,\Psi} = 0, \end{aligned} \tag{13.2}$$

where at this stage Λ and b are arbitrary parameters. Details about the derivation of the Einstein–dilaton equations and analysis of the known solutions can be found in Garcia–Diaz and Gutierrez–Cano (2014b).

13.2 Static Black Hole Coupled to a Scalar $\Psi(r) = k\ln(r)$

The static cyclic symmetric metric in the $(2+1)$ Schwarzschild coordinate frame is given by

$$g = -N(r)^2 dt^2 + \frac{dr^2}{L(r)^2} + r^2 d\phi^2. \tag{13.3}$$

The scalar field equation for the dilaton $\Psi(r) = k\ln(r)$ becomes

$$FEQ := \frac{L^2}{r\,N}\frac{d\,N}{dr} + \frac{L}{r}\frac{d\,L}{dr} + \frac{1}{4k}\,b\Lambda\,r^{bk} = 0. \tag{13.4}$$

The simplest Einstein equation is ${EQ_3}^3$

$${EQ_3}^3 = -\frac{L^2}{Nr}\frac{d\,N}{dr} - \frac{L}{r}\frac{dL}{dr} + 2\,\Lambda\,r^{bk} = 0, \tag{13.5}$$

thus, from ${EQ_3}^3 + FEQ$, one gets the constant relation

$$r^{bk}\Lambda\;(b + 8\;k) = 0 \rightarrow b = -8k. \tag{13.6}$$

On the other hand, equation ${EQ_1}^1 - {EQ_2}^2 - {EQ_3}^3$ gives a first-order equation for L^2, namely

$$\frac{d}{dr}L^2 + 4\,k^2\,\frac{L^2}{r} - 2\,\Lambda r^{1-8\,k^2} = 0 \tag{13.7}$$

with integral

$$\begin{aligned} L(r)^2 &= r^{-4\;k^2}\,C_1 + r^{2-8\;k^2}\frac{\Lambda}{1-2\;k^2} \\ &= \left(r^{4\;k^2}\,C_1 + \frac{\Lambda}{1-2\;k^2}r^2\right)r^{-8\;k^2}. \end{aligned} \tag{13.8}$$

The remaining Einstein equation arises from $-({EQ_1}^1 - {EQ_2}^2 + {EQ_3}^3)\,r/(2\,L^2)$, and yields

$$\frac{1}{N}\frac{dN}{dr} - \frac{\Lambda}{L^2}r^{(1-8\,k^2)} - 2\,\frac{k^2}{r} = 0. \tag{13.9}$$

Finally, substituting $L(r)^2$ from (13.8) into the above equation (13.9) for $N(r)$, and integrating one obtains

$$N(r)^2 = {C_N}^2\left(r^{4\;k^2}\,C_1 + \frac{\Lambda}{1-2\;k^2}r^2\right),\; N(r)^2 = {C_N}^2\,r^{8\;k^2}\,L(r)^2. \tag{13.10}$$

The remaining equation with second derivatives of $\frac{d^2}{dr^2}N$, $-{EQ_1}^1 = \frac{L^2}{Nr}\frac{d}{dr}N + \frac{L}{N}\frac{dL}{dr}\frac{dN}{dr} + \frac{L^2}{N}\frac{d^2N}{dr^2} - 2\;\Lambda\,r^{bk} = 0$, is fulfilled for the determined structural functions. Thus, the general static solution can be given as

$$\begin{aligned} g &= -\,{C_N}^2\,r^{8\;k^2}\;L(r)^2 dt^2 + \frac{dr^2}{L(r)^2} + r^2 d\phi^2, \\ L(r)^2 &= \left(r^{4\;k^2}\,C_1 + \frac{\Lambda}{1-2\;k^2}r^2\right)r^{-8\;k^2}, \\ \Psi(r) &= k\;\ln(r). \end{aligned} \tag{13.11}$$

endowed with three significant parameters: mass $-C_1$, cosmological constant $\Lambda \to \pm\frac{1}{l^2}$, and the dilaton parameter k. The constant C_N can be absorbed by scaling the coordinate t.

The constant Λ can be equated to minus the standard $\Lambda_s = \pm\frac{1}{l^2}$; indeed, by setting in (13.11)

$$\Lambda = \pm\frac{1}{l^2}\alpha^2,\, r \to r\alpha^{1/(4\,k^2)},\, \phi \to \phi\,\alpha^{-1/(4\,k^2)},$$
$$C_1 \to C_1\,\alpha^{1+1/(2\,k^2)},\, C_N \to C_N\,\alpha^{-(1+1/(4\,k^2))}, \tag{13.12}$$

one arrives at the metric (13.11) with structural functions with $\Lambda = \pm\frac{1}{l^2}$.

Notice that the Λ used by in Chan–Mann, when considered as a cosmological constant, differs from the standard cosmological constant $\Lambda_s = \pm\frac{1}{l^2} = -\Lambda$, where $+$ and $-$ stand correspondingly for de Sitter and anti-de Sitter (AdS).

If $\Lambda\,(1 - 2\,k^2) > 0$ then one has

a) dS horizonless: $\Lambda < 0 \wedge \{k < -\frac{1}{\sqrt{2}}, k > \frac{1}{\sqrt{2}}\}, C_1 > 0$,
b) dS cosmological singularity: $\Lambda < 0 \wedge \{k < -\frac{1}{\sqrt{2}}, k > \frac{1}{\sqrt{2}}\}, C_1 < 0$,
c) AdS horizonless: $\Lambda > 0 \wedge \{-\frac{1}{\sqrt{2}} < k < \frac{1}{\sqrt{2}}\}, C_1 > 0$,
d) AdS black hole: $\Lambda > 0 \wedge \{-\frac{1}{\sqrt{2}} < k < \frac{1}{\sqrt{2}}\}, C_1 < 0$,

If $\Lambda\,(1 - 2\,k^2) < 0$ then one has

e) dS cosmological singularity: $\Lambda < 0 \wedge \{-\frac{1}{\sqrt{2}} < k < \frac{1}{\sqrt{2}}\}, C_1 > 0$,
f) AdS event horizon: $\Lambda > 0 \wedge \{k < -\frac{1}{\sqrt{2}}, k > \frac{1}{\sqrt{2}}\}, C_1 > 0$.

Notice that for vanishing dilaton parameter $k = 0$, the derived solution reduces to the BTZ one for $C_1 = -M$, $C_N = 1$, and $\Lambda = 1/l^2$, on the other hand, for opposite in sign constants one gets the cosmological dS solution.

Accomplishing in the general above solution (13.11) the transformations and constant parameterizations:

$$r \to r^{N/2}, \phi \to \beta\,\theta,\, k = \pm\frac{1}{2}\sqrt{\frac{2-N}{N}},\; C_N = 2\beta/N,\, 2\,\beta^2\,C_1 = -M\,N,$$

one gets the Chan–Mann static solution: Chan (1997); Chan and Mann (1996)

$$ds^2 = -U(r)dt^2 + \beta^2\frac{dr^2}{U(r)} + \beta^2 r^N d\theta^2,$$
$$U(r) = \beta^2\,F(r) = -\frac{2M}{N}r^{1-\frac{N}{2}} + \frac{8\Lambda\beta^2}{(3\,N-2)N}r^N,$$
$$\Psi(r) = \pm\frac{1}{4}\sqrt{N(2-N)}\,\ln(r),\, 0 < N < 2,\, N \neq 2/3. \tag{13.13}$$

13.2.1 Quasi Local Momentum, Energy, and Mass

To characterize non-asymptotically flat solutions one uses the Brown–York formalism, see Brown and York Jr. (1993); Brown et al. (1994), of quasi–local momentum, energy, and mass quantities. The case to be treated in detail corresponds to the black hole solution with parameters $\Lambda > 0$, $C_1 = -M, 1 - 2k^2 = B^2, 0 < B < 1$, $C_N = 1$. Evaluating the quasi–local characteristics as functions of r, one obtains for the energy and mass

$$\epsilon(r) = -r^{2\,B^2-2}\sqrt{1 - \frac{MB^2}{r^{2\,B^2}\Lambda}}\frac{\sqrt{\Lambda}}{\pi\,B} - \epsilon_0,$$
$$E(r) = -2\,r^{2\,B^2-1}\sqrt{1 - \frac{MB^2}{r^{2\,B^2}\Lambda}}\frac{\sqrt{\Lambda}}{B} - 2\pi r\epsilon_0,$$
$$M(r) = 2\,M - 2\,\frac{r^{2B^2}}{B^2}\Lambda - 2\,r\frac{\sqrt{\Lambda}}{B}\sqrt{1 - \frac{MB^2}{r^{2B^2}\Lambda}}\pi\,\epsilon_0. \tag{13.14}$$

The series expansion of energy and mass functions independent of ϵ_0 behave at infinity, $r \to \infty$, as

$$\epsilon(r \to \infty, \epsilon_0 = 0) \approx \frac{B\,M}{2\sqrt{\Lambda}r^2\pi} - \frac{\sqrt{\Lambda}}{r^2 B\pi}r^{2B^2},$$
$$E(r \to \infty, \epsilon_0 = 0) \approx \frac{B\,M}{r\sqrt{\Lambda}} - 2\,\frac{\sqrt{\Lambda}}{r\,B}r^{2B^2},$$
$$M(r \to \infty, \epsilon_0 = 0) = 2\,M - 2\,\frac{\Lambda}{B^2}r^{2B^2}. \tag{13.15}$$

The reference energy density to be used in this evaluation is the one corresponding to the anti-de Sitter metric with parameter M_0,

$$\epsilon_0(M_0) = -\frac{1}{\pi\,r}\sqrt{\frac{r^2}{l^2} - M_0},\ \epsilon_{0|\infty}(M_0) \approx -\frac{1}{\pi\,l} + \frac{l\,M_0}{2\pi\,r^2} = -\frac{\sqrt{\Lambda}}{\pi} + \frac{M_0}{2\sqrt{\Lambda}r^2\pi}.$$

The series expansions of the corresponding quantities at $r \to \infty$ result in

$$\epsilon(r \to \infty, \epsilon_{0|\infty}(M_0)) \approx \frac{B\,M - M_0}{2\pi\sqrt{\Lambda}r^2} - \frac{\sqrt{\Lambda}r^{2\,B^2-2}}{B\pi} + \frac{\sqrt{\Lambda}}{\pi},$$
$$E(r \to \infty, \epsilon_{0|\infty}(M_0)) \approx \left(-2\,\frac{r^{2\,B^2-1}}{B} + 2\,r\right)\sqrt{\Lambda} + \frac{B\,M - M_0}{r\sqrt{\Lambda}},$$
$$M(r \to \infty, \epsilon_{0|\infty}(M_0)) \approx -2\,\frac{\Lambda\,r^{2\,B^2}}{B^2} + 2\,\frac{\Lambda\,r^2}{B} - \frac{M_0}{B} + 2\,M$$
$$-B\,Mr^{(2-2\,B^2)}. \tag{13.16}$$

For $B = 1$, which corresponds to vanishing dilaton parameter k, one arrives at the static BTZ metric, and at its quasi local energy and mass; compere with the energy–mass characteristics of the BTZ solution. Therefore, one concludes that role of the mass parameter is played by M. At spatial infinity all physical quantities are infinite; in particular, the field Ψ increases logarithmically.

13.2.2 Classification of the Energy–Momentum and Cotton Tensors

The Ricci tensor, determining in (2+1) dimensions the Riemann curvature tensor, or equivalently, for this class of dilatons, the energy–momentum tensor matrix $(T^{\mu}{}_{\nu})$ is described by

$$\begin{bmatrix} -2\,\Lambda\,r^{-8\,k^2} & 0 & 0 \\ 0 & 4\,k^2 r^{-2-4\,k^2}\_C1 - 2r^{-8\,k^2}\Lambda\,\frac{(4\,k^2-1)}{(2\,k^2-1)} & 0 \\ 0 & 0 & -2\,\Lambda\,r^{-8\,k^2} \end{bmatrix},$$

and is algebraically characterized by the following eigenvectors:

$$\begin{aligned} &\lambda_{1,2} = -2\,\Lambda\,r^{-8\,k^2} : \mathbf{V1,\ 2} = [V^1, 0, V^3], V_\mu V^\mu = -{V^1}^2 N^2 + {V^3}^2 r^2, \\ &V^3 = \frac{\delta\,V^1\,N}{r}, V_\mu V^\mu = \left(V^1\right)^2 N^2\,\left(\delta^2 - 1\right); \\ &\delta^2 > 1, \mathbf{V1} = \mathbf{S1}, \delta = 1, \mathbf{V1} = \mathbf{N1}, \delta^2 < 1, \mathbf{V1} = \mathbf{T1} \\ &\lambda_3 = 4\,k^2 r^{-2-4\,k^2}\,C_1 - 2\Lambda\,r^{-8\,k^2}\,\frac{(4\,k^2-1)}{(2\,k^2-1)} = T^2{}_2 : \mathbf{V3} = [0, V^2, 0], \\ V_\mu V^\mu = \frac{V^{2^2}}{L^2}, &\mathbf{V3} = \mathbf{S3}. \end{aligned} \tag{13.17}$$

Hence, depending on the sign of the norm, one will have spacelike, $\delta^2 > 1$, null, $\delta^2 = 1$, or timelike, $\delta^2 < 1$, eigenvectors: $\mathbf{V1,\ 2} = \{\mathbf{S}, \mathbf{N}, \mathbf{T}\}$. Therefore, in the case of a double root one may choose different vector components determining, for instance, one spacelike vector S and the other timelike T or null N one. Therefore, one may have the algebraic Ricci types: $\{S, 2S\}, \{S, 2N\}, \{S, 2T\}$ and $\{S, (S,T)\}, \{S, (T,T)\}, \ldots, \{S, (T,N)\}$.

The matrix of the Cotton tensor is given by

$$\begin{aligned} (C^{\mu}{}_{\nu}) &= \begin{bmatrix} 0 & 0 & C^1{}_3 \\ 0 & 0 & 0 \\ C^3{}_1 & 0 & 0 \end{bmatrix}, \\ C^1{}_3 &= \frac{C_1}{C_N} k^2\,\left(1 - 2\,k^2\right)\,r^{-2-8\,k^2}, \\ C^3{}_1 &= -C_1\,k^2\,\left(1 - 2\,k^2\right) r^{-4} L(r)^2, \\ C^1{}_3 C^3{}_1 &= -\frac{{C_1}^2}{C_N}\,k^4\,\left(1 - 2\,k^2\right)^2 r^{-6-8\,k^2}\,L(r)^2 < 0. \end{aligned} \tag{13.18}$$

The characteristic equation allows for one real and a pure imaginary eigenvalues. Therefore, the Cotton tensor is characterized algebraically by

$$\begin{aligned} &\lambda_1 = 0 : \mathbf{V1} = [0, V^2, 0], V_\mu V^\mu = \frac{V^2}{L^2}, \mathbf{V1} = \mathbf{S1}, \\ &\lambda_2 = \bar{\lambda}_3 = \sqrt{C^1{}_3 C^3{}_1} : \end{aligned}$$

$$\mathbf{V2} = \mathbf{Z} = [V^1, 0, V^3 = \frac{\sqrt{C^3{}_1}}{\sqrt{C^1{}_3}} V^1], \mathbf{V3} = \bar{\mathbf{Z}}. \tag{13.19}$$

Thus, the algebraic type of the Cotton tensor is Type I: $\{S, Z, \bar{Z}\}$.

13.3 General Static Chan–Mann Solution

With the purpose of establishing the uniqueness of the Chan–Mann solution, its derivation is accomplished in a coordinate frame such that the static cyclic symmetric line element is given by

$$g = -F(r)dt^2 + \frac{dr^2}{F(r)} + H(r)d\phi^2.$$

The Einstein-field equations (13.2) for this metric are

$$\begin{aligned}
EQ_{tt} &= -2\,\Lambda\,\mathrm{e}^{b\Psi} F + \frac{F}{4\,H}\left(\frac{d}{dr}H\right)\frac{d}{dr}F + \frac{1}{2}F\left(\frac{d^2}{dr^2}F\right),\\
EQ_{rr} &= 2\,\frac{\Lambda\,\mathrm{e}^{b\Psi}}{F} + \frac{1}{4}\frac{\left(\frac{d}{dr}H\right)^2}{(H)^2} - \frac{1}{2}\frac{\frac{d^2}{dr^2}H}{H} - \frac{1}{2}\frac{\frac{d^2}{dr^2}F}{F}\\
&\quad -\frac{1}{4}\frac{\left(\frac{d}{dr}F\right)\frac{d}{dr}H}{HF} - 4\left(\frac{d}{dr}\Psi\right)^2,\\
EQ_{\phi\phi} &= 2\,\Lambda\,\mathrm{e}^{b\Psi} H - \frac{1}{2}\left(\frac{d}{dr}F\right)\frac{d}{dr}H + \frac{1}{4}\frac{F\left(\frac{d}{dr}H\right)^2}{H} - \frac{1}{2}\left(\frac{d^2}{dr^2}H\right)F
\end{aligned} \tag{13.20}$$

while the dynamical equation for the scalar Ψ amounts to

$$EQF = \left(\frac{d}{dr}F\right)\frac{d}{dr}\Psi + \left(\frac{d^2}{dr^2}\Psi\right)F + \frac{1}{2}\frac{F\left(\frac{d}{dr}H\right)\frac{d}{dr}\Psi}{H} + \frac{1}{4}b\Lambda\,\mathrm{e}^{b\Psi}. \tag{13.21}$$

The combination $EQ_{tt}/F(r) + EQ_{rr}F(r)$ yields

$$2\,H\,\frac{d^2\,H}{dr^2} - \left(\frac{d\,H}{dr}\right)^2 + 16\left(\frac{d\Psi}{dr}\right)^2 H^2 = 0, \tag{13.22}$$

which, introducing the new function $Z = \sqrt{H}$, and substituting $\Psi = k\ln(r)$, gives rise to an Euler equation

$$\frac{d^2}{dr^2}Z\,(r) + 4\,\frac{k^2 Z\,(r)}{r^2} = 0 \tag{13.23}$$

with solutions of the form r^α, $\alpha^2 - \alpha + 4\,k^2 = 0$, $\alpha_\pm = 1/2 \pm 1/2\,\sqrt{1 - 16\,k^2}$. Therefore

$$H(r) = \left(Z_+ r^{\alpha_+} + Z_- r^{\alpha_-}\right)^2, Z_\pm = \text{const.} \tag{13.24}$$

The field equation (13.21), introducing a new function $Y(r)$ through $F = Y(r)/\sqrt{H}$ and replacing $\Psi = k\ln(r)$, becomes

$$4\,k\frac{d}{dr}(\frac{Y(r)}{r}) + b\Lambda\,r^{b\,k}\,\sqrt{H} = 0, \tag{13.25}$$

with first integral

$$4\,k\,Y(r) = -r\int b\Lambda\,r^{b\,k}\,\sqrt{H}dr + C_0\,r, \tag{13.26}$$

and using here $H(r)$ from (13.24), one gets

$$\begin{aligned} Y(r) = C_0\,r - \frac{b\Lambda\,Z_+}{2k}\,\frac{r^{bk+5/2+1/2\,\sqrt{1-16\,k^2}}}{\left(2\,bk+3+\sqrt{1-16\,k^2}\right)} \\ -\frac{b\Lambda\,Z_-}{2k}\,\frac{r^{bk+5/2-1/2\,\sqrt{1-16\,k^2}}}{\left(2\,bk+3-\sqrt{1-16\,k^2}\right)}. \end{aligned} \tag{13.27}$$

Consequently, the structural function $F(r)$ is

$$F(r) = \frac{Y(r)}{\left(Z_+\,r^{1/2+1/2\,\sqrt{1-16\,k^2}} + Z_-\,r^{1/2-1/2\,\sqrt{1-16\,k^2}}\right)}. \tag{13.28}$$

The remaining equations to be satisfied impose some constraints on $Z_\pm$, and on the values of b, namely $b_- := -\frac{1+\sqrt{1-16\,k^2}}{k}$ for $Z_+ = 0, Z_- \neq 0$ and $b_+ := \frac{-1+\sqrt{1-16\,k^2}}{k}$ for $Z_- = 0, Z_+ \neq 0$. Correspondingly, one gets the solutions in the form

$$\begin{aligned} g_\mp &= -dt^2\,F(r)_\mp + \frac{dr^2}{F(r)_\mp} + d\phi^2\,H(r)_\mp, \\ F(r)_\mp &= \frac{C_0}{Z_\mp}\,r^{\frac{1}{2}\pm\frac{1}{2}\,\sqrt{1-16\,k^2}} + 8\,\frac{r^{1\mp\sqrt{1-16\,k^2}}\,\Lambda}{4\mp4\,\sqrt{1-16\,k^2}-48\,k^2}, \\ H(r)_\mp &= {Z_\mp}^2 r^{1\mp\sqrt{1-16\,k^2}},\ \Psi(r) = k\ln(r). \end{aligned} \tag{13.29}$$

13.3.1 Regular $F(r)_+$ Function for the Metric g_+

The function $F(r)_+$ would be regular (with respect to the set of parameters) for certain values of the dilaton parameter k: if k is restricted to the range $0 < k < 1/4$, in F_+ the denominator $D_+ := 4 + 4\sqrt{1-16\,k^2} - 48\,k^2$ does not approach zero in this whole range of k, $D_+ > 0$, therefore $F(r)_+$ is regular for $0 < k < 1/4$. At $k = 0$ the metric is well-behaved and recognizable as the static BTZ metric with parameters $C_0 = -M, \Lambda = 1/l^2$, and $Z_+ = 1$. On the other hand, introducing a new constant ν related to the dilaton, $\nu = \sqrt{1-16\,k^2}$, $0 < \nu < 1$, one rewrites the structural functions as

$$
\begin{aligned}
F(r)_+ &= \frac{C_0}{Z_+}\, r^{(1-\nu)/2} + 8\,\frac{r^{1+\nu}\Lambda}{(\nu+1)(3\nu+1)},\\
r_h &= \left(-\frac{8\,\Lambda\, Z_+}{C_0(3\nu+1)(\nu+1)}\right)^{(3\nu+1)/2},\\
H(r)_+ &= {Z_+}^2 r^{1+\nu},\ \Psi(r) = \pm\frac{\sqrt{1-\nu^2}}{4}\ln(r),\ 0<\nu<1. \qquad (13.30)
\end{aligned}
$$

It is apparent that these structural functions for $\nu = 1$ yield the functions of the static BTZ metric, under minor arrangements. Other interesting solution contained in this metric representation is the one determined by $\nu = 0$, namely

$$
\begin{aligned}
F(r)_+ &= \frac{C_0}{Z_+}\, r^{1/2} + 8\,\Lambda\, r,\ r_h = \left(-\frac{8\,\Lambda\, Z_+}{C_0}\right)^{1/2},\\
H(r)_+ &= {Z_+}^2 r,\ \Psi(r) = \pm\frac{1}{4}\ln(r), \qquad (13.31)
\end{aligned}
$$

which is identifiable with the string solution.

On the other hand, if one were adopting the choice

$$
\nu = 1+\sqrt{1-16\,k^2} \to k = \frac{1}{4}\sqrt{\nu(2-\nu)},\ 1<\nu<2, \qquad (13.32)
$$

then the structural functions would be

$$
\begin{aligned}
F(r)_+ &= \frac{C_0}{Z_+}\, r^{1-\nu/2} + \frac{8\Lambda}{(3\,\nu-2)\nu}\, r^{\nu},\ H(r)_+ = Z_+^2\, r^{\nu},\\
\Psi(r) &= \pm\frac{1}{4}\sqrt{\nu(2-\nu)}\,\ln(r),\ 1<\nu<2. \qquad (13.33)
\end{aligned}
$$

13.3.2 Chan–Mann Solution

The function $F(r)_-$ exhibits a singularity in its dependence on k: the reality of the radical in the powers of r, $\sqrt{1-16\,k^2}$, imposes to k to ranges $0<k<1/4$, moreover in F_- the denominator $D_- := 4-4\sqrt{1-16\,k^2}-48\,k^2$ riches 0 at $k=\pm\sqrt{2}/6$, and at $k=0$, therefore depending on the range of k, $F(r)_-$ splits into two branches:

a) $F(r)_-$ for k within $-\sqrt{2}/6<k<0$, and $0<k<\sqrt{2}/6$, has $D_-<0$, consequently, for $\Lambda>0$ and $C_0>0$, $(Z_\mp=1)$, the function $F(r)_-$ becomes zero at $r=(-\frac{C_0 D_-}{8\Lambda Z_\mp})^{(1\mp\sqrt{1-16\,k^2})/2}$.

b) $F(r)_-$ for k within $-1/4<k<-\sqrt{2}/6$, and $\sqrt{2}/6<k<1/4$, has $D_->0$.

One may introduce a new dilaton constant, say N, through

$$
N = 1-\sqrt{1-16\,k^2} \to k = \pm\sqrt{N(2-N)}/4,
$$

consequently, this correspondence establishes the following relationship in the ranges of values of the dilaton constants

$$0 < k < \frac{\sqrt{2}}{6} \to 0 < N < \frac{2}{3}, \frac{\sqrt{2}}{6} < k < 1/4 \to \frac{2}{3} < N < 1, \tag{13.34}$$

with the structural metric functions and dilaton field in the form

$$F(r) = \frac{C_0}{Z_-} r^{1-\frac{N}{2}} + \frac{8\Lambda}{(3\,N-2)N} r^N, \; H(r) = Z_-^2\, r^N,$$

$$\Psi(r) = \pm\frac{1}{4}\sqrt{N(2-N)}\,\ln(r). \tag{13.35}$$

Further, by setting $C_0 = -2M/(N\,\beta)$, $Z_- = \beta$, and scaling the coordinate $t \to \beta t$, and adding the ranges of values of the dilaton constants N and ν, from (13.34) and (13.32) respectively, one arrives at the solution

$$ds^2 = -U(r)dt^2 + \beta^2 \frac{dr^2}{U(r)} + \beta^2 r^N d\theta^2,$$

$$U(r) = \beta^2\, F(r) = -\frac{2M}{N} r^{1-\frac{N}{2}} + \frac{8\Lambda\beta^2}{(3\,N-2)N} r^N,$$

$$\Psi(r) = \pm\frac{1}{4}\sqrt{N(2-N)}\,\ln(r),\; 0 < N < 2,\; N \neq 2/3,$$

known as the Chan–Mann black hole solution for minimally coupled scalar field; see Chan (1997); Chan and Mann (1996). In this way, the uniqueness of the solution for a dilaton of the form $\Psi(r) = k\,\ln(r)$ coupled to a exponential potential $\exp(b\,\Psi)$ has been established avoiding guesses about a special structure of the metric. It describes a black hole with horizon at

$$r_h = \left(\frac{M\,(3N-2)}{4\beta^2\Lambda}\right)^{\frac{2}{3N-2}},$$

for $N \neq 2/3$, $0 < N < 2$, $M > 0$ and $\Lambda > 0$.

13.4 Stationary Solution Coupled to $\Psi(r) = k\,\ln(r)$

The (2+1)-dimensional stationary cyclic symmetric metric in the Schwarzschild coordinate frame $\{t, r, \phi\}$ can be given as

$$g = -N(r)^2 dt^2 + \frac{dr^2}{L(r)^2} + r^2(d\phi + W(r)dt)^2. \tag{13.36}$$

In this frame the scalar field equation becomes

$$\frac{1}{N}\frac{dN}{dr} + \frac{1}{L}\frac{dL}{dr} + \frac{b\Lambda\, r^{bk+1}}{4\,k\,L^2} = 0. \tag{13.37}$$

The simplest Einstein equation is ${EQ_3}^1$:

$$\frac{\frac{d^2W}{dr^2}}{\frac{dW}{dr}} = -\frac{3}{r} - \frac{1}{L}\frac{dL}{dr} + \frac{1}{N}\frac{dN}{dr}, \tag{13.38}$$

with first integral

$$\frac{dW}{dr} = W_0 \frac{N}{r^3 L}, \tag{13.39}$$

Replacing $\frac{dL}{dr}$ from (13.37) and $\frac{dW}{dr}$ from (13.39) into the Einstein equations, from ${EQ_3}^3$ one gets

$$\frac{b\Lambda\, r^{bk}}{k} - 2\,\frac{{W_0}^2}{r^4} + 8\ \Lambda\, r^{bk} = 0, \tag{13.40}$$

which is fulfilled if

$$b = -\frac{4}{k} \neq 0,\ \ {W_0}^2 = 2\,\frac{\Lambda\,\left(2\ k^2 - 1\right)}{k^2}, \tag{13.41}$$

thus, for this class the dilaton field ought always to exist, $b \neq 0$ and $k \neq 0$, there is no room for a vacuum plus cosmological constant field. Moreover, one has to distinguish two branches: $\Lambda > 0 \wedge \left(2\ k^2 - 1\right) > 0$, and $\Lambda < 0 \wedge \left(2\ k^2 - 1\right) < 0$.

Replacing b into the field equation one gets

$$\frac{1}{N}\frac{d}{dr}N + \frac{1}{L}\frac{d}{dr}L - \frac{\Lambda}{k^2 r^3 L^2} = 0. \tag{13.42}$$

The remaining ${EQ_1}^1 - {EQ_2}^2$ Einstein equation yields

$$-\frac{1}{N}\frac{d}{dr}N + \frac{1}{L}\frac{d}{dr}L + 4\,\frac{k^2}{r} = 0. \tag{13.43}$$

Adding these two last equations, one arrives at a linear first-order equation for L^2

$$\frac{d}{dr}L^2 + 4\,\frac{L^2 k^2}{r} = \frac{\Lambda}{k^2 r^3}, \tag{13.44}$$

with integral

$$L(r)^2 = r^{-4\ k^2}\, C_1 + \frac{\Lambda}{2\,k^2\,(2\ k^2 - 1)}\frac{1}{r^2}. \tag{13.45}$$

Finally, substituting this result for $L(r)^2$ into the first-order linear equation (13.43) for $N(r)$ and integrating one obtains

$$N(r)^2 = {C_N}^2 r^{8\ k^2}\ L(r)^2. \tag{13.46}$$

The remaining Einstein equation with second derivatives of $\frac{d^2}{dr^2}N$ is fulfilled for this set of functions. On the other hand, the equation for W (13.39) becomes

$$\frac{d}{dr}W = W_0\, C_N\, r^{4\ k^2 - 3} \tag{13.47}$$

with general integral

$$\begin{aligned} W &= \frac{W_0\, C_N}{2(2\ k^2 - 1)} r^{4\ k^2 - 2} + W_1 \\ \to W(r) &= \sqrt{\frac{\Lambda}{2\ k^2 - 1}}\frac{C_N}{\sqrt{2}\,k} r^{4\ k^2 - 2} + W_1. \end{aligned} \tag{13.48}$$

The parameter W_1 can be equated to zero by a coordinate translation. Thus, this stationary solution coupled to a dilaton can be given by

$$\begin{aligned}
g &= -{C_N}^2\, r^{8\,k^2}\, L(r)^2\, dt^2 + \frac{dr^2}{L(r)^2} + r^2\,(d\phi + W(r)dt)^2\,,\\
L^2 &= r^{-4\,k^2}\, C_1 + \frac{\Lambda}{2\,k^2\,(2\,k^2-1)}\frac{1}{r^2},\\
W &= \sqrt{\frac{\Lambda}{2\,k^2-1}}\frac{C_N}{\sqrt{2}\,k}r^{4\,k^2-2},\\
\Psi(r) &= k\,\ln(r).
\end{aligned} \tag{13.49}$$

Recall that in the derivation of this solution one has to fit the condition $\Lambda\,\left(2\,k^2-1\right) > 0$, together with $k \neq 0$, $k \neq 1/\sqrt{2}$. Moreover, having in mind that $g_{tt} = -{C_N}^2 C_1 r^{4\,k^2}$, the signature $\{-,+,+\}$ imposes some constraints on the range of values of the constant:

assuming that Λ is related with the standard cosmological constant Λ_s then $\Lambda_s = \pm\frac{1}{l^2} = -\Lambda$, where + and − stand correspondingly for de Sitter and anti-de Sitter (AdS) cases.

If $\Lambda < 0$, then $\left(2\,k^2-1\right) < 0$ with

- dS horizonless: $\Lambda < 0 \wedge \{-\frac{1}{\sqrt{2}} < k < \frac{1}{\sqrt{2}}\} \wedge k \neq 0, C_1 > 0$,

If $\Lambda > 0$, then $\left(2\,k^2-1\right) > 0$ and arises

- AdS horizonless $\Lambda > 0 \wedge \{k < -\frac{1}{\sqrt{2}}, k > \frac{1}{\sqrt{2}}\}, C_1 > 0$.

Thus the allowed dilaton solutions are horizonless.

13.4.1 Momentum, Energy, and Mass for a Rotating Dilaton

To characterize non-asymptotically flat solutions one uses the Brown–York formalism of quasi-local momentum, energy, and mass quantities. The expressions for the momentum density and the quasi-local momentum are

$$j = \frac{\sqrt{2\,\Lambda\,(2\,k^2-1)}}{2\pi\,k\,r},\, J(r) = \frac{\sqrt{2\,\Lambda\,(2\,k^2-1)}}{k} = W_0. \tag{13.50}$$

Evaluating the quasi-local characteristics as functions of r one obtains for the energy and mass

$$\begin{aligned}
\epsilon(r) &= -\frac{L(r)}{\pi\,r} - \epsilon_0,\; E(r) = -2\,L(r) - 2\,\pi\,r\,\epsilon_0,\\
M(r) &= -2\,C_N\,C_1 - 2\,C_N\frac{\Lambda}{2\,k^2-1}\,r^{4\,k^2-2} - 2\,C_N r^{4\,k^2}\,L(r)\pi\,\epsilon_0.
\end{aligned} \tag{13.51}$$

The series expansions of energy and mass functions independent of ϵ_0 behave at infinity $r \to \infty$ as

$$\epsilon(r \to \infty, \epsilon_0 = 0) \approx -\frac{\sqrt{2}k\,C_1\,\sqrt{2\,k^2-1}}{2\,\pi\sqrt{\Lambda}\,\left(r^{\,k^2}\right)^4} - \frac{\sqrt{2\,\Lambda}}{2\,\pi\,k\,r^2\,\sqrt{2\,k^2-1}},$$
$$E(r \to \infty, \epsilon_0 = 0) \approx -\left(1 + \frac{C_1\,k^2\,(2\,k^2-1)}{\Lambda\,r^{4\,k^2-2}}\right)\frac{\sqrt{2\Lambda}}{r\,k\,\sqrt{2\,k^2-1}},$$
$$M(r \to \infty, \epsilon_0 = 0) = -2\,C_N\,C_1 - 2\,C_N\frac{\Lambda}{2\,k^2-1}\,r^{4\,k^2-2}. \tag{13.52}$$

The constant value in $M(r \to \infty)$ one may identify with a mass, namely $M = -C_1 C_N$, for $C_N = -1$. Notice that by scaling the coordinate t, without any loss of generality, one can set $C_N = \pm 1$. Nevertheless, as has been established above, the range that the constant may assume imposes certain constraints, in particular the positiveness of C_1, the obtained stationary solution cannot have the appropriate mass $M = -C_1$, instead one has to recur to $C_N = -1$ to overcome the difficulty.

At spatial infinity all physical quantities are infinite, except for the constant quasi global momentum J, in particular, the mass function increases as a power of r, while the dilaton field Ψ logarithmically increases.

13.4.2 Classification of the Energy–Momentum and Cotton Tensors

The Ricci tensor, determining in (2+1) dimensions the Riemann curvature tensor, or equivalently, for this class of dilatons, the energy–momentum tensor is described by the matrix

$$(T^\mu{}_\nu) = \begin{bmatrix} -\frac{2\,\Lambda}{r^4} & 0 & 0 \\ 0 & \frac{4\,k^2\,L(r)^2}{r^2} - \frac{2\Lambda}{r^4} & 0 \\ 0 & 0 & -\frac{2\,\Lambda}{r^4} \end{bmatrix}, \tag{13.53}$$

and is algebraically characterized by the following eigenvectors:

$$\begin{aligned} \lambda_{1,2} &= -2\,\frac{\Lambda}{r^4} : \mathbf{V1,\ 2} = [V^1, 0, V^3], V_\mu V^\mu = r^2\left(V^3 + W V^1\right)^2 - {V^1}^2 N^2; \\ &V_\mu V^\mu > 1, \mathbf{V1} = \mathbf{S1}, V_\mu V^\mu = 0, \mathbf{V1} = \mathbf{N1}, V_\mu V^\mu < 1, \mathbf{V1} = \mathbf{T1}, \\ &V_\mu V^\mu > 1, \mathbf{V2} = \mathbf{S2}, V_\mu V^\mu = 0, \mathbf{V2} = \mathbf{N2}, V_\mu V^\mu < 1, \mathbf{V2} = \mathbf{T2}, \\ \lambda_3 &= 4\,\frac{k^2 L^2}{r^2} - 2\,\frac{\Lambda}{r^4} = {T^2}_2 : \mathbf{V3} = [0, V^2, 0], V_\mu V^\mu = \frac{(V^2)^2}{L^2}, \mathbf{V3} = \mathbf{S3}. \end{aligned} \tag{13.54}$$

In the case of a double root, one may choose different vector components determining, for instance, one spacelike vector S and the other timelike T or null N one, and other possible combinations. Therefore, one may have the algebraic Ricci–energy–momentum types: $\{S, 2S\}, \{S, 2N\}, \{S, 2T\}$ and $\{S, (S, T)\}, \{S, (T, T)\}, \ldots, \{S, (T, N)\}$.

The matrix of the Cotton tensor is given by

$$(C^{\mu}{}_{\nu}) = \begin{bmatrix} C^{1}{}_{1} & 0 & C^{1}{}_{3} \\ 0 & 0 & 0 \\ C^{3}{}_{1} & 0 & -C^{1}{}_{1} \end{bmatrix},$$

$$C^{1}{}_{1} = -C^{3}{}_{3} = \frac{C_1}{2}\sqrt{2\Lambda}k\,\sqrt{2\,k^2-1}\,r^{-4\,k^2-4},$$

$$C^{1}{}_{3} = -\frac{k^2(2\,k^2-1)}{C_N}\left(\frac{C_1}{\left(r^{k^2}\right)^4} + \frac{\Lambda}{(2\,k^2-1)\,r^2k^2}\right) r^{-4\,k^2-2},$$

$$C^{3}{}_{1} = k^2\left(2\,k^2-1\right)\,C_N {C_1}^2\,r^{-4\,k^2-4}. \tag{13.55}$$

The characteristic equation allows for one real and pure imaginary eigenvalues. Consequently, the Cotton tensor is algebraically characterized by

$$\lambda_1 = 0 : \mathbf{V1} = [0, V^2, 0], V_{\mu}V^{\mu} = \frac{(V^2)^2}{L^2}, \mathbf{V1} = \mathbf{S1},$$

$$\lambda_2 = \bar{\lambda}_3 = \sqrt{(C^{1}{}_{1})^2 + C^{1}{}_{3}C^{3}{}_{1}} = i\,k^2\,C_1\left(2\,k^2-1\right)\,r^{-4\,k^2-3}\,L(r):$$

$$\mathbf{V2} = \mathbf{Z} = [V^1, 0, V^3], \mathbf{V3} = \bar{\mathbf{Z}} = [\bar{V}^1, 0, \bar{V}^3]. \tag{13.56}$$

Hence, the algebraic type of the Cotton tensor is Type I: $\{S, Z, \bar{Z}\}$.

13.5 Stationary Dilaton Solutions Generated via $SL(2, R)$ Transformations

Subjecting the metric (13.11) to a general $SL(2, R)$ transformation

$$t = \alpha T + \beta\,\Phi,$$
$$\phi = \gamma T + \delta\,\Phi,\ \Delta := \alpha\delta - \beta\gamma. \tag{13.57}$$

one arrives at the stationary metric equipped with rotation

$$g = -\left(\alpha^2\,N(r)^2 - \gamma^2 r^2\right)dT^2 - 2\left(\beta\,\alpha\,N(r)^2 - \delta\,\gamma\,r^2\right)dr\,d\Phi$$
$$+\frac{dr^2}{L(r)^2} + \left(\delta^2 r^2 - \beta^2\,N(r)^2\right)d\Phi^2,$$

$$L(r)^2 = \left(r^{4\,k^2}\,C_1 + \frac{\Lambda}{1-2\,k^2}r^2\right)r^{-8\,k^2},\ N(r)^2 = C_N^2\,L(r)^2,$$

$$\Psi(r) = k\,\ln(r). \tag{13.58}$$

or, in the standard representation used to evaluate energy, mass, and momentum, one has

$$
\begin{aligned}
g &= -\mathsf{N}(r)^2 dT^2 + \frac{dr^2}{\mathsf{L}(r)^2} + \mathsf{K}(r)^2 (d\Phi + \mathsf{W}(r) dT)^2, \\
\mathsf{N}(r)^2 &= \frac{N^2 r^2}{\delta^2 r^2 - \beta^2 N^2} \left(\alpha\,\delta - \beta\,\gamma\right)^2, \\
\mathsf{K}(r)^2 &= \delta^2 r^2 - \beta^2 N^2, \\
\mathsf{L}(r)^2 &= r^{-4\,k^2}\, C_1 + \frac{\Lambda}{1 - 2\,k^2} r^{-8\,k^2+2}, \\
\mathsf{W}(r) &= \frac{\delta\,\gamma\,r^2 - \beta\,\alpha\,N^2}{\delta^2 r^2 - \beta^2 N^2}, \ \beta \neq 0 \neq \delta, \\
N^2 &:= C_N^2 \left(r^{4\,k^2}\, C_1 + \frac{\Lambda}{1 - 2\,k^2} r^2\right) = C_N^2\, r^{8\,k^2} \mathsf{L}(r)^2.
\end{aligned} \tag{13.59}
$$

For the AdS ($\Lambda = 1/l^2$)–black hole branch, the constants appearing in the structural functions will be replaced by $C_N \to \sqrt{1 - 2\,k^2} = B$, $k = \sqrt{1 - B^2}/\sqrt{2}$, $0 < k < 1/\sqrt{2}, 1 > B > 0$, thus

$$
N(r)^2 = N_B^2 = r^2 \left(r^{-2\,B^2}\, C_1\, B^2 + \Lambda\right), \ \mathsf{L}(r)^2 = L_B^2 = \frac{N_B^2}{B^2} r^{-4+4\,B^2}. \tag{13.60}
$$

The quasi-local momentum $J(r)$ is constant in the whole spacetime, namely

$$
J(r) = J(r \to \infty) = 2\,\beta\,\delta\,B^3\, C_1 = W_0 = J_0. \tag{13.61}
$$

The quasi-local energy is given by

$$
\begin{aligned}
E(r) = -2\, \frac{r^{2\,B^2-1}\sqrt{\Lambda + \_C1\, r^{-2\,B^2} B^2}}{B\sqrt{\delta^2 - \beta^2\Lambda - \beta^2 \_C1\, B^2 r^{-2\,B^2}}} & \left(\delta^2 - \beta^2\Lambda - \beta^2 \_C1\, B^2 r^{-2\,B^2}\right. \\
& \left. + \beta^2 \_C1\, r^{-2\,B^2} B^4\right),
\end{aligned} \tag{13.62}
$$

while the quasi-local mass amounts to

$$
\begin{aligned}
M(r) = &- \frac{2\Delta\left(\Lambda + C_1\, B^2\, r^{-2\,B^2}\right)}{B\left(\delta^2 - \Lambda\beta^2 - \beta^2\, C_1\, B^2\, r^{-2\,B^2}\right)} \left[\beta^2\, C_1 B^2 (B^2 - 1)\right. \\
&\left. + r^{2\,B^2}(\delta^2 - \Lambda\beta^2)\right] + J_0\, \frac{\beta\,\alpha\,\Lambda - \delta\,\gamma + \beta\,\alpha\, C_1\, B^2\, r^{-2\,B^2}}{\delta^2 - \beta^2\Lambda - \beta^2\, C_1\, B^2\, r^{-2\,B^2}} \\
&- 2\,\epsilon_0\,\pi\,\Delta\, r^2 \sqrt{\Lambda + C_1\, B^2\, r^{-2\,B^2}}.
\end{aligned} \tag{13.63}
$$

The reference energy density to be used in this evaluation is that which corresponds to the anti-de Sitter metric with parameter M_0,

$$
\epsilon_0(M_0) = -\frac{\sqrt{-M_0 + r^2\Lambda}}{\pi\, r}, \ \epsilon_{0|\infty}(M_0) \approx -\frac{\sqrt{\Lambda}}{\pi} + 1/2\, \frac{M_0}{\sqrt{\Lambda}\pi\, r^2}.
$$

Then the series expansion of the mass at spatial infinity, $r \to \infty$, amounts to

$$M(r \to \infty) \approx -2\,\frac{\Delta\, C_1\, B\left(\delta^2 - \beta^2\Lambda + B^2\beta^2\Lambda\right)}{\delta^2 - \beta^2\Lambda} - 2\,\frac{\Delta\, r^{2\,B^2}\,\Lambda}{B}$$
$$+J_0\,\frac{(\beta\,\alpha\,\Lambda - \delta\,\gamma)}{\delta^2 - \beta^2\Lambda} + M(\epsilon_{0|\infty}),$$
$$M(\epsilon_{0|\infty}) = M(r \to \infty, \epsilon_{0|\infty}(M_0)) \approx \Delta\, C_1\, B^2\, r^{2-2\,B^2}$$
$$+2\,\Delta\,\Lambda\, r^2 - \Delta\, M_0.$$

When the dilaton field and the rotation vanish, $B = 1$, $\beta = 0$, the generated solution reduces to the AdS metric and one identifies $C_1 = -M$, or comparing with the quasi-local BTZ mass $M_{BTZ}(r \to \infty) = M - M_0$, the same conclusion is achieved. Frequently, in the literature one encounters the $SL(2,R)$ transformation

$$t = \frac{T}{\sqrt{1 - \frac{\omega^2}{l^2}}} - \omega\,\frac{\Phi}{\sqrt{1 - \frac{\omega^2}{l^2}}},\ \phi = -\frac{\omega}{l^2}\,\frac{T}{\sqrt{1 - \frac{\omega^2}{l^2}}} + \frac{\Phi}{\sqrt{1 - \frac{\omega^2}{l^2}}}, \tag{13.64}$$

correspondingly, the structural functions (13.59) assume the form

$$\mathsf{N}(r) = -\frac{\left(l^2 - \omega^2\right) r^2 N^2}{l^2\left(\omega^2 N^2 - r^2\right)},\ \mathsf{L}(r) = L(r),\ N = N(r),$$
$$\mathsf{W}(r) = -\frac{\omega\,\left(N^2 l^2 - r^2\right)}{l^2\left(\omega^2 N^2 - r^2\right)},\mathsf{K}(r) = -\frac{l^2\left(\omega^2 N^2 - r^2\right)}{l^2 - \omega^2}. \tag{13.65}$$

In this parametrization, the quasi-local momentum becomes

$$J(r) = -2\,\omega\,\frac{l^2 B^3\, C_1}{l^2 - \omega^2} = J_0,$$

and vanishes as soon the rotation ω becomes zero.

13.5.1 Sub-Class of Rotating Dilaton Black Holes

Requiring the fulfillment of the relationship

$$\gamma\,\delta = \beta\,\alpha\,\frac{{C_N}^2\Lambda}{(1 - 2\,k^2)}, \tag{13.66}$$

one obtains a sub-family of rotating dilaton solutions

$$g = -\alpha^2\,{C_N}^2\left(r^{8\,\kappa\,k^2}\,L^2 - \frac{\beta^2\,{C_N}^2\Lambda^2}{\delta^2\left(1 - 2\,k^2\right)^2}r^2\right)dt^2 - 2\beta\,\alpha\,{C_N}^2\,C_1\,r^{4\,k^2}\,dr\,d\phi$$
$$+\frac{dr^2}{L^2} + \left(\delta^2 r^2 - \beta^2 r^{8\,k^2}\,{C_N}^2 L^2\right)d\phi^2,$$
$$L(r)^2 = \left(r^{4\,k^2}\,C_1 + \frac{\Lambda}{1 - 2\,k^2}r^2\right)r^{-8\,k^2},\ \Psi(r) = k\,\ln(r), \tag{13.67}$$

in which the term with power r^2 in W or equivalently in $g_{T\Phi}$, in the metric (13.58), cancels out. On the other hand, in the limit of vanishing dilaton parameter $k=0$, under the coordinate transformation and new parameterizations

$$r^2 = \frac{\beta^2 C_1 C_N{}^2 + \rho^2}{\delta^2 - \beta^2 C_N{}^2 \Lambda} = \frac{2\left(M + \sqrt{M^2 - J_0{}^2\Lambda}\right)\rho^2 - J_0{}^2}{4\,\delta^2\sqrt{M^2 - J_0{}^2\Lambda}},$$
$$\alpha = \frac{\delta}{C_N},\ \beta = \frac{J_0}{2\,\delta\, C_N\, C_1},\ C_1 = -\frac{M + \sqrt{M^2 - J_0{}^2\Lambda}}{2\,\delta^2}, \tag{13.68}$$

this sub-branch becomes the rotating BTZ metric

The transformations above (13.68) may be used in the metric (13.67) to obtain the studied case in terms of the physical BTZ parameters M and J_0 and the dilaton parameter k. Nevertheless, the expressions are quite involved due to the $r \rightarrow \rho$ transformation.

13.5.2 Rotating Chan–Mann Dilaton Black Hole

Moreover, the metric (13.67) gives rise to the Chan–Mann rotating dilaton solution. Straightforwardly by subjecting the static seed metric (13.11) to the following $SL(2,R)$ transformation together with a transformation of the radial coordinate

$$t = \frac{\sqrt{3\,N - 2}\sqrt{N}\,(2\,A\,T + \omega\,\Phi)}{2\,B\sqrt{3\,N^2A^2 - 2\,NA^2 - 2\,\Lambda\,\omega^2}},\ r = \rho^{N/2},$$
$$\phi = \frac{B\left(A\sqrt{N}\sqrt{3\,N - 2}\,\Phi + 4\,\frac{\Lambda\,\omega}{\sqrt{N}\sqrt{3\,N-2}}\,T\right)}{\sqrt{3\,N^2A^2 - 2\,N\,A^2 - 2\,\Lambda\,\omega^2}}, \tag{13.69}$$

where N is the new dilaton parameter. Do not mix up with the structural function $N(r)$, together with the introduction of new constants through

$$k = \pm\frac{1}{2}\sqrt{\frac{2-N}{N}} = 1, C_N = 2\frac{B^2}{N},$$
$$C_1 = \frac{\left(3N^2A^2 - 2\,N\,A^2 - 2\,\Lambda\,\omega^2\right)N}{4\,(3\,N - 2)\,AB^2}, \tag{13.70}$$

one arrives at the Chan and Mann (1996) metric, Eq. (3.1a), namely

$$g = -\left(A\rho^{1-\,N/2} + 8\,\frac{\rho^N\Lambda\,B^2}{(3\,N - 2)\,N}\right)dT^2$$
$$+\left[8\,\frac{\rho^N\Lambda}{(3\,N - 2)\,N} + \frac{\rho^{1-\,N/2}}{B^2}\left(A - 2\,\frac{\Lambda\,\omega^2}{A\,(3\,N - 2)\,N}\right)\right]^{-1} d\rho^2$$
$$-\omega\,\rho^{1-\,N/2}\,dT\,d\Phi + \left(\rho^N B^2 - \frac{\omega^2\rho^{1-\,N/2}}{4\,A}\right)d\Phi^2 \tag{13.71}$$

where r of Eq. (3.1a) is replaced by ρ, and B stands for β of Eq. (3.1a) of Chan and Mann (1996), to avoid confusion.

It calls one's attention to the sophisticated modification of the function $g_{\rho\rho}$, which under the $SL(2,R)$ remains invariant and could be used (the original in relation to the static solution) for the adequate evaluation of the energy and mass.

13.6 Dilaton Coupled to Einstein–Maxwell Fields

The previous approach to determine Einstein–dilaton solutions in $(2+1)$ gravity is extended to the Einstein–Maxwell–dilaton case; thus, the Schwarzschild coordinate frame is used to determine static cyclic symmetric metrics to $(2+1)$ Einstein equations coupled to an electric Maxwell field and a dilaton logarithmically depending on the radial coordinate in the presence of an exponential potential. The general solution is derived and identified with the Chan–Mann charged dilaton solution. Via a general $SL(2,R)$ transformation, applied on the obtained charged dilaton metric, a family of stationary dilaton solutions has been generated; these solutions possess five parameters: dilaton and cosmological constants, charge, momentum, and mass for some values of them. All the exhibited solutions have been characterized by their quasi-local energy, mass, and momentum through their series expansions at spatial infinity. The structural functions determining these solutions increase with the radial coordinate; hence, they do not exhibit an dS–AdS behavior at infinity. Moreover, the algebraic structure of the Maxwell field, energy–momentum, and Cotton tensors is given explicitly.

13.6.1 Einstein–Maxwell-Scalar Field Equations

The action to be considered in this work dealing (2+1)-dimensional gravity is given by

$$\mathcal{S} = \int d^3x\sqrt{-g}\left[R - \frac{B}{2}\nabla_\mu\Psi\,\nabla^\mu\Psi + 2\,e^{b\Psi}\Lambda - e^{-4\,a\,\Psi}\,F^2\right], \tag{13.72}$$

where Λ, b are arbitrary at this stage parameters, Ψ is the massless minimally coupled scalar field, R is the scalar curvature, and $F^2 = F_{\mu\,\nu}\,F^{\mu\,\nu}$ the electromagnetic invariant. The variations of this action yield the dynamical equations

$$\begin{aligned}
&R_{\mu\nu} = \frac{B}{2}\nabla_\mu\Psi\,\nabla_\nu\Psi - 2g_{\mu\nu}e^{\,b\,\Psi}\Lambda + 2\,e^{-4\,a\,\Psi}\left(F_\mu{}^\alpha\,F_{\nu\alpha} - g_{\mu\,\nu}\,F^2\right),\\
&\frac{B}{2}\nabla^\mu\nabla_\mu\Psi + b\,e^{\,b\,\Psi}\Lambda + 2\,a\,e^{-4\,a\,\Psi}\,F^2 = 0,\\
&\nabla^\mu\left(e^{-4\,a\,\Psi}\,F_{\mu\,\nu}\right) = 0.
\end{aligned} \tag{13.73}$$

13.7 Static Charged Solution Coupled to $\Psi(r) = k\,\ln(r)$

The static cyclic symmetric metric in the (2+1) Schwarzschild coordinate frame is given by

$$g = -N(r)^2 dt^2 + \frac{dr^2}{L(r)^2} + r^2 d\phi^2. \tag{13.74}$$

The electromagnetic field equations for the tensor field $F_{\mu\nu} = 2F_{tr}\delta^t_{[\mu}\delta^r_{\nu]}$, and the dilaton $\Phi(r) = k\,\ln(r)$ becomes

$$EQ_F = \frac{d}{dr}\frac{F_{tr}\,L\,r^{-4\,a\,k+1}}{N} \rightarrow F_{tr} = Q\frac{N}{L}r^{4\,ak-1}. \tag{13.75}$$

The simplest Einstein equation is $R_{11} + R_{22}\,L^2\,N^2$, which yields

$$\frac{1}{N}\frac{d}{dr}N - \frac{1}{L}\frac{d}{dr}L - \frac{1}{2}\frac{Bk^2}{r} = 0, \tag{13.76}$$

thus one gets

$$N(r) = C_N\,L\,(r)\,r^{B\,k^2/2}. \tag{13.77}$$

On the other hand, the equation R_{33} gives a first-order equation for $L^2 = Y\,(r)$, namely

$$\frac{d}{dr}Y\,(r) + \frac{1}{2}\frac{Bk^2 Y\,(r)}{r} + 2\frac{r^{4\,ak}\,Q^2}{r} - 2\,\Lambda\,r^{bk+1} = 0 \tag{13.78}$$

integrating one obtains

$$L(r)^2 = Y\,(r) = -4\frac{r^{4\,ak}\,Q^2}{Bk^2 + 8\,ak} + 4\frac{\Lambda\,r^{2+bk}}{4 + Bk^2 + 2bk} + r^{-1/2\,Bk^2}\,C_1. \tag{13.79}$$

Substituting this expression of $Y\,(r)$ into the remaining scalar field equation

$$\frac{d}{dr}Y\,(r) + \frac{1}{2}\frac{Bk^2 Y\,(r)}{r} - 8\frac{r^{4\,ak}\,aQ^2}{B\,k\,r} + 2\frac{b\Lambda\,r^{bk+1}}{Bk} = 0, \tag{13.80}$$

one arrives at relationships between constants, namely

$$a = -\frac{1}{4\,B\,k},\; b = -B\,k. \tag{13.81}$$

Therefore, the general charged dilaton static solution can be given as

$$g = -{C_N}^2\,r^{B\,k^2}\,L(r)^2 dt^2 + \frac{dr^2}{L(r)^2} + r^2 d\phi^2,$$

$$L(r)^2 = \left(r^{B\,k^2/2}\,C_1 + 4\frac{r^2\Lambda}{4 - B\,k^2} + 4\frac{Q^2}{Bk^2}\right) r^{-B\,k^2},\; k^2 \neq \frac{4}{B},$$

$$F_{\mu\nu} = 2F_{tr}\delta^t_{[\mu}\delta^r_{\nu]},\; F_{tr} = QC_N\,r^{-1/2\,Bk^2-1} = -A_{t,r},\; A_t = 2\frac{QC_N}{B\,k^2}\,r^{-B\,k^2/2},$$

$$\Psi(r) = k\,\ln(r), \tag{13.82}$$

endowed with four relevant parameters: in particular, one may identify the mass $M = -C_1$, cosmological constant $\Lambda \rightarrow \pm\frac{1}{l^2}$, dilaton parameter k, and the charge

Q. The constant C_N can be absorbed by scaling the coordinate t, thus it can be equated to unit, $C_N \to 1$. Moreover, one has to set the charge Q to zero, $Q = 0$, when looking for the limiting solutions for vanishing dilaton $k = 0$, which are just the dS and AdS solutions with parameters $C_1 = \pm M$ respectively, and $C_N = 1$. There is no static electrically charged limit of this solution for a vanishing dilaton field.

The constant Λ can be equated to minus the standard cosmological constant $\Lambda_s = \pm\frac{1}{l^2}$; indeed, by setting in (13.82)

$$\Lambda = \pm\frac{1}{l^2}\alpha^2,\, r \to r\alpha^{2/(B\,k^2)},\, \phi \to \phi\,\alpha^{-2/(B\,k^2)},\, Q \to Q\,\alpha^{(1+2/(B\,k^2))},$$
$$C_1 \to C_1\,\alpha^{1+4/(B\,k^2)},\, C_N \to C_N\,\alpha^{-(1+2/(B\,k^2))}, \tag{13.83}$$

one arrives at the metric (13.82) with $\Lambda = \pm\frac{1}{l^2}$. Notice that the Λ used in Chan–Mann works, when considered as a cosmological constant, differs from the standard cosmological constant $\Lambda_s = \pm\frac{1}{l^2} = -\Lambda$, where $+$ and $-$ stand correspondingly for de Sitter and anti-de Sitter (AdS).

If $B > 0$, $\Lambda\,(4 - B\,k^2) > 0$, then one has

a) dS horizonless: $\Lambda < 0 \wedge \{k < -\frac{\sqrt{B}}{\sqrt{2}}, k > \frac{\sqrt{B}}{\sqrt{2}}\}, C_1 > 0$,
b) dS cosmological singularity: $\Lambda < 0 \wedge \{k < -\frac{\sqrt{B}}{\sqrt{2}}, k > \frac{\sqrt{B}}{\sqrt{2}}\}, C_1 < 0$,
c) AdS horizonless: $\Lambda > 0 \wedge \{-\frac{\sqrt{B}}{\sqrt{2}} < k < \frac{\sqrt{B}}{\sqrt{2}}\}, C_1 > 0$,
d) AdS black hole: $\Lambda > 0 \wedge \{-\frac{\sqrt{B}}{\sqrt{2}} < k < \frac{\sqrt{B}}{\sqrt{2}}\}, C_1 < 0$.

If $B > 0$, $\Lambda\,(4 - B\,k^2) > 0$, then one has

e) dS cosmological singularity: $\Lambda < 0 \wedge \{-\frac{\sqrt{B}}{\sqrt{2}} < k < \frac{\sqrt{B}}{\sqrt{2}}\}, C_1 > 0$,
f) AdS event horizon: $\Lambda > 0 \wedge \{k < -\frac{\sqrt{B}}{\sqrt{2}}, k > \frac{\sqrt{B}}{\sqrt{2}}\}, C_1 > 0$.

The norm of the normal vector $n_\mu = \nabla_\mu r$ to the surface $r = r_0 =$ const. is given by $g^{rr}(r_0) = L(r_0)^2$. It becomes a null vector at the horizon r_h of the black hole solution with $B > 0$, $k^2 < 4/B$, $C_1 < 0$, and $\Lambda = \frac{1}{l^2}$ which is determined as the outer root of $g^{rr}(r) = L(r)^2 = 0$, namely

$$r^{B\,k^2/2}C_1 + 4\frac{r^2\Lambda}{4 - B\,k^2} + 4\frac{Q^2}{Bk^2} = 0. \tag{13.84}$$

The electromagnetic invariant of this solution amounts to

$$F_{\mu\nu}F^{\mu\nu} = -2\frac{Q^2\,{C_1}^2}{{C_N}^2}r^{-2(1+\,Bk^2)}.$$

Accomplishing in the general above solution (13.82) the r-transformation and constant parameterizations:

$$r \to r^{N/2},\, k = \pm\frac{\sqrt{2}}{\sqrt{B}}\sqrt{\frac{2-N}{N}},$$

one gets, modulo minor constants arrangements, the Chan and Mann (1994, 1996) static solution:

$$\begin{aligned}
ds^2 &= -\,C_N^2 U(r)dt^2 + \frac{N^2}{4U(r)}dr^2 + r^N d\phi^2,\\
U(r) &= C_1 r^{1-N/2} + \frac{2N\Lambda}{(3N-2)}r^N + 4\frac{N}{2-N}Q^2,\\
F_{\mu\nu} &= 2F_{tr}\delta^t_{[\mu}\delta^r_{\nu]},\ F_{tr} = \frac{N}{2}QC_N r^{N/2-2} = -A_{t,r}\\
\to A_t &= \frac{N}{2-N}\frac{QC_N}{}r^{N/2-1},\\
\Psi(r) &= \pm\sqrt{\frac{N(2-N)}{2B}}\ \ln(r),\ 0<N<2,\ N\neq 2/3.
\end{aligned} \tag{13.85}$$

13.7.1 Quasi-Local Mass, Momentum, Energy for Charged Dilaton

To characterize non-asymptotically flat solutions one uses the Brown–York formalism of quasi-local momentum, energy, and mass quantities; see 1.3.1. The evaluation of the mass for the studied charged dilaton solution yields

$$\begin{aligned}
M(r,\epsilon_0) &:= -2C_N r^{1/2\,Bk^2}L^2 - 2\pi\ r^{1+1/2\,Bk^2}L\,\epsilon_0,\\
M(r,\epsilon_0=0) &= -2\ C_N\ C_1 - 8\,\frac{r^{(4-\,Bk^2)/2}C_N\,\Lambda}{4-Bk^2}\\
&\quad -8\,\frac{r^{-1/2\,Bk^2}\,C_N\,Q^2}{Bk^2}.
\end{aligned} \tag{13.86}$$

Comparing with the energy characteristics of the BTZ solution, one concludes that role of the mass parameter is played by $C_1 C_N = -M$. Recall that C_N can be equated to 1. At spatial infinity all physical quasi-local quantities are infinite, in particular, the field Ψ logarithmically increases.

13.7.2 Algebraic Classification of the Field, Energy–Momentum, and Cotton Tensors

The Maxwell field tensor is given by

$$(F^{\mu}{}_{\nu}) = \begin{bmatrix} 0 & -\frac{Q}{C_N L^2}r^{-1-1/2\,Bk^2} & 0\\ -r^{-1+1/2\,Bk^2}L^2 Q\,C_N & 0 & 0\\ 0 & 0 & 0 \end{bmatrix}, \tag{13.87}$$

and is algebraically characterized by the following eigenvectors:

$$\begin{aligned}
&\lambda_1 = r^{-1-Bk^2}Q: \mathbf{V1{,}2} = [V^1, -r^{1/2\,Bk^2}L^2 C_N V^1, 0],\ V_\mu V^\mu = 0,\ \mathbf{V1} = \mathbf{N1},\\
&\lambda_2 = -r^{-1-Bk^2}Q: \mathbf{V2} = [V^1, r^{1/2\,Bk^2}L^2 C_N V^1, 0],\ V_\mu V^\mu = 0,,\ \mathbf{V2} = \mathbf{N2},\\
&\lambda_3 = 0: \mathbf{V3} = [0,0,V^3],\ V_\mu V^\mu = {V^3}^2 r^2,\ \mathbf{V3} = \mathbf{S3},
\end{aligned} \tag{13.88}$$

therefore, its type is $\{N, N, S\}$. For this class of charged dilatons, the energy–momentum tensor is described by

$$T^{\mu}{}_{\nu} = 2Q^2 r^{-2(1+Bk^2)}\delta^{\mu}{}_{\phi}\delta^{\phi}{}_{\nu}, \tag{13.89}$$

and is algebraically characterized by the following eigenvectors:

$$\begin{aligned}
&\lambda_{1,2} = 0 : \mathbf{V1,2} = [V^1, V^2, 0], \mathcal{N} := V_{\mu}V^{\mu} = -{V^1}^2 {C_N}^2 r^{B\,k^2}\,L(r)^2 + \frac{{V^2}^2}{L^2},\\
&\quad \mathbf{V1,2 = T1,2, N1,2, S1,2},\\
&\lambda_3 = 2\,Q^2 r^{-2-2\,Bk^2} : \mathbf{V3} = [0,0,V^3], V_{\mu}V^{\mu} = {V^3}^2 r^2, \mathbf{V3 = S3}.
\end{aligned} \tag{13.90}$$

Hence depending on the sign of the norm $\mathcal{N}$ one will have spacelike, $\mathcal{N} > 0$, null, $\mathcal{N} = 0$, or timelike, $\mathcal{N} < 0$, eigenvectors: $\mathbf{V1,2} = \{\mathbf{S,N,T}\}$. Therefore, in the case of a double root one may choose different vector components determining, for instance, one spacelike vector S and the other timelike T or null N one. Therefore, for the Maxwell energy tensor one may have the algebraic types: $\{2S, S\}, \{2N, S\}, \{2T, S\}$ and $\{(S,T), S\}$,$\{(T,T), S\}$,...

The Cotton tensor amounts to

$$\begin{aligned}
C^{\mu}{}_{\nu} &= C^1{}_3\delta^{\mu}{}_t\delta^{\phi}{}_{\nu} + C^3{}_1\delta^{\mu}{}_{\phi}\delta^t{}_{\nu},\\
C^1{}_3 &= \frac{1}{32\,C_N r^2}\left(B\,C_1\,k^2\left(4-Bk^2\right)r^{-Bk^2} + 16\,Q^2\left(2+Bk^2\right)r^{-3/2\,Bk^2}\right)\\
C^3{}_1 &= -\frac{C_N}{32\,r^4 Bk^2}\left(r^{-Bk^2/2}{C_1}^2B^2k^4\left(4-Bk^2\right)\right.\\
&\quad +12\,r^{-Bk^2}Q^2\,C_1\,Bk^2\left(4+Bk^2\right)\\
&\quad +4\,B^2 r^{2-Bk^2}\Lambda\,C_1\,k^4 + 64\,r^{-3\,Bk^2/2}Q^4\left(2+Bk^2\right)\\
&\quad \left. +64\,Br^{2-3\,Bk^2/2}\Lambda\,Q^2k^2\frac{\left(2+Bk^2\right)}{\left(4-Bk^2\right)}\right).
\end{aligned} \tag{13.91}$$

Depending on the signs of C_1 and $\Lambda = \pm 1/l^2$ the components of $C^1{}_3$ and $C^3{}_1$ may be positive or negative; consequently their product could be positive. Therefore, the Cotton tensor, for both $C^1{}_3$ and $C^3{}_1$, positive or negative, is characterized algebraically by

$$\begin{aligned}
&\lambda_1 = 0 : \mathbf{V1} = [0, V^2, 0], V_{\mu}V^{\mu} = \frac{{V^2}^2}{L^2}, \mathbf{V1 = S1},\\
&\lambda_2 = \sqrt{C^1{}_3C^3{}_1} : \mathbf{V2} = [V^1, 0, V^1\left(\frac{C^3{}_1}{C^1{}_3}\right)^{1/2}],\\
&\mathcal{N} = V_{\mu}V^{\mu} = -{V^1}^2(N^2\,C^1{}_3 - r^2C^3{}_1)/C^1{}_3,\\
&\mathbf{V2 = T2, N2, S2},\\
&\lambda_2 = -\sqrt{C^1{}_3C^3{}_1} : \mathbf{V3} = [V^1, 0, -V^1\left(\frac{C^3{}_1}{C^1{}_3}\right)^{1/2}],\\
&V_{\mu}V^{\mu} = -{V^1}^2(N^2\,C^1{}_3 - r^2C^3{}_1)/C^1{}_3,\\
&\mathbf{V3 = T3, N3, S3}.
\end{aligned} \tag{13.92}$$

Depending on the sign of the norm $\mathcal{N}$ one has spacelike, $\mathcal{N} > 0$, null, $\mathcal{N} = 0$, or timelike, $\mathcal{N} < 0$, eigenvectors, denoted by: $\mathbf{V1,2} = \{\mathbf{S},\mathbf{N},\mathbf{T}\}$. Therefore, when the eigenvalues are real, the Cotton tensor allows for types I: $\{2S,S\}$,$\{2N,S\}$,$\{2T,S\}$ and $\{(S,T),S\}$, $\{(T,T),S\}$,...

In the spacetime region where ${C^1}_3$ and ${C^3}_1$ are of opposite signs, one of the eigenvalues becomes imaginary and the following scheme arises

$$\begin{aligned}
&\lambda_1 = 0 : \mathbf{V1} = [0,V,0], V_\mu V^\mu = \frac{V^2}{L^2}, \mathbf{V1} = \mathbf{S1},\\
&\lambda_2 = \bar{\lambda}_3 = \sqrt{{C^1}_3 {C^3}_1} : \mathbf{V2} = \mathbf{Z} = [V^1, 0, \frac{\sqrt{{C^3}_1}}{\sqrt{{C^1}_3}} V^1], \mathbf{V3} = \bar{\mathbf{Z}},
\end{aligned} \tag{13.93}$$

thus, the algebraic type of the Cotton tensor is $\{S, Z, \bar{Z}\}$.

13.8 Stationary Charged Dilaton Generated via $SL(2,R)$

Subjecting the metric (13.82) to a general $SL(2,R)$ transformation

$$\begin{aligned}
t &= \alpha T + \beta\,\Phi,\\
\phi &= \gamma T + \delta\,\Phi,\ \Delta := \alpha\delta - \beta\gamma,
\end{aligned}$$

one arrives at the stationary metric equipped with rotation

$$\begin{aligned}
g = &-\left(\alpha^2 {C_N}^2 r^{Bk^2} L^2 - \gamma^2 r^2\right) dT^2 + \frac{dr^2}{L^2}\\
&-2\left(\beta\,\alpha\, {C_N}^2 r^{Bk^2} L^2 - \delta\,\gamma\, r^2\right) dT d\,\Phi + \left(\delta^2 r^2 - \beta^2 {C_N}^2 r^{Bk^2} L^2\right) d\Phi^2,\\
L(r)^2 = &\left(r^{B\,k^2/2} C_1 + 4\,\frac{r^2\Lambda}{4 - B\,k^2} + 4\,\frac{Q^2}{Bk^2}\right) r^{-B\,k^2}.
\end{aligned} \tag{13.94}$$

Using the standard notation, this charged rotating dilaton solution is given as

$$\begin{aligned}
g &= -\mathsf{N}(r)^2 dT^2 + \frac{dr^2}{\mathsf{L}(r)^2} + \mathsf{K}(r)^2 (d\Phi + \mathsf{W}(r) dT)^2,\\
\mathsf{N}(r)^2 &= \frac{N^2 r^2}{\delta^2 r^2 - \beta^2 N^2} \left(\alpha\,\delta - \beta\,\gamma\right)^2,\\
N^2 &:= C_N^2 \left(r^{B\,k^2/2} C_1 + 4\,\frac{r^2\Lambda}{4 - B\,k^2} + 4\,\frac{Q^2}{Bk^2}\right) = C_N^2\, r^{B\,k^2} \mathsf{L}(r)^2,\\
\mathsf{L}(r)^2 &= \left(r^{B\,k^2/2} C_1 + 4\,\frac{r^2\Lambda}{4 - B\,k^2} + 4\,\frac{Q^2}{Bk^2}\right) r^{-B\,k^2},\\
\mathsf{K}(r)^2 &= \delta^2 r^2 - \beta^2 N^2,\\
\mathsf{W}(r) &= \frac{\delta\,\gamma\, r^2 - \beta\,\alpha\, N^2}{\delta^2 r^2 - \beta^2 N^2},\ \beta \neq 0 \neq \delta,\\
\Psi &= k \ln(r),\\
F_{\mu\nu} &= 2F_{Tr}\delta^T_{[\mu}\delta^r_{\nu]} + 2F_{\Phi\,r}\delta^\Phi_{[\mu}\delta^r_{\nu]},
\end{aligned}$$

$$
\begin{aligned}
F_{Tr} &= -A_{T,r} = \alpha\, F_{tr} = -\alpha \frac{d}{dr}\left(2\frac{QC_N}{B\,k^2}\, r^{-B\,k^2/2}\right),\\
F_{tr} &= QC_N\, r^{-1/2\,Bk^2-1},\\
F_{\Phi\, r} &= -A_{\Phi,r} = \beta\, F_{tr} = -\beta \frac{d}{dr}\left(2\frac{QC_N}{B\,k^2}\, r^{-B\,k^2/2}\right),\\
A_t &= 2\frac{QC_N}{B\,k^2}\, r^{-B\,k^2/2}.
\end{aligned}
\tag{13.95}
$$

It is worth noticing that the structure of the electromagnetic energy tensor amounts to

$$
(T_{\mu\nu}) = 2\, Q^2 r^{-2\,Bk^2} \begin{bmatrix} \gamma^2 & 0 & \gamma\,\delta \\ 0 & 0 & 0 \\ \gamma\,\delta & 0 & \delta^2 \end{bmatrix}. \tag{13.96}
$$

For the AdS ($\Lambda = 1/l^2$)–black hole branch, the constants appearing in the structural functions will be replaced by $C_N \to \sqrt{1-2\,k^2} = B$, $k = \sqrt{1-B^2}/\sqrt{2}$, $0 < k < 1/\sqrt{2}, 1 > B > 0$, thus

$$
N(r)^2 = N_B^2 = r^2\left(r^{-2\,B^2}\, C_1\, B^2 + \Lambda\right),\ \mathsf{L}(r)^2 = L_B^2 = \frac{N_B^2}{B^2} r^{-4+4\,B^2}. \tag{13.97}
$$

13.8.1 Quasi-Local Mass and Momentum

The evaluation of the quasi-local momentum $J(r)$ yields

$$
J(r) = \frac{1}{2}\,\beta\,\delta\, C_N\, C1\,\left(4 - Bk^2\right) + 8\,\frac{\delta\, C_N\,\beta\, Q^2}{Bk^2} r^{-\,Bk^2/2}, \tag{13.98}
$$

hence, for positive $B > 0$, the contribution of the electromagnetic field $Q \neq 0$ to the momentum, at spatial infinity, disappears and one has

$$
J(r \to \infty) = \frac{1}{2}\,\beta\,\delta\, C_N\, C1\,\left(4 - Bk^2\right) =: J_0. \tag{13.99}
$$

The evaluation of the quasi local mass yields

$$
\begin{aligned}
M(r) &= \frac{1}{2}\,\frac{C_N\,\Delta}{r^3 Bk^2\,(Bk^2-4)}\,\frac{M_L}{D} - \frac{1}{2}\,\frac{\beta\,\delta\, C_N}{B^2k^4r^2}\,\frac{M_J}{D},\\
M_L &:= -4\,Bk^2\, C_1\, r^3\,\left(Bk^2-4\right)\left(P + \Lambda\,\beta^2\, {C_N}^2 Bk^2\right)\\
&\quad +4\,r\,Q^2\beta^2\, {C_N}^2\, C_1\,\left(Bk^2-4\right)^2 Bk^2 + 16\,Bk^2 r^{5-1/2\,Bk^2}\,\Lambda\, P\\
&\quad + r^{1+1/2\,Bk^2}\, {C_1}^2\beta^2\, {C_N}^2 B^2k^4\left(Bk^2-4\right)^2 - 16\, Q^2 r^{3-1/2\,Bk^2}\left(Bk^2-4\right) P,\\
M_J &= \left(16\,Bk^2\, Q^2 r^{-Bk^2/2} - B^2\, C_1\, k^4\right)\left[\delta\,\gamma\,\left(Bk^2-4\right)\right.\\
&\quad \left. +4\,\Lambda\, {C_N}^2\alpha\,\beta\right]\left(Bk^2-4\right)\, r^2\\
&\quad -64\beta\,\alpha\, r^{-1/2\,Bk^2}\, Q^4\, {C_N}^2\left(Bk^2-4\right) + \beta\,\alpha B^2k^4\, {C_1}^2\, {C_N}^2 r^{1/2\,Bk^2}\left(Bk^2-4\right)^2
\end{aligned}
$$

$$+4\,\beta\,\alpha\,Bk^2\,C_1\,{C_N}^2 Q^2\left(Bk^2-4\right)\left(Bk^2-8\right),$$

$$D=P-4\,\frac{\beta^2\,{C_N}^2 Q^2\left(Bk^2-4\right)}{Br^2k^2}-\beta^2\,{C_N}^2 r^{(\,Bk^2-4)/2}\,C_1\left(Bk^2-4\right), \tag{13.100}$$

where $P=\delta^2 Bk^2-4\,\delta^2+4\,\beta^2\,{C_N}^2\Lambda$. The order zero in the series expansion of $M(r)$ is given by

$$M_0=2\,M\,\Delta\,\frac{\Lambda\,\beta^2\,{C_N}^2 Bk^2+\delta^2 Bk^2-4\,\delta^2}{\delta^2 Bk^2-4\,\delta^2+4\,\beta^2\,{C_N}^2\Lambda}-J_0\frac{Bk^2\delta\,\gamma+4\,\beta\,\alpha\,{C_N}^2\Lambda-4\,\delta\,\gamma}{\delta^2 Bk^2-4\,\delta^2+4\,\beta^2\,{C_N}^2\Lambda}, \tag{13.101}$$

where $C_N\,C1$ has been replaced by $-M$. Comparing with the quasi-local BTZ mass $M_{BTZ}=2\,M-2\,r^2/l^2$; when the rotation vanishes, $\beta=0$, $\Delta=\alpha\delta\rightarrow 1$, implies that $M_0\rightarrow -C_N\,C1=M$. It should be pointed out that, due to the presence of terms with positive powers of r in the series expansion of M(r), it increases as $r\rightarrow\infty$, like the M_{BTZ} does.

Frequently, in the literature one encounters the $SL(2,R)$ transformation

$$t=\frac{T}{\sqrt{1-\frac{\omega^2}{l^2}}}-\omega\,\frac{\Phi}{\sqrt{1-\frac{\omega^2}{l^2}}},\ \phi=-\frac{\omega}{l^2}\,\frac{T}{\sqrt{1-\frac{\omega^2}{l^2}}}+\frac{\Phi}{\sqrt{1-\frac{\omega^2}{l^2}}},$$

correspondingly, the structural functions (13.95) assume the form

$$\mathsf{N}(r)^2=\frac{\left(l^2-\omega^2\right)r^2N^2}{l^2\left(r^2-\omega^2N^2\right)},\ \mathsf{L}(r)=\mathsf{L}(r),\ N=N,$$

$$\mathsf{W}(r)=\frac{\omega\,\left(N^2l^2-r^2\right)}{l^2\left(r^2-\omega^2N^2\right)},\ \mathsf{K}(r)^2=\frac{l^2\left(r^2-\omega^2N^2\right)}{l^2-\omega^2}. \tag{13.102}$$

In this parametrization, the quasi-local momentum becomes

$$J_0=-\frac{1}{2}\,\frac{\omega\,l^2}{l^2-\omega^2}\,C_1\,C_N(4-B\,k^2),$$

and vanishes as soon the rotation ω becomes zero.

The generated stationary metric (13.59) gives rise, among others, to a charged generalization of Chan–Mann rotating dilaton solution; see below, Section 13.8.3.

With this result we are giving an answer to the remark contained in the Conclusions of Chan and Mann (1996): "Although the static charged black solutions of (1.1) exist, Chan and Mann (1994), at present we are unable to generalize our spinning solution to charged cases. This endeavor is complicated by the fact that when one adds Maxwell fields to a spinning solution, both electric and magnetic fields must be present..."

13.8.2 Algebraic Classification of the Field, Energy–Momentum, and Cotton Tensors

The Maxwell electromagnetic field tensor matrix $(F^{\mu}{}_{\nu})$ amounts to

$$\begin{bmatrix} 0 & -\frac{\delta\, r^{-1-3/2\,Bk^2}\,Q}{\Delta\, C_N\, L^2} & 0 \\ -\alpha\, L^2 Q r^{-1-1/2\,Bk^2}\, C_N & 0 & -\beta\, L^2 Q\, r^{-1-1/2\,Bk^2}\, C_N \\ 0 & \frac{\gamma\, r^{-1-3/2\,Bk^2}\,Q}{\Delta\, C_N\, L^2} & 0 \end{bmatrix},$$

and has the following eigenvectors

$$\begin{aligned} &\lambda_1 = 0 : \mathbf{V1} = [-\frac{\beta\, V^3}{\alpha}, 0, V^3], V_\mu V^\mu = \frac{(V^3)^2 r^2 \Delta^2}{\alpha^2}, \\ &\lambda_2 = Q r^{-1-Bk^2} : \mathbf{V2} = [-r^{-\,Bk^2/2}\frac{\delta\, V^2}{\Delta\, L^2\, C_N}, V^2, r^{-\,Bk^2/2}\frac{\gamma\, V^2}{\Delta\, L^2\, C_N}], \\ &V_\mu V^\mu = 0, \mathbf{V2} = \mathbf{N2}, \\ &\lambda_3 = -Q r^{-1-Bk^2} : \mathbf{V3} = [r^{-\,Bk^2/2}\frac{\delta\, V^2}{\Delta\, L^2\, C_N}, V^2, -r^{-\,Bk^2/2}\frac{\gamma\, V^2}{\Delta\, L^2\, C_N}], \\ &V_\mu V^\mu = 0, \mathbf{V3} = \mathbf{N3}. \end{aligned} \tag{13.103}$$

Correspondingly, its type is $\{S, N, N\}$.

The electromagnetic energy–momentum tensor is given by

$$(T^{\mu}{}_{\nu}) = -2\,\frac{Q^2 r^{-(2+2\,Bk^2)}}{\Delta} \begin{bmatrix} \beta\,\gamma & 0 & \beta\,\delta \\ 0 & 0 & 0 \\ -\alpha\,\gamma & 0 & -\alpha\,\delta \end{bmatrix}, \tag{13.104}$$

and is algebraically characterized by:

$$\begin{aligned} &\lambda_{1,2} = 0 : \mathbf{V1,2} = [V^1, V^2, -V^1\gamma/\delta], \\ &\mathcal{N} = V_\mu V^\mu = -\frac{{C_N}^2 L^2 r^{Bk^2}\Delta^2}{\delta^2}{V^1}^2 + \frac{{V^2}^2}{L^2}, \\ &\mathcal{N} > 0, \mathbf{V1,2} = \mathbf{S1,2}, \mathcal{N} = 0, \mathbf{V1,2} = \mathbf{N1,2}, \mathcal{N} < 0, \mathbf{V1,2} = \mathbf{T1}, \\ &\lambda_3 = 2\,Q^2 r^{-(2+2\,Bk^2)} : \mathbf{V3} = [-\frac{\beta\, V^3}{\alpha}, 0, V_3], \\ &V_\mu V^\mu = \frac{(V^3)^2 r^2 \Delta^2}{\alpha^2}, \mathbf{V3} = \mathbf{S3}. \end{aligned} \tag{13.105}$$

In the case of a double root, depending on the sign of the norm $V_\mu V^\mu$, one will have spacelike, $\mathcal{N} > 0$, null, $\mathcal{N} = 0$, or timelike, $\mathcal{N} < 0$, eigenvectors: $\mathbf{V1,2} = \{\mathbf{S1,2}, \mathbf{N1,2}, \mathbf{T1,2}\}$. Therefore, one may choose different vector components determining, for instance, one spacelike vector S and the other timelike T or null N one. Therefore, one may have the algebraic Ricci types: $\{2S, S\}, \{2N, S\}, \{2T, S\}$ and $\{(S,T), S\}, \{(T,T), S\}, \ldots, \{(T,N), S\}$, where parenthesis is used to stand out the multiplicity of the root under consideration. The matrix of the Cotton tensor for the rotating charged solution is given by

$$(C^{\mu}{}_{\nu}) = \begin{bmatrix} \frac{-\alpha\,\beta\,C^3{}_1+\gamma\,\delta\,C^1{}_3}{\Delta} & 0 & \frac{-\beta^2C^3{}_1+\delta^2C^1{}_3}{\Delta} \\ 0 & 0 & 0 \\ -\frac{-\alpha^2C^3{}_1+\gamma^2C^1{}_3}{\Delta} & 0 & -\frac{-\alpha\,\beta\,C^3{}_1+\gamma\,\delta\,C^1{}_3}{\Delta} \end{bmatrix}, \tag{13.106}$$

where

$$\begin{aligned} C^1{}_3 &= \frac{1}{32\,C_N\,r^2}\left(B\,C_1\,k^2\,(4-Bk^2)\,r^{-Bk^2}\right. \\ &\quad \left.+16\,Q^2\,(2+Bk^2)\,r^{-3/2\,Bk^2}\right), \\ C^3{}_1 &= -\frac{C_N}{32\,r^4Bk^2}\left(r^{-Bk^2/2}\,{C_1}^2B^2k^4\,(4-Bk^2)\right. \\ &\quad +12\,r^{-Bk^2}\,Q^2\,C_1\,Bk^2\,(4+Bk^2) \\ &\quad +4\,B^2r^{2-Bk^2}\Lambda\,C_1\,k^4+64\,r^{-3\,Bk^2/2}\,Q^4\,(2+Bk^2) \\ &\quad \left.+64\,Br^{2-3\,Bk^2/2}\Lambda\,Q^2k^2\frac{(2+Bk^2)}{(4-Bk^2)}\right). \end{aligned} \tag{13.107}$$

Because of the presence of C_1 and $\Lambda=\pm1/l^2$ in $C^1{}_3$ and $C^3{}_1$, these tensor components may be positive or negative quantities in some spacetime regions. Therefore, in general, the Cotton tensor can be characterized algebraically by three real eigenvalues

$$\begin{aligned} \lambda_1 &= 0:\mathbf{V1}=[0,V^2,0],\,V_\mu V^\mu=\frac{{V^2}^2}{L^2},\,\mathbf{V1}=\mathbf{S1}, \\ \lambda_2 &= \sqrt{C^1{}_3C^3{}_1}:\mathbf{V2}=[V^1,0,V^3], \\ -V^3 &= \frac{V^1\left(-\alpha\,\beta\,C^3{}_1+\gamma\,\delta\,C^1{}_3-\sqrt{C^3{}_1\,C^1{}_3}\Delta\right)}{\delta^2C^1{}_3-\beta^2C^3{}_1}, \\ V_\mu V^\mu &= \frac{{V^1}^2\Delta^2\left(C^3{}_1\,r^2-C^1{}_3\,N^2\right)\left(\delta^2C^1{}_3+\beta^2C^3{}_1+2\,\delta\,\beta\sqrt{C^3{}_1\,C^1{}_3}\right)}{\left(\delta^2C^1{}_3-\beta^2C^3{}_1\right)^2}, \\ \mathbf{V2} &= \mathbf{T2},\mathbf{N2},\mathbf{S2}, \\ \lambda_3 &= -\sqrt{C^1{}_3C^3{}_1}:\mathbf{V\ 3}=[V^1,0,V^3] \\ -V^3 &= \frac{V^1\left(-\alpha\,\beta\,C^3{}_1+\gamma\,\delta\,C^1{}_3+\sqrt{C^3{}_1\,C^1{}_3}\Delta\right)}{\delta^2C^1{}_3-\beta^2C^3{}_1}, \\ V_\mu V^\mu &= \frac{{V^1}^2\Delta^2\left(C^3{}_1\,r^2-C^1{}_3\,N^2\right)\left(\delta^2C^1{}_3+\beta^2C^3{}_1-2\,\delta\,\beta\sqrt{C^3{}_1\,C^1{}_3}\right)}{\left(\delta^2C^1{}_3-\beta^2C^3{}_1\right)^2}, \\ \mathbf{V3} &= \mathbf{T3},\mathbf{N3},\mathbf{S3}. \end{aligned} \tag{13.108}$$

In this case the Cotton tensor is Type I with particular eigenvector subclasses:

$$\{T, T, S\}, \{T, S, S\}, \ldots, \{S, S, S\}.$$

For one real and two complex conjugate eigenvalues, one has

$$\begin{aligned}
&\lambda_1 = 0 : \mathbf{V1} = [0, V, 0], V_\mu V^\mu = \frac{V^2}{L^2}, \mathbf{V1} = \mathbf{S1},\\
&\lambda_2 = \bar{\lambda}_3 = \sqrt{C^1{}_3 C^3{}_1} :\\
&\mathbf{V\ 2} = \mathbf{Z} = [V^1, 0, V^3],\ V^3 = -\frac{V^1\left(-\alpha\,\beta\,C^3{}_1 + \gamma\,\delta\,C^1{}_3 - \sqrt{C^3{}_1 C^1{}_3 \Delta}\right)}{\delta^2 C^1{}_3 - \beta^2 C^3{}_1},\\
&\mathbf{V\ 3} = \bar{\mathbf{Z}}.
\end{aligned} \tag{13.109}$$

Thus, the algebraic type of the Cotton tensor is Type I: $\{S, Z, \bar{Z}\}$.

13.8.3 Particular Stationary Charged Dilaton via $SL(2,R)$ Transformation

Requiring the fulfillment of the relation

$$\gamma\,\delta = -4\frac{{C_N}^2\Lambda}{(B\ k^2 - 4)}\alpha\,\beta \tag{13.110}$$

the term with power r^2 in W or equivalently in $g_{T\Phi}$ disappears. The generalized rotating charged Chan–Mann metric is given by

$$\begin{aligned}
g = &-\left({C_N}^2\alpha^2 L^2 r^{Bk^2} - 16\,\beta^2\frac{{C_N}^4\alpha^2\Lambda^2 r^2}{\delta^2\left(Bk^2 - 4\right)^2}\right) dT^2 + \frac{dr^2}{L^2}\\
&-2\alpha\,\beta\frac{{C_N}^2}{Bk^2}\left(r^{\,B\,k^2/2}C_1\,Bk^2 + 4\,Q^2\right) dT d\,\Phi\\
&+\left(\delta^2 r^2 - \beta^2\,{C_N}^2 r^{Bk^2}\,L^2\right) d\Phi^2,\\
L^2 = &\left(r^{B\,k^2/2}C_1 + 4\,\frac{r^2\Lambda}{4 - B\,k^2} + 4\,\frac{Q^2}{Bk^2}\right) r^{-B\,k^2},\ \Psi = k\ln(r),\\
F_{\mu\nu} = &\ 2F_{Tr}\delta^T_{[\mu}\delta^r_{\nu]} + 2F_{\Phi\,r}\delta^\Phi_{[\mu}\delta^r_{\nu]},\ F_{Tr} = \alpha\,F_{tr},\ F_{\Phi\,r} = \beta\,F_{tr},\\
F_{tr} = &\ QC_N r^{-1-B\,k^2/2}.
\end{aligned} \tag{13.111}$$

Switching the rotation $\beta = 0$, the above-presented metric reduces to the Chan–Mann charged dilaton metric.

13.9 Summary of Dilaton Minimally Coupled to Gravity

Using the Schwarzschild coordinate frame for a static cyclic symmetric metric in (2 + 1) gravity coupled minimally to a dilaton logarithmically depending on the radial coordinate in the presence of a exponential potential and an electrical Maxwell field the general solutions of the Einstein–dilaton and Einstein–Maxwell–dilaton equations are derived.

In the Einstein–dilaton case, by solving first-order linear equations, the general solution is derived and identified with the Chan–Mann dilaton solution. Moreover, for completeness, the Chan–Mann solution is also derived in its original coordinate frame avoiding any ansatz. A new stationary dilaton solution is obtained in the Schwarzschild coordinate frame; it does not allows for an AdS limit at spatial infinity, where its structural functions increase indefinitely; it is horizonless and singular at the origin of coordinates; moreover, one can identify at infinity a mass parameter and, in the whole space, a constant (quasi-local) angular momentum.

In the Einstein–Maxwell–dilaton branch, the static solution is equivalent to the Chan–Mann charged dilaton static solution. Via a $SL(2,R)$ transformation of the Killing coordinates, applied on the derived dilaton and charged static cyclic symmetric metric, families of stationary dilaton solutions has been generated; they are equipped with five relevant parameters interpretable as dilaton parameter, charge, momentum, cosmological constant, and mass for some values of them. A particular $SL(2,R)$ transformation is identified, which gives raise to the charged generalization of the rotating Chan–Mann dilaton solution. At spatial infinity all these solutions do not allow for an AdS–dS limit; their structural functions increase indefinitely as the radial coordinate increases. There exists a horizon, structurally common to the full class of solutions, determining their black hole character for a range of physical parameters. This families of solutions, the static and stationary ones, have been characterized by their quasi-local energy, mass, and momentum through their series expansions at spatial infinity. The algebraic classifications of the electromagnetic field, Maxwell energy–momentum, and Cotton tensors are established. The electromagnetic field tensor belongs to the type $\{S,N,N\}$. The Maxwell energy–momentum tensor is of types: $\{2S,S\}$,$\{2N,S\}$,{2T,S},$\{(S,T),S\}$,... The Cotton tensor exhibits various possibilities. For real roots it falls into type I:$\{T,T,S\}$, $\{T,S,S\}$,$\cdots$ For a complex root the Cotton tensor is of the Type I: $\{S,Z,\bar{Z}\}$.

14

Scalar Field Non-Minimally Coupled to (2+1) Gravity

A stationary cyclic symmetric black hole solution for a scalar field non-minimally coupled to (2+1)-dimensional gravity is derived and analyzed; its quasi-local momentum, energy and mass are evaluated. They are asymptotically similar to the ones of the rotating anti–de Sitter black hole solution with a contribution due to the angular momentum parameter. The algebraic structure of the Ricci, energy-momentum, and Cotton tensors is established.

14.1 Einstein Equations for Non-Minimally Coupled Scalar Field

The action for a dilaton non-minimally coupled to (2+1)-dimensional gravity is given by

$$\mathcal{S} = \frac{1}{2}\int d^3x\sqrt{-g}\left[\frac{1}{\kappa}(R+\frac{2}{l^2}) - \nabla_\mu\Psi\,\nabla^\mu\Psi - \zeta R\,\Psi^2\right], \tag{14.1}$$

where $\Lambda = -1/l^2$ stands for the (negative) cosmological constant, Ψ is the massless non-minimally coupled scalar field, ζ is the coupling constant, and R the scalar curvature. The variations of this action yield the dynamical equations

$$\begin{aligned} G_\mu{}^\nu - \frac{1}{l^2}\delta_\mu{}^\nu = \kappa\,\Big[&\nabla_\mu\Psi\,\nabla^\nu\Psi - \frac{1}{2}\delta_\mu{}^\nu\nabla_\sigma\Psi\,\nabla^\sigma\Psi \\ &+ \zeta\left(\Box\Psi^2 - \nabla_\mu\nabla^\nu\Psi^2 + G_\mu{}^\nu\Psi^2\right)\Big], \\ \Box\Psi^2 = \zeta\, R\,\Psi&, \end{aligned} \tag{14.2}$$

where $\Box = \nabla^\mu\nabla_\mu$ stands for the Laplace–Beltrami operator.

Within (2+1)-dimensional Einstein theory, the general form of a stationary cyclic symmetric line element – allowing for a timelike Killing vector $\boldsymbol{\partial_t}$ and a spacelike Killing vector field $\boldsymbol{\partial_\phi}$ – can be given as

$$g = -N(\rho)^2dt^2 + \frac{d\rho^2}{L(\rho)^2} + K(\rho)^2(d\phi + W(\rho)dt)^2.$$

14.1.1 Martinez–Zanelli Black Hole Solution with $T_\mu{}^\mu = 0$

In the static case, $W = 0$, and for the class of metrics allowing for the polar variable ρ, $K = \rho$, and restricted also to $N(\rho) = L(\rho)$, the Einstein equations (14.2) and the dynamical equation for the scalar Ψ allow for a unique solution; see Ayón-Beato et al. (2000),

$$N(\rho)^2 = L(\rho)^2 = \frac{(\rho+m)^2\,(\rho-2\,m)}{l^2\,\rho},\ K(\rho) = \rho,\ W(\rho) = 0,$$
$$\Psi_{MZ}(\rho) = \sqrt{8}\sqrt{\frac{m}{\kappa\,(\rho+m)}}, \tag{14.3}$$

known as the Martínez and Zanelli (1996) black hole solution for a scalar field non-minimally coupled to gravity . This solution arises for a coupling constant $\zeta = 1/8$, which yields the vanishing of the trace $T_\mu{}^\mu = 0$ and consequently the scalar curvature becomes constant $R = 6\,l^{-2}$. Moreover, it describes a black hole with horizon at $\rho_+ = 2\,m$, and presents a curvature singularity at the origin $\rho = 0$.

14.2 Stationary Black Hole for a Non-Minimally Coupled Scalar Field

Subjecting the above metric to the $SL(2,R)$ transformation

$$t = \frac{l}{\sqrt{l^2-\omega^2}}T - \frac{l\,\omega}{\sqrt{l^2-\omega^2}}\Phi,$$
$$\phi = -\frac{l\,\omega}{\sqrt{l^2-\omega^2}}T + \frac{l}{\sqrt{l^2-\omega^2}}\Phi,$$
$$\rho = \rho, \tag{14.4}$$

one arrives at the metric

$$g = -\frac{(l^2-\omega^2)\,\rho^3 - 3\,l^2\rho\,m^2 - 2\,l^2m^3}{(l^2-\omega^2)\,\rho\,l^2}dt^2 - 2\frac{\omega\,m^2\,(3\,\rho+2\,m)}{(l^2-\omega^2)\,\rho}dt\,d\phi$$
$$+\frac{3\,\omega^2\rho\,m^2 + 2\,\omega^2m^3 + (l^2-\omega^2)\,\rho^3}{(l^2-\omega^2)\,\rho}d\phi^2 + \frac{\rho\,l^2}{(\rho-2\,m)\,(\rho+m)^2}d\rho^2. \tag{14.5}$$

Choosing the parameters as

$$\omega = -\frac{lR_-}{J} = -\frac{l\,J}{R_+},\ R_\pm = Ml \pm \sqrt{M^2l^2+J^2},\ m = \frac{M^{1/2}}{3^{1/2}}\,l, \tag{14.6}$$

one brings (14.5) to a representation close to the one of the BTZ solution at spatial infinity $\rho \to \infty$, namely

$$g = -N(\rho)^2dt^2 + \frac{dr^2}{L(\rho)^2} + K(\rho)^2(d\phi + W(\rho)dt)^2,$$
$$L(\rho)^2 = \frac{\rho^2}{l^2} - M - 2\,\frac{M^{3/2}}{3^{3/2}}\,\frac{l}{\rho},$$

$$K(\rho)^2 = \rho^2 - \frac{1}{2}\,l R_- - \frac{M^{1/2}\,l^2 R_-}{3^{3/2}\quad\rho},\; N(\rho) = \rho\frac{L(\rho)}{K(\rho)},$$
$$W(\rho) = -\frac{J}{2\,K(\rho)^2}\left(1 + 2\frac{M^{1/2}}{3^{3/2}}\frac{l}{\rho}\right),$$
$$\Psi^2 = 8\,\frac{\sqrt{M}l}{\kappa\left(\sqrt{3}\rho + \sqrt{M}l\right)}. \tag{14.7}$$

This solution represents a black hole with horizon at $\rho = 2\,m = 2\,\frac{M^{1/2}}{3^{1/2}}\,l$.

14.2.1 Quasi-Local Momentum, Energy, and Mass

To characterize non asymptotically flat solutions one uses the Brown–York formalism of quasi-local momentum, energy, and mass quantities, see Section 1.3 and the summary (1.3.1), which for the stationary cyclic symmetric metric (14.7) gives the physical quasi-local momenta

$$j_\phi(\rho) = \frac{\left(3\,\rho + \sqrt{3}\sqrt{M}l\right)J}{\pi\sqrt{2}\sqrt{\rho}\sqrt{18\,\rho^3 - 9\,lR_-\,\rho - 2\,\sqrt{3}\sqrt{M}l^2 R_-}},$$
$$j(\rho\to\infty) \approx \frac{J}{2\pi\,\rho} + \frac{\sqrt{M}l\,J}{2\sqrt{3}\pi\,\rho^2},$$
$$J(\rho) = J + \frac{\sqrt{M}l\,J}{\sqrt{3}\rho},\; J(\rho\to\infty)\approx J, \tag{14.8}$$

while for the energy and mass one gets

$$\epsilon(\rho) = -\frac{\sqrt{9\,\rho^3 - 9\,Ml^2\rho - 2\,\sqrt{3}M^{3/2}l^3}\left(18\,\rho^3 + \sqrt{3}\sqrt{M}l^2 R_-\right)}{3\,l\,\pi\rho^{3/2}\left(18\,\rho^3 - 9\,lR_-\,\rho - 2\,\sqrt{3}\sqrt{M}l^2 R_-\right)},$$
$$E(\rho) = -\frac{\sqrt{2}\sqrt{9\,\rho^3 - 9\,Ml^2\rho - 2\,\sqrt{3}M^{3/2}l^3}\left(18\,\rho^3 + \sqrt{3}\sqrt{M}l^2 R_-\right)}{9\,l\rho^2\sqrt{18\,\rho^3 - 9\,lR_-\,\rho - 2\,\sqrt{3}\sqrt{M}l^2 R_-}},$$
$$M(\rho) = -\frac{4\left(9\,\rho^3 - 9\,Ml^2\rho - 2\,\sqrt{3}M^{3/2}l^3\right)\left(18\,\rho^3 + \sqrt{3}\sqrt{M}l^2 R_-\right)}{9\,l^2\,\rho\left(18\,\rho^3 - 9\,lR_-\,\rho - 2\,\sqrt{3}\sqrt{M}l^2 R_-\right)}$$
$$+\frac{4J^2\left(9\,\rho + 2\,\sqrt{3}\sqrt{M}l\right)\left(3\,\rho + \sqrt{3}\sqrt{M}l\right)}{3\rho\left(18\,\rho^3 - 9\,lR_-\,\rho - 2\,\sqrt{3}\sqrt{M}l^2 R_-\right)}. \tag{14.9}$$

The series expansion of energy and mass functions independent of ϵ_0 behave at infinity $\rho\to\infty$ as

$$\begin{aligned}
\epsilon(\rho \to \infty, \epsilon_0 = 0) &\approx -\frac{1}{l\pi} + \frac{\sqrt{M^2 l^2 + J^2}}{2\pi\,\rho^2},\\
E(\rho \to \infty, \epsilon_0 = 0) &\approx -2\,\frac{\rho}{l} + \frac{Ml}{2\rho} + \frac{\sqrt{M^2 l^2 + J^2}}{2\rho},\\
M(\rho \to \infty, \epsilon_0 = 0) &= -2\,\frac{\rho^2}{l^2} + M + \frac{\sqrt{M^2 l^2 + J^2}}{l}.
\end{aligned} \tag{14.10}$$

The reference energy density to be used is the one corresponding to the AdS metric with parameter M_0, $\epsilon_0(M_0) = -\frac{1}{\pi\rho}\sqrt{\frac{\rho^2}{l^2} - M_0}$, $\epsilon_{0|\infty}(M_0) \approx -\frac{1}{\pi\, l} + \frac{l\, M_0}{2\pi\,\rho^2}$, the series expansions of the corresponding quantities at $\rho \to$ infinity result in

$$\begin{aligned}
\epsilon(\rho \to \infty, \epsilon_{0|\infty}(M_0)) &\approx \frac{\sqrt{M^2 l^2 + J^2} - l M_0}{2\pi\,\rho^2} \approx \frac{M - M_0}{2\pi\rho^2} + \frac{1}{4\,\pi\,\rho^2}\frac{J^2}{M\,l},\\
E(\rho \to \infty, \epsilon_{0|\infty}(M_0)) &\approx \frac{\sqrt{M^2 l^2 + J^2} - l M_0}{\rho} \approx \frac{M - M_0}{\rho} + \frac{1}{2\,\rho}\frac{J^2}{M\,l}\\
M(\rho \to \infty, \epsilon_{0|\infty}(M_0)) &\approx \frac{\sqrt{M^2 l^2 + J^2}}{l} - M_0 \approx M - M_0 + \frac{J^2}{2\,M\,l^2}.
\end{aligned} \tag{14.11}$$

Where the third column in the above approximations hold for $Ml \gg J$. Therefore, comparing with the energy characteristics of the BTZ solution – see Section 4.1.1, (4.10) – one concludes that role of the mass parameter is played by M, and of the momentum parameter by J. At spatial infinity all physical quantities are finite; in particular, the field Ψ vanishes.

14.2.2 Algebraic Classification of the Ricci, Energy–Momentum, and Cotton Tensors

The Ricci tensor, determining in (2+1) dimensions the Riemann curvature tensor, is described by the matrix $(R^{\mu}{}_{\nu})$

$$\left[\begin{array}{ccc}
-\frac{-2\,l^2\rho^3 + l^2 m^3 + 2\,\omega^2\rho^3 + 2\,\omega^2 m^3}{\rho^3\,(l^2 - \omega^2)\,l^2} & 0 & \frac{3\omega\, m^3}{\rho^3\,(l^2 - \omega^2)}\\
0 & \frac{2\,\rho^3 - m^3}{l^2\rho^3} & 0\\
-\frac{3\,\omega\, m^3}{\rho^3\,(l^2 - \omega^2)\,l^2} & 0 & \frac{2\,l^2\rho^3 + 2\,l^2 m^3 - 2\,\omega^2\rho^3 + \omega^2 m^3}{\rho^3\,(l^2 - \omega^2)\,l^2}
\end{array}\right], \tag{14.12}$$

and is algebraically characterized by the following eigenvectors:

$$\begin{aligned}
\lambda_1 &= 2\,\frac{\rho^3 + m^3}{l^2\rho^3} : \mathbf{V1} = [\omega\, V^3, 0, V^3], V_\mu V^\mu = \frac{(V^3)^2\rho^2\,(l^2 - \omega^2)}{l^2}, \mathbf{V1} = \mathbf{S1},\\
\lambda_{2,3} &= \frac{2\,\rho^3 - m^3}{l^2\rho^3} : \mathbf{V2,3} = [V^1, V^2, \frac{V^1\,\omega}{l^2}],\\
V_\mu V^\mu &= \frac{-(V^1)^2\,(\rho - 2\,m)^2\,(\rho + m)^4\,(l^2 - \omega^2) + (V^2)^2\rho^2 l^6}{l^4\rho\,(\rho - 2\,m)\,(\rho + m)^2},
\end{aligned} \tag{14.13}$$

hence depending on the sign of this norm one will have spacelike ($V^1 = 0$), null ($V^2 = \pm\sqrt{l^2-\omega^2}\,(\rho+m)^2\,(\rho-2\,m)\,V^1/(l^3\rho)$), or timelike ($V^2 = 0$) eigenvectors, and also when

$$\mathbf{V2,3} = [V^1, \frac{\delta\sqrt{l^2-\omega^2}\,(\rho+m)^2\,(\rho-2\,m)\,V^1}{l^3\rho}, \frac{V^1\,\omega}{l^2}] = \{\mathbf{S,N,T}\}, \quad (14.14)$$

where δ may assume the values greater than unit $\delta > 1$, equal to unit $\delta = 1$ and less than unit $\delta < 1$, determining correspondingly spacelike, null, or timelike eigenvectors. Thus, in the case of a double root one may choose different vector components determining, for instance, one spacelike vector S and the other timelike T or null N one. Therefore, one may have the algebraic Ricci types: $\{S, 2S\}, \{S, 2N\}, \{S, 2T\}$ and $\{S, (S, T)\}, \{S, (T, T)\}, \ldots, \{S, (T, N)\}$.

For the energy–momentum tensor matrix $(T^{\mu}{}_{\nu})$

$$\begin{bmatrix} \frac{(l^2 m^2+2\,\omega^2 m^2+\omega^2\rho^2-l^2\rho^2)m}{\kappa\,l^2\rho^2(\rho+m)(l^2-\omega^2)} & 0 & -\frac{3\omega\,m^3}{\kappa\,\rho^2(\rho+m)(l^2-\omega^2)} \\ 0 & \frac{(m-\rho)m}{\rho^2\kappa\,l^2} & 0 \\ \frac{3\omega\,m^3}{\kappa\,l^2\rho^2(\rho+m)(l^2-\omega^2)} & 0 & -\frac{(2\,l^2 m^2+\omega^2 m^2-\omega^2\rho^2+l^2\rho^2)m}{\kappa\,l^2\rho^2(\rho+m)(l^2-\omega^2)} \end{bmatrix}$$

one has the eigenvectors

$$\lambda_1 = -\frac{m\,(2\,m^2+\rho^2)}{\kappa\,l^2\rho^2\,(\rho+m)} : \mathbf{V1} = [-\frac{\rho^2\omega\,V_3}{l^2}, 0, \rho^2\,V_3], V_\mu V^\mu = \frac{{V_3}^2\rho^2\,(l^2-\omega^2)}{l^2},$$
$$\mathbf{V1} = \mathbf{S1},$$
$$\lambda_2 = \frac{(m-\rho)\,m}{\kappa\,\rho^2 l^2} : \mathbf{V2,3} = [V^1, V^2, \frac{V^1\,\omega}{l^2}],$$
$$V_\mu V^\mu = \frac{-(V^1)^2\,(\rho-2\,m)^2\,(\rho+m)^4\,(l^2-\omega^2)+(V^2)^2\rho^2 l^6}{l^4\rho\,(\rho-2\,m)\,(\rho+m)^2}. \quad (14.15)$$

Therefore, as in the case of the algebraic structure of the Ricci tensor, one may distinguishes spacelike ($V^1 = 0$), null

$$V^2 = \pm\sqrt{l^2-\omega^2}\,(\rho+m)^2\,(\rho-2\,m)\;V^1/(l^3\rho),$$

or timelike ($V^2 = 0$) eigenvectors, and also when

$$\mathbf{V2,3} = [V^1, \frac{\delta\sqrt{l^2-\omega^2}\,(\rho+m)^2\,(\rho-2\,m)\;V^1}{l^3\rho}, \frac{V^1\,\omega}{l^2}] = \{\mathbf{S,N,T}\}, \quad (14.16)$$

for $\delta > 1, = 1, < 1$. The same comment about the double root and its vectors applies in this case. Consequently, one may have the algebraic energy–momentum types: $\{S, 2S\}, \{S, 2T\}, \{S, 2N\}$ and $\{S, (S, T)\}, \ldots, \{S, (T, N)\}$.

The existence of this relationship between the eigenvectors of the Ricci and energy–momentum tensors is due to the tensor relation

$$R_\mu{}^\nu + \frac{\kappa}{\rho} T_\mu{}^\nu - \frac{(2\,\rho - m)}{l^2 \rho} \delta_\mu{}^\nu = 0.$$

For the Cotton tensor matrix $(C^\mu{}_\nu)$

$$\begin{bmatrix} -\frac{3\omega\, m^3\left(-2\,\rho^3+3\,\rho\, m^2+2\, m^3\right)}{2 l^2 \rho^6 (l^2-\omega^2)} & 0 & \frac{3m^3\left(-l^2\rho^3-\omega^2\rho^3+3\,\omega^2\rho\, m^2+2\,\omega^2 m^3\right)}{2l^2\rho^6(l^2-\omega^2)} \\ 0 & 0 & 0 \\ -\frac{3m^3\left(-l^2\rho^3+3\,l^2\rho\, m^2+2\,l^2 m^3-\omega^2\rho^3\right)}{2l^4\rho^6(l^2-\omega^2)} & 0 & \frac{3\omega\, m^3\left(-2\,\rho^3+3\,\rho\, m^2+2\, m^3\right)}{2l^2\rho^6(l^2-\omega^2)} \end{bmatrix},$$

the characteristic equation allow for one real and a pure imaginary eigenvalues, therefore, the Cotton tensor is characterized algebraically by

$$\lambda_1 = 0 : \mathbf{V1} = [0,\, V^2, 0],\, V_\mu V^\mu = \frac{(V^2)^2 \rho\, l^2}{(\rho - 2\,m)\,(\rho + m)^2},\, \mathbf{V1} = \mathbf{S1},$$

$$\lambda_2 = \bar{\lambda}_3 = \frac{3\,i\sqrt{\rho}\,(\rho - 2\,m)\,(\rho + m)\,m^3}{2\rho^5 l^3} :$$

$$\mathbf{V2} = \mathbf{Z} = [V^1, 0,\, \frac{m^3\omega\,(6\,\rho^3 - 9\,m^2\,\rho - 6\,m^3) - 2\,\lambda\, l^2 \rho^6 (l^2 - \omega^2)}{3\,[(l^2 + \omega^2)\rho^3 - \omega^2\, m^2 (3\,\rho\, + 2\,m)]\, m^3} V^1],$$

$$\mathbf{V3} = \bar{\mathbf{Z}}. \tag{14.17}$$

Thus, the algebraic type of the Cotton tensor is given by Type I: $\{S, Z, \bar{Z}\}$.

15
Low-Energy (2+1) String Gravity

In this chapter the dynamical equations for an n-dimensional heterotic string theory of the Horowitz type are shown explicitly in the string frame and in the Einstein frame too. In particular, the dynamical equations of the three-dimensional string theory are also given. A detailed derivation of these equations has been accomplished by Garcia–Diaz and Gutierrez–Cano (2014a); this work is followed in part. The relation of the Horowitz–Welch and Horne–Horowitz string black hole solution is exhibited. The Chan–Mann charged dilaton solution is derived and the subclass of string solutions is explicitly identified. The stationary generalization, via $SL(2,R)$ transformations, of the static (2+1) Horne–Horowitz string black hole solution is given.

15.1 n-Dimensional Heterotic String Dynamical Equations

In this section we reproduce the field equations "for a part of the low energy action" of Horowitz (1992), to an n-dimensional heterotic string theory described by a metric $g_{\mu\nu}$, a scalar field Φ, a Maxwell field $F_{\mu\nu}$, and a three-form $H_{\mu\nu\lambda}$. The three-form $\boldsymbol{H}$ is related to the two-form potential $\boldsymbol{B}$ and a gauge field A_μ through $\boldsymbol{H} = \boldsymbol{dB} - a\,\boldsymbol{A} \wedge \boldsymbol{dF}$, where a is a constant to be adjusted at the end for final results. In this text, n is used instead of D to denote the number of dimensions. Moreover, Λ is reserved for the standard cosmological constant, whereas Λ_H, and $\Lambda_{CM} = -\Lambda$, denote the Λs used by Horowitz (1992) and Chan and Mann (1996), respectively.

15.1.1 String Frame

The heterotic string action for dimension n, $S = \int d^n x \mathfrak{L}$, is given by

$$S = \int d^n x \sqrt{-g} e^{-2\Phi} \left[R - 2\Lambda + U(\Phi) + 4(\nabla\Phi)^2 - F^2 - \frac{1}{12} H^2 \right]. \tag{15.1}$$

Variations with respect to the metric, the scalar Φ field, the totally antisymmetric H field, and the electromagnetic F field lead to the dynamical field equations

$$G_{\mu\nu} - \frac{1}{2}g_{\mu\nu}\,(-2\Lambda + U) + 2\nabla_\mu\nabla_\nu\Phi - 2g_{\mu\nu}\nabla^2\Phi + 2g_{\mu\nu}(\nabla\,\Phi)^2$$
$$-2\,(F_{\mu\alpha}F_{\nu\beta}g^{\alpha\beta} - \frac{1}{4}g_{\mu\nu}F^2) - \frac{1}{4}(H_{\mu\alpha\beta}H_\nu{}^{\alpha\beta} - \frac{1}{6}g_{\mu\nu}H^2) = 0, \quad (15.2a)$$

$$4\nabla^2\,\Phi - 4(\nabla\,\Phi)^2 + R - 2\,\Lambda - F^2 - \frac{1}{6}H^2 + U(\Phi) - \frac{1}{2}\frac{dU}{d\,\Phi} = 0, \quad (15.2b)$$

$$\nabla_\sigma(\mathrm{e}^{-2\Phi}H^{\mu\nu\sigma}) = 0, \quad (15.2c)$$

$$\nabla_\lambda\left(\mathrm{e}^{-2\Phi}F^{\lambda\epsilon}\right) + a\frac{1}{12}\mathrm{e}^{-2\Phi}F_{\nu\lambda}\,H^{\epsilon\nu\lambda} = 0. \quad (15.2d)$$

The last equation differs in sign from Horowitz's corresponding equation (2.10b).

By contracting (15.2a) one evaluates R,

$$R = \frac{2}{n-2}\left(n\Lambda - \frac{n}{2}U - 2(n-1)\nabla^2\Phi + 2n(\nabla\Phi)^2 + \frac{n-4}{2}F^2 + \frac{n-6}{24}H^2\right), \quad (15.3)$$

which, replaced again into (15.2a) and (15.2b), allows one to rewrite the set of dynamical equations as

$$R_{\mu\nu} = -2\nabla_\mu\nabla_\nu\Phi + 2F_{\mu\alpha}F_{\nu\beta}g^{\alpha\beta} + \frac{1}{4}\,H_{\mu\alpha\beta}H_\nu{}^{\alpha\beta}$$
$$+\frac{2}{n-2}g_{\mu\nu}\left(\Lambda - \frac{1}{2}U - \nabla^2\Phi + 2(\nabla\Phi)^2 - \frac{1}{2}F^2 - \frac{1}{12}H^2\right), \quad (15.4a)$$

$$2\nabla^2\Phi - 4(\nabla\Phi)^2 - 2\Lambda + F^2 + \frac{1}{6}H^2 + U + \frac{n-2}{4}\frac{dU}{d\,\Phi} = 0, \quad (15.4b)$$

$$\nabla_\sigma(\mathrm{e}^{-2\Phi}H^{\mu\nu\sigma}) = 0, \quad (15.4c)$$

$$\nabla_\lambda\left(\mathrm{e}^{-2\Phi}F^{\lambda\epsilon}\right) + a\frac{1}{12}\mathrm{e}^{-2\Phi}F_{\nu\lambda}\,H^{\epsilon\nu\lambda} = 0. \quad (15.4d)$$

15.1.2 Einstein Frame

On the other hand, to pass to the Einstein frame description of this low energy string theory, one accomplishes a conformal transformation of the form

$$\tilde{g}_{\mu\nu} = \mathrm{e}^{2\sigma}g_{\mu\nu}, \tilde{g}^{\mu\nu} = \mathrm{e}^{-2\sigma}g^{\mu\nu} \to \tilde{g} = \mathrm{e}^{2\,n\sigma}g, \mathrm{dim} = \mathrm{n}, \quad (15.5)$$

which transforms the action S (15.1), considered as the barred one, taking into account that $\tilde{F}^2 = \tilde{g}^{\mu\alpha}\tilde{g}^{\nu\beta}F_{\mu\nu}F_{\alpha\beta} = \mathrm{e}^{-4\sigma}g^{\mu\alpha}g^{\nu\beta}F_{\mu\nu}F_{\alpha\beta} = \mathrm{e}^{-4\sigma}F^2$, and $\tilde{H}^2 = \mathrm{e}^{-6\sigma}H^2$, to the form

$$S = \int d^n x\sqrt{-g}\mathrm{e}^{(n\,\sigma - 2\,\sigma - 2\Phi)}\left[\mathrm{e}^{2\sigma}\tilde{R} + \mathrm{e}^{2\,\sigma}(-2\Lambda + U) + 4(\nabla\Phi)^2\right.$$
$$\left.-\mathrm{e}^{-2\sigma}F^2 - \frac{1}{12}\mathrm{e}^{-4\sigma}H^2\right]. \quad (15.6)$$

Thus, one may use the conformal transformed curvature scalar

$$e^{2\sigma}\tilde{R} = R - 2(n-1)g^{\mu\nu}\nabla_\nu\nabla_\mu\sigma - (n-1)(n-2)g^{\mu\nu}\nabla_\mu\sigma\nabla_\nu\sigma, \tag{15.7}$$

and choose

$$n\,\sigma - 2\,\sigma - 2\Phi = 0 \to \sigma = \frac{2}{n-2}\Phi,\ \Phi = \frac{n-2}{2}\sigma. \tag{15.8}$$

Substituting these relations in the action above (15.6), one arrives at

$$S = \int d^n x\sqrt{-g}\left[R - \frac{4}{n-2}(\nabla\Phi)^2 - 2e^{4\,\Phi/(n-2)}\Lambda + V(\Phi) \right.$$
$$\left. -e^{-4\,\Phi/(n-2)}F^2 - \frac{1}{12}e^{-8\,\Phi/(n-2)}H^2\right], \tag{15.9}$$

where the divergence has been dropped from this action

$$-2(n-1)\sqrt{-g}g^{\mu\nu}\nabla_\nu\nabla_\mu\sigma = -2(n-1)(\sqrt{-g}\sigma^{,\nu})_{,\nu},$$

and denoted

$$e^{2\sigma}U(\Phi) = V(\Phi).$$

The extremum of S is achieved along the dynamical equations:

$$G_{\mu\nu} = -g_{\mu\nu}\,e^{4\Phi/(n-2)}\Lambda + \frac{1}{2}g_{\mu\nu}\,V(\Phi) + \frac{4}{n-2}\left(\nabla_\mu\Phi\,\nabla_\nu\Phi - \frac{1}{2}g_{\mu\nu}\nabla_\alpha\Phi\,\nabla^\alpha\Phi\right)$$
$$+ 2\,e^{-4\Phi/(n-2)}(F_{\mu\sigma}F_\nu{}^\sigma - \frac{1}{4}g_{\mu\nu}F^2) - \frac{1}{4}\,e^{-8\Phi/(n-2)}(H_{\mu\alpha\beta}H_\nu{}^{\alpha\beta} - \frac{1}{6}g_{\mu\nu}H^2), \tag{15.10a}$$

$$8\,\nabla^\nu\nabla_\nu\Phi - 8\,\Lambda e^{4\Phi/(n-2)} + 4\,e^{-4\Phi/(n-2)}\,F^2 + \frac{2}{3}\,e^{-8\Phi/(n-2)}\,H^2 + \frac{n-2}{2}\frac{dV}{d\Phi} = 0, \tag{15.10b}$$

$$\nabla_\sigma\left(e^{-8\Phi/(n-2)}H^{\alpha\beta\sigma}\right) = 0, \tag{15.10c}$$

$$\nabla_\lambda\left(e^{-4\Phi/(n-2)}F^{\lambda\epsilon}\right) + \frac{a}{12}\,e^{-8\Phi/(n-2)}F_{\alpha\beta}H^{\alpha\beta\epsilon} = 0. \tag{15.10d}$$

Replacing in (15.10a) the scalar curvature R,

$$R = \frac{4}{n-2}(\nabla\Phi)^2 + \frac{n-4}{n-2}\,e^{-4\Phi/(n-2)}F^2 + \frac{1}{12}\frac{n-6}{n-2}\,e^{-8\Phi/(n-2)}H^2$$
$$- \frac{n}{n-2}(-2\Lambda\,e^{4\Phi/(n-2)} + V),$$

one rewrites (15.10a) as

$$
\begin{aligned}
R_{\mu\nu} &= \frac{2\Lambda}{n-2}\, g_{\mu\nu}\, \mathrm{e}^{4\Phi/(n-2)} - \frac{1}{n-2} g_{\mu\nu}\, V(\Phi) + \frac{4}{n-2}\, \nabla_\mu \Phi\, \nabla_\nu \Phi \\
&+ \mathrm{e}^{-4\Phi/(n-2)} \left(2\, F_{\mu\sigma} F_\nu{}^\sigma - \frac{1}{n-2} g_{\mu\nu}\, F^2\right) \\
&+ \frac{1}{4} \mathrm{e}^{-8\Phi/(n-2)} \left(2\, H_{\mu\alpha\beta} H_\nu{}^{\alpha\beta} - \frac{2}{3} \frac{1}{n-2} g_{\mu\nu}\, H^2\right). \qquad (15.11)
\end{aligned}
$$

15.2 Dynamical Equations in (2+1) String Gravity

In the three-dimensional case, the above Einstein action (15.9) reduces to

$$
S = \int d^3x \sqrt{-g} \left[R - 2\mathrm{e}^{4\Phi}\Lambda - 4(\nabla\Phi)^2 - \mathrm{e}^{-4\Phi} F^2 - \frac{1}{12} \mathrm{e}^{-8\Phi} H^2 + V(\phi)\right], \qquad (15.12)
$$

and the Einstein frame dynamical equations (15.10) become

$$
\begin{aligned}
R_{\mu\nu} &= 2\Lambda\, g_{\mu\nu}\, \mathrm{e}^{4\Phi} + 4\, \nabla_\mu \Phi\, \nabla_\nu \Phi + \mathrm{e}^{-4\Phi} \left(2\, F_{\mu\sigma} F_\nu{}^\sigma - g_{\mu\nu}\, F^2\right) \\
&+ \frac{1}{2} \mathrm{e}^{-8\Phi} \left(H_{\mu\alpha\beta} H_\nu{}^{\alpha\beta} - \frac{1}{3} g_{\mu\nu}\, H^2\right) - g_{\mu\nu}\, V(\Phi). \qquad (15.13a)
\end{aligned}
$$

$$
8\, \nabla^2\Phi - 8\, \Lambda \mathrm{e}^{4\Phi} + 4\, \mathrm{e}^{-4\Phi}\, F^2 + \frac{2}{3}\, \mathrm{e}^{-8\Phi}\, H^2 + \frac{1}{2} \frac{dV}{d\Phi} = 0, \qquad (15.13b)
$$

$$
\nabla_\sigma \left(\mathrm{e}^{-8\Phi} H^{\alpha\beta\sigma}\right) = 0, \qquad (15.13c)
$$

$$
\nabla_\lambda \left(\mathrm{e}^{-4\Phi} F^{\lambda\epsilon}\right) + \frac{a}{12}\, \mathrm{e}^{-8\Phi} F_{\alpha\beta} H^{\alpha\beta\epsilon} = 0. \qquad (15.13d)
$$

One can recover the string dynamical equations by using the conformal inverse relations (see Eisenhart, 1966),

$$
\begin{aligned}
\sigma_{,i;j} &= \tilde{\sigma}_{,i;j} + 2\sigma_{,i}\sigma_{,j} - \tilde{g}_{ij}(\tilde{\nabla}\sigma)^2, \quad \mathrm{e}^{-2\,\sigma} \tilde{\sigma}^{;k}_{;k} = \sigma^{,k}{}_{;k} - (n-2)(\tilde{\nabla}\sigma)^2, \\
\overset{W}{R}_{ij} &= \overset{\tilde{W}}{R}_{ij} + \tilde{g}_{ij} \tilde{\sigma}^{;k}_{;k} + (n-2)\left(\tilde{\sigma}_{,i;j} + \sigma_{,i}\sigma_{,j} - \tilde{g}_{ij}(\tilde{\nabla}\sigma)^2\right), \\
\mathrm{e}^{-2\,\sigma} \overset{W}{R} &= \overset{\tilde{W}}{R} + 2(n-1)\tilde{\sigma}^{;k}_{;k} - (n-1)(n-2)(\tilde{\nabla}\sigma)^2, \qquad (15.14)
\end{aligned}
$$

where tilde is used to denote that covariant differentials are constructed with $\tilde{\Gamma}$s or contravariant tensor components are built with $\tilde{g}^{\mu\nu}$. For $\sigma = 2\Phi$ and $n = 3$ one gets

$$
\begin{aligned}
\Phi_{,\mu;\nu} &= \tilde{\Phi}_{,\mu;\nu} + 4\Phi_{,\mu}\Phi_{,\nu} - 2\tilde{g}_{\mu\nu}(\tilde{\nabla}\Phi)^2, \\
\Phi^{;\alpha}{}_{;\alpha} &= \mathrm{e}^{4\Phi} \left(\tilde{\Phi}^{;\alpha}_{;\alpha} - 2(\tilde{\nabla}\Phi)^2\right), \\
R_{\mu\nu} &= \tilde{R}_{\mu\nu} + 2\tilde{\Phi}_{,\mu;\nu} + 2\tilde{g}_{\mu\nu}\tilde{\Phi}^{;\alpha}_{;\alpha} + 4\Phi_{,\mu}\Phi_{,\nu} - 4\tilde{g}_{\mu\nu}(\tilde{\nabla}\Phi)^2, \\
R &= \mathrm{e}^{4\Phi} \left(\tilde{R} + 8\, \tilde{\Phi}^{;\alpha}_{;\alpha} - 8\, (\tilde{\nabla}\Phi)^2\right), \qquad (15.15)
\end{aligned}
$$

and conformally transforming the above dynamical equations (15.13), using relations (15.15), one gets the barred dynamical equations of the (2+1) string theory under consideration

$$S = \int d^3x\sqrt{-\tilde{g}}e^{-2\Phi}\left[\tilde{R} - 2\Lambda + 4(\tilde{\nabla}\Phi)^2 - \tilde{F}^2 - \frac{1}{12}\tilde{H}^2 + U(\phi)\right], \tag{15.16a}$$

$$\tilde{R}_{\mu\nu} + 2\nabla_\mu\nabla_\nu\Phi - 2F_{\mu\alpha}F_{\nu\beta}\tilde{g}^{\alpha\beta} - \frac{1}{4}H_{\mu\alpha\beta}H_{\nu\gamma\lambda}\tilde{g}^{\gamma\alpha}\tilde{g}^{\lambda\beta}$$
$$+ \tilde{g}_{\mu\nu}\left(-2\Lambda + U + \tilde{F}^2 + \frac{\tilde{H}^2}{6} + 2\tilde{\nabla}^2\Phi - 4(\tilde{\nabla}\Phi)^2\right) = 0, \tag{15.16b}$$

$$2\tilde{\nabla}^2\Phi - 4(\tilde{\nabla}\Phi)^2 + \tilde{F}^2 - 2\Lambda + \frac{1}{6}\tilde{H}^2 + U(\Phi) + \frac{1}{4}\frac{dU}{d\Phi} = 0, \tag{15.16c}$$

$$\tilde{\nabla}_\sigma(e^{-2\Phi}\tilde{H}^{\mu\nu\sigma}) = 0, \tag{15.16d}$$

$$\tilde{\nabla}_\lambda\left(e^{-2\Phi}\tilde{F}^{\lambda\epsilon}\right) + a\frac{1}{12}e^{-2\Phi}\tilde{F}_{\nu\lambda}\tilde{H}^{\epsilon\nu\lambda} = 0. \tag{15.16e}$$

15.3 Horne–Horowitz Black String

Horne and Horowitz (1992) published an exact string black hole solution in three dimensions for the full string theory of Section 15.1, endowed with mass, axion charge per unit length, the asymptotic value of the dilaton, and a cosmological constant. This string solution is given by

$$\overset{S}{g} = -(1 - \frac{M}{r})dt^2 + (1 - \frac{Q^2}{M\,r})dx^2 + (1 - \frac{M}{r})^{-1}(1 - \frac{Q^2}{M\,r})^{-1}\frac{k\,dr^2}{8r^2},$$
$$H_{rtx} = \frac{Q}{r^2}, \quad \phi = \ln r + \frac{1}{2}\ln\frac{k}{2}. \tag{15.17}$$

The identification of the functions appearing in the action (15.1) and equations (15.2) with the ones of Horne and Horowitz (1992) corresponds to $\phi = -2\Phi$, and $\frac{8}{k} = -2\Lambda = \frac{4}{l^2}$, $k = 2\,l^2$, where l has the dimension of length.

Accomplishing in the above mentioned metric a conformal transformation $\overset{E}{g}_{\mu\nu} = e^{-2\sigma}\overset{S}{g}_{\mu\nu} = \frac{k\,r^2}{2}\overset{S}{g}_{\mu\nu}$, where superscripts E and S stand for Einstein and String respectively, one arrives at the corresponding solution in the Einstein frame, namely

$$\overset{E}{g} = \frac{k\,r^2}{2}(1 - \frac{Q^2}{M\,r})dx^2 + (1 - \frac{M}{r})^{-1}(1 - \frac{Q^2}{M\,r})^{-1}\frac{k^2\,dr^2}{16},$$
$$- \frac{k\,r^2}{2}(1 - \frac{M}{r})dt^2, \quad H_{rtx} = \frac{Q}{r^2}, \quad \Phi = -\frac{1}{2}\ln r - \frac{1}{4}\ln\frac{k}{2}, \tag{15.18}$$

fulfilling the dynamical equations (15.10) for dimension $n = 3$.

By accomplishing a $SL(2,R)$ transformation of the Killing coordinates

$$t = \alpha T + \beta \phi, \ \Delta := \alpha\delta - \beta\gamma.$$
$$x = \gamma T + \delta \phi,$$

one arrives at a stationary HH-string solution

$$\begin{aligned}
\overset{E}{g} = & -\frac{k r^2}{2}\left[(1-\frac{M}{r})\alpha^2 - (1-\frac{Q^2}{M r})\gamma^2\right] dT^2 \\
& + k r^2\left[(1-\frac{Q^2}{M r})\gamma\delta - (1-\frac{M}{r})\alpha\beta\right] dT\, d\phi \\
& + \frac{k r^2}{2}\left[(1-\frac{Q^2}{M r})\delta^2 - (1-\frac{M}{r})\beta^2\right] d\phi^2 + (1-\frac{M}{r})^{-1}(1-\frac{Q^2}{M r})^{-1}\frac{k^2\, dr^2}{16},
\end{aligned}$$

$$H_{rtx} = \frac{Q}{r^2},\ \Phi = -\frac{1}{2}\ln r - \frac{1}{4}\ln\frac{k}{2},\ \frac{8}{k} = -2\Lambda = \frac{4}{l^2}. \tag{15.19}$$

In the literature one frequently encounters the $SL(2,R)$ transformation

$$t = \frac{T}{\sqrt{1-\frac{\omega^2}{l^2}}} - \omega\frac{\phi}{\sqrt{1-\frac{\omega^2}{l^2}}},\ x = -\frac{\omega}{l^2}\frac{T}{\sqrt{1-\frac{\omega^2}{l^2}}} + \frac{\phi}{\sqrt{1-\frac{\omega^2}{l^2}}},$$

which yields

$$\begin{aligned}
g = & -\frac{(r-M)\left(rM-Q^2\right)rk\left(l^2-\omega^2\right)}{2\,l^2\left(-\omega^2 rM+\omega^2M^2+rM-Q^2\right)}dT^2 \\
& + (1-\frac{M}{r})^{-1}(1-\frac{Q^2}{M r})^{-1}\frac{k^2\, dr^2}{16} \\
& + \frac{l^2kr\left(-\omega^2 rM+\omega^2M^2+rM-Q^2\right)}{2M\left(l^2-\omega^2\right)} \\
& \times\left(d\phi - \frac{\omega\left(-l^2Mr+l^2M^2+rM-Q^2\right)}{l^2\left(-\omega^2 rM+\omega^2M^2+rM-Q^2\right)}dT\right)^2,
\end{aligned}$$

$$H_{rtx} = \frac{Q}{r^2},\ \Phi = -\frac{1}{2}\ln r - \frac{1}{4}\ln\frac{k}{2}. \tag{15.20}$$

The evaluation of the quasilocal mass, energy and momentum is done using the Brown–York approach; this yields

$$J(r\to\infty) = 2\omega\, l^2\,\frac{\left(M^2-Q^2\right)}{\left(l^2-\omega^2\right)M}, \tag{15.21a}$$

$$\epsilon(r\to\infty) = -1/2\,\frac{1}{r\pi\, l^2} - 1/4\,\frac{(M-\omega\, Q)\,(M+\omega\, Q)}{Ml^2\pi\,(\omega-1)\,(\omega+1)\,r^2} - \epsilon_0, \tag{15.21b}$$

$$E(r\to\infty) = \frac{2\sqrt{1-\omega^2}}{\sqrt{l^2-\omega^2}} - 2\pi\, K(r)\epsilon_0, \tag{15.21c}$$

$$M_{BY}(r \to \infty) = 4r + 4\,l\,(-2r^2 + M_0 l^2), \tag{15.21d}$$

$$\epsilon_0 = -\frac{1}{\pi\, l} + \frac{1}{2}\frac{M_0\, l}{\pi\, r^2}. \tag{15.21e}$$

ω is interpreted as a rotating parameter and the mass function increases as the radial coordinate approach spatial infinity. Moreover, various pathologies take place at this location.

15.4 Horowitz–Welch Black String

Horowitz and Welch (1993) published an exact string black hole solution in three dimensions for the low-energy string theory

$$S = \int d^3x \sqrt{-g} e^{-2\Phi} \left[R - 2\Lambda + 4(\nabla\Phi)^2 - \frac{1}{12} H^2 \right],\ \Lambda = -2/k_{HW}, \tag{15.22}$$

of Section 15.1, endowed with mass, angular momentum, axion charge per unit length, and a negative cosmological constant. This string solution is given by a modified BTZ black hole to a $(2+1)$ string theory with vanishing scalar Φ and electromagnetic $F_{\alpha\beta}$ fields; in this last case the field equations (15.4) become

$$R_{\mu\nu} - \frac{1}{4} H_{\mu\alpha\beta} H_\nu{}^{\alpha\beta} = 0, \tag{15.23a}$$

$$-\,2\Lambda + \frac{1}{6} H^2 = 0, \tag{15.23b}$$

$$\nabla_\sigma (H^{\mu\nu\sigma}) = 0. \tag{15.23c}$$

The totally anti-symmetric tensor $H_{\mu\nu\alpha}$ has to be proportional to the volume three form $\epsilon_{\mu\nu\alpha}$, because of the equation (15.23c), the proportionality factor ought to be a constant, hence one may choose

$$H_{\mu\nu\sigma} = \alpha \frac{1}{l} \epsilon_{\mu\nu\sigma}. \tag{15.24}$$

Taking into account the properties of ϵ

$$\epsilon_{\alpha\nu\sigma}\epsilon^{\beta\nu\sigma} = -2\delta^\beta{}_\alpha, \tag{15.25}$$

therefore

$$H_{\alpha\nu\sigma} H^{\beta\nu\sigma} = -2\frac{\alpha^2}{l^2}\delta^\beta{}_\alpha,\ H^2 = -6\frac{\alpha^2}{l^2}. \tag{15.26}$$

Consequently, from (15.23a) one has

$$R_{\mu\nu} - \frac{1}{4} H_{\mu\alpha\beta} H_\nu{}^{\alpha\beta} = 0 \to R_{\mu\nu} = -\frac{2}{l^2} g_{\mu\nu},\ \alpha = 2 \tag{15.27}$$

$$-\,2\Lambda + \frac{1}{6} H^2 = 0 \to \Lambda = -\frac{\alpha^2}{2l^2} = -\frac{2}{l^2},\ k_{HW} = l^2. \tag{15.28}$$

On the other hand, $\boldsymbol{H} = \boldsymbol{dB}$, thus

$$H_{\alpha\beta\gamma} = 3B_{[\alpha\beta,\gamma]} = \frac{2}{l}\epsilon_{\alpha\beta\gamma}. \tag{15.29}$$

As is concluded in Horowitz and Welch (1993): "Thus every solution to three dimensional general relativity with negative cosmological constant is a solution to low energy string theory with: $\Phi = 0$, $H_{\alpha\beta\gamma} = \frac{2}{l}\epsilon_{\alpha\beta\gamma}$, and $\Lambda = -\frac{2}{l^2}$."

In particular, in Horowitz and Welch (1993) it is established that the BTZ black hole metric

$$g = -(\frac{r^2}{l^2} - M)dt^2 - J\,dt\,d\phi + r^2 d\phi^2 + (\frac{r^2}{l^2} - M + \frac{J^2}{4l^2})^{-1}dr^2, \tag{15.30}$$

in the presence of an anti-symmetric B field

$$B_{\phi\,t} = \frac{r^2}{l},\ \boldsymbol{H} = \boldsymbol{dB}, \tag{15.31}$$

is a solution of the string theory with a zero scalar field Φ.
By a target space duality transformation, Eq. (13) of Horowitz and Welch (1993), which referred to Buscher (1993), means that from a given solution $(g_{\mu\nu}, B_{\mu\nu}, \Phi)$ independent on one variable, say x, one generates a new solution $(\tilde{g}_{\mu\nu}, \tilde{B}_{\mu\nu}, \tilde{\Phi})$ with

$$\begin{aligned}
\tilde{g}_{xx} &= \frac{1}{g_{xx}},\ \tilde{g}_{x\alpha} = \frac{B_{x\alpha}}{g_{xx}},\\
\tilde{g}_{\alpha\beta} &= g_{\alpha\beta} - \frac{1}{g_{xx}}(g_{x\alpha}\,g_{x\beta} - B_{x\alpha}\,B_{x\beta}),\\
\tilde{B}_{x\alpha} &= \frac{g_{x\alpha}}{g_{xx}},\ \tilde{B}_{\alpha\beta} = B_{\alpha\beta} - 2\frac{g_{x[\alpha}B_{\beta]\,x}}{g_{xx}},\\
\tilde{\Phi} &= \Phi - \frac{1}{2}\ln g_{xx},
\end{aligned} \tag{15.32}$$

where α and β run over all directions except x. Applying this transformation to expressions (15.30) and (15.31), along the coordinate symmetry ϕ, one gets Eq. (14) of Horowitz and Welch (1993), namely

$$\overset{S}{g} = (M - \frac{J^2}{4r^2})dt^2 + \frac{2}{l}dt\,d\phi + \frac{1}{r^2}\,d\phi^2 + (\frac{r^2}{l^2} - M + \frac{J^2}{4r^2})^{-1}dr^2, \tag{15.33}$$

which, once diagonalized by means of a $SL(2, R)$ coordinate transformation

$$\begin{aligned}
t &= -\frac{l}{\sqrt{r_+^2 - r_-^2}}\tilde{t} + \frac{l}{\sqrt{r_+^2 - r_-^2}}\tilde{x},\\
\phi &= \frac{r_+^2}{\sqrt{r_+^2 - r_-^2}}\tilde{t} - \frac{r_-^2}{\sqrt{r_+^2 - r_-^2}}\tilde{x},
\end{aligned} \tag{15.34}$$

and the r-coordinate transformation

$$r^2 = l\tilde{r} \tag{15.35}$$

yields the string solution derived in Horne and Horowitz (1992); see also Section 15.3. Dropping primes, it becomes

$$\overset{S}{g} = -(1-\frac{\mathcal{M}}{r})dt^2 + (1-\frac{\mathcal{Q}^2}{\mathcal{M}\,r})dx^2 + (1-\frac{\mathcal{M}}{r})^{-1}(1-\frac{\mathcal{Q}^2}{\mathcal{M}\,r})^{-1}\frac{l^2\,dr^2}{4r^2},$$

$$B_{xt} = \frac{\mathcal{Q}}{r},\ \phi = -\frac{1}{2}\ln(r\,l), \tag{15.36}$$

where $\mathcal{M} = r_+^2/l$ and $\mathcal{Q} = J/2$.

The identification of the functions appearing in the action (15.1) and equations (15.2) with the ones of Horne and Horowitz (1992),

$$S = \int d^3x\sqrt{-g}e^{\phi}\left[R - 2\Lambda + (\nabla\Phi)^2 - \frac{1}{12}H^2\right],\ \Lambda = -4/k_{HH}, \tag{15.37}$$

requires that $\phi = -2\Phi_{HW}$, and $k_{HH} = 2\,k_{HW}$, $k_{HW} = l^2$, $\Lambda = -2/l^2$, where the subscripts are in correspondence with the initial of the author's family name.

15.5 Chan–Mann String Solution

Chan and Mann (1994, see also Chan and Mann, 1996), derived a class of solutions to dilaton minimally coupled to (2+1) Einstein–Maxwell gravity. There is a subclass of solutions allowing an interpretation from the viewpoint of the low energy (2+1) string theory for specific values of the charged dilaton solution. First, we derive the dilaton solution. Next, assigning specific values to constants characterizing the charged dilaton, the correspondence with the string theory developed in the previous section is established.

15.5.1 Einstein–Maxwell-Scalar Field Equations

The Chan and Mann (1994) action for a (2+1)-dimensional gravity is given by

$$\mathcal{S} = \int d^3x\sqrt{-g}\left[R - \frac{B}{2}\nabla_\mu\Psi\,\nabla^\mu\Psi + 2\,\mathrm{e}^{b\Psi}\Lambda_{CM} - \mathrm{e}^{-4\,a\,\Psi}\,F^2\right],$$

where the parameters Λ_{CM}, b are arbitrary at this stage, Ψ is the massless minimally coupled scalar field, R is the scalar curvature, and $F^2 = F_{\mu\,\nu}\,F^{\mu\,\nu}$ the electromagnetic invariant. The variations of this action yield the dynamical equations

$$R_{\mu\nu} = \frac{B}{2}\nabla_\mu\Psi\,\nabla_\nu\Psi - 2g_{\mu\nu}\mathrm{e}^{b\,\Psi}\Lambda_{CM} + 2\,\mathrm{e}^{-4\,a\,\Psi}\left(F_\mu{}^\alpha\,F_{\nu\alpha} - g_{\mu\,\nu}\,F^2\right),$$

$$\frac{B}{2}\nabla^\mu\nabla_\mu\Psi + b\,\mathrm{e}^{b\,\Psi}\Lambda_{CM} + 2\,a\,\mathrm{e}^{-4\,a\,\Psi}\,F^2 = 0,$$

$$\nabla^\mu\left(\mathrm{e}^{-4\,a\,\Psi}\,F_{\mu\,\nu}\right) = 0.$$

15.5.2 Static and Stationary Black String Solutions

The static cyclic symmetric metric in the $(2+1)$ Schwarzschild coordinate frame was derived in Section 13.7. It is given explicitly by (13.82), namely

$$\begin{aligned} g_E &= -r^2\, L(r)^2 dt^2 + \frac{dr^2}{L(r)^2} + r^2 d\phi^2, \\ L(r)^2 &= \left(r\, C_1 - 2\, r^2 \Lambda_s + 2\, Q^2\right) r^{-2}, \\ F_{\mu\nu} &= 2 F_{tr} \delta^t_{[\mu} \delta^r_{\nu]},\ F_{tr} = \frac{Q}{r^2} = -A_{t,r} \rightarrow A_t = \frac{Q}{r}, \\ \Psi(r) &= -1/2\, \ln(r). \end{aligned} \tag{15.38}$$

This string solution was derived for the first time in Chan and Mann (1994), for $B = 8, k = -1/2, a = 1, b = 4$; it fulfills the Einstein string equations (15.13) for $\Lambda = \Lambda_s = \pm\frac{1}{l^2} = -\Lambda_{CM}$, in the case of vanishing H.

Under the conformal transformation

$$\tilde{g}_{\mu\nu} = \mathrm{e}^{4\,\Psi(r)} g_{\mu\nu} = \frac{1}{r^2} g_{\mu\nu} \tag{15.39}$$

it becomes

$$\begin{aligned} g_S &= -\left(r\, C_1 - 2 r^2 \Lambda_s + 2\, Q^2\right) r^{-2}\, dt^2 + \frac{dr^2}{\left(r\, C_1 - 2\, r^2 \Lambda_s + 2\, Q^2\right)} + d\phi^2, \\ F_{\mu\nu} &= 2 F_{tr} \delta^t_{[\mu} \delta^r_{\nu]},\ F_{tr} = \frac{Q}{r^2} = -A_{t,r} \rightarrow A_t = \frac{Q}{r}, \\ \Psi(r) &= -1/2\, \ln(r); \end{aligned} \tag{15.40}$$

this is a solution of the equations (15.4) of the $(2+1)$ string theory.

Moreover, subjecting the metric (15.40) to $SL(2,R)$ transformations of the Killing coordinates

$$t = \alpha\, \tau + \beta\, \theta,\ \phi = \gamma\, \tau + \delta\, \theta$$

one arrives at a rotating charged string solution, namely

$$\begin{aligned} g_S &= -\left(\alpha^2\, \mathcal{L}^2/r^2 - \gamma^2\right) d\tau^2 - 2\left(\alpha\beta\, \mathcal{L}^2/r^2 - \gamma\, \delta\right) d\theta\, d\tau \\ &\quad + \left(\delta^2 - \beta^2\, \mathcal{L}^2/r^2\right) d\theta^2 + \frac{dr^2}{\mathcal{L}^2}, \\ F_{\mu\nu} &= 2\alpha\, F_{tr} \delta^\tau_{[\mu} \delta^r_{\nu]} - 2\beta\, F_{tr} \delta^\theta_{[\mu} \delta^r_{\nu]},\ F_{tr} := \frac{Q}{r^2} = -A_{t,r} \rightarrow A_t = \frac{Q}{r}, \\ \Psi(r) &= -1/2\, \ln(r),\ \mathcal{L}^2 := r\, C_1 - 2\, r^2 \Lambda_s + 2\, Q^2. \end{aligned} \tag{15.41}$$

In particular when $\Lambda_{CM} = 1/l^2$, AdS branch, and the usual choice of the $SL(2,R)$ transformations is adopted

$$\begin{aligned} t &= \frac{\tau}{\sqrt{1 - \omega^2/l^2}} - \omega \frac{\theta}{\sqrt{1 - \omega^2/l^2}}, \\ \phi &= -\frac{\omega}{l^2} \frac{\tau}{\sqrt{1 - \omega^2/l^2}} + \frac{\theta}{\sqrt{1 - \omega^2/l^2}}, \end{aligned} \tag{15.42}$$

where ω stands for the rotation parameter, this metric can be written as

$$\begin{aligned} g_S = & -\frac{\mathcal{L}^2}{r^2}\frac{1-\omega^2/l^2}{1-\omega^2+\mathcal{L}^2/r^2}d\tau^2 \\ & +\frac{1-\omega^2\,\mathcal{L}^2/r^2}{1-\omega^2/l^2}\left(d\theta-\frac{\omega}{l^2}\frac{1-\omega^2/l^2}{1-\omega^2\,\mathcal{L}^2/r^2}\,d\tau\right)^2+\frac{dr^2}{\mathcal{L}^2}, \end{aligned} \tag{15.43}$$

and is endowed with four parameters: the mass $M = -C_1$, charge Q, rotation ω, and cosmological constant $\Lambda_s = \pm 1/l^2$. It is worth pointing out that string solutions can be found in various dimensions; see, for instance, the works of Witten (1991); Mandal et al. (1991); Maki and Shiraishi (1993), among others.

16
Topologically Massive Gravity

In this part of the book we deal with exact solutions to the Einstein topologically massive gravity equations. However, since the material to be included only represents twenty per cent of the whole book's subject matter, we prefer to present this content in the form of chapters devoted to a concise but, we hope, complete (in the range of the possibilities) exposition of the exact solutions in topologically massive gravity (TMG) in three dimensions in the case of vacuum in the presence of a cosmological constant Λ of both signs. Thus, this chapter has an introductory character, while the next three chapters deal with very specific families of the existing Petrov-type Cotton solutions in TMG.

The extension of the $3D$ Einstein gravity to other field theories to provide them with certain degrees of freedom (a massive spin 2 graviton) was proposed more than thirty-five years ago by Deser, Jackiw and Templeton; see Deser et al. (1982a): "Three-dimensional massive gauge theories," which is known as the TMG. It includes a Chern–Simons term constructed from connections (Cotton tensor) with broken parity invariance; see also its extended version, with a detailed analysis, in Deser et al. (1982b). A cosmological constant was introduced in these three-dimensional theories by Deser (1984). The sign in front of the curvature scalar has been chosen opposite to the standard one of the $4D$ Einstein gravity to yield, in the limit of the linearized theory, to the existence of a spin 2 graviton with positive energy. A modern treatment of these aspects in cosmological massive gravity appeared recently in Carlip et al. (2009).

As far as the determination of exact solutions to vacuum equations with a cosmological constant in TMG is concerned, through this long period, various classes of solutions have been reported apart from the trivial Minkowski flat spacetime, the de Sitter $(A)dS_3$ cosmology, and the BTZ black hole – conformally flat (zero Cotton tensor) solutions – for any value of the coupling mass parameter μ. Among these classes one can mention homogeneous spaces, pp-waves metrics, cosmological solutions, and Kerr–Schild metrics. Fortunately, a systematic up-to-date classification of the existing families of vacuum solutions in TMG has been done recently by Chow, Pope, and Sezgin (CPS): see Chow et al. (2010a), referred to as CPSa, and Chow et al. (2010b), referred to as CPSb, for Kundt spacetimes. Thus, I shall take advantage of this progress and incorporate their results in this book; I shall work out the derivation of the various classes reported in the literature using a presentation pattern similar to CPS and exhibit their characterization by means of the García et al. (2004) Petrov–Segré classification of the Cotton tensor and other relevant features.

There are other kinds of generalizations of the TMG that pursue other purposes. For instance, implications in the AdS_3/CFT_2 correspondence. The construction of a new massive gravity theory, NMG, preserving the parity invariance and possessing a single massive spin 2 field has been achieved by Bergshoeff, Hohm and Townsend (see Bergshoeff et al., 2009); this theory is constructed with combinations of quadratic invariants $F(R, R_{\mu\nu}\, R^{\mu\nu})$. The generalization of this last theory took place with the publication of "minimal massive gravity," which has positive energy spin 2 graviton and positive central charges for the "asymptotic AdS-boundary algebra." These theories experience nowadays a lot of activity; many researchers are engaged in their developments, and the list of references on these topics is quite vast: a Living review concerned these developments not only in three dimensions but also in higher dimensions, "Massive gravity," has been published recently by de Rham (2014).

16.1 Chern–Simons Action and Field Equations of TMG

The Chern–Simons action

$$I = \frac{1}{\kappa}\int dx^3\sqrt{-g}\left[R - 2\Lambda + \frac{1}{\mu}\epsilon^{\lambda\mu\nu}\Gamma^{\rho}{}_{\lambda\sigma}\left(\partial_\mu\Gamma^{\sigma}{}_{\rho\nu} + \frac{2}{3}\Gamma^{\sigma}{}_{\mu\tau}\Gamma^{\tau}{}_{\nu\rho}\right)\right], \tag{16.1}$$

where $\kappa = 16\pi G$, in the presence of a matter distribution, under variations, yields the Einstein topological massive gravity (E-TMG) equations

$$E_{\alpha\beta} := R_{\alpha\beta} - \frac{1}{2}R\, g_{\alpha\beta} + \Lambda\, g_{\alpha\beta} + \frac{1}{\mu}C_{\alpha\beta} = \kappa\, T_{\alpha\beta},$$
$$C_{\alpha\beta} = \epsilon_{\alpha}{}^{\mu\nu}\nabla_\mu(R_{\nu\beta} - \frac{1}{4}R\, g_{\nu\beta}), \tag{16.2}$$

where $\epsilon_{\alpha\mu\nu}$ is the Levi–Civita pseudotensor, and $C_{\alpha\beta}$ is the symmetric, $C_{\alpha\beta} = C_{\beta\alpha}$, traceless, $C^{\alpha}_{\alpha} = 0$, Cotton (pseudo-)tensor.

Notice that in the vacuum case the E-TMG equations (16.2) can be represented equivalently by the system of equations

$$R = 6\,\Lambda, \tag{16.3a}$$

$$S_{\alpha\,\beta} + \frac{1}{\mu}C_{\alpha\,\beta} = 0, \tag{16.3b}$$

where $S_{\alpha\beta}$ is the traceless Ricci tensor,

$$S_{\alpha\beta} := R_{\alpha\beta} - \frac{R}{3} g_{\alpha\beta}, \; S^{\alpha}{}_{\alpha} = 0. \tag{16.4}$$

Practically, as was stated previously, in this text we follow, with minor changes, the CPS reviews, Chow et al. (2010a and 2010b); the cosmological constant is mostly denoted by its standard notation Λ. Nevertheless, when using the CPS representations of the metrics reported in their reviews, they are presented as in the CPS works, where they first kept the same coordinates as in the originals that appeared in the literature, with possible minimal rearrangements, and later found coordinate transformations that brought the solutions to a canonical form. In the CPS publications, any cosmological constant only appears via its parameter m, which has dimensions of mass. Moreover, the leitmotifs pursued in the original publications are quoted at the beginnings of the corresponding paragraphs to show the diversity of problems yielding a similar answer.

16.2 Exact Vacuum Solutions of TMG with Λ

In the CPS reviews on exact solutions to E-TMG equations, namely Chow et al. (2010a and 2010b), it was established that almost all local metrics reduce to four (five in our subclassification) particular families of solutions:

1. General triaxially squashed vacuum solutions ($\Lambda = 0$)
2. Timelike biaxially squashed AdS_3 solutions
3. Spacelike biaxially squashed AdS_3 solutions
4. AdS_3 non-covariantly and covariantly constant type N wave solutions
5. Kundt dS_3–AdS_3 solutions

1\. Details in Chapter 17. A systematic search of squashed solutions in TMG dates back to the works by Nutku and Baekler (1989), and, independently, by Ortiz (1990), (NBO), dealing with the derivation of homogeneous spacetimes of various Bianchi types (BT) in vacuum without a cosmological constant – the so-called triaxially squashed spacetimes, although $3D$ Bianchi-type VIII spaces were determined by Vuorio (1985) from a different point of view, and studied in detail some years before the NBO publications by Percacci, Sodano, and Vuorio; see Percacci et al. (1987). In the context of Einstein-TMG solutions, one of the first works, to our knowledge, where the term “squashed space” appeared is the publication by Nutku and Baekler (1989). Moreover, in this work one finds all the building material to construct Bianchi spacetimes to vacuum Einstein-TMG equations; as an introduction to this topic one has to address them; they posed the problem of finding homogeneous, anisotropic solutions of the Bianchi types to E-TMG equations.

Later Nutku (1993), in the same Bianchi-type VIII framework but now with a cosmological constant, derived timelike biaxially squashed and spacelike spacetimes.

All these squashed spacetimes are Petrov type D, and exhibit the constant scalar invariant (CSI) property, i.e., all scalar polynomial curvature invariants are constant.

2. Details in Section 17.5. Timelike biaxially squashed AdS_3 spacetime can be regarded as a timelike fibration over $\mathbb{H}$, or as a stationary cyclic symmetric spacetime decomposed with respect to the timelike Killing vector. This spacetime is Petrov type D.

3. Details in Section 17.6. Spacelike biaxially squashed AdS_3 spacetime can be regarded as a spacelike fibration over AdS_2, or as a stationary cyclic symmetric spacetime decomposed with respect to the spacelike Killing vector. This spacetime is Petrov type D.

4. Details in Chapter 18. In the spirit of CPS, the AdS_3 pp-wave solutions are generalizations of pp-waves to include a negative cosmological constant; these solutions are Petrov type N. Nevertheless, we prefer to restrict the name of pp-waves to those solutions possessing a covariantly constant $k_{\mu;\nu} = 0$ null congruence k as they are defined in the standard $4D$ Einstein gravity. In TMG there is room for a wider class of solutions with a non-covariantly constant $k_{\mu;\nu} \neq 0$ null congruence k of Cotton type N, to be denoted as TN-wave solutions.

5. Details in Chapter 19. In general, Kundt spacetimes are defined by metrics admitting a null geodesic vector field that is shear-free, twist-free, and expansion-free. Therefore, in three dimensions, because the twist and shear trivially vanish, a "Kundt spacetime is simply one that admits an expansion-free null geodesic congruence." These spacetimes allow for Petrov type II, D, III, N, and O solutions. Moreover, all the existing Kundt solutions of TMG are CSI – constant scalar invariants – spacetimes for which all polynomial scalar (Ricci) curvature invariants are constant; these late metric structures were studied for the first time by Coley, Hervik, and Pelavas (see Coley et al., 2006 and Coley et al., 2008) in three dimensions in a wider context.

Many of these solutions "have been independently rediscovered as solutions of TMG several times. The literature is rather fragmented, using different coordinate systems that are not obviously related." Thus, adopting the aim of CPS, a review establishing the relationships, if any, between families of solutions becomes pertinent; the metrics are presented in the same coordinates as the original literature, although certain rearrangements have been made. Each metric is then transformed to a canonical form.

17

Bianchi-Type (BT) Spacetimes in TMG; Petrov Type D

17.1 Generalities on Bianchi-Type (BT) 3*D* Spaces

The classification of the Bianchi-type 3*D* spaces – from *I* to *IX* – was accomplished more than hundred years ago on the basis of the Lie groups that characterize them. The classification of homogeneous 3*D* Bianchi-type spaces can be found in Stephani et al. (2003), Chapter 13 and 14, Table 13.4.

A quite complete outline of the main concepts needed when studying the symmetries of a manifold is detailed in Ortiz (1990), §2: vector field on a manifold, exponential map, Lie derivative, invariance of the metric tensor, Killing vector field, Lie group formed by the set of all symmetry generators, commutators of generators, Lie algebras, structure constants, right invariant vector field ↔ right invariant 1-form, left invariant vector field ↔ left invariant 1-form, transitive group, isometry group, homogeneous spacetime, space of left invariant 1-forms, decomposition of the matrix of structure constants, and classification of 3*D* Lie algebras – Bianchi types.

Let the set of r independent Killing vectors $\{\xi_A{}^{\alpha}, A = 1, \cdots, r\}$ forms a basis of a G_r group, then the generators of this symmetry group, defined as

$$K_A = \xi_A{}^{\alpha} \frac{\partial}{\partial x^{\alpha}},$$

fulfill the commutation rule

$$[K_A, K_B] = -C^N{}_{A\,B} K_N,\; C^N{}_{A\,B} = -C^N{}_{B\,A},$$

and the Jacobi identities,

$$[K_A, [K_B, K_C]] + [K_B, [K_C, K_A]] + [K_C, [K_A, K_B]] = 0.$$

Moreover, the quantities $C^N{}_{A\,B}$, called constants of structure of the group, fulfill the Lie identity

$$C^N{}_{M[\,L}C^M{}_{A\,B]} = 0.$$

If by linear transformations T of the r basis Killing operators K_A one can convert one set of structure constants into one another, then the corresponding groups that they represent are isomorphic. On the contrary, if there is no such transformation, the groups are non-isomorphic. If a basis of generators is given by the set $\{K_A\}$, then a linear combination of them by means of an invertible constant matrix $T^A{}_{\tilde{B}}$ also determines a basis;

$$K_{\tilde{A}} = T^N{}_{\tilde{A}} K_N,$$

such that the new structure constants obey the following transformation law

$$C^{\tilde{L}}{}_{\tilde{M}\tilde{N}} = T^A{}_{\tilde{M}} T^B{}_{\tilde{N}} T^{\tilde{L}}{}_D C^D{}_{AB},$$

where $T^{\tilde{L}}{}_D$ is inverse to $T^A{}_{\tilde{M}}$.

Returning to the classification of $3D$ homogeneous spaces, to describe these spaces one uses the left invariant 1–forms σ^α, $\alpha = 1,2,3$ which fulfill the Maurer–Cartan structure equations,

$$d\sigma^\alpha = \frac{1}{2} C^\alpha{}_{\beta\gamma} \sigma^\beta \wedge \sigma^\gamma \tag{17.1}$$

where the structure constants $C^\alpha{}_{\beta\gamma} = -C^\alpha{}_{\gamma\beta}$ obey the Lie identity

$$C^\alpha{}_{\beta[\mu} C^\beta{}_{\nu\gamma]} = 0. \tag{17.2}$$

Each Bianchi type is characterized by a specific set of structure constants, which in turn determines the corresponding group. The classification of homogeneous $3D$ Bianchi-type spaces can be found also in the textbook by Stephani (1990).

Following Nutku and Baekler (1989), the triad description of the homogeneous, anisotropic cosmological models in TMG keep a close parallelism to the Maurer–Cartan structure description. By means of a linear transformation between the 1-forms σ^a and the triad 1-forms ω^a

$$\begin{aligned} \omega^a &= \lambda^a{}_\alpha \sigma^\alpha, \\ \lambda^a{}_\alpha &= \lambda_a \delta^a{}_\alpha \text{ no sum in } a,\ \lambda_a = \text{constants}, \end{aligned} \tag{17.3}$$

one determines the 3D spacetime metric with Lorentzian signature defined through

$$\begin{aligned} ds^2 &= \eta_{ab}\,\omega^a \omega^b,\ \omega^a = \lambda_a \sigma^a, \text{no sum in } a, \\ \eta_{ab} &= \operatorname{diag}(-1,1,1). \end{aligned} \tag{17.4}$$

From the first Cartan structure equations

$$d\omega^a + \Gamma^a{}_b \wedge \omega^b = 0, \tag{17.5}$$

one determines the connection 1-forms, $\Gamma^a{}_b = \Gamma^a{}_{bs}\omega^s$, which, used in the second Cartan structure equations

$$\Theta^a{}_b := d\omega^a{}_b + \omega^a{}_s \wedge \omega^s{}_b = \frac{1}{2} R^a{}_{bcd}\omega^c \wedge \omega^d, \tag{17.6}$$

gives rise to the Riemann curvature tensor

$$R^s{}_{abc} = \partial_b \Gamma^s{}_{ac} - \partial_c \Gamma^s{}_{ab} + \Gamma^v{}_{ac}\Gamma^s{}_{bv} - \Gamma^v{}_{ab}\Gamma^s{}_{cv}. \tag{17.7}$$

Because of the constancy of the structure constants $C^\alpha{}_{\beta\mu}$ in the case of 3D Bianchi-type spacetimes, the structure constants are related according to

$$C^m{}_{ns} = \lambda^m{}_\alpha \lambda^\beta{}_n \lambda^\gamma{}_s C^\alpha{}_{\beta\gamma},\ \lambda^m{}_\alpha \lambda^\alpha{}_n = \delta^m{}_n,\ \lambda^m{}_\alpha \lambda^\beta{}_m = \delta^\beta{}_\alpha, \tag{17.8}$$

thus the connections are

$$\Gamma^a{}_{bc} = \frac{1}{2}\left(-C^a{}_{bc} + \eta_{bm}\eta^{an}C^m{}_{cn} - \eta_{cm}\eta^{na}C^m{}_{nb}\right) \tag{17.9}$$

and consequently in the Riemann tensor components the derivatives of Γs drop out. Moreover, its contraction yields the Ricci tensor and a subsequent contraction leads the crucial important formula for the scalar curvature R in terms of polynomials of the scale factors λ_a, as explicitly shown by Nutku and Baekler (1989) and Ortiz (1990) (NBO for short), and Nutku (1993). We derive R in the forthcoming paragraphs for specific metrics. It is clear that the curvature invariants constructed on constant Riemann tensor are also constant objects, i.e., scalar constant invariants. Moreover, the invariants built on covariant derivatives of the Riemannian curvature are constant scalars too.

Nutku and Baekler (1989) and Ortiz (1990) established that the only Bianchi types allowed by the E-TMG equations are the Bianchi Type VIII, and III spacetimes. These classes are derived here in some detail. In the original papers, different Bianchi types were integrated for both Euclidean and Lorentzian signatures.

17.2 Nutku–Baekler–Ortiz "Timelike" BT VIII Spacetime

The first variant of Bianchi type VIII spacetime, the so-called NBO "timelike" triaxilly squashed BT VIII spacetime, arises for the metric, in coordinates $\{\psi, \theta, \phi\}$, and orthonormal triad given by

$$\begin{aligned}
ds^2 &= -\lambda_0^2\, {\sigma^{(0)}}^2 + \lambda_1^2\, {\sigma^{(1)}}^2 + \lambda_2^2\, {\sigma^{(2)}}^2 = \eta_{ab}\,\omega^{(a)}\omega^{(b)},\\
\omega^{(a)} &= \lambda_a \sigma^{(a)},\ \omega^{(a)} = \omega^{(a)}{}_\alpha dx^\alpha,\ \{\lambda_0, \lambda_1, \lambda_2\} = \text{constants},\\
\sigma^{(0)} &= d\psi + \sinh\theta\, d\phi,\\
\sigma^{(1)} &= -\sin\psi d\theta + \cos\psi \cosh\theta d\phi,\\
\sigma^{(2)} &= \cos\psi d\theta + \sin\psi \cosh\theta d\phi,\\
d\sigma^{(1)} &= \sigma^{(2)} \wedge \sigma^{(0)},\ d\sigma^{(2)} = \sigma^{(0)} \wedge \sigma^{(1)}, d\sigma^{(0)} = -\sigma^{(1)} \wedge \sigma^{(2)}.
\end{aligned} \tag{17.10}$$

We denote the triad components enclosed within a parenthesis to avoid confusion with coordinate components.

The E–TMG equations determine the constraints on the scale factors: among them, they yield the scalar curvature

$$R = \frac{(\lambda_1 + \lambda_0 - \lambda_2)(\lambda_1 + \lambda_0 + \lambda_2)(\lambda_2 - \lambda_1 + \lambda_0)(-\lambda_2 - \lambda_1 + \lambda_0)}{2{\lambda_0}^2{\lambda_2}^2{\lambda_1}^2}. \tag{17.11}$$

In vacuum TMG, the vanishing of the curvature scalar R requires the fulfillment of the relations

$$\lambda_0 \pm \lambda_1 \mp \lambda_2 = 0 \rightarrow \lambda_0 + \lambda_1 + \lambda_2 = 0; \tag{17.12}$$

this last relation can be always adopted because of the triad $\{\omega^{(a)}\}$ is defined up the signs of λ_a. The equations $E^{(a)}{}_{(b)} = 0$ lead, taking into account the relation above, to the condition

$$\mu = -\frac{{\lambda_0}^2 + {\lambda_1}^2 + {\lambda_2}^2}{\lambda_0\,\lambda_1\,\lambda_2} = 2\,\frac{{\lambda_1}^2 + \lambda_2\,\lambda_1 + {\lambda_2}^2}{\lambda_1\,\lambda_2\,(\lambda_1 + \lambda_2)}. \tag{17.13}$$

These Bianchi type VIII solutions are called triaxially squashed spacetimes of the TMG theory; it is apparent that they are unique to their class; they were derived in Nutku and Baekler (1989), Eq. (4.1), and Ortiz (1990) (type(a) with $a = 0$). They are characterized by a Petrov type D Cotton tensor: its eigenvectors are given by

$$\begin{aligned} L_1 &= -\mathcal{L}_+\,\lambda_1;\; V_{I\,\alpha} = \omega^{(1)}{}_\alpha, \text{ spacelike},\\ L_2 &= -\mathcal{L}_+\,\lambda_2;\; V_{II\,\alpha} = \omega^{(2)}{}_\alpha, \text{ spacelike},\\ L_0 &= -\mathcal{L}_+\,\lambda_0;\; V_{III\,\alpha} = \omega^{(0)}{}_\alpha, \text{ timelike}, \end{aligned} \tag{17.14}$$

where the positive $\mathcal{L}_+$ is given by

$$\mathcal{L}_+ = -2\,\frac{\mu}{\lambda_0\,\lambda_1\,\lambda_2} = 2\,\frac{{\lambda_0}^2 + {\lambda_1}^2 + {\lambda_2}^2}{{\lambda_0}^2{\lambda_2}^2{\lambda_1}^2} = 4\,\frac{{\lambda_1}^2 + \lambda_2\,\lambda_1 + {\lambda_2}^2}{(\lambda_1 + \lambda_2)^2\,{\lambda_2}^2{\lambda_1}^2}. \tag{17.15}$$

In terms of its eigenvectors, the Cotton tensor is expressed as

$$C_{\alpha\,\beta} = \mathcal{L}_+\,\lambda_0\,\omega^{(0)}{}_\alpha\omega^{(0)}{}_\beta - \mathcal{L}_+\,\lambda_1\,\omega^{(1)}{}_\alpha\omega^{(1)}{}_\beta - \mathcal{L}_+\,\lambda_2\,\omega^{(2)}{}_\alpha\omega^{(2)}{}_\beta, \tag{17.16}$$

hence, its constant triad components $C_{(a)\,(b)} = \omega^\alpha{}_{(a)}C_{\alpha\,\beta}\omega^\beta{}_{(b)}$ are

$$C_{(a)\,(b)} = \mathcal{L}_+\,\lambda_0\,\delta^{(0)}{}_{(a)}\delta^{(0)}{}_{(b)} - \mathcal{L}_+\,\lambda_1\,\delta^{(1)}{}_{(a)}\delta^{(1)}{}_{(b)} - \mathcal{L}_+\,\lambda_2\,\delta^{(2)}{}_{(a)}\delta^{(2)}{}_{(b)}. \tag{17.17}$$

17.2.1 Nutku Timelike Biaxially Squashed Metric

Nutku extended the results of NBO to the presence of a cosmological constant Λ. In such case the scalar curvature is equal to 6Λ, $R = 6\,\Lambda$; in Nutku (1993), $\Lambda_{Nutku} = -\Lambda_{standard} = -\Lambda$. Now one leaves (17.12), and instead requires $\lambda_i = \pm\lambda_j, i \neq j$. In particular, without loss of generality one may set

$$\lambda_2 = \lambda_1, \tag{17.18}$$

which, substituted into $E^\alpha{}_\beta$, yields

$$E^\psi{}_\phi = 2\,\mu\,{\lambda_1}^2{\lambda_0}^2 - 2\,\mu\,{\lambda_1}^4 + 3\,{\lambda_0}^3 - 3\,\lambda_0\,{\lambda_1}^2 = 0. \tag{17.19}$$

Solving for λ_0 one obtains

$$\lambda_0 = -\frac{2}{3}\,\mu\,{\lambda_1}^2, \tag{17.20}$$

together with the trivial solutions $\lambda_0 = \pm\lambda_1$. The substitution of $\lambda_2 = \lambda_1$ into the scalar curvature $R = 6\Lambda$ leads to

$$(\lambda_0 - 2\,\lambda_1)\,(\lambda_0 + 2\,\lambda_1) - 12\,{\lambda_1}^4\,\Lambda = 0, \tag{17.21}$$

which, substituting λ_0 from (17.20), gives rise to the equations

$$\mu^2{\lambda_1}^2 - 9 - 27\,\Lambda\,{\lambda_1}^2 = 0 \to \lambda_1 = 3\,\frac{1}{\sqrt{\mu^2 - 27\,\Lambda}},\ \lambda_0 = -6\,\frac{\mu}{\mu^2 - 27\,\Lambda}. \tag{17.22}$$

Finally, this Nutku (1993) solution, Eq. (18), can be given by the metric

$$ds^2 = \frac{9}{\mu^2 - 27\Lambda}\left[-\frac{4\mu^2}{\mu^2 - 27\Lambda}\,(d\psi + \sinh\theta d\phi)^2 + d\theta^2 + \cosh^2\theta d\phi^2\right], \tag{17.23}$$

which is characterized by a Petrov type D Cotton tensor

$$(C^\alpha{}_\beta) = \mathcal{L}\begin{bmatrix} -1 & 0 & 0 \\ 0 & 2 & 3\sinh\theta \\ 0 & 0 & -1 \end{bmatrix},\ \mathcal{L} := \frac{1}{9}\mu\,\left(\mu^2 + 9\,\Lambda\right). \tag{17.24}$$

Vuorio Timelike Biaxially Squashed Metric

Moreover, it contains the timelike biaxially squashed solution in the case of vanishing cosmological constant. To derive this subclass of solutions from the vanishing condition of the scalar curvature (17.11) one uses here the relation $\lambda_2 = \lambda_1$, thus

$$R(\lambda_2 = \lambda_1) = 0 \to \lambda_0 = \mp 2\lambda_1. \tag{17.25}$$

Replacing these λs in equation (17.13) with μ, one arrives at

$$\mu = \pm\frac{3}{\lambda_1} \to \lambda_1 = \pm\frac{3}{\mu}, \tag{17.26}$$

which determines a metric of the form (17.23)

$$ds^2 = \frac{9}{\mu^2}\,\left[-4\,(d\psi + \sinh\theta d\phi)^2 + d\theta^2 + \cosh^2\theta d\phi^2\right], \tag{17.27}$$

for $\Lambda = 0$ known as the Vuorio (1985) solution.

17.3 Nutku–Baekler–Ortiz "Spacelike" Squashed BT VIII Spacetime

In this paragraph, a second branch of Bianchi type VIII solutions is derived – the NBO "spacelike" triaxilly squashed BT VIII spacetime: the metric and orthonormal triad used in the coordinates $\{\psi, \theta, \phi\}$ are

$$ds^2 = -\lambda_0^2\,{\sigma^{(0)}}^2 + \lambda_1^2\,{\sigma^{(1)}}^2 + \lambda_2^2\,{\sigma^{(2)}}^2 = \eta_{ab}\,\omega^{(a)}\omega^{(b)},\ \omega^{(a)} = \lambda_a\sigma^{(a)},$$
$$\sigma^{(0)} = \cosh\psi d\theta + \sinh\psi\cos\theta d\phi,$$

$$\begin{aligned}\sigma^{(1)} &= \sinh\psi d\theta + \cosh\psi\cos\theta d\phi,\\ \sigma^{(2)} &= d\psi + \sin\theta\, d\phi,\\ \{\lambda_0, \lambda_1, \lambda_2\} &= \text{constants.}\end{aligned} \tag{17.28}$$

where the left invariant 1-forms $\sigma^{(a)}$ fulfill the Maurer–Cartan equations

$$d\sigma^{(1)} = \sigma^{(2)} \wedge \sigma^{(0)},\ d\sigma^{(2)} = \sigma^{(0)} \wedge \sigma^{(1)}, d\sigma^{(0)} = -\sigma^{(1)} \wedge \sigma^{(2)}.$$

It should be pointed out that this metric structure (17.28) corresponds to another real cut of the metric (17.10), when considered as a complex one: making there the complex transformations and identifications

$$\begin{aligned}&\theta \to i\theta,\ \psi \to i\psi,\ \phi \to \phi,\ \sigma^{(0)} \to i\sigma^{(2)},\ \sigma^{(1)} \to \sigma^{(1)},\ \sigma^{(2)} \to i\sigma^{(0)},\\ &\omega^{(0)} \to i\omega^{(2)},\ \omega^{(1)} \to \omega^{(1)},\ \omega^{(2)} \to i\omega^{(0)},\ \lambda_0 \to \lambda_2,\ \lambda_1 \to \lambda_1,\ \lambda_2 \to \lambda_0,\end{aligned}$$

one arrives at the metric structure (17.28).

The E-TMG equations, among others, yield the scalar curvature

$$R = \frac{(\lambda_0 + \lambda_1 + \lambda_2)(\lambda_0 + \lambda_1 - \lambda_2)(-\lambda_2 + \lambda_0 - \lambda_1)(\lambda_2 + \lambda_0 - \lambda_1)}{2{\lambda_2}^2{\lambda_0}^2{\lambda_1}^2}. \tag{17.29}$$

In vacuum TMG, the vanishing of the curvature scalar R requires the fulfillment of the relations $\lambda_0 \pm \lambda_1 \mp \lambda_2 = 0$, which one fixes to be

$$\lambda_0 + \lambda_1 + \lambda_2 = 0, \tag{17.30}$$

because the triad is defined up the signs of λ_a. The equation $E^{\psi}{}_{\psi} = 0$ leads, taking into account the relation above, to the coupling constant

$$\mu = \frac{{\lambda_0}^2 + {\lambda_1}^2 + {\lambda_2}^2}{\lambda_0\,\lambda_1\,\lambda_2} = -2\frac{{\lambda_1}^2 + \lambda_1\lambda_2 + {\lambda_2}^2}{\lambda_1\lambda_2(\lambda_1 + \lambda_2)}. \tag{17.31}$$

These λs (17.30) and μ (17.31) determine a second family of Bianchi type VIII solutions, called triaxilly squashed spacetimes of the TMG theory; it is clear that they are unique. They are characterized by a Petrov type D Cotton tensor. Searching for its eigenvectors one solves the equations

$$C^{\alpha}{}_{\beta}V^{\beta}{}_{A} - L\,V_{A}{}^{\alpha} = 0,\ \det(C^{\alpha}{}_{\beta} - L\,\delta^{\alpha}{}_{\beta}) = 0,$$

which lead to

$$\begin{aligned}L_1 &= -\lambda_1\mathcal{L}_-,\ V_{I\alpha} = \omega^{(1)}{}_{\alpha}, \text{spacelike},\\ L_2 &= -\lambda_2\mathcal{L}_-,\ V_{II\alpha} = \omega^{(2)}{}_{\alpha}, \text{spacelike},\\ L_0 &= -\lambda_0\mathcal{L}_-,\ V_{0\alpha} = \omega^{(0)}{}_{\alpha},\ \text{timelike},\end{aligned} \tag{17.32}$$

where the negative constant $\mathcal{L}_-$ is equal to

$$\mathcal{L}_- = -2\,\frac{\mu}{\lambda_1\,\lambda_2\,\lambda_0} = -4\,\frac{{\lambda_2}^2 + \lambda_2\,\lambda_1 + {\lambda_1}^2}{{\lambda_2}^2\,(\lambda_2 + \lambda_1)^2\,{\lambda_1}^2}. \tag{17.33}$$

They allow to express $C_{\alpha\beta}$ as

$$C_{\alpha\beta} = \lambda_0 \mathcal{L}_-\, \omega^{(0)}{}_\alpha \omega^{(0)}{}_\beta - \lambda_1 \mathcal{L}_-\, \omega^{(1)}{}_\alpha \omega^{(1)}{}_\beta - \lambda_2 \mathcal{L}_-\, \omega^{(2)}{}_\alpha \omega^{(2)}{}_\beta; \qquad (17.34)$$

hence, its constant triad components $C_{(a)\,(b)} = \omega^\alpha{}_{(a)} C_{\alpha\beta} \omega^\beta{}_{(b)}$ are

$$C_{(a)\,(b)} = \lambda_0 \mathcal{L}_-\, \delta^{(0)}{}_{(a)} \delta^{(0)}{}_{(b)} - \lambda_1 \mathcal{L}_-\, \delta^{(1)}{}_{(a)} \delta^{(1)}{}_{(b)} - \lambda_2 \mathcal{L}_-\, \delta^{(2)}{}_{(a)} \delta^{(2)}{}_{(b)}. \qquad (17.35)$$

17.3.1 Spacelike Biaxially Squashed Metric; Nutku Solution Counterpart

Nutku (1993) extended the results of NBO to the presence of a cosmological constant Λ and derived the timelike biaxially squashed metric (17.23), nevertheless no mention is made of this second spacelike possibility, although in the publication by Nutku and Baekler (1989) the metric structure (17.28) is reported under the number (4.6). For that reason we propose to call this second branch "Nutku solution counterpart".

In the presence of a cosmological constant Λ, the scalar curvature fulfills $R = 6\,\Lambda$; now one has to abandon (17.30) and require $\lambda_i = \pm\lambda_j, i \neq j$. Without loss of generality, one may demand

$$\lambda_0 = \lambda_1 \qquad (17.36)$$

which, substituted into $E^\alpha{}_\beta$, yields the component $E^\psi{}_\phi$ of the form

$$E^\psi{}_\phi : 3\,{\lambda_1}^2 \lambda_2 - 3\,{\lambda_2}^3 - 2\,\mu\,{\lambda_2}^2 {\lambda_1}^2 + 2\,\mu\,{\lambda_1}^4 = 0 \to \lambda_1 = \sqrt{-\frac{3\,\lambda_2}{2\mu}}. \qquad (17.37)$$

Its substitution into the remaining TMG equations leads to a single constraint, namely

$$\lambda_2\,\mu^2 - 27\,\lambda_2\,\Lambda + 6\,\mu = 0 \to \lambda_2 = -6\,\frac{\mu}{\mu^2 - 27\,\Lambda},\ \lambda_1 = 3\,\frac{1}{\sqrt{\mu^2 - 27\,\Lambda}}, \qquad (17.38)$$

hence, the Nutku solution counterpart can be given by the metric

$$ds^2 = \frac{9}{\mu^2 - 27\Lambda}\left[-d\theta^2 + \cos^2\theta d\phi^2 + \frac{4\mu^2}{\mu^2 - 27\Lambda}\,(d\psi + \sin\theta d\phi)^2\right]. \qquad (17.39)$$

Although this solution is not explicitly given in Nutku (1993), it can straightforwardly be derived by complex coordinate transformations from Nutku (1993), Eq. (18). This solution is characterized by a Petrov type D Cotton tensor.

Spacelike Biaxially Squashed Vuorio Solution Counterpart

Moreover, it contains the biaxially squashed spacelike solution in the case of vanishing cosmological constant, i.e., the metric (17.39) for $\Lambda = 0$. For completeness, this subclass of solutions is derived from the vanishing condition of the scalar curvature (17.29); as the first step one uses the equality $\lambda_2 = \lambda_1$, next its substitution in R gives

$$R(\lambda_2 = \lambda_1) = 0 \rightarrow \lambda_0 = \mp 2\lambda_1, \tag{17.40}$$

finally, replacing these λs in the equation (17.31) for μ one arrives at

$$\mu = \pm\frac{3}{\lambda_1} \rightarrow \lambda_1 = \pm\frac{3}{\mu}, \tag{17.41}$$

which determines a metric of the form (17.39) for $\Lambda = 0$,

$$ds^2 = \frac{9}{\mu^2}\left[-d\theta^2 + \cos^2\theta d\phi^2 + 4\left(d\psi + \sin\theta d\phi\right)^2\right], \tag{17.42}$$

which can be named "Vuorio solution counterpart."

17.4 Nutku–Baekler–Ortiz Solutions of Bianchi Type III

Another branch of solutions, see Nutku and Baekler (1989), Eq. (4.10), and Ortiz (1990) (type(a) with $a \neq 0$), is derived using the Bianchi type III left 1-forms from the metric

$$\begin{aligned}
ds^2 &= -\lambda_1^2\,{\sigma^{(1)}}^2 + \lambda_2^2\,{\sigma^{(2)}}^2 + \lambda_3^2\,{\sigma^{(3)}}^2 = \eta_{ab}\,\omega^{(a)}\omega^{(b)},\ \omega^{(a)} = \lambda_a\sigma^{(a)},\\
\sigma^{(1)} &= \mathrm{e}^{\alpha\,\theta}(\cosh\theta\,dx + \sinh\theta\,dy),\\
\sigma^{(2)} &= \mathrm{e}^{\alpha\,\theta}(\sinh\theta\,dx + \cosh\theta\,dy),\\
\sigma^{(3)} &= d\theta,
\end{aligned}\tag{17.43}$$

where $\sigma^{(a)}$ are Bianchi III left invariant 1-forms that satisfy

$$\begin{aligned}
&d\sigma^{(1)} = \alpha\sigma^{(3)}\wedge\sigma^{(1)} + \sigma^{(3)}\wedge\sigma^{(2)},\ d\sigma^{(2)} = \alpha\sigma^{(3)}\wedge\sigma^{(2)} + \sigma^{(3)}\wedge\sigma^{(1)},\\
&d\sigma^{(3)} = 0,\ {C^{(1)}}_{(3)(1)} = \alpha = {C^{(2)}}_{(3)(2)}, {C^{(1)}}_{(3)(2)} = 1 = {C^{(2)}}_{(3)(1)},
\end{aligned}\tag{17.44}$$

which corrects the relations in the text line following Eqs. (A.28) and (A.56) of Chow et al. (2010a), denoted for short CPSa. The scalar curvature for the vacuum TMG is given by

$$R = -\frac{2\,{\lambda_1}^2{\lambda_2}^2 + 12\,{\lambda_1}^2\alpha^2{\lambda_2}^2 - {\lambda_1}^4 - {\lambda_2}^4}{2\,{\lambda_1}^2{\lambda_2}^2{\lambda_3}^2} = 0$$

from which one isolates α^2

$$\alpha^2 = \frac{1}{12}\,\frac{\left(\lambda_1 - \lambda_2\right)^2\left(\lambda_1 + \lambda_2\right)^2}{{\lambda_1}^2{\lambda_2}^2}$$

with two possible branches

$$\alpha = \pm\frac{1}{6}\,\frac{\sqrt{3}\,(\lambda_1 + \lambda_2)\,(\lambda_1 - \lambda_2)}{\lambda_2\,\lambda_1}.$$

Nevertheless, one can always choose the sign of α to be positive, $\alpha = |\alpha|$; if α were negative, $\alpha = -|\alpha|$, then by changing $\theta \to -\theta$, $y \to -y$ one gets the case with positive α. From the equation $E^{\theta}{}_{\theta}$ one obtains the expression of μ

$$\mu = \frac{3}{2}\frac{{\lambda_1}^2 + {\lambda_2}^2}{\lambda_3\,\lambda_1\,\lambda_2}.$$

Substituting μ into $E^{z}{}_{y}$ one gets

$$15\,{\lambda_1}^2{\lambda_2}^4 - 15\,{\lambda_1}^4{\lambda_2}^2 - {\lambda_2}^6 + {\lambda_1}^6 = 0,$$

with solutions

$$\lambda_1 = \left(2 + \sqrt{3}\right)\lambda_2, \tag{17.45a}$$

or

$$\lambda_1 = \left(2 - \sqrt{3}\right)\lambda_2. \tag{17.45b}$$

Evaluating α and μ one gets

$$\alpha = \pm 1 \to \alpha = 1,\ \mu = \frac{6}{\lambda_3} \to \lambda_3 = \frac{6}{\mu}. \tag{17.46}$$

Therefore, one distinguishes only two possible branches of Bianchi type III solutions without a cosmological constant.

17.4.1 Nutku–Baekler–Ortiz BT III Timelike Solution with $\Lambda = 0$

Part of the name used above has been borrowed from CPSa to designate in a definite manner this class of solutions; in a similar way we proceed to denote the second spacelike branch of solutions. For the choice

$$\alpha = 1,\ \lambda_1 = \left(2 + \sqrt{3}\right)\lambda,\ \lambda_2 = \lambda,\ \lambda_3 = \frac{6}{\mu}, \tag{17.47a}$$

one obtains the metric given in Chow et al. (2010a) as (A.28), namely

$$\begin{aligned} ds^2 &= \lambda^2\left[-(2+\sqrt{3})^2\,{\sigma^{(1)}}^2 + {\sigma^{(2)}}^2\right] + \frac{36}{\mu^2}{\sigma^{(3)}}^2 = \frac{36}{\mu^2}\,d\theta^2 \\ &\quad + \lambda^2 e^{2\theta}\left[-(2+\sqrt{3})^2(\cosh\theta\,dx + \sinh\theta\,dy)^2 + (\sinh\theta\,dx + \cosh\theta\,dy)^2\right] \\ &= -\frac{3+2\sqrt{3}}{6}\lambda^2\left[\sqrt{3}(dx - dy) + 2e^{2\theta}(dx + dy)\right]^2 \\ &\quad + \frac{3+2\sqrt{3}}{6}\lambda^2 e^{4\theta}(dx + dy)^2 + \frac{36}{\mu^2}\,d\theta^2. \end{aligned} \tag{17.47b}$$

Making the coordinate transformations CPSa(A.29),

$$\frac{\sqrt{3+2\sqrt{3}}}{6}\frac{\mu}{\sqrt{2}}\lambda(x-y)\rightarrow\tau,\ \frac{\sqrt{3+2\sqrt{3}}}{3}\frac{\mu}{\sqrt{6}}\lambda(x+y)\rightarrow x',\ \mathrm{e}^{-2\theta}\rightarrow y',\tag{17.47c}$$

dropping primes, one gets

$$ds^2_{tl}=\frac{9}{\mu^2}\left[-\lambda^2(d\tau+\frac{dx}{y})^2+\frac{dx^2+dy^2}{y^2}\right],\tag{17.47d}$$

which coincides with the solution given by Eq. (A.11) of Chow et al. (2010a) with $\Lambda=0$, CPSa(A.11).

17.4.2 Nutku–Baekler–Ortiz BT III Spacelike Solution with $\Lambda = 0$

On the other hand, for the second available choice

$$\alpha=1,\ \lambda_1=\left(2-\sqrt{3}\right)\lambda,\ \lambda_2=\lambda,\ \lambda_3=\frac{6}{\mu},\tag{17.48a}$$

one obtains the solution given by

$$\begin{aligned}ds^2&=\lambda^2\left[-(2-\sqrt{3})^2\sigma^{(1)2}+\sigma^{(2)2}\right]+\frac{36}{\mu^2}\sigma^{(3)2}=\frac{36}{\mu^2}d\theta^2\\&+\lambda^2\mathrm{e}^{2\theta}\left[-(2-\sqrt{3})^2(\cosh\theta\,dx+\sinh\theta\,dy)^2+(\sinh\theta\,dx+\cosh\theta\,dy)^2\right]\\&=\frac{2\sqrt{3}-3}{6}\lambda^2\left[\sqrt{3}(dx-dy)-2\mathrm{e}^{2\theta}(dx+dy)\right]^2\\&-\frac{2\sqrt{3}-3}{6}\lambda^2\mathrm{e}^{4\theta}(dx+dy)^2+\frac{36}{\mu^2}d\theta^2,\end{aligned}\tag{17.48b}$$

coinciding with CPSa (A.56). Making the coordinate transformations CPSa-(A.57),

$$\frac{\sqrt{2\sqrt{3}-3}}{3\sqrt{3}}\frac{\mu}{\sqrt{2}}\lambda(x+y)\rightarrow t,\ \frac{\sqrt{2\sqrt{3}-3}}{6}\frac{\mu}{\sqrt{2}}\lambda(x-y)\rightarrow z,\ \mathrm{e}^{-2\theta}\rightarrow x,\tag{17.48c}$$

one arrives at $\Lambda=0$.

$$ds^2_{sl}=\frac{9}{\mu^2}\left[\lambda^2(dz+\frac{dt}{x})^2+\frac{dx^2-dt^2}{x^2}\right],\tag{17.49}$$

which corresponds to (A.46) with $\Lambda=0$ of Chow et al. (2010a).

All these homogeneous Bianchi type VIII and III solutions – the unique three dimensional spacetimes of this kind in TMG – are presented in a succinct manner in the following table:

Table 17.4.1 *Bianchi Type (BT) VIII and BT III TMG solutions*

References Bianchi T	Eqs.text	Eqs. in Ref.	Rel. to CPSa
Nutku and Baekler (1989)	TriAtl:(17.10)	(4.1)	(A.27)
BT VIII	TriAsl:(17.28)	(4.6)	(A.54)
Ortiz (1990)	TriAtl:(17.10)	$(a = 0)$	(A.28)
BT VIII	TriAsl:(17.28)	↓ (5.3)	(A.56)
Nutku and Baekler (1989)	sl:(17.47b)		(A.27)
BT III	tl:(17.48b)	(4.10)	(A.27)
Ortiz (1990)	sl:(17.47b)	$(a \neq 0)$	(A.28)
BT III	tl:(17.48b)	↓ (5.5)	(A.28)
Nutku (1993); Λ	BiA sl: (17.39)		(A.48)
BT VIII; Λ	BiA tl: (17.23)	(18)	(A.12)
CPSa; Chow et al. (2010a)	17.5.4		
tl – timelike, sl – spacelike	TriA – triaxial	BiA – biaxial	T – type

17.5 Timelike Biaxially Squashed Metrics

17.5.1 Representation of the Vacuum Biaxially Squashed Solutions

Most of the known exact solutions in TMG belong to the biaxially squashed – timelike or spacelike – homogeneous anisotropic Bianchi type VIII families of spacetimes. Fortunately, the identification and sorting of the known solutions in the timelike and spacelike classes have been accomplished by the familiar Chow, Pope and Sezgin; see CPSa: Chow et al. (2010a), who made a complete classification of the existing solutions. Moreover, they classify in that work the pp-wave solutions too.

In $(2+1)$-dimensional gravity any stationary cyclic symmetric metric can be given as

$$g = g_{tt}dt^2 + 2g_{t\phi}dt\, d\phi + g_{\phi\phi}\, d\phi^2 + g_{rr}dr^2, \tag{17.50}$$

where $g_{\alpha\beta}$ depend on the coordinate r, $g_{\alpha\beta}(r)$.

Without loss of generality one can choose a "spacelike" representation of the stationary cyclic symmetric $(2+1)$ metric developed with respect to the cyclic symmetry $\boldsymbol{\partial}_\phi$,

$$m^\alpha = \delta^\alpha{}_\phi/\sqrt{g_{\phi\,\phi}},\ \boldsymbol{m} = \sqrt{g_{\phi\,\phi}}\left(\boldsymbol{d}\phi + \frac{g_{t\,\phi}}{g_{\phi\,\phi}}\boldsymbol{d}t\right) \tag{17.51}$$

obtaining

$$g = g_{\phi\,\phi}\left(d\phi + \frac{g_{t\,\phi}}{g_{\phi\,\phi}}dt\right)^2 - \frac{g_{t\,\phi}^2 - g_{tt}g_{\phi\,\phi}}{g_{\phi\,\phi}}dt^2 + g_{rr}dr^2, \tag{17.52}$$

or, in a more familiar way,

$$g = -\frac{F(r)}{H(r)}dt^2 + \frac{dr^2}{F(r)} + H(r)\left[d\phi + W(r)dt\right]^2. \tag{17.53}$$

On the other hand, if one chooses the stationary symmetry $\boldsymbol{\partial}_t$

$$k^\alpha = \delta^\alpha{}_t/\sqrt{|g_{t\,t}|},\ \boldsymbol{k} = \sqrt{|g_{t\,t}|}\left(\boldsymbol{d}t + \frac{g_{t\,\phi}}{g_{t\,t}}\boldsymbol{d}\phi\right), \tag{17.54}$$

as the fundamental Killing field, the stationary cyclic symmetric $(2+1)$ metric can be written in the "timelike" representation

$$g = -|g_{t\,t}|\left(d\,t + \frac{g_{t\,\phi}}{g_{t\,t}}d\phi\right)^2 + \frac{g_{tt}g_{\phi\,\phi} - g_{t\,\phi}^2}{g_{t\,t}}d\phi^2 + g_{rr}dr^2, \tag{17.55}$$

or in the more standard way,

$$g = -\frac{F(r)}{h(r)}\left[dt - \omega(r)\,d\phi\right]^2 + h(r)\,d\phi^2 + \frac{dr^2}{F(r)}. \tag{17.56}$$

Vuorio (1985), one of the first researchers looking for exact solutions in TMG, used the stationary "rotationally" symmetric *ansatz* for the metric, i.e., the metric (17.56) up to coordinate redefinition.

Although the general timelike biaxially squashed Nutku solution has been derived in the previous paragraph 17.2.1, it is enlightening to exhibit its derivation from the point of view of the above-detailed stationary cyclic symmetric metric structure. Consider the middle of the road metric element in coordinates $\{\phi, \theta, \tau\}$

$$ds^2 = -\frac{\lambda_2}{4}(d\tau + \cosh\theta\, d\phi)^2 + \frac{\lambda_1}{4}(d\theta^2 + \sinh^2\theta\, d\phi^2), \tag{17.57}$$

where the scale factors λ_1 and λ_2 are constants, and seek for solutions to the Einstein-TMG equations (16.2). The relevant E-TMG equations amount to:

$$E^\tau_\phi : (3\,{\lambda_2}^{3/2} - \mu\,\lambda_1\,\lambda_2)\,(\lambda_2 - \lambda_1) = 0 \rightarrow \mu\,\lambda_1 = 3\,\sqrt{\lambda_2}, \tag{17.58}$$

substituting λ_1 into the equation $E^\phi_\phi = E^\theta_\theta$,

$$E^\phi_\phi : -\left(\lambda_2 + \Lambda\,{\lambda_1}^2\right)\mu\,\lambda_1 + 4\,\sqrt{\lambda_2}\,(\lambda_2 - \lambda_1) = 0, \tag{17.59}$$

one arrives at

$$\begin{aligned}
&\left(\mu^2 - 27\,\Lambda\right)\sqrt{\lambda_2} = 12\,\mu \rightarrow \lambda_2 = 144\,\frac{\mu^2}{\left(\mu^2 - 27\,\Lambda\right)^2} = 4\,\frac{9}{\mu^2 - 27\,\Lambda}\,\lambda^2,\\
&\lambda^2 := \frac{4\mu^2}{\mu^2 - 27\,\Lambda},\ \lambda_1 = \frac{36}{\mu^2 - 27\,\Lambda}.
\end{aligned} \tag{17.60}$$

To obtain these solutions for λs one could use the equation arising from the vanishing of the contraction $E^\alpha{}_\alpha = 0$ of the vacuum $T^\alpha{}_\beta = 0$ E–TMG equations: $4\,\lambda_1 + 3\,\Lambda\,{\lambda_1}^2 = \lambda_2$.

The timelike biaxially squashed *dS–AdS* solution can be given by the metric

$$\begin{aligned}
ds^2_{tl} &= \frac{9}{\mu^2 - 27\Lambda}\left[-\lambda^2(d\tau + \cosh\theta\, d\phi)^2 + d\theta^2 + \sinh^2\theta d\phi^2\right],\\
\lambda^2 &:= \frac{4\mu^2}{\mu^2 - 27\,\Lambda}.
\end{aligned} \tag{17.61}$$

It is characterized by the constant scalar curvature

$$R = 6\Lambda, \tag{17.62}$$

the Cotton tensor

$$C^{\alpha}{}_{\beta} = -\frac{\mu(\mu^2 + 9\Lambda)}{9}\left[\delta^{\alpha}{}_{\theta}\delta^{\theta}{}_{\beta} + \delta^{\alpha}{}_{\phi}\delta^{\phi}{}_{\beta} - 2\delta^{\alpha}{}_{\tau}\delta^{\tau}{}_{\beta} - 3\cosh\theta\,\delta^{\alpha}{}_{\tau}\delta^{\phi}{}_{\beta}\right], \tag{17.63}$$

and the traceless Ricci tensor $S_{\mu\nu} = R_{\mu\nu} - \frac{1}{3}R\,g_{\mu\nu}$, which, as it should be, for all vacuum E-TMG solutions plus Λ the relation between the Cotton and the traceless Ricci tensor

$$S^{\alpha}{}_{\beta} + \frac{1}{\mu}C^{\alpha}{}_{\beta} = 0$$

holds. These tensors allow for an alternative representation through the metric tensor and a unit vector, say K^{α}, of the form

$$S^{\alpha}_{\beta} = \frac{\mu^2 + 9\Lambda}{9}(\delta^{\alpha}{}_{\beta} + 3\,K^{\alpha}\,K_{\beta}) \tag{17.64}$$

where the unit timelike vector K^{α}, is given by

$$K^{\alpha} = \frac{\mu^2 - 27\,\Lambda}{6\,\mu}\,\delta^{\alpha}{}_{\tau},\; K_{\alpha} = -\frac{6\mu}{\mu^2 - 27\,\Lambda}\cosh\theta\delta_{\alpha}{}^{\phi} - \frac{6\mu}{\mu^2 - 27\,\Lambda}\delta_{\alpha}{}^{\tau}. \tag{17.65}$$

17.5.2 Eigenvectors of the Cotton Tensor; Triad Formulation

This paragraph is devoted to the search of the eigenvectors of the Cotton tensor, and, consequently, of the traceless Ricci tensor because of (17.5.1). Using the orthonormal triad 1-forms

$$\Theta^{(a)} = \Theta^{(a)}{}_{\alpha}\,dx^{\alpha} \equiv h^{(a)}{}_{\alpha}\,dx^{\alpha}, a = 1, 2, 3,\;\; \alpha = \theta, \phi, \tau,$$

the metric is given as

$$ds^2 = g_{(a)(b)}\Theta^{(a)}\Theta^{(b)} = {\Theta^{(1)}}^2 + {\Theta^{(2)}}^2 - {\Theta^{(3)}}^2,$$

where the constant triad metric $\left(g_{(a)(b)}\right) = \mathrm{diag}(1, 1 - 1) = \left(g^{(a)(b)}\right)$.

For the timelike squashed metric (17.61) under consideration, the basis 1-forms are

$$\begin{aligned}\Theta^{(1)} &= \frac{3}{\sqrt{\mu^2 - 27\,\Lambda}}\,d\theta,\; \Theta^{(2)} = \frac{3\sinh\theta}{\sqrt{\mu^2 - 27\,\Lambda}}\,d\phi,\\ \Theta^{(3)} &= \frac{6\mu\cosh\theta}{\mu^2 - 27\,\Lambda}\,d\phi + \frac{6\mu}{\mu^2 - 27\,\Lambda}\,d\tau,\end{aligned} \tag{17.66}$$

therefore the matrix of the tensor $h^{(a)}{}_{\alpha}$, $\Theta^{(a)}{}_{\alpha} = h^{(a)}{}_{\alpha}\,dx^{\alpha}$, amounts to

$$\left(h^{(a)}{}_{\alpha}\right) = \begin{bmatrix} 0 & 3\,\frac{1}{\sqrt{\mu^2-27\,\Lambda}} & 0 \\ 3\,\frac{\sinh\theta}{\sqrt{\mu^2-27\,\Lambda}} & 0 & 0 \\ 6\,\frac{\mu\,\cosh\theta}{\mu^2-27\,\Lambda} & 0 & 6\,\frac{\mu}{\mu^2-27\,\Lambda} \end{bmatrix}, \tag{17.67}$$

with inverse matrix $h_{(a)}{}^{\alpha}$, $\partial_{(a)} = \partial_{(a)}{}^{\alpha}\frac{\partial}{\partial x^{\alpha}} = h_{(a)}{}^{\alpha}\frac{\partial}{\partial x^{\alpha}}$, given by

$$\left(h_{(a)}{}^{\alpha}\right) = \begin{bmatrix} 0 & \frac{\sqrt{\mu^2-27\,\Lambda}}{3} & 0 \\ \frac{\sqrt{\mu^2-27\,\Lambda}}{3\sinh\theta} & 0 & -\frac{\sqrt{\mu^2-27\,\Lambda}\cosh\theta}{3\sinh\theta} \\ 0 & 0 & \frac{\mu^2-27\,\Lambda}{6\mu} \end{bmatrix}. \tag{17.68}$$

These tensors are used to construct the triad tensor (constant) components. In the case of the Cotton tensor given by coordinate components (17.63), using

$$C^{(a)}{}_{(b)} = h^{(a)}{}_{\alpha} C^{\alpha}{}_{\beta}\, h_{(b)}{}^{\beta}, \tag{17.69}$$

one gets its triad components

$$\left(C^{(a)}{}_{(b)}\right) = \frac{\mu}{9}\left(9\,\Lambda + \mu^2\right)\begin{bmatrix} -1 & 0 & 0 \\ 0 & -1 & 0 \\ 0 & 0 & 2 \end{bmatrix}, \tag{17.70}$$

from which it becomes apparent which are the eigenvalues of the Cotton tensor, $\lambda_S = \frac{2}{9}\,\mu\,\left(\mu^2 + 9\,\Lambda\right) = -2\lambda_D$, and its Petrov type D character.

Accordingly, one may write the triad components of the Cotton tensor in a very simple way

$$C_{(a)(b)} = \lambda_D \delta^{(1)}_{(a)}\delta^{(1)}_{(b)} + \lambda_D \delta^{(2)}_{(a)}\delta^{(2)}_{(b)} - \lambda_S \delta^{(3)}_{(a)}\delta^{(3)}_{(b)}, \tag{17.71}$$

hence its coordinate components $C_{\alpha\beta} = C_{(a)(b)}\Theta^{(a)}{}_{\alpha}\Theta^{(b)}{}_{\beta}$ become

$$C_{\alpha\beta} = \lambda_D \Theta^{(1)}{}_{\alpha}\Theta^{(1)}{}_{\beta} + \lambda_D \Theta^{(2)}{}_{\alpha}\Theta^{(2)}{}_{\beta} - \lambda_S \Theta^{(3)}{}_{\alpha}\Theta^{(3)}{}_{\beta}. \tag{17.72}$$

Proceeding in the standard way in the search of the eigenvectors of the Cotton tensor certainly one should arrive at the same result.

17.5.3 Complex Extension Toward the Spacelike Squashed Metric

Accomplishing in the metric (17.61) the transformation

$$\cosh\theta = x \rightarrow \theta = \mathrm{arccosh} x$$

one gets the timelike metric in the form

$$ds^2_{tl} = \frac{9}{\mu^2 - 27\Lambda}\left[-\lambda^2(d\tau + x\, d\phi)^2 + \frac{dx^2}{x^2-1} + (x^2-1)d\phi^2\right]. \tag{17.73}$$

From this metric (17.73), by "complex transformations"

$$x = i\,y,\ \tau = i\,z,\ ,\phi = t, \tag{17.74}$$

one arrives at the spacelike biaxially squashed metric

$$ds^2_{sl} = \frac{9}{\mu^2 - 27\Lambda}\left[\lambda^2(dz + y\,dt)^2 + \frac{dy^2}{y^2+1} - (y^2+1)dt^2\right]. \tag{17.75}$$

Moreover, the transformation

$$y = \sinh\rho$$

brings the above metric to the spacelike biaxially squashed form

$$ds^2_{sl} = \frac{9}{\mu^2 - 27\Lambda}\left[\lambda^2(dz + \sinh\rho\, dt)^2 + d\rho^2 - \cosh^2\rho\, dt^2\right]. \tag{17.76}$$

From the complex point of view, these timelike and spacelike biaxially squashed metric structures can be considered as two real cuts of a complex metric. Therefore, from this perspective, any solution of a specific class, should it be timelike or spacelike, would have a counterpart. This procedure is well known in Maxwell electrodynamics: from electric solutions one gets their magnetic counterpart via complex relationships.

At this stage, it is worth noticing the real timelike cut arising from (17.76) by means of

$$z \to i\tau,\; t \to i\phi, \tag{17.77}$$

namely

$$ds^2_{tl} = \frac{9}{\mu^2 - 27\Lambda}\left[-\lambda^2(d\tau + \sinh\rho\, d\phi)^2 + d\rho^2 + \cosh^2\rho\, d\phi^2\right]. \tag{17.78}$$

Similarly, from (17.61), via

$$\tau \to iz,\; \phi \to it, \theta \to \rho \tag{17.79}$$

one gets another spacelike squashed metric representation

$$ds^2_{sl} = \frac{9}{\mu^2 - 27\Lambda}\left[\lambda^2(d\tau + \cosh\rho\, dt)^2 + d\rho^2 - \sinh^2\rho\, dt^2\right]. \tag{17.80}$$

17.5.4 Alternative Metric Representation of ${ds_{tl}}^2$

To present in "canonical" coordinates the timelike biaxially squashed metrics, even when this family is reducible to a single solution of TMG, there are given various forms of the spacelike $2D$ metric. Following Chow et al. (2010a), these representations are related to the coordinates used in the $2D$ "metric ds_2^2 of the hyperbolic space $\mathbb{H}^2$ with squared radius $L^2 = 9/(\mu^2 - 27\Lambda^2)$, $L > 0$, which is the upper leaf $X^0 \geq L$ of the hyperboloid

$$-(X^0)^2 + (X^1)^2 + (X^2)^2 = -L^2$$

in the $3D$ flat spacetime with metric

$$ds^2 = -(dX^0)^2 + (dX^1)^2 + (dX^2)^2.\text{''}\ \mathrm{CPSa(A.2)}.$$

There are several coordinate systems to be chosen:
For the choice CPSa(A.4):

$$X^0 = L\cosh\theta,\ X^1 = L\sinh\theta\,\cos\phi,\ X^1 = L\sinh\theta\,\sin\phi, \tag{17.81a}$$

one gets $\mathbb{H}^2$ in polar coordinates, CPSa(A.5):

$$ds_2{}^2 = L^2\left(d\theta^2 + \sinh^2\theta d\phi^2\right), \tag{17.81b}$$

which, by means of the transformation:

$$\cosh\theta = Ar^2 + B, \phi \to \phi/(2AL^2),$$

can be brought to $\mathbb{H}^2$ in polynomial representation CPSa(A.13),

$$ds_2{}^2 = \frac{4L^2A^2r^2}{(Ar^2+B)^2-1}dr^2 + \frac{(Ar^2+B)^2-1}{4L^2A^2}d\phi^2. \tag{17.81c}$$

For the choice CPSa(A.6):

$$X^0 = \frac{y}{2} + \frac{L^2+x^2}{2y},\ X^1 = L\frac{x}{y},\ X^2 = -\frac{y}{2} + \frac{L^2-x^2}{2y}, \tag{17.81d}$$

one gets $\mathbb{H}^2$ in Poincaré coordinates, CPSa(A.7):

$$ds_2{}^2 = \frac{L^2}{y^2}\left(dx^2 + dy^2\right), \tag{17.81e}$$

which, by means of the transformation:

$$x = \frac{\phi}{2AL^2},\ y = \frac{1}{(Ar^2+B)}$$

can be brought to the polynomial representation CPSa(A.14)

$$ds_2{}^2 = \frac{4L^2A^2r^2}{(Ar^2+B)^2}dr^2 + \frac{(Ar^2+B)^2}{4L^2A^2}d\phi^2. \tag{17.81f}$$

The choice CPSa(A.8):

$$X^0 = L\cosh\theta\cosh\phi,\ X^1 = L\sinh\theta,\ X^2 = L\cosh\theta\,\sinh\phi, \tag{17.81g}$$

leads to $\mathbb{H}^2$ in polar coordinates CPSa(A.9):

$$ds_2{}^2 = L^2\left(d\theta^2 + \cosh^2\theta d\phi^2\right), \tag{17.81h}$$

which, by means of the transformation:

$$\sinh\theta = Ar^2 + B, \phi \to \phi/(2AL^2),$$

can be brought to the polynomial representation CPSa(A.15)

$$ds_2{}^2 = \frac{4L^2A^2r^2}{(Ar^2+B)^2+1}dr^2 + \frac{(Ar^2+B)^2+1}{4L^2A^2}d\phi^2. \tag{17.81i}$$

In Chow et al. (2010a), to facilitate comparison with the solutions reported in the literature, the timelike squashed solution is represented in various alternative metric forms, correspondingly CPSa(A.10), CPSa(A.11), and CPSa(A.12) for

$$\lambda^2 := \frac{4\mu^2}{\mu^2 - 27\Lambda}, \tag{17.82a}$$

explicitly

$$ds_{tl}^2 = \frac{9}{\mu^2 - 27\Lambda}\left[-\lambda^2(d\tau + \cosh\theta\, d\phi)^2 + d\theta^2 + \sinh^2\theta d\phi^2\right], \tag{17.82b}$$

$$ds_{tl}^2 = \frac{9}{\mu^2 - 27\Lambda}\left[-\lambda^2(d\tau + \frac{dx}{y})^2 + \frac{dx^2 + dy^2}{y^2}\right], \tag{17.82c}$$

$$ds_{tl}^2 = \frac{9}{\mu^2 - 27\Lambda}\left[-\lambda^2(d\tau + \sinh\theta\, d\phi)^2 + d\theta^2 + \cosh^2\theta d\phi^2\right]. \tag{17.82d}$$

Another possibility is achieved by transformations of the type cosh $= Ar^2 + B$, which yields CPSa(A.16), namely

$$ds_{tl}^2 = -(dt - \frac{\mu}{3}r^2\, d\phi)^2 + \frac{r^2}{F(r)}dr^2 + F(r)d\phi^2,$$

$$F(r) = \frac{1}{36}(\mu^2 - 27\Lambda)r^4 + k_1 r^2 + k_0. \tag{17.82e}$$

In the forthcoming subsections the identification of the solutions reported by different authors is carried out following the pattern given in Chow et al. (2010a), citing their equation for reference.

Vuorio Solution; $\Lambda = 0$

Vuorio (1985) reported the first stationary rotationally symmetric solution (2.21), CPSa(A.25), to the E-TMG equations, namely:

$$ds^2 = \frac{9}{\mu^2}\left[-(dt + 2(1 - \cosh\sigma)\, d\theta)^2 + d\sigma^2 + \sinh^2\sigma\, d\theta^2\right]. \tag{17.83}$$

Carrying out the coordinates transformations and constants redefinitions

$$t \to 2\tau + 2\phi,\ \theta \to -\phi,\ \sigma \to \theta,\ \Lambda \to 0,$$

one brings the above solution to CPSa(A.10) with $\Lambda = 0$.

Percacci–Sodano–Vuorio Solution; $\Lambda = 0$

Percacci et al. (1987) considered stationary solutions possessing a timelike Killing vector with constant scalar twist; their solution (3.20), CPSa(A.26), is given by

$$ds^2 = -3\left[dx^2 + \frac{2}{3}e^{(\mu x^1/3)}dx^0\right]^2 + (dx^1)^2 + \frac{1}{3}e^{(2\mu x^1/3)}(dx^0)^2. \tag{17.84a}$$

Doing the coordinates transformations

$$\frac{\mu}{2\sqrt{3}}x^2 \to \tau,\ \frac{\mu}{3\sqrt{3}}x^0 \to x,\ \exp\left(-\frac{\mu}{3}x^1\right) \to y, \tag{17.84b}$$

one arrives at CPSa(A.11) with $\Lambda = 0$.

Nutku–Baekler and Ortiz BT III Timelike Solutions; $\Lambda = 0$

Nutku and Baekler (1989), and Ortiz (1990) determined solutions formed from left-invariant 1-forms of Bianchi type III spaces. A detailed derivation of them is given here in 17.4, 17.4.1. The metric of these solutions – (4.10) of Nutku and Baekler (1989) and (type(a) with $a \neq 0$) of Ortiz (1990) – is given by (17.47b); it can be reduced to CPSa(A.11) with $\Lambda = 0$.

Clément Solutions; $\Lambda = 0$

Clément (1992a), and Clément (1992b) too, considered stationary rotationally symmetric solutions. His solution (4.4), CPSa(A.30), is given by the metric

$$ds^2 = -\left[dt + 2c d\theta - 2c\cosh(\frac{\mu}{3}r)d\theta\right]^2 + dr^2 + c^2\,\sinh^2(\frac{\mu}{3}r)\,d\theta^2. \tag{17.85a}$$

Making the coordinates transformations and constants redefinitions

$$\frac{\mu}{6}(t + 2c\,\theta) \to \tau,\ \frac{\mu}{3}r \to \theta',\ -\frac{\mu\,c}{3}\theta \to \phi, \tag{17.85b}$$

dropping prime, one arrives at (17.82b), CPSaA(24), with $\Lambda = 0$.

The Clément's solution (4.5), CPSa(A.31),

$$ds^2 = -\left[dt - 2c\sinh(\frac{\mu}{3}r)d\theta\right]^2 + dr^2 + c^2\,\cosh^2(\frac{\mu}{3}r)\,d\theta^2, \tag{17.86a}$$

subjected to the coordinates transformations

$$\frac{\mu}{6}t \to \tau,\ \frac{\mu}{3}r \to \theta',\ -\frac{\mu\,c}{3}\theta \to \phi, \tag{17.86b}$$

dropping prime, gives rise to (17.82d), CPSa (A.12), for $\Lambda = 0$.

On the other hand, the solution (4.7) by Clément (1992a) has various possible choices of signs:

Two choices give

$$ds^2 = -\frac{c}{d}\left[dt \mp 2d\,\mathrm{e}^{\pm\mu\,r/3}d\theta\right]^2 + dr^2 + c\,d\,\mathrm{e}^{\pm 2\mu\,r/3}\,d\theta^2, \tag{17.87a}$$

which is CPSa(A.32). Subjecting (17.87a) to the coordinates transformations CPSa(A.32)

$$\sqrt{\frac{c}{d}}t \to \tau,\ \mp\frac{\mu}{3}\sqrt{c}\,d\theta \to x,\ \mathrm{e}^{\mp\mu\,r/3} \to y, \tag{17.87b}$$

one arrives at (17.82c), CPSa(A.11), with $\Lambda = 0$.

Another two choices of signs lead to

$$ds^2 = -3\,c\,d\left[d\theta \pm \frac{2}{3d}\,e^{\pm\mu\,r/3}dt\right]^2 + dr^2 + \frac{c}{3d}\,e^{\pm 2\mu\,r/3}\,dt^2, \tag{17.88a}$$

see CPSa(A.33). Making the coordinates transformations

$$\sqrt{3c\,d}\,\theta \to \tau,\ \pm\frac{\mu}{3}\sqrt{\frac{c}{3d}}\,t \to x,\ e^{\mp\mu\,r/3} \to y, \tag{17.88b}$$

one arrives at (17.82c), CPSa(A.11), with $\Lambda = 0$.

Nutku Solutions

Nutku (1993) published a black hole solution (25), CPSa(A.21), to the E–TMG equations given by

$$ds^2 = -\frac{2J-M}{6}\left[dt - \frac{2\mu\,r^2 - 3J/\mu}{2J-M}d\theta\right]^2 + \frac{r^2}{F(r)}dr^2 + \frac{6}{2J-M}\,F(r)d\theta^2,$$
$$F(r) = \frac{1}{36}(\mu^2 - 27\Lambda)r^4 - \frac{1}{6}M\,r^2 + \frac{1}{4}\frac{J^2}{\mu^2}. \tag{17.89a}$$

Making the coordinates transformations CPSa(A.22)

$$\sqrt{\frac{2J-M}{6}}t + \sqrt{\frac{3}{2}}\frac{1}{\sqrt{2J-M}}\frac{J}{\mu}\theta \to t',\ \frac{\sqrt{6}}{\sqrt{2J-M}}\theta \to \phi,$$
$$-\frac{M}{6} \to k_1,\ \frac{1}{4}\frac{J^2}{\mu^2} \to k_0, \tag{17.89b}$$

dropping prime, one arrives at the solution (17.82e), CPSa(A.16).

Moreover, Nutku generalized the Λ zero Vuorio (1985) solution (see 17.5.4) to include the cosmological constant; the Nutku (1993) solution (3.11), after relabelling $\psi_N \to \tau$ corresponds to (17.82d), CPSa(A.12). On the other hand, Nutku succeeded in deriving his solution (3.17), which corresponds to (17.82e), CPSa(A.16), for $k_1 = 1$ and $k_0 = 0$ and $\theta_N \to \phi$.

Gürses "Gödel Type" Solution

Gürses (2008) considered "solutions that are of Gödel type, which are defined by his equations: (42), (43), (44), (46), and part of the unlabelled equation following (39)," as discussed, (A.17), in Chow et al. (2010a), namely

$$ds^2 = -\left[\sqrt{a_0}dt + u_2(r,\theta)d\theta + u_1(r,\theta)dr\right]^2 + \frac{e_0^2}{a_0}r^2\psi(r)d\theta^2 + \frac{1}{\psi(r)}dr^2,$$
$$\psi(r) = \frac{\mu^2 - 27\Lambda}{36}r^2 + b_0 + \frac{b_1}{r^2}. \tag{17.90a}$$

CPSa(A.18), where u_1 and u_2 are constrained by the equation CPSa(A.19)

$$\frac{\partial u_1}{\partial\theta} = \frac{\partial u_2}{\partial r} + \frac{2}{3}\frac{\mu\,e_0}{\sqrt{a_0}}r, \tag{17.90b}$$

and hence of the form CPSa(A.19),

$$u_1 = \frac{\partial}{\partial r}U(r,\theta), u_2 = -\frac{1}{3}\frac{\mu\, e_0}{\sqrt{a_0}}r^2 + \frac{\partial}{\partial \theta}U(r,\theta). \tag{17.90c}$$

Doing the coordinates transformations and constants redefinitions

$$\sqrt{a_0}\,t + U(r,\theta) \to t', \ \frac{e_0}{\sqrt{a_0}}\theta \to \phi,$$
$$b_0 \to k_1, \ b_1 \to k_0, \tag{17.90d}$$

dropping prime, one arrives at the solution (17.82e), CPSa(A.16).

The Gürses solution (17.91a) below is an special case of the above solution (17.90) for $u_1 = 0$, $u_2 = -\frac{1}{\sqrt{a_0}}(c_0 + \frac{1}{3}e_0\,\mu\, r^2)$, $c_0 =$ constant.

Gürses Solution

In the work by Gürses (1994) dealing with a perfect fluid (5) of constant pressure and density as solution of the Einstein plus Λ equations, its interpretation is established as a vacuum TMG plus Λ solution given by CPSa(A.18):

$$ds^2 = -\left[\sqrt{a_0}dt - \frac{3c_0 + \mu\, e_0\, r^2}{3\sqrt{a_0}}d\theta\right]^2 + \frac{e_0^2}{a_0}r^2\psi(r)d\theta^2 + \frac{1}{\psi(r)}dr^2,$$
$$\psi(r) = \frac{\mu^2 - 27\Lambda}{36}r^2 + b_0 + \frac{b_1}{r^2}. \tag{17.91a}$$

Accomplishing the coordinates transformations and constants redefinitions

$$\sqrt{a_0}\,t - \frac{c_0}{\sqrt{a_0}}\,\theta \to t', \ \frac{e_0}{\sqrt{a_0}}\theta \to \phi, b_0 \to k_1, \ b_1 \to k_0, \tag{17.91b}$$

dropping prime, one arrives at the solution (17.82e), CPSa(A.16).

Clément Solution

Clément (1994) considered a "Killing symmetry reduction procedure to obtain stationary rotationally symmetric solutions". Clément's solution (18), CPSa(A.23), is given as

$$ds^2 = -\left[\sqrt{2\,a}dt + \frac{1}{\sqrt{2\,a}}\left(\frac{3b}{\mu} - \frac{2\,\mu}{3}\rho\right)d\theta\right]^2 + \frac{d\rho^2}{F(r)} + \frac{1}{2\,a}F(\rho)\,d\theta^2,$$
$$F(\rho) = \frac{1}{9}(\mu^2 - 27\Lambda)\rho^2 + 4(a-b)\,\rho + 9\frac{b^2}{\mu^2}. \tag{17.92a}$$

Accomplishing ing the coordinates transformations and constants redefinitions CPSa(A.24)

$$\sqrt{2\,a}\,t + \frac{3b}{\mu\sqrt{2\,a}}\theta \to t', \ \rho \to r^2, \ \frac{\sqrt{2}}{\sqrt{a}}\theta \to \phi, a-b \to k_1, \ \frac{9\,b^2}{4\,\mu^2} \to k_0, \tag{17.92b}$$

dropping prime, one arrives at (17.82e), CPSa(A.16).

Anninos–Li–Padi–Song–Strominger Solution

Anninos et al. (2009) reported "warped" AdS_3 black hole solutions of TMG; their solution (3.4), after accomplishing the redefinitions $\sigma \to \theta$, $u \to \phi$, becomes the solution (17.82d), CPSa(A.12).

The solutions reported in this subsection are gathered in the following table:

Table 17.5.1 *Timelike biaxially squashed solutions*

References	Secs. or (Eqs.)	Eqs. in Ref.	Rel. to CPSa	Λ
Vuorio (1985)	(17.83)	(2.21)	(A.25)	–
Percacci et al. (1987)	17.84	(3.20)	(A.26)	–
Nutku and Baekler (1989)	17.4.1	(4.10)	(A.27)	–
and Ortiz (1990)	(17.47b)		(A.28)	–
Clément (1992a), (1992b)	(17.85)	(4.4–7)	(A.30–33)	–
Nutku (1993)	(17.89)	(25)	(A.21)	$\surd$
Gürses (2008)	(17.90)	(42–46)	(A.17)	$\surd$
Gürses (1994)	(17.91)	(5)	(A.18)	$\surd$
Clément (1994)	(17.92)	(18)	(A.23)	$\surd$
Anninos et al. (2009)	(17.82d)	(3.4)	(A.12)	$\surd$
CPSa: Chow et al. (2010a)	17.5.4		(A.10–15)	$\surd$

17.6 Spacelike Biaxially Squashed Metrics

In this paragraph the integration of the Bianchi spacelike biaxially squashed solutions is accomplished, starting from the middle-of-the-road metric element

$$ds^2 = \frac{\lambda_1}{4}(dz + \sinh\rho\, d\tau)^2 + \frac{\lambda_2}{4}(d\rho^2 - \cosh^2\rho\, d\tau^2), \tag{17.93}$$

and searching for solutions to the Einstein-TMG equations. The trace of the field equation for TMG in vacuum reads

$$E^{\mu}{}_{\mu} = \frac{4\,\lambda_2 + 3\,\Lambda\,{\lambda_2}^2 - \lambda_1}{{\lambda_2}^2} = 0 \to \lambda_1 = 4\,\lambda_2 + 3\,\Lambda\,{\lambda_2}^2, \tag{17.94a}$$

the equation component

$$E^{z}{}_{\tau} = \lambda_1\left(3\sqrt{\lambda_1} - \mu\,\lambda_2\right)(-\lambda_2 + \lambda_1) = 0 \to \lambda_1 = 1/9\,\mu^2{\lambda_2}^2, \tag{17.94b}$$

hence one arrives from (17.94a) at

$$1/9\,\mu^2{\lambda_2}^2 = 4\,\lambda_2 + 3\,\Lambda\,{\lambda_2}^2 = 0 \to \lambda_2 = \frac{36}{\mu^2 - 27\,\Lambda}, \lambda_1 = \frac{144\,\mu^2}{(\mu^2 - 27\,\Lambda)^2}. \tag{17.94c}$$

The derived solution can be given by the metric in coordinates $x^\alpha = \{z, \rho, \tau\}$

$$ds^2 = \frac{36\,\mu^2}{(\mu^2 - 27\,\Lambda)^2}(dz + \sinh\rho\, d\tau)^2 + \frac{9}{(\mu^2 - 27\,\Lambda)}(d\rho^2 - \cosh^2\rho\, d\tau^2), \tag{17.95}$$

and characterized by the Cotton tensor

$$C^{\alpha}{}_{\beta} = \frac{1}{9}\,\mu\,(\mu^2 + 9\,\Lambda)\,[2\delta^{\alpha}{}_{z}\delta^{z}{}_{\beta} - \delta^{\alpha}{}_{\rho}\delta^{\rho}{}_{\beta} - \delta^{\alpha}{}_{\tau}\delta^{\tau}{}_{\beta} + 3\,\sinh\rho\delta^{\alpha}{}_{z}\delta^{\tau}{}_{\beta}]\,, \tag{17.96}$$

and a traceless Ricci tensor $\kappa\, S^{\alpha}{}_{\beta} = R^{\alpha}{}_{\beta} - \frac{R}{3}\delta^{\alpha}{}_{\beta}$, which can be given also as

$$S^{\alpha}_{\beta} = \frac{\mu^2 + 9\Lambda}{9}(\delta^{\alpha}{}_{\beta} - 3\,K^{\alpha}\,K_{\beta}),\; K^{\alpha} = \pm\frac{1}{6}\frac{\mu^2 - 27\,\Lambda}{\mu}\,\delta^{\alpha}{}_{z}, \tag{17.97}$$

where K^{α} is a spacelike unit vector.

17.6.1 Eigenvectors of the Cotton Tensor; Triad Formulation

As in the previous timelike case, one can represent the studied solution by means of the triad formalism, in which the metric is given by

$$ds^2 = g_{(a)(b)}\Theta^{(a)}\Theta^{(b)} = {\Theta^{(1)}}^2 + {\Theta^{(2)}}^2 - {\Theta^{(3)}}^2 \tag{17.98}$$

where the orthonormal basis triad $\Theta^{(a)} = h^{(a)}{}_{\alpha}\,dx^{\alpha}$ is defined as

$$\begin{aligned}
\Theta^{(1)}{}_{\alpha} &= [\frac{6\,\mu}{\mu^2 - 27\,\Lambda}, 0, \frac{6\,\mu\,\sinh(\rho)}{\mu^2 - 27\,\Lambda}],\ \text{spacelike},\\
\Theta^{(2)}{}_{\alpha} &= [0, 3\,\frac{1}{\sqrt{\mu^2 - 27\,\Lambda}}, 0],\ \text{spacelike},\\
\Theta^{(3)}{}_{\alpha} &= [0, 0, 3\,\frac{\cosh(\rho)}{\sqrt{\mu^2 - 27\,\Lambda}}],\ \text{timelike},
\end{aligned} \tag{17.99}$$

hence, the triad tensor components $h^{(a)}{}_{\alpha}$ can be represented by the matrix

$$\left(h^{(a)}{}_{\beta}\right) = \begin{bmatrix} 6\,\frac{\mu}{\mu^2-27\,\Lambda} & 0 & 6\,\frac{\mu\,\sinh(\rho)}{\mu^2-27\,\Lambda} \\ 0 & 3\,\frac{1}{\sqrt{\mu^2-27\,\Lambda}} & 0 \\ 0 & 0 & 3\,\frac{\cosh(\rho)}{\sqrt{\mu^2-27\,\Lambda}} \end{bmatrix}. \tag{17.100}$$

The directional derivative vectors are:

$$\begin{aligned}
\partial_{(a)} &= \partial_{(a)}{}^{\alpha}\frac{\partial}{\partial\, x^{\alpha}} = h_{(a)}{}^{\alpha}\frac{\partial}{\partial\, x^{\alpha}},\\
\partial_{(1)}{}^{\alpha} &= [\frac{\mu^2 - 27\,\Lambda}{6\,\mu}, 0, 3\,\frac{\cosh(\rho)}{\sqrt{\mu^2 - 27\,\Lambda}}],\\
\partial_{(2)}{}^{\alpha} &= [0, \frac{1}{3}\,\sqrt{\mu^2 - 27\,\Lambda}, 0],\\
\partial_{(3)}{}^{\alpha} &= [-\frac{\sinh(\rho)\,\sqrt{\mu^2 - 27\,\Lambda}}{3\cosh(\rho)}, 0, \frac{\sqrt{\mu^2 - 27\,\Lambda}}{3\cosh(\rho)}],
\end{aligned} \tag{17.101}$$

thus, their components can be gathered in the matrix form

$$\left(h_{(a)}{}^{\alpha}\right) = \frac{1}{3}\sqrt{\mu^2 - 27\,\Lambda}\begin{bmatrix} \frac{1}{2}\frac{\sqrt{\mu^2-27\,\Lambda}}{\mu} & 0 & 0 \\ 0 & 1 & 0 \\ -\frac{\sinh(\rho)}{\cosh(\rho)} & 0 & \frac{1}{\cosh(\rho)} \end{bmatrix}. \tag{17.102}$$

From the coordinate components of the Cotton tensor (17.96), via

$$C^{(a)}{}_{(b)} = h^{(a)}{}_{\alpha} C^{\alpha}{}_{\beta}\, h_{(b)}{}^{\beta},$$

one determines its triad components

$$C^{(a)}{}_{(b)} = \frac{1}{9}\,\mu\,\left(\mu^2 + 9\,\Lambda\right)\left[2\delta^{(a)}{}_{(1)}\delta^{(1)}{}_{(b)} - \delta^{(a)}{}_{(2)}\delta^{(2)}{}_{(b)} - \delta^{(a)}{}_{(3)}\delta^{(3)}{}_{(b)}\right], \tag{17.103}$$

and reciprocally the Cotton tensor coordinate components are

$$\begin{aligned} C_{\alpha\beta} &= C_{(a)(b)}\Theta^{(a)}{}_{\alpha}\Theta^{(b)}{}_{\beta} = \lambda_s\Theta^{(1)}{}_{\alpha}\Theta^{(1)}{}_{\beta} + \lambda_d\Theta^{(2)}{}_{\alpha}\Theta^{(2)}{}_{\beta} - \lambda_d\Theta^{(3)}{}_{\alpha}\Theta^{(3)}{}_{\beta}, \\ \lambda_s &= 2\frac{\mu}{9}\left(\mu^2 + 9\,\Lambda\right),\ \lambda_d = -\frac{\mu}{9}\left(\mu^2 + 9\,\Lambda\right). \end{aligned} \tag{17.104}$$

17.6.2 Alternative Metric Representation of ds^2_{sl}

These representations are related to the coordinates used in the $2D$ "metric ds^2_2 of AdS_2 space, with squared radius

$$L^2 = 9/(\mu^2 - 27\Lambda^2),\ L > 0.$$

$2D$ AdS_2 with AdS radius $L > 0$ is the hyperboloid

$$-(X^0)^2 + (X^1)^2 - (X^2)^2 = -L^2$$

in the flat $3D$ spacetime with metric

$$ds^2 = -(dX^0)^2 + (dX^1)^2 - (dX^2)^2.\text{''}\ \text{CPSa(A.36)}.$$

There are several coordinate systems to be chosen:

For the choice, CPSa(A.37):

$$X^0 = L\cosh\rho\cos\tau,\ X^1 = L\sinh\rho\,, X^2 = L\cosh\rho\,\sin\tau. \tag{17.105a}$$

one gets ${ds_2}^2$ in polar coordinates, CPSa(A.38):

$${ds_2}^2 = L^2\left(d\rho^2 - \cosh^2\rho\, d\tau^2\right). \tag{17.105b}$$

For the choice CPSa(A.39):

$$X^0 = L\frac{t}{\rho},\ X^1 = \frac{L^2}{2\rho} - \frac{\rho^2 - t^2}{2\rho},\ X^2 = \frac{L^2}{2\rho} + \frac{\rho^2 - t^2}{2\rho}. \tag{17.105c}$$

one gets ${ds_2}^2$ in conformally flat coordinates that cover the Poincaré patch CPSa(A.40):

$$ {ds_2}^2 = \frac{L^2}{\rho^2}\left(d\rho^2 - dt^2\right). \tag{17.105d} $$

The choice CPSa(A.41):

$$ X^0 = L\cosh\rho\,,\; X^1 = L\sinh\rho\cosh\tau,\; X^2 = L\sinh\rho\sinh\tau. \tag{17.105e} $$

leads to ${ds_2}^2$ in coordinates that cover $X^0 \geq L$, CPSa(A.42):

$$ {ds_2}^2 = L^2\left(d\rho^2 - \sinh^2\rho d\tau^2\right). \tag{17.105f} $$

The choice CPSa(A.43):

$$ X^0 = L\sin\tau,\; X^1 = L\cos\tau\sinh\phi,\; X^2 = L\cos\tau\,\cosh\phi, \tag{17.105g} $$

gives ${ds_2}^2$ in coordinates that cover only $-L \leq X^0 \leq L$, CPSa(A.44):

$$ {ds_2}^2 = L^2\left(\cos^2\tau d\phi^2 - d\tau^2\right). \tag{17.105h} $$

To facilitate comparison with the solutions reported in the literature, the space-like squashed (SLS) solution, following Chow et al. (2010a), is presented in various alternative forms with minor changes from our side, namely, restoring the negative cosmological constant Λ and the λ constant, explicitly

$$ m^2 = -\Lambda,\; \lambda^2 = \frac{4\mu^2}{\mu^2 - 27\Lambda}. \tag{17.106} $$

These solutions, corresponding respectively to CPSa(A.45), CPSa(A.47), and CPSa(A.46), are given by

$$ ds^2_{sl} = \frac{9}{\mu^2 - 27\Lambda}\left[\lambda^2(dz + \sinh\rho\,d\tau)^2 + d\rho^2 - \cosh^2\rho d\tau^2\right], \tag{17.107a} $$

$$ ds^2_{sl} = \frac{9}{\mu^2 - 27\Lambda}\left[\lambda^2(dz + \cosh\rho\,d\tau)^2 + d\rho^2 - \sinh^2\rho d\tau^2\right], \tag{17.107b} $$

$$ ds^2_{sl} = \frac{9}{\mu^2 - 27\Lambda}\left[\lambda^2(dz + \frac{dt}{x})^2 + \frac{-dt^2 + dx^2}{x^2}\right], \tag{17.107c} $$

and the time-dependent real cut of the complex extension of (17.107a), $\rho \to it, \tau \to -i\phi$, which yields

$$ ds^2_{sl} = \frac{9}{\mu^2 - 27\Lambda}\left[\lambda^2(dz + \sin t\,d\phi)^2 - dt^2 + \cos^2 t d\phi^2\right], \tag{17.107d} $$

comparable with CPSa(A.48).

Hall–Morgan–Perjés SLS Solution; $\Lambda = 0$

Hall et al. (1987) searched for solutions of TMG of a particular Petrov–Segré type; type D to be specific. Their solution (61), CPSa(A.51), is

$$ds^2 = 2du\,dv - \frac{\mu^2}{9}\,v^2\,du^2 + \left(dy - \frac{2}{3}\mu\,v\,du\right)^2 + 2\,v\,f(u)\,du^2. \tag{17.108a}$$

Accomplishing the coordinate transformations

$$u' = \int \mathrm{e}^{-F(u)}\,du,\; v' = \mathrm{e}^{F(u)}\,v,\;\; F(u) := \int f(u)du, \tag{17.108b}$$

one sees that $f(u)$ is redundant, and one can equate it to zero, $f(u) \to 0$. Under such a choice, the second coordinate transformations

$$\frac{\mu^2}{18}u' \to \hat{u},\; \frac{1}{v'} + \frac{\mu^2}{18}u' \to \hat{v} \tag{17.108c}$$

bring the metric to the form CPSa(A.52)

$$ds^2 = -\frac{36}{\mu^2}\frac{d\hat{u}\,d\hat{v}}{(\hat{u}-\hat{v})^2} + \left(dy + \frac{12}{\mu}\frac{d\hat{u}}{\hat{u}-\hat{v}}\right)^2. \tag{17.108d}$$

Finally, the coordinate changes

$$\frac{1}{2}(\hat{u}+\hat{v}) \to t,\; \frac{1}{2}(\hat{v}-\hat{u}) \to x,\; \frac{\mu}{6}y + \log\left|\frac{1}{2}(\hat{v}-\hat{u})\right| \to -z, \tag{17.108e}$$

leads to (17.107c), CPSa(A.46), with $\Lambda = 0$.

Nutku–Baekler and Ortiz BT III Spacelike Solutions; $\Lambda = 0$

Nutku and Baekler (1989) and independently Ortiz (1990) considered solutions of Bianchi type III formed from left-invariant 1-forms without a cosmological constant. The derivation of these solutions is given in 17.4, and 17.4.2, while the metric is explicitly reported in (17.48b), it can be reduced to (17.107c), CPSa(A.46), with $\Lambda = 0$.

Moussa–Clément–Leygnac SLS Solution; $\Lambda = 0$

Moussa et al. (2003) obtained black hole solutions by analytically continuing the Vuorio (1985) solution, 17.5.4. Their solution (4), CPSa(A.58), is given by

$$ds^2 = \frac{9}{\mu^2}\left[-\frac{\rho^2-\rho_0^2}{3}\,d\phi^2 + \frac{d\rho^2}{\rho^2-\rho_0^2} + 3\left(dt - (\omega + \frac{2}{3}\rho)\,d\phi\right)^2\right]. \tag{17.109a}$$

Making the coordinates transformations

$$\frac{\rho_0}{\sqrt{3}}\phi \to \tau,\; \frac{\rho}{\rho_0} \to \cosh\rho',\; \frac{\sqrt{3}}{2}(\omega\,\phi - t) \to z, \tag{17.109b}$$

dropping prime, one arrives at (17.107b), CPSa(A.47), with $\Lambda = 0$; $\omega =$ constant.

Bouchareb–Clément SLS Solution

Bouchareb and Clément (2007) reported their black hole solution (4.1), CPSa(A.49),

$$ds^2 = -\frac{\mu^2 - 27\Lambda}{3(\mu^2 + 9\Lambda)}(\rho^2 - \rho_0^2)d\phi^2 + \frac{9}{\mu^2 - 27\Lambda}\frac{d\rho^2}{\rho^2 - \rho_0^2}$$
$$+ 3\frac{\mu^2 + 9\Lambda}{4\mu^2}\left[dt - \left(\frac{1}{3}\frac{4\mu^2}{\mu^2 + 9\Lambda}\rho + \omega\right)d\phi\right]^2. \qquad (17.110a)$$

Making the coordinates transformations

$$\frac{\rho_0}{3\sqrt{3}}\frac{\mu^2 - 27\Lambda}{\sqrt{\mu^2 + 9\Lambda}}\phi \to \tau,\ \frac{\rho}{\rho_0} \to \cosh\rho',\ \frac{\sqrt{\mu^2 + 9\Lambda}}{\sqrt{3}}\frac{\mu^2 - 27\Lambda}{4\mu^2}(\omega\,\phi - t) \to z, \qquad (17.110b)$$

dropping prime, one arrives at (17.107b), CPSa(A.47); $\omega =$ constant.

Anninos–Li–Padi–Song–Strominger SLS Solution

Anninos et al. (2009) consider "warped" AdS_3 black hole solutions of TMG. After making the coordinate relabellings $\sigma \to \rho$, $u \to z$, their solution (3.3) becomes (17.107a), CPSa(A.45).

All the above-reported solutions can be gathered for quick reference in the form of a table:

Table 17.6.1 *Spacelike biaxilly squashed solutions*

References	Secs. or (Eqs.)	Eqs. in Ref.	Rel. to CPSa	Λ
Hall et al. (1987)	(17.108)	(61)	(A.51)	–
Nutku and Baekler (1989)	17.4.2		(A.53)	–
and Ortiz (1990)	17.6.2		(A.56)	
Moussa et al. (2003)	(17.109)	(4)	(A.58)	–
Bouchareb and Clément (2007)	(17.110)	(4.1)	(A.49)	√
Anninos et al. (2009)	17.6.2	(3.3)	(A.45)	√
CPSa:Chow et al. (2010b)	(17.107)		(A.34–48)	√

18

Petrov Type N Wave Metrics

18.1 Brinkmann-Like 3D Metric

Solutions with a covariantly constant null vector field k_a (which is their invariant characterization), $k_a k^a = 0$,

$$k_{a;b} = 0 \rightarrow k_{a;b}\, k^b = 0,\; R_{ab}\, k^a\, k^b = 0, \tag{18.1}$$

are called plane-fronted gravitational waves with parallel propagated null rays and denoted as *pp*-waves. They have been known in 4D Einstein relativity since long ago; Brinkmann (1923) discovered them in 1923; see Kramer et al. (1980), §21.5. The condition on the Ricci tensor point on the existence of solutions for null electromagnetic fields, radiation and vacuum, excluding other field solutions, among them dS–AdS cosmological constant solutions. This fact establishes a difference with the 3D wave solutions in TMG.

To get an insight of how the wave solutions with a non-vanishing cosmological constant arise in the framework of the TMG, let us consider the Brinkmann-like 3D Killing wave metric

$$ds^2 = d\rho^2 + 2F(\rho, u)dudv + H(\rho, u)du^2. \tag{18.2}$$

One chooses the null vector field k_a in the direction ∂_v, thus

$$k^\mu = \delta^\mu{}_v,\; k_\mu = \delta^u{}_\mu F(\rho, u),\; k^\mu\, k_\mu = 0. \tag{18.3}$$

Its covariant differential is

$$k_{\mu;\nu} = (\delta^u{}_\mu \delta^\rho{}_\nu - \delta^\rho{}_\mu \delta^u{}_\nu)\frac{\partial F(\rho, u)}{\partial \rho}, \tag{18.4}$$

consequently, the congruence is geodesic and Killingian one, but it is not covariantly constant to shear the name of pp-wave; this takes place only for the structural function

$$F(\rho, u) = f(u) \rightarrow 1, \tag{18.5}$$

but if this were the case, the metric could not allow for a cosmological constant even in TMG; this subclass of solutions with vanishing cosmological constant can properly be named *pp-wave solutions*; see Section 18.3 for specific solutions.

For the class of wave solutions with a Killingian null congruence k, such that $k_{\mu;\nu} \neq 0$, (18.4), all the scalar invariants representing respectively the twist, shear, and expansion:

$$\omega^2 = k^{\mu;\nu} k_{[\mu;\nu]}, \quad \sigma^2 = k^{\mu;\nu} k_{\mu;\nu} - (k^\mu{}_{;\mu})^2, \quad \theta = \frac{1}{2} k^\mu{}_{;\mu}, \tag{18.6}$$

vanish; these wave solutions can be called *type N Killing waves* and denoted as *TN-waves* instead of pp-waves. This class of TN-solutions is different from type N Kundt family of metrics, arising as limit of type III Kundt metric, to be treated in detail in Chapter 19; it is worth pointing out that the null vector for this Kundt class is a non-Killingian one; thus this class of wave solutions with a shear-free, twist-free, expansionless, and non-Killingian null congruence can be called *type N non-Killingian waves* and denoted *TNnK-wave solutions*. Having made these remarks, we can proceed with the search for the corresponding solutions.

18.2 AdS_3 Non-Covariantly Constant TN-Waves

To be consistent with the comments stressed above, we first integrate the cosmological case for the metric (18.2), and vector field (18.3). The $E^\rho{}_\rho$ component of the TMG equations is

$$\left(\frac{\partial}{\partial \rho} F(\rho, u)\right)^2 + 4\,\Lambda\, F^2 = 0 \tag{18.7}$$

because of the nonlinearity, the equation splits in two branches of AdS solutions, $\Lambda = -m^2$:

$$F(\rho) = F_0(u)\, \mathrm{e}^{\pm 2\, m\, \rho} \to F(\rho) = \mathrm{e}^{2\, m\, \rho}. \tag{18.8}$$

The factor $F_0(u)$ can be set equal to the unit, $F_0(u) \to 1$, by a transformation of the coordinate u, moreover, it is enough to treat the case with m, the branch of solutions with $-m$ is derivable from the first one simply by replacing there $m \to -m$. The $E^v{}_v$ component yields a dependent equation, which, for the solutions of (18.8), identically holds.

The only remaining equation is the $E^v{}_u$ one, which for the $F(\rho)$ solution (18.8) becomes

$$\frac{\partial^3}{\partial \rho^3} H(\rho, u) + (\mu - 3\, m)\, \frac{\partial^2}{\partial \rho^2} H(\rho, u) - 2\, m\, (\mu - m)\, \frac{\partial}{\partial \rho} H(\rho, u) = 0. \tag{18.9}$$

There arise three different cases depending on the $\mu(m)$ relation.

Case $\mu \neq \pm m$:
The general solution of (18.9) for $H(\rho, u)$ is given by

$$H(\rho, u) = f_1(u) + f_2(u)\,e^{2m\rho} + f_3(u)\,e^{(-\mu+m)\rho}, \tag{18.10}$$

which is characterized by the Cotton tensor with components

$$C_{\alpha\beta} = \frac{1}{2}\mu\left(\mu^2 - m^2\right) e^{-(\mu-m)\rho} f_3(u)\,\delta^u{}_\alpha \delta^u{}_\beta. \tag{18.11}$$

Using the null vector k_α, (18.3), which is common for all the cases we are dealing with: $k_\alpha = F(\rho)\,\delta^u{}_\alpha$, is equipped with a non-vanishing covariant derivative

$$k_{\alpha;\beta} = 2\,m e^{2m\rho} \delta^u{}_{[\alpha}\delta^\rho{}_{\beta]}, \tag{18.12}$$

thus one concludes that the shear tensor and the divergence vanish, $k_{(\alpha;\beta)} = 0$ and $k^\alpha{}_{;\alpha} = 0$, while the rotation tensor (twist) is different from zero $k_{[\alpha;\beta]} \neq 0$, although the shear, rotation, and expansion invariant scalars are all zero. Consequently, this geeodesic null vector is also a Killing vector $k_{(\alpha;\beta)} = 0$. With the help of this null Killing vector k, one may write the Cotton tensor (18.11) as

$$C_{\alpha\beta} = \frac{1}{2}\mu\left(\mu^2 - m^2\right) e^{-(\mu+3m)\rho} f_3(u)\,k_\alpha k_\beta,\; k^\alpha = \delta^\alpha_v, \tag{18.13}$$

which manifestly shows the Petrov type N character of this Cotton tensor possessing a triple zero eigenvalue, $C_{\alpha\beta}\,k^\alpha = 0$.

Case $\mu = m$:
For this branch of solution, the $E^v{}_u$ equation (18.9) becomes

$$\frac{\partial^3}{\partial\rho^3} H(\rho, u) - 2m\frac{\partial^2}{\partial\rho^2} H(\rho, u) = 0 \tag{18.14}$$

hence it general solution can be given as

$$H(\rho, u) = f_1(u) + f_2(u)\,\rho + f_3(u)\,e^{2m\rho}. \tag{18.15}$$

It is characterized by a Cotton tensor

$$C_{\alpha\beta} = -f_2(u)\,m^2\,\delta^u{}_\alpha \delta^u{}_\beta, \tag{18.16}$$

or, using the null vector k from (18.3), one can represent it as

$$C_{\alpha\beta} = -f_2(u)\,m^2 e^{-4m\rho}\,k_\alpha k_\beta,\; k^\alpha = \delta^\alpha_v, \tag{18.17}$$

from which its Petrov type N property is apparent.

Case $\mu = -m$:
For this branch of solutions, the $E^v{}_u$ equation (18.9) amounts to

$$\frac{\partial^3}{\partial\rho^3} H(\rho, u) - 4m\frac{\partial^2}{\partial\rho^2} H(\rho, u) + 4m^2\frac{\partial}{\partial\rho} H(\rho, u) = 0, \tag{18.18}$$

its general solution can be given as

$$H(\rho, u) = f_1(u) + f_2(u)\, e^{2m\rho} + f_3(u)\, e^{2m\rho}\rho \tag{18.19}$$

This type N-wave gravitational field is characterized by the Petrov type N Cotton tensor

$$C_{\alpha\beta} = -f_3(u)\, m^2 e^{2m\rho}\, \delta^u{}_\alpha \delta^u{}_\beta = -f_3(u)\, m^2 e^{-2m\rho}\, k_\alpha k_\beta,\ k^\alpha = \delta^\alpha_v, \tag{18.20}$$

where the null vector k from (18.3) has been used.

18.2.1 AdS_3 TN-Waves with $\Lambda \neq 0$

Summarizing, the general type N wave solutions with a geodesic Killing null vector shear-free, twist-free, expansionless, and *non-covariantly constant* congruence, denoted as **TN-wave**, are given by the metrics

$$\mu \neq \pm m : ds^2 = d\rho^2 + 2e^{2m\rho} du dv + [e^{(m-\mu)\rho} f_1 + e^{2m\rho} f_2 + f_3] du^2, \tag{18.21a}$$

$$\mu = m : ds^2 = d\rho^2 + 2e^{2m\rho} du dv + [\rho\, f_1 + e^{2m\rho} f_2 + f_3] du^2, \tag{18.21b}$$

$$\mu = -m : ds^2 = d\rho^2 + 2e^{2m\rho} du dv + [\rho\, e^{2m\rho} f_1 + e^{2m\rho} f_2 + f_3] du^2, \tag{18.21c}$$

where $m = \sqrt{-\Lambda}$, with Λ being a non-vanishing negative cosmological constant; some of the arbitrary functions $f_i := f_i(u)$ can be removed via coordinate transformations. Nevertheless, it is convenient to include them when making comparisons with the solutions existing in the literature; these representations are equivalent to Eqs. (4.1)–(4.3) of Chow et al. (2010a).

In the literature about exact solutions for vacuum E-TMG equations in the presence of a cosmological constant one find the same solution discovered by different authors and given in different coordinate systems. Fortunately, the task of identifying them has been accomplished in Chow et al. (2010a) (CPSa), and their results are reproduced in this text.

18.2.2 Nutku TN-Wave Solution

Nutku (1993) considered several solutions with a cosmological constant. The Nutku's wave solution (16–17) is given by the metric

$$ds^2 = \frac{1}{[m\,x + h]^2}\left[-2dudv + d\,x^2 - 2\left(\frac{x}{m}\ddot{h} + c\,[m\,x + h]^{1+\mu/m}\right) du^2\right], \tag{18.22}$$

CPSa(A.59), where $c =$ constant and $h(u)$ is an arbitrary function. Accomplishing the coordinates transformations

$$-\frac{1}{m}\log(m\,x + h(u)) \to \rho,\ -v + \frac{\dot{h}}{m^2}(m\,x + h(u)) \to v',$$

dropping prime, one arrives at (18.21a), CPSa(4.1), with

$$f_1 = -c,\ f_2 = \frac{1}{m^2}(2h\ddot{h} + \dot{h}^2),\ f_3 = 0.$$

18.2.3 Clément TN-Wave Solution

Clément (1994) considered a Killing symmetry reduction procedure to obtain stationary rotationally symmetric solutions. The Clément solution (21) can be written as

$$ds^2 = \frac{1}{4m^2}\frac{d\rho^2}{\rho^2} + 2m^2\rho\left(dt^2 - \frac{d\theta^2}{m^2}\right) + \frac{M}{2}\left(1 + c\,\rho^{(1-\mu/m)/2}\right)\left(dt - \frac{d\theta}{m}\right)^2, \tag{18.23}$$

CPSa(A.60), where c and M are constant. Making the coordinates transformations

$$\frac{1}{2m}\log\rho \to \rho',\ t - \frac{\theta}{m} \to u,\ \frac{1}{m^2}\left(t + \frac{\theta}{m}\right) \to 2\,v,$$

one arrives at (18.21a), or CPSa(4.1), with

$$f_1 = \frac{Mc}{2},\ f_2 = 0,\ f_3 = \frac{M}{2}.$$

18.2.4 Ayón–Hassaïne TN-Wave Solution

Ayón-Beato and Hassaïne (2005) derived AdS waves by using the general AdS wave ansatz. Their solutions given by equations (A3), (A5), and (A4) are reported in CPSa correspondingly as CPSa(A.61)–CPSa(A.63):

$$ds^2 = \frac{1}{m^2\,y^2}\left[dy^2 - 2dudv - (m\,y)^{1+\mu/m}F_1(u)du^2\right],\ \mu \neq \pm m, \tag{18.24a}$$

$$ds^2 = \frac{1}{m^2\,y^2}\left[dy^2 - 2dudv - y^2\log(-m\,y)F_1(u)du^2\right],\ \mu = m, \tag{18.24b}$$

$$ds^2 = \frac{1}{m^2\,y^2}\left[dy^2 - 2dudv - \log(-m\,y)F_1(u)du^2\right],\ \mu = -\,m, \tag{18.24c}$$

where $F_1(u)$ is an arbitrary function. Doing the coordinates transformations

$$-\frac{1}{m}\log(-m\,y) \to \rho,\ -v \to v',$$

dropping prime, one arrives at (18.21a), (18.21b), (18.21c), or CPSa(4.1–4.3), CPSa(A.11), with

$$f_1 = -F_1(u),\ f_2 = 0 = f_3.$$

Considering a correspondence between Cotton gravity with a conformally coupled scalar field and TMG, see Deser et al. (2005) in this respect; Ayón-Beato and Hassaïne (2006) had previously found AdS waves.

18.2.5 Ölmez–Sarioğlu–Tekin TN-Wave Solution

Ölmez et al. (2005) considered supersymmetric solutions. Their solution (9) can be given as

$$ds^2 = d\rho^2 + 2e^{\mp 2m\rho}du\,dv + \left[\beta_2 e^{\mp(\mu+m)\rho} + \beta_1 e^{\mp 2m\rho} + \beta_0\right] dv^2, \qquad (18.25)$$

CPSa(A.64), where $\{\beta_0(v), \beta_1(v), \beta_2(v)\}$ are arbitrary functions. Carrying out the coordinates transformations

$$u \to v', \, v \to u', \, \mp\rho \to \rho',$$

dropping primes, one arrives at (18.21a), CPSa(4.1), with

$$f_1 = \beta_2, \, f_2 = \beta_1, \, f_3 = \beta_0.$$

18.2.6 Dereli-Sarioğlu TN-Wave Solution

Dereli and Sarioğlu (2001) considered supersymmetric solutions. Their solution (3.27–3.36) is a special case of the above solution of (18.25) with β_i = constants or (18.21a), CPSa(4.1), with constants f_i.

18.2.7 Carlip–Deser–Waldron–Wise TN-Wave Solution

Carlip et al. (2008) derived an AdS wave solution (3.14), for $\mu \neq \pm m$, CPSa(A.66) given by

$$ds^2 = \frac{1}{m^2 z^2}\left[dz^2 + 2\,dx^+ dx^- + 2h(x^+)\,(m\,z)^{1+\mu/m}(dx^+)^2\right], \qquad (18.26)$$

where $h(x^+)$ is an arbitrary function. Subjecting (18.26) to the coordinates transformations

$$-\frac{1}{m}\log(m\,z) \to \rho, \; x^+ \to u, \; x^- \to v,$$

for $\mu \neq \pm m$, one arrives at (18.21a), CPSa(4.1), with

$$f_1 = \frac{2h}{m^2}, \; f_2 = 0 = f_3.$$

18.2.8 Gibbons–Pope–Sezgin TN-Wave Solution

Gibbons et al. (2008) considered supersymmetric solutions of TMG. They found that all such solutions are AdS waves; their solutions are just the ones reported above (18.21a), (18.21b), and (18.21c), or CPSa((4.1), (4.2), (4.3)).

18.2.9 Anninos–Li–Padi–Song–Strominger TN-Wave Solution

Anninos et al. (2009) considered "warped" AdS_3 black hole solutions of TMG. Their solution (3.7), for $\mu = -3m$, can be written as CPSa(A.65)

$$ds^2 = \frac{1}{m^2\,u^2}\left(du^2 + dx^+ dx^- + \frac{1}{u^2}(dx^-)^2\right). \tag{18.27}$$

Making the coordinates transformations

$$-\frac{1}{m}\log(m\,u) \to \rho,\; m^2\,x^- \to u',\; \frac{x^+}{2m^2} \to v,$$

dropping prime, one arrives at (18.21a), CPSa(4.1) with

$$\mu = -3m,\; f_1 = 1,\; f_2 = 0 = f_3.$$

18.2.10 Garbarz–Giribet–Vásquez TN-Wave Solution

Garbarz et al. (2009) obtained solutions for the special values of $\mu = \pm m$. Their solution (1.5), for $\mu = m$, is given by

$$\begin{aligned} ds^2 &= \frac{r^2}{m^2}\left(r^2 - \frac{\kappa^2 M}{2m^2}\right)^{-2} dr^2 - m^2\left(r^2 - \frac{\kappa^2 M}{2m^2}\right)\left(dt^2 - \frac{1}{m^2}d\phi^2\right) \\ &+ \left[k\log\left(\frac{r^2}{r_0^2} - \frac{\kappa^2 M}{2m^2 r_0^2}\right) + \frac{\kappa^2 M}{2}\right]\left(dt - \frac{d\phi}{m}\right)^2, \end{aligned} \tag{18.28}$$

CPSa(A.67), where M, r_0, κ, k are constants. Accomplishing the coordinates transformations

$$m^2\left(r^2 - \frac{\kappa^2 M}{2m^2}\right) \to e^{2m\rho},\; t - \frac{1}{m}\phi \to u,\; t + \frac{1}{m}\phi \to -2\,v,$$

one arrives at (18.21b), CPSa(4.2), with

$$f_1 = 2km,\; f_2 = 0,\, f_3 = \frac{\kappa^2 M}{2} - 2k\log(m r_0).$$

Their second solution (3.18), for $\mu = -m$, is given as

$$\begin{aligned} ds^2 &= \tfrac{r^2}{m^2}\left(r^2 - \tfrac{\kappa^2 M}{2m^2}\right)^{-2} dr^2 - m^2\left(r^2 - \tfrac{\kappa^2 M}{2m^2}\right)\left(dt^2 - \tfrac{1}{m^2}d\phi^2\right) \\ &+ \left[k\left(r^2 - \tfrac{\kappa^2 M}{2m^2}\right)\log\left(\tfrac{r^2}{r_0^2} - \tfrac{\kappa^2 M}{2m^2 r_0^2}\right) + \tfrac{\kappa^2 M}{2}\right]\left(dt + \tfrac{d\phi}{m}\right)^2, \end{aligned} \tag{18.29}$$

CPSa(A.68), where M, r_0, κ, k are constants. Carrying out the coordinates transformations

$$m^2\left(r^2 - \frac{\kappa^2 M}{2m^2}\right) \to e^{2m\rho},\; t - \frac{1}{m}\phi \to u,\; t + \frac{1}{m}\phi \to -2\,v,$$

one arrives at (18.21c), CPSa(4.3), with

$$f_1 = \frac{2k}{m},\; f_2 = -\frac{2k}{m^2}\log(m r_0),\, f_3 = \frac{\kappa^2 M}{2}.$$

These type N waves solutions, with a Killing geodesic non-covariantly constant null congruence, can be shown in the form of a table:

Table 18.2.1 *Type N solutions with a non-covariantly constant null vector*

TN-wave solutions; Authors	Equations	Ref.	Λ
Chow et al. (2010a)	(18.21)	(4.1)–(4.3)	√
Nutku (1993)	(18.22)	(16–17)	√
Clément (1994)	(18.23)	(21)	√
Ayón-Beato and Hassaïne (2005)	(18.24)	(A.3–5)	√
Ölmez et al. (2005)	(18.25)	(9)	√
Dereli and Sarioğlu (2001)	(18.25) β_i const.		√
Carlip et al. (2008)	(18.26)	(3.14)	√
Gibbons et al. (2008)	(18.21a)–(18.21c)		√
Anninos et al. (2009)	(18.27)	(3.7)	√
Garbarz et al. (2009)	(18.28), (18.29)	(1.5), (3.18)	√

18.3 pp-Wave Solutions; $\Lambda = 0$

For a covariantly constant null congruence k, (18.1), for the wave metric (18.2), the TMG equations considerably simplify: first, the cosmological constant Λ vanishes, Λ=0; second, from the $E^{\rho}{}_{\rho}$ component equation, (18.7), one establishes that

$$\frac{d}{d\rho}F\left(\rho\right) = 0 \rightarrow F(\rho) = F_0 \rightarrow 1 \tag{18.30}$$

while the $E^{v}{}_{u}$ component (18.9) yields

$$\frac{\partial^3}{\partial\rho^3}H\left(\rho, u\right) + \mu\,\frac{\partial^2}{\partial\rho^2}H\left(\rho, u\right) = 0, \tag{18.31}$$

with general solution

$$H\left(\rho, u\right) = f_3\left(u\right) + f_2\left(u\right)\rho + f_1\left(u\right)\mathrm{e}^{-\mu\,\rho}. \tag{18.32}$$

Therefore the metric for pp-wave solutions is given by

$$ds^2 = d\rho^2 + 2\,du\,dv + \left(f_1(u)\mathrm{e}^{-\mu\,\rho} + \rho\,f_2(u) + f_3(u)\right)du^2, \tag{18.33}$$

coinciding with CPSa(A.70); the functions $f_2(u)$ and $f_3(u)$ can be equated to zero via transformations, but here are maintained as they stand for comparison with solutions reported in the literature. This solution is characterized by the Petrov type N Cotton tensor

$$C_{\alpha\beta} = \frac{1}{2}f_1(u)\mu^3\mathrm{e}^{-\mu\,\rho}\,k_\alpha k_\beta,\; k^\alpha = \delta^\alpha_v,\; k_\alpha = \delta^u_\alpha. \tag{18.34}$$

where k is a null vector.

18.3.1 Martinez–Shepley pp-Wave Solution; $\Lambda = 0$

Martinez and Shepley (1986) happen to be the first researches to look for pp-wave solutions of TMG, as one realizes from an unpublished preprint of them with zero cosmological constant.

18.3.2 Aragone pp-Wave Solution; $\Lambda = 0$

Aragone (1987) reported his (11) solution by the metric

$$ds^2 = \frac{4N_0^2}{(c(u) - \mu\, v)^2} dx^2 + \frac{2}{\mu} \dot{c}(u)\, du^2 - 2du\, dv - 2N_0 du\, dx, \tag{18.35}$$

CPSa(A.71). Subjecting (18.35) to the coordinates transformations

$$v - \frac{c(u)}{\mu} \to e^{\mu\rho/2},\ N_0\, x + N_0 \left(v - \frac{c(u)}{\mu}\right) \to u',$$

$$- u + \frac{4}{\mu^2} \left(v - \frac{c(u)}{\mu}\right)^{-1} \to v',$$

dropping primes, one arrives at CPSa(A.70) with

$$f_1 = \frac{4}{\mu^2},\ f_2 = 0 = f_3.$$

18.3.3 Percacci–Sodano–Vuorio pp-Wave Solution; $\Lambda = 0$

Percacci et al. (1987) considered stationary solutions for which the timelike Killing vector has a constant scalar twist. The solution given by their equation (3.19), is

$$ds^2 = \left(\frac{2}{\mu x^1}\right)^2 (dx^1)^2 - \frac{1}{2} dx^0\, dx^2 - \left(\frac{\mu x^1}{8}\right)^2 (dx^0)^2, \tag{18.36}$$

CPSa(A.72). In the original expression of the solution (3.19), there are functions ω_i not specified explicitly, but subjected to a certain equation; in CPSa(A.72) these functions are chosen as $\omega_1 = 0$ and $\omega_2 = 16/\mu^3 (x^2)^2$; any other choice is equivalent by redefinition of x^0.

Making the coordinates transformations

$$\frac{x^0}{4} \to u,\ x^1 \to e^{-\mu\rho/2},\ -x^2 \to v,$$

one arrives at (18.33), CPSa(A.70), with

$$f_1 = -\frac{\mu^2}{4},\ f_2 = 0 = f_3.$$

18.3.4 Hall–Morgan–Perjés pp-Wave Solution; $\Lambda = 0$

Hall et al. (1987) reported a pp-wave solution (46) of the form

$$ds^2 = du^2 + 2\, dx\, dr - 2e^{-\mu\, u} f(x)\, dx^2, \tag{18.37}$$

CPSa(A.73), where $f(x)$ is an arbitrary function. Doing the coordinate transformations

$$u \to \rho,\ x \to u',\ r \to v,$$

dropping prime, one arrives at (18.33), CPSa(A.70), with

$$f_1 = -2f,\ f_2 = 0 = f_3.$$

18.3.5 Dereli–Tucker pp-Wave Solution; $\Lambda = 0$

Dereli and Tucker (1988) considered solutions with a pp-wave-like ansatz. The solution in their equations (2.14) and (2.22) can be given by

$$ds^2 = dx^2 + 2du\,dv + 2\left[\frac{1}{\mu^2}e^{\mu\,x} f_1 + \left(f_3 - \frac{1}{\mu} f_1\right) x + f_2 - \frac{1}{\mu^2} f_1\right] du^2, \quad (18.38)$$

CPSa(A.74), where $f_1(u)$, $f_2(u)$, and $f_3(u)$ are arbitrary functions. Carrying out the coordinates transformations $x \to -\rho$, one arrives at (18.33), CPSa(A.70), but with the replacements of $f_1 \to 2f_1/\mu^2$, $f_2 \to 2(f_1/\mu - f_3)$, $f_3 \to 2(f_2 - f_1/\mu^2)$ there.

18.3.6 Deser–Steif pp-Wave Solution; $\Lambda = 0$

Deser and Steif (1992) considered an impulsive pp-wave solution. Their solution (2.1) of the equation

$$G_{\mu\nu} + \frac{1}{\mu} C_{\mu\nu} = -\kappa^2 T_{\mu\nu},\ T_{uu} = E\delta(y)\delta(u) \quad (18.39)$$

is given as

$$ds^2 = dy^2 - du\,dv + \left[C + B\,y + 2\,E\kappa^2\left(y - \frac{1}{\mu}(1 - e^{-\mu\,y})\right)\theta(y)\delta(u)\right] du^2, \quad (18.40)$$

CPSa(A.76), where $C(u)$, and $B(u)$ are arbitrary functions.

For $y > 0$ and $\delta(u) \to 1$, making the coordinates transformations

$$y \to \rho,\ v \to -v'/2,$$

dropping prime, one arrives at (18.33), CPSa(A.70), with

$$f_1 = 2\kappa^2 E/\mu,\ f_2(u) = 2\kappa^2 E + B(u),\ f_3 = -2\kappa^2 E/\mu + C(u).$$

18.3.7 Clément pp-Wave Solution; $\Lambda = 0$

Clément (1994) considered, in the class of stationary rotationally symmetric solutions, a pp-wave metric in his equation (4.15), which amounts to

$$ds^2 = dr^2 \pm 2\sigma_0(dt - \omega_0 d\theta)\,d\theta - \sigma_0\,(a + br + c\,e^{\mp\mu\,r})\,(dt - \omega_0 d\theta)^2, \quad (18.41)$$

see CPSa(A.75), where $a, b, c, \omega_0, \sigma_0$ are constants. Accomplishing the coordinates transformations

$$\pm r \to \rho,\ t - \omega_0\theta \to u,\ \pm\omega_0\theta \to v,$$

one arrives at (18.33), CPSa(A.70), with

$$f_1 = -c\,\omega_0,\ f_2 = \mp b\,\omega_0,\ f_3 = -a\,\omega_0.$$

18.3.8 Cavaglia pp-Wave Solution; $\Lambda = 0$

Cavaglia (1999) reported various classes of pp-waves. The solutions in equations (23)–(26) of Cavaglia (1999) are given by the metric

$$ds^2 = -(h(u-v))^2 du\, dv - H(u+v)(du+dv)\, d\phi, \tag{18.42}$$

$$h(u-v) = \begin{cases} 1) \ \ \frac{\alpha}{\mu}\tanh\left[\frac{1}{4}(u-v)\alpha - \beta\right], \\ 2) - \frac{\alpha}{\mu}\tanh\left[\frac{1}{4}(u-v)\alpha - \beta\right], \\ 3) \ \ \frac{\alpha}{\mu}\left[\frac{1}{4}(u-v)\alpha - \beta\right]^{-1}, \end{cases}$$

where α and β are constants, CPSa(A.77). Making the coordinates transformations

$$\int H(u+v)d(u+v) \to u', \ \phi \to -2v'$$

together with the transformation involving ρ,

$$\text{for 1)} : \cosh\left[\frac{1}{4}(u-v)\alpha - \beta\right] \to e^{\mu\rho/2}, \ f_1 = -f_3 = \left(\frac{\alpha}{2\mu H}\right)^2, \ f_2 = 0,$$

$$\text{for 2)} : \cos\left[\frac{1}{4}(u-v)\alpha - \beta\right] \to e^{\mu\rho/2}, \ -f_1 = f_3 = \left(\frac{\alpha}{2\mu H}\right)^2, \ f_2 = 0,$$

$$\text{for 3)} : \frac{1}{4}(u-v)\alpha - \beta \to e^{\mu\rho/2}, \ f_1 = -\left(\frac{\alpha}{2\mu H}\right)^2, \ f_2 = 0 = f_3,$$

dropping primes, one arrives at (18.33), CPSa(A.70).

18.3.9 Dereli–Sarioğlu pp-Wave Solution; $\Lambda = 0$

Dereli and Sarioğlu (2001) considered supersymmetric solutions. The solution, in their equations (43),(44), and (45), is obtained as a limit of their more general solution with a non-vanishing cosmological constant reviewed in Ölmez et al. (2005).

18.3.10 García–Hehl–Heinicke–Macías pp-Wave Solution; $\Lambda = 0$

García et al. (2004) constructed a pp-wave solution (2.137)–(2.141)

$$ds^2 = dy^2 + dx^2 - dt^2 - (C + Ay + Be^{\mu\, y})(dt - dx)^2, \tag{18.43}$$

CPSa(A.78). where A, B, C are constants. Carrying out the coordinates transformations

$$y \to -\rho, \ t - x \to u, \ t + x \to -v,$$

one arrives at (18.33), CPSa(A.70), with

$$f_1 = -B, \ f_2 = A, \ f_3 = -C.$$

18.3.11 Macías–Camacho pp-Wave Solution; $\Lambda = 0$

Macías and Camacho (2005) considered Kerr–Schild solutions of TMG without a cosmological constant. They gave two solutions explicitly. The solution in their equation (63) is reported in Chow et al. (2010a) as CPSa(A.79), namely

$$ds^2 = d\xi^2 - 2du\,dv + 2\left[c + b(\xi + y_0\,u) + \frac{a}{\mu}e^{\mu(\xi + y_0\,u)}\right]\left[dv + y_0 d\xi + \frac{y_0^2}{2}\,du\right]^2, \tag{18.44}$$

where y_0 is a constant. Making the coordinates transformations

$$\xi + y_0\,u \to -\rho,\; v + y_0\,\xi + y_0^2\,u/2 \to u',\; u \to -v',$$

dropping primes, one arrives at (18.33), CPSa(A.70), with

$$f_1 = \frac{2a}{\mu},\; f_2 = -2b,\; f_3 = 2c.$$

The second solution of their equation (72) amounts to

$$ds^2 = d\xi^2 - 2du\,dv - \frac{1}{\mu}\left[c + \alpha(\xi + 1/\mu) + \gamma e^{\mu\,\xi}\right]\left(dv + y_0 d\xi + y_0^2 du/2\right)^2, \tag{18.45}$$

CPSa(A.80), where $c = c(v)$ and $\gamma = \gamma(v)$ are arbitrary functions, and α is a constant. Accomplishing the coordinates transformations

$$\xi \to -\rho,\; -v \to u',\; u \to v',$$

dropping primes, one arrives at (18.33), CPSa(A.70), with

$$f_1 = -\frac{\gamma(u)}{\mu},\; f_2 = \frac{\alpha}{\mu},\; f_3 = -\frac{C(u) + \alpha/\mu}{\mu}.$$

All the references of these pp-wave solutions can be gathered in a table:

Table 18.3.1 *Type N solutions with a covariantly constant null vector; pp-wave solutions*

References	Eqs.	Eqs. to Ref.	Λ
Martinez and Shepley (1986)	18.3.1		–
Aragone (1987)	(18.35)	(11)	–
Percacci et al. (1987)	(18.36)	(3.19)	–
Hall et al. (1987)	(18.37)	(46)	–
Dereli and Tucker (1988)	(18.38)	(2.14), (2.22)	–
Deser and Steif (1992)	(18.40)	(7)	–
Clément (1994)	(18.41)	(4.15)	–
Cavaglia (1999)	(18.42)	(23)–(26)	–
Dereli and Sarioğlu (2001)	18.3.9	(43)–(45)	–
García et al. (2004)	(18.43)	(20.137–141)	–
Macías and Camacho (2005)	(18.44), (18.45)	(63), (72)	–
CPSa: Chow et al. (2010a)	(18.33)	(A.70)	–

19
Kundt Spacetimes in TMG

In general, Kundt spacetimes are defined as those spacetimes that admit a null geodesic vector field which is shear-free, twist-free, and expansion-free. In three dimensions, because of this dimensionality, the twist and shear trivially vanish, therefore a Kundt spacetime is simply one that admits an expansion-free null geodesic congruence. These spacetimes allow for Petrov type II, D, III, N, and O solutions as we shall see in the next paragraphs. Moreover, except for a general type II branch of metrics, all the existing Kundt solutions are CSI-constant scalar invariants – spacetimes for which all polynomial scalar (Ricci) curvature invariants are constant.

19.1 Null Geodesic Vector Field

In general relativity Ehlers (1961) introduced the frame of reference description of the kinematics of a fluid u^α; the frame is constructed on timelike trajectories to which the timelike vector field u^α is tangent, and a local spacelike hypersurface orthogonal to those trajectories. In $(2+1)$ dimensions the equivalent description was worked out by Barrow et al. (1986).

Similarly, when dealing with light trajectories, one has at one's disposal a null vector field k^α – the tangent vector to the light rays – and one can decompose the spacetime with respect to a frame constructed on the basis of the null vector k^α but, this time, using a second null vector l^α such that $k^\alpha\, l_\alpha = -1$, to define the metric projection tensor

$$\begin{aligned} b_{\mu\nu} &= g_{\mu\nu} + k_\mu\, l_\nu + k_\nu\, l_\mu;\; k^\mu\, k_\mu = 0 = l^\mu\, l_\mu,\; k^\mu\, l_\mu = -1, \\ b_{\mu\nu} k^\nu &= 0 = b_{\mu\nu} l^\nu, \end{aligned} \tag{19.1}$$

hence, the sub-manifold with metric $b_{\alpha\beta}$ is orthogonal to the directions k^α and l^α, and consequently spanned by spacelike vectors; for a $(3+1)$-dimensional spacetime the sub-manifold is a $2D$ orthogonal hypersurface, while for a

(2 + 1) spacetime it is a one-dimensional curve. There is a full parallelism in the formulation of scalar invariants in (3 + 1) and (2 + 1) spacetime geometries, except for minor changes in numerical factors due to the difference in dimensions. Thus let us consider a null vector field k^{μ}, which is required to be geodesic and affinely parametrised, therefore

$$k_{\mu;\nu}\, k^{\nu} = 0, \;\; k^{\mu}\, k_{\mu} = 0. \tag{19.2}$$

Instead of considering the decomposition of $k_{\mu;\nu}$ into its kinematical parts with respect to the spacetime metric $g_{\alpha\,\beta}$, as one commonly does for the timelike fluid vector field, defining the shear $\sigma_{\mu\nu}$ and rotation $\omega_{\mu\nu}$ tensors, and the expansion scalar θ according to

$$\begin{aligned} k_{\mu;\nu} &= \omega_{\mu\nu} + \sigma_{\mu\nu} + \frac{2\,\theta}{D}\, g_{\mu\nu}, \;\; \theta = \frac{1}{2} k^{\mu}{}_{;\mu}, \\ \omega_{\mu\nu} &:= \frac{1}{2}(k_{\mu;\nu} - k_{\nu;\mu}), \; \sigma_{\mu\nu} := \frac{1}{2}\left(k_{\mu;\nu} + k_{\nu;\mu} - \frac{4\,\theta}{D} g_{\mu\nu}\right), \end{aligned} \tag{19.3}$$

one can do it equivalently using its projection onto the sub-manifold by means of $b_{\alpha\beta}$

$$\begin{aligned} k_{\overline{\alpha};\overline{\beta}} := b^{\mu}{}_{\alpha}\, b^{\nu}{}_{\beta}\, k_{\mu;\nu} &= k_{\alpha;\beta} + k_{\alpha}\left(k_{\sigma;\beta} l^{\sigma} + \frac{1}{2} k_{\beta} k_{\gamma;\sigma} l^{\gamma} l^{\sigma}\right) \\ &\quad + k_{\beta}\left(k_{\alpha;\sigma} l^{\sigma} + \frac{1}{2} k_{\alpha} k_{\gamma;\sigma} l^{\gamma}\, l^{\sigma}\right), \end{aligned}$$

to define the expansion θ, the twist $\omega_{\overline{\mu};\overline{\nu}}$, and the shear $\sigma_{\overline{\mu};\overline{\nu}}$ tensors according with

$$\begin{aligned} k_{\overline{\mu};\overline{\nu}} &= \omega_{\overline{\mu};\overline{\nu}} + \sigma_{\overline{\mu};\overline{\nu}} + \frac{2\,\theta}{D-2} b_{\mu\nu}, \; \theta = \frac{1}{2} g^{\mu\nu}\, k_{\overline{\mu};\overline{\nu}} = \frac{1}{2} k^{\mu}{}_{;\mu}, \\ \omega_{\overline{\mu}\overline{\nu}} &:= \frac{1}{2}(k_{\overline{\mu};\overline{\nu}} - k_{\overline{\nu};\overline{\mu}}), \; \sigma_{\overline{\mu}\overline{\nu}} := \frac{1}{2}(k_{\overline{\mu};\overline{\nu}} + k_{\overline{\nu};\overline{\mu}} - \frac{4\theta}{D-2} b_{\mu\nu}), \\ b^{\mu}{}_{\mu} &= g^{\mu}{}_{\mu} - 2 = D - 2. \end{aligned} \tag{19.4}$$

These definitions we shall adopt from now on. They depend on the choice of the vector field l^{α}. Nevertheless the scalars of expansion θ, rotation ω^2, and shear σ^2 depend only on k_{μ}; these last two invariants are defined as follows

$$\omega^2 = \omega_{\mu\nu}\omega^{\mu\nu} = k^{\mu;\nu} k_{[\mu;\nu]}, \;\; \sigma^2 = \sigma^{\mu\nu}\sigma_{\mu\nu} = k^{\mu;\nu} k_{\mu;\nu} - \frac{(k^{\mu}{}_{;\mu})^2}{D-2}. \tag{19.5}$$

In the particular case of $3D$ spacetime geometry, as has been stated previously, the hypersurface to which k and l are orthogonal is one-dimensional and spacelike. Thus there exits a spacelike vector field m_{α}, $m^{\alpha}\, m_{\alpha} = 1$, such that one may identify the projection tensor with its tensor product, namely $b_{\alpha\beta} = m_{\alpha}\, m_{\beta}$. Because of the dimension 1 of the direction to which m is tangent, there is no room for the rotation and shear; as stated in CPS (Chow et al., 2010b) the shear and the twist (rotation) trivially vanish, consequently, "every null geodesic congruence of a three-dimensional spacetime is shear-free and twist-free." Hence, in three dimensions, "a Kundt spacetime is simply one that admits an expansion-free null geodesic congruence."

D-dimensional Kundt spacetimes are defined by metrics admitting a null geodesic vector field that is shear-free $\sigma^2 = 0$, and twist-free $\omega^2 = 0$, and expansion-free, $\theta = 0$.

Similarly, D-dimensional Robinson–Trautman are defined as those spaces admitting a null geodesic vector field that is shear-free $\sigma^2 = 0$, and twist-free $\omega^2 = 0$, but possessing expansion, $\theta \neq 0$. Consequently, a three-dimensional Robinson–Trautman spacetime admits an expanding, $k^{\mu}{}_{;\mu} \neq 0$, null geodesic congruence, $k_{\mu;\nu}k^{\nu} = 0$.

19.2 General Kundt Metrics

Despite TMG being a higher-derivative theory, for Kundt metrics, as quoted in Chow et al. (2010b), the field equations are still linear in curvature, and "so the calculations and results bear some similarity to those for Kundt metrics of four-dimensional Einstein gravity."

In this text, the term *metric* is used to denote a geometrical structure – line element – equipped with a metric tensor, i.e., a set of metric structural functions, while the term *solution* refers to that set of metric structural functions which fulfill a set of field equations, and consequently there is no freedom in the structural functions. Nevertheless, referring to a specific solution, it is given by the metric and a set of particular functions.

The four-dimensional Kundt metric, borrowed from the Kramer–Stephani et al. books (see Kramer et al., 1980 and Stephani et al., 2003, §27), is given by:

$$ds^2 = \frac{2}{P^2} d\zeta\, d\bar{\zeta} - 2du\left(dv + W d\zeta + \bar{W} d\bar{\zeta} + H du\right). \tag{19.6}$$

where $P(\zeta, \bar{\zeta}, u)$, and $H(\zeta, \bar{\zeta}, u)$ are real functions, and $W = W(\zeta, \bar{\zeta}, u)$ is a complex function. This metric is invariant under the following coordinate transformations and the associated transformations of the functions P, W, and H functions:

$$\begin{aligned} \zeta' &= f(\zeta, u) : \\ P'^2 &= P^2 f_{,\zeta} \bar{f}_{,\bar{\zeta}},\, W' = W/f_{,\zeta},\, +\bar{f}_{,u}/(P^2 f_{,\zeta} \bar{f}_{,\bar{\zeta}}), \\ H' &= H - \frac{1}{f_{,\zeta} \bar{f}_{,\bar{\zeta}}} \left(f_{,u} \bar{f}_{,u}/P^2 + W f_{,u} \bar{f}_{,\bar{\zeta}} + \bar{W}\, \bar{f}_{,u} f_{,\zeta}\right), \end{aligned} \tag{19.7a}$$

$$\begin{aligned} v' &= v + g(\zeta, \bar{\zeta}, u) : \\ P' &= P,\, W' = W - g_{,\zeta},\, H' = H - g_{,u}, \end{aligned} \tag{19.7b}$$

$$u' = h(u),\, v' = \frac{v}{h_{,u}}:$$
$$P' = P,\, W' = \frac{W}{h_{,u}},\, H' = \frac{1}{{h_{,u}}^2}\left(H + \frac{h_{,u,u}}{h_{,u}}\right). \tag{19.7c}$$

A special subclass of metrics – Kundt vacuum type III and N metrics ($R_{12} = \partial_\zeta\partial_{\bar\zeta}\ln P + \Psi_2 + \bar\Psi_2 = 0$) – arises for

$$\partial_\zeta\partial_{\bar\zeta}\ln P = 0 \to P P_{,\zeta,\bar\zeta} - P_{,\zeta}P_{,\bar\zeta} = 0.$$

In such a case, always, via the transformation (19.7a), P can be brought to $P = 1$; the demonstration goes as follows

$$\partial_\zeta\partial_{\bar\zeta}\ln P = 0 \to \ln P = F(\zeta,u) + \bar F(\bar\zeta,u) \to P = \mathrm{e}^{F(\zeta,u)}\mathrm{e}^{\bar F(\bar\zeta,u)}$$
$$\to P'^2 = \mathrm{e}^{2F(\zeta,u)}\mathrm{e}^{2\bar F(\bar\zeta,u)} f_{,\zeta}\bar f_{,\bar\zeta},$$

hence, if one requires $P'^2 \to 1$, then

$$\mathrm{e}^{2F(\zeta,u)} f_{,\zeta} \to 1,\ \frac{\partial f(\zeta,u)}{\partial\zeta} = \mathrm{e}^{-2F(\zeta,u)},$$
$$f(\zeta,u) \equiv \zeta'(\zeta,u) = \int_{\zeta_0}^{\zeta} \mathrm{e}^{-2F(\zeta,u)} d\zeta,$$

achieving in this way $P' = 1$.

19.2.1 3D Kundt Metric

A dimensional reduction of the $4D$ Kundt metric (19.6) is achieved by considering the sub-manifold with coordinates $\{v, u, \mathrm{Re}\zeta\}$, i.e., the slices $z = 0$ of the complex coordinate $\zeta = (r + iz)/\sqrt{2}$

$$\zeta = \to r/\sqrt{2},\ u \to -u, \text{ together with},\ H(v,u,\zeta) \to -f(v,u,r)/2,$$
$$W(v,u,\zeta) + \bar W(v,u,\bar\zeta) \to \sqrt{2}W(v,u,r), \tag{19.8}$$

(instead of the coordinate ρ used in CPSb, Chow et al. (2010b), we prefer to denote it by r), gives rise to the general $3D$ Kundt metric in the form

$$ds^2 = \frac{1}{P^2(v,u,r)}dr^2 + 2\,du dv + f(v,u,r)\,du^2 + 2\,W(v,u,r)\,du\,dr. \tag{19.9}$$

The transformation (19.7a) becomes

$$r' = r'(r,u):$$
$$P' = P\left(r'_{,r}\right),\ W' = W/r'_{,r} - r'_{,u}/(P\left(r'_{,r}\right))^2,$$
$$f' = f + \left(r'_{,u}\right)^2/(P\left(r'_{,r}\right))^2 - W\,r'_{,u}/(r'_{,r}). \tag{19.10}$$

In particular, the change $r'(u,r) = \int^r 1/P(u,r)\,dr$, can be used to bring, dropping primes, the Kundt metric to

$$ds^2 = dr^2 + 2du\,dv + f(v,u,r)du^2 + 2W(v,u,r)du\,dr. \tag{19.11a}$$

The null geodesic vector k^μ is given by

$$k^\mu = \delta^\mu{}_v, k_\mu = \delta^u{}_\mu \tag{19.11b}$$

with its covariant derivative being

$$k_{\mu;\nu} = \frac{1}{2}\frac{\partial}{\partial v}f\,(v,u,r)\,\delta^u{}_\mu\delta^u{}_\nu + \frac{1}{2}\frac{\partial}{\partial v}W\,(v,u,r)\,(\delta^u{}_\mu\delta^r{}_\nu + \delta^r{}_\mu\delta^u{}_\nu), \tag{19.11c}$$

hence, k is geodesic and expansion-free as can be established straightforwardly form the above equation, additionally, it is shear–free and rotationless. Moreover this congruence is, in general, non-Killingian; for a single pp-wave case it becomes a Killing null field.

When using numbers to denote coordinates, the correspondence is $\{v,u,r\}$ $= \{1,2,3\}$. Most of the functions are denoted with Latin characters.

For the general Kundt metric (19.11a) the TMG equations read:

$$E^2{}_1 = -\frac{1}{2\mu}\frac{\partial^3}{\partial v^3}W\,(v,u,r) = 0, \tag{19.11d}$$

integrating one gets

$$W(v,u,r) = v^2W_2\,(u,r) + vW_1\,(u,r) + W_0\,(u,r), \tag{19.11e}$$

Replacing this $W(v,u,r)$ into

$$E^3{}_1 = E^2{}_3 = \frac{1}{4\mu}\left(2\,\mu\frac{\partial^2 W}{\partial v^2} + \frac{\partial^3 f}{\partial v^3}\right) = 0 \tag{19.11f}$$

one can integrate $f\,(v,u,r)$ as

$$\begin{aligned} &f\,(v,u,r) = F_3\,(u,r)\;v^3 + F_2\,(u,r)\,v^2 + F_1\,(u,r)\,v + F_0\,(u,r)\,, \\ &F_3\,(u,r) := -\frac{2\,\mu}{3}\,W_2\,(u,r)\,. \end{aligned} \tag{19.11g}$$

On the other hand the evaluation of scalar curvature R, which for this vacuum TMG theory is equal to 6Λ, $R = 6\Lambda$, gives

$$R = \frac{\partial^2}{\partial v^2}f - \frac{3}{2}\left(\frac{\partial}{\partial v}W\right)^2 - 2\,W\frac{\partial^2}{\partial v^2}W + 2\,\frac{\partial^2}{\partial v\partial r}W = 6\Lambda. \tag{19.11h}$$

Substituting into (19.11h) the obtained metric functions one gets

$$\begin{aligned} &-\frac{45}{2\mu^2}\,v^2\,(F_3)^2 + v\left(6\,F_3 + \frac{6}{\mu}\frac{\partial}{\partial r}F_3 - \frac{15}{\mu}\,F_3W_1\right) - \frac{6}{\mu}\,W_0F_3 \\ &+2\,F_2 + 2\,\frac{\partial}{\partial r}W_1 - 3/2\,(W_1)^2 - 6\Lambda = 0, \end{aligned} \tag{19.11i}$$

therefore, because of the variable independence on v, one has

$$F_3 = 0 \rightarrow W_2 = 0, \tag{19.11j}$$

consequently, solving for F_2, one obtains

$$F_2\left(u,r\right) = 3\,\Lambda + \frac{3}{4}\left(W_1\right)^2 - \frac{\partial}{\partial r}W_1, \tag{19.11k}$$

which substituted in the remaining TMG equation $E^2{}_2$ yields

$$\begin{aligned} E^2{}_2 : \frac{\partial^2}{\partial r^2}W_1 + \left(\mu - \frac{5}{2}\,W_1\right)\frac{\partial}{\partial r}W_1 \\ + \left(3\,\Lambda + \frac{3}{4}\left(W_1\right)^2 - \frac{1}{2}\,\mu W_1\right)W_1 - 2\,\Lambda\,\mu = 0 \end{aligned} \tag{19.11l}$$

which, worked out in the CPSb fashion, can be written in the "subjective" form

$$\frac{\partial}{\partial r}\left(\frac{\partial}{\partial r}W_1 - \frac{1}{2}W_1^2 - 2\Lambda\right) + (\mu - \frac{3}{2}W_1)(\frac{\partial}{\partial r}W_1 - \frac{1}{2}W_1^2 - 2\Lambda) = 0. \tag{19.11m}$$

As we shall see in a forthcoming paragraph, this equation plays a fundamental role in the algebraic type properties of the resulting solutions.

Notice that one can always set $W_0(u,r) = 0$ by means of a coordinate transformation of the coordinate v; in fact

$$v \rightarrow v + \Phi(u,r); \tag{19.11n}$$

the g_{ur}-component becomes

$$g_{ur} \rightarrow g_{ur} = \partial_r\Phi(u,r) + \Phi(u,r)\,W_1 + W_0 + v\,W_1, \tag{19.11o}$$

thus, requiring

$$\partial_r\Phi(u,r) + \Phi(u,r)\,W_1 + W_0 = 0, \tag{19.11p}$$

which occurs if

$$\Phi\left(u,r\right) = \left(\int -\mathrm{e}^{\int W_1(u,r)dr}\,W_0\left(u,r\right)dr + \phi\left(u\right)\right)\mathrm{e}^{-\int W_1(u,r)dr}, \tag{19.11q}$$

one achieves the purpose

$$W_0(u,r) \rightarrow 0,\; g_{ur} = v\,W_1\left(u,r\right). \tag{19.11r}$$

19.3 3*D* Canonical Kundt Metric

In this section, the main geometrical features of the Kundt metrics derived by Chow et al. (2010b) in the framework of the vacuum topologically massive gravity are exhibited and analyzed in detail.

Summarizing the results derived previously, one begins with the general canonical Kundt metric which can be written as

$$ds^2 = dr^2 + 2du\,dv + (v^2F_2 + vF_1 + F_0)du^2 + 2\,v\,W_1\,du\,dr, \tag{19.12a}$$

where $F_i = F_i(u,r)$, $i = 0,1,2$, and $W_1 = W_1(u,r)$. These structural functions ought to fulfill the TMG equations: the "key" equation

$$\frac{\partial}{\partial r}\left(\frac{\partial}{\partial r}W_1 - \frac{1}{2}W_1^2 - 2\Lambda\right) + (\mu - \frac{3}{2}W_1)(\frac{\partial}{\partial r}W_1 - \frac{1}{2}W_1^2 - 2\Lambda) = 0; \quad (19.12b)$$

next, the algebraic equation to determine F_2

$$F_2(u,r) = 3\Lambda + \frac{3}{4}(W_1(u,r))^2 - \frac{\partial}{\partial r}W_1(u,r), \quad (19.12c)$$

a linear ordinary second-order equation for F_1 or first-order equation for $\frac{\partial}{\partial r}F_1$

$$\frac{\partial^2}{\partial r^2}F_1 + \left(\mu - \frac{1}{2}W_1\right)\frac{\partial}{\partial r}F_1 + \left(\frac{5}{2}W_1 - \mu\right)\frac{\partial}{\partial u}W_1 - 3\frac{\partial^2}{\partial u \partial r}W_1 = 0, \quad (19.12d)$$

and finally, a linear ordinary third-order equation to determine F_0

$$\frac{\partial^3}{\partial r^3}F_0 + \left(\mu + \frac{3}{2}W_1\right)\frac{\partial^2}{\partial r^2}F_0 - \left(3\Lambda - \frac{7}{2}\frac{\partial}{\partial r}W_1 + \frac{1}{4}{W_1}^2 - W_1\mu\right)\frac{\partial}{\partial r}F_0 + \left[(4\Lambda + {W_1}^2)\left(\mu - \frac{3}{2}W_1\right) - \left(\mu - \frac{9}{2}W_1\right)\frac{\partial}{\partial r}W_1\right]F_0 = -\frac{\partial^2}{\partial u \partial r}F_1 - \frac{1}{2}F_1\frac{\partial}{\partial r}F_1 + \frac{1}{2}F_1\frac{\partial W_1}{\partial u}\frac{\partial^2 W_1}{\partial u^2}. \quad (19.12e)$$

The covariant derivative of the geodesic expansionless null vector $k^\alpha = \delta^\alpha{}_v$ in this representation amounts to

$$k_{\alpha;\beta} = \left[v\left(\frac{3}{4}{W_1}^2 - \frac{\partial}{\partial r}W_1 + 3\Lambda\right) + \frac{1}{2}F_1\right]\delta^u{}_\alpha\delta^u{}_\beta + W_1\delta^u{}_{(\alpha}\delta^r{}_{\beta)}. \quad (19.13)$$

Notice that Kundt metrics in general possess a geodesic non-Killingian null congruence k, only in the pp-wave limit, which takes place for $\{W_1 = 0,\ F_1 = 0,\ \Lambda = 0\}$, the covariant derivative $k_{\alpha;\beta}$ vanishes.

The evaluation of the traceless Ricci tensor $S^\mu{}_\nu$, modulo the field equations (19.12), yields

$$S^\mu{}_\nu = S^1{}_1(\delta^\mu{}_v\delta^v{}_\nu + \delta^\mu{}_u\delta^u{}_\nu - 2\delta^\mu{}_r\delta^r{}_\nu) + S^1{}_2\delta^\mu{}_v\delta^u{}_\nu + S^1{}_3\delta^\mu{}_v\delta^r{}_\nu + S^3{}_2\delta^\mu{}_r\delta^u{}_\nu \quad (19.14a)$$

$$S^1{}_1 = S^2{}_2 = -\frac{1}{2}EVC = -\frac{1}{2}\left(\frac{\partial}{\partial r}W_1 - \frac{1}{2}(W_1)^2 - 2\Lambda\right),$$

$$S^3{}_3 = EVC = -2S^1{}_1,$$

$$S^1{}_3 = -\frac{1}{2}EVCv(3W_1 - 2\mu) + \frac{1}{2}\left(\frac{\partial}{\partial r}F_1 - \frac{\partial}{\partial u}W_1\right),$$

$$S^3{}_2 = v\mu\,EVC + \frac{1}{2}\left(\frac{\partial}{\partial r}F_1 - \frac{\partial}{\partial u}W_1\right),$$

$$S^1{}_2 = -\frac{1}{4}\,EVC\mu\,(3\,W_1 - 2\,\mu)\,v^2 - \frac{1}{4}\left(\frac{\partial}{\partial r}F_1 - \frac{\partial}{\partial u}W_1\right)(W_1 - 2\,\mu)\,v$$
$$-\frac{1}{2}\left(\frac{\partial^2}{\partial r^2}F_0 + W_1\frac{\partial}{\partial r}F_0 + F_0\frac{\partial}{\partial r}W_1\right),$$
$$EVC = \frac{\partial}{\partial r}W_1 - \frac{1}{2}\,(W_1)^2 - 2\,\Lambda. \qquad (19.14b)$$

19.3.1 Petrov Classification of the Cotton and Traceless Ricci Tensors

It is apparent that $S^1{}_1$ plays the role of the eigenvalue of the matrix tensor $(S^\alpha{}_\beta)$; if $S^1{}_1 \neq 0$ then the tensor $S^\alpha{}_\beta$ is Petrov type II with a possible degeneration to Petrov type D: the traceless Ricci tensor matrix $\mathcal{S}$, via elementary (similarity) algebraic transformations – addition of one column to a second one multiplied by a factor (scalar or function other than zero and addition of a row to another one multiplied by a factor and multiplication of a column by a factor other than zero) – can be brought into its equivalent Jordan matrix form, namely

$$\mathcal{S} := \begin{bmatrix} S^1{}_1 & S^1{}_2 & S^1{}_3 \\ 0 & S^1{}_1 & 0 \\ 0 & S^3{}_2 & -2\,S^1{}_1 \end{bmatrix} \sim \mathcal{J}_{S_{II}} = \mathcal{B}\mathcal{S}\mathcal{A} = \begin{bmatrix} S^1{}_1 & \tilde{S}^1_2 & 0 \\ 0 & S^1{}_1 & 0 \\ 0 & 0 & -2\,S^1{}_1 \end{bmatrix} \sim$$
$$\mathcal{J}_{S_{II}} = \begin{bmatrix} S & 1 & 0 \\ 0 & S & 0 \\ 0 & 0 & -2\,S \end{bmatrix}, \quad \tilde{S}^1_2 := \frac{S^3{}_2\,S^1{}_3 + 2\,S^1{}_1\,S^1{}_2}{2S^1{}_1}, \qquad (19.15)$$

where the multiplication by the matrix $\mathcal{A}$ adds in $\mathcal{S}$ the third column multiplied by $S^3{}_2/(2S^1{}_1)$ to the second one to eliminate the term in the intersection of the second column and the third row, $T^2{}_3 \to 0$, while the multiplication by the matrix $\mathcal{B}$ adds in the resulting matrix the third row multiplied by $S^1{}_3/(2\,S^1{}_1)$ to the first one to eliminate $T^1{}_3 \to 0$, these matrices are given by

$$\mathcal{A} = \begin{bmatrix} 1 & 0 & 0 \\ 0 & 1 & 0 \\ 0 & S^3{}_2/(2S^1{}_1) & 1 \end{bmatrix}, \quad \mathcal{B} = \begin{bmatrix} 1 & 0 & S^1{}_3/(2S^1{}_1) \\ 0 & 1 & 0 \\ 0 & 0 & 1 \end{bmatrix}. \qquad (19.16)$$

Above, it has been assumed $S^1{}_1 \neq 0$, thus the equivalent Jordan matrix $\mathcal{J}_S$ to $\mathcal{S}$ is of Petrov type II for $\tilde{S}^1_2 \neq 0$, and Petrov type D if $\tilde{S}^1_2$ vanishes, $\tilde{S}^1_2 = 0$, with Jordan form

$$\mathcal{J}_{S_D} = \mathcal{B}\mathcal{S}\mathcal{A} = \begin{bmatrix} S^1{}_1 & 0 & 0 \\ 0 & S^1{}_1 & 0 \\ 0 & 0 & -2\,S^1{}_1 \end{bmatrix}. \qquad (19.17)$$

Table 19.3.1 *Petrov types of general Kundt metrics*

$S = \partial_r W_1 - {W_1}^2/2 - 2\Lambda$ $S^\alpha{}_\beta$:$\{S^1{}_1 = S^2{}_2 = S, S^3{}_3 = -2S, S^1{}_2, S^1{}_3, S^3{}_2\}$			
Petrov types No type I	Eigenvalues	Jordan Matrix	Λ
II	$S \neq 0$	$\mathcal{J}_{S_{II}}, \{S, S^1{}_2\}$	✓
D	$S \neq 0$	$\mathcal{J}_{S_D}, \{S, S^1{}_2 = 0\}$	✓
III	$S = 0$	$\mathcal{J}_{S_{III}}, \{S = 0 \neq S^1{}_2, S^1{}_3, S^3{}_2\}$	✓
N	$S = 0$	$\mathcal{J}_{S_N}, \{S = 0 \neq S^1{}_2, S^1{}_3, S^3{}_2 = 0\}$	✓
O	$S = 0$	$\{0\}$	✓

On the other hand, if the eigenvalue vanishes, $S^1{}_1 = 0$, then the tensor $S^\alpha{}_\beta$ is Petrov type III with possible degeneration to Petrov type N or O (conformally flat).

$$\mathcal{S} := \begin{bmatrix} 0 & S^1{}_2 & S^1{}_3 \\ 0 & 0 & 0 \\ 0 & S^3{}_2 & 0 \end{bmatrix} \sim \mathcal{J}_{S_{III}} = \mathcal{T}\mathcal{S} = \begin{bmatrix} 0 & S^1{}_2 & 0 \\ 0 & 0 & \tilde{S}^2_{\ 3} \\ 0 & 0 & 0 \end{bmatrix}$$

$$\tilde{S}^2_{\ 3} := -\frac{S^3{}_2\, S^1{}_3}{S^1{}_2}, \tag{19.18}$$

where the transformation matrix $\mathcal{T}$,

$$\mathcal{T} = \begin{bmatrix} 0 & 0 & S^1{}_2/S^3{}_2 \\ -S^3{}_2/S^1{}_2 & 0 & 1 \\ 0 & 1 & 0 \end{bmatrix}, \tag{19.19}$$

adds rows multiplied by certain specific factors and permute the resulting second by the third row. It is assumed that $S^1{}_2 \neq 0$. If $\tilde{S}^2_{\ 3} := -S^3{}_2\, S^1{}_3/S^1{}_2$ is different from zero, $\tilde{S}^2_{\ 3} \neq 0$ then the metric is of Petrov type III. On the contrary, if $\tilde{S}^2_{\ 3} = 0$, then the spacetime is of Petrov type N,

$$\mathcal{J}_{S_N} = \mathcal{T}\mathcal{S} = \begin{bmatrix} 0 & S^1{}_2 & 0 \\ 0 & 0 & 0 \\ 0 & 0 & 0 \end{bmatrix} \sim \begin{bmatrix} 0 & 1 & 0 \\ 0 & 0 & 0 \\ 0 & 0 & 0 \end{bmatrix}. \tag{19.20}$$

Finally, if additionally $S^1{}_2 = 0$ the metric becomes conformally flat or Petrov type O.

Summing up, for the $3D$ Kundt metrics the Petrov types of the traceless Ricci and Cotton tensors are presented in the table above.

For the Cotton tensor – a symmetric and traceless tensor – a similar algebraic classification to the one presented above for the traceless Ricci tensor takes place;

instead of the components $S^{\alpha}{}_{\beta}$ replace them by the Cotton components $C^{\alpha}{}_{\beta}$ with eigenvalue $C^{1}{}_{1}$.

As is well known, for vacuum spacetimes the E-TMG equations are equivalent to the scalar curvature equation and the vanishing relation between the traceless Ricci and Cotton tensors:

$$R = 6\,\Lambda,\ S_{\alpha\,\beta} + \frac{1}{\mu}C_{\alpha\,\beta} = 0.$$

These tensors shear the same eigenvalues up to the factor $-\mu$ and consequently the same Petrov types. In particular, these relations hold for the canonical metric (19.12a).

The general Petrov type II Kundt metric with a variable $W_1(u,r)$ fulfilling $EVC \neq 0$ and equation (19.12b) is unknown. The class of Petrov type II with $W_1 = 2\mu/3$ has been completely integrated. For Petrov type III Kundt metrics EVC ought to vanish, $EVC = 0$; all solutions has been derived and fully characterized.

Invariants. As far as to the Riemann invariants is concerned, one has: Riemann–Ricci invariants

$$R = R^{\alpha}{}_{\beta}, \tag{19.21a}$$

$$R_{\alpha\beta}R^{\alpha\beta} = \frac{3}{2}\,EVC^{2} + 12\,\Lambda^{2}, \tag{19.21b}$$

$$R_{\alpha\beta;\gamma}R^{\alpha\beta;\gamma} = \left[\frac{5}{8}(3\,W_1(u,r) - 2\mu)^2 - \mu^2\right] EVC^2 \tag{19.21c}$$

traceless Ricci tensor $S^{\alpha}{}_{\beta}$ invariants

$$S_{\alpha\beta}S^{\alpha\beta} = \frac{3}{2}\ EVC^{2}, \tag{19.22a}$$

$$S^{\sigma}{}_{\alpha}S_{\sigma\beta}S^{\alpha\beta} = 3/4\,EVC^{3}, \tag{19.22b}$$

$$S^{\sigma}{}_{\alpha}S_{\sigma\beta}S^{\nu\alpha}S_{\nu}{}^{\beta} = \frac{9}{8}\ EVC^{4}, \tag{19.22c}$$

$$S_{\alpha\beta;\gamma}S^{\alpha\beta;\gamma} = \frac{3}{4}\,W1\,(u,r)\,(3\,W_1\,(u,r) - 4\,\mu)\ \ EVC^{2}, \tag{19.22d}$$

$$S_{\alpha\beta;\gamma}S^{\alpha\beta;\epsilon}S^{\gamma}{}_{\epsilon} = -\frac{3}{8}\,W_1\,(u,r)\,(3\,W_1\,(u,r) - 4\,\mu)\ EVC^{3} \tag{19.22e}$$

Constant Scalar Invariant Spacetimes

Coley and collaborators accomplished a systematic study of "spacetimes for which all polynomial scalar invariants constructed from the Riemann curvature and its covariant derivatives are constant and are called constant scalar invariant (CSI) spacetimes," see Coley et al. (2006). In Coley et al. (2008) it was established that a $3D$ CSI spacetime is either locally homogeneous or is a Kundt spacetime, and all $3D$ CSI spacetimes were determined. The proof of this assertion is based on a theorem stating, roughly speaking, "that on an open neighborhood in which the Segré type does not change and all Ricci invariants are constant (consequently its eigenvalues are constant too) there exits a frame such that all the components of the Ricci tensor are constants and of the canonical (Segré) form in any dimension." An extensive use of the Segré forms and of **triad** frames in which the Ricci tensor components are constant, together with the constancy (vanishing) of the invariants constructed from covariant derivatives of the Ricci tensor allow to establish this powerful result in three dimensions. Extensions to higher dimensions have been carried out by Coley et al. (2009a and 2009b).

The invariants of $S_{\alpha\beta}$, equivalently, of the Cotton tensor $C_{\alpha\beta}$, involving covariant differentiations, contain, as factor, the function $W_1(u,r)\,EVC$ to some power. Thus, there are type II metrics, unknown at present, with $EVC \neq 0$ and with a function $W_1(u,r)$ depending on its variables and constrained to fulfill the key equation (19.12b). Moreover, type II metrics with constant EVC fulfilling (19.12b) arise for $W1(u,r) = 2\mu/3$; the invariants become constants and the solutions become constant scalar invariant (CSI) spacetimes.

If the Cotton tensor is of type III, its eigenvalues are zero, $EVC = 0$, together with its invariants. Moreover, the Cotton tensor invariants, or the curvature Ricci invariants, become constants, hence all Kundt type III metrics are constant scalar invariant spacetimes.

19.3.2 Sub-Branch $W_1(r)$ of the General Kundt Metric in TMG

In Chow et al. (2010b) the $W_1(r)$ case is studied in detail. It is argued that by means of a transformation $r \to r + g(u)$ one can eliminate one of the integration constants from the two that appear in the solution of the second-order equations (19.12b); this is true but in no way this means that this solution will be independent of the coordinate u, considered as a parameter, for a partial differential equation of the form of the equation (19.12b). As an example, consider the third-order differential equation (19.11e), explicitly

$$\frac{\partial^3}{\partial v^3} W(v,u,r) = 0 \to W(v,u,r) = v^2 W_2(u,r) + v\,W_1(u,r) + W_0(u,r)$$

which, by means of coordinate transformations and function redefinitions, one can achieve $W_0(u,r) = 0$, but still $W_1(u,r)$ and $W_2(u,r)$ remain functions of their arguments.

An important general integral, within this framework, is given by the integration of the function $F_1(u,r)$, equation (19.12d),

$$\frac{\partial^2}{\partial r^2} F_1 + \left(\mu - \frac{1}{2} W_1\right) \frac{\partial}{\partial r} F_1 = \left(\mu - \frac{5}{2} W_1\right) \frac{\partial}{\partial u} W_1 + 3 \frac{\partial^2}{\partial u \partial r} W_1,$$

namely

$$F_1(u,r) = \int \{ \int \left[\left(\mu - \frac{5}{2}\,W_1 \right) \frac{\partial W_1}{\partial u} + 3\,\frac{\partial^2 W_1}{\partial u \partial r} \right] \frac{dr}{EI(r)} \} EI(r)dr$$

$$+ f_{11}\,(u) \int EI(r)dr + f_{10}\,(u)\,,\; EI(r) := \mathrm{e}^{\int(-\mu+\frac{1}{2}\,W_1)dr}, \qquad (19.23)$$

where the function $W_1(u,r)$ participates actively in the integrals, and, in general for $W_1(u,r)$ depending on r and u too, it should contribute substantially with nonzero terms to the function $F_1(u,r)$.

Therefore, extending these considerations to the case of the key equation (19.12b) for $W_1(u,r)$, a second-order equation, from the two parameters of integration depending on u one would be able, by transformations, to exclude one of them, but not both at the same time, thus the remaining integration function will depend on the coordinate u. The general solution of the key equation (19.12b) still remains unknown; it is a challenge to overcome.

The case $W_1(r)$ is a sub-branch of Petrov type II Kundt metrics; even in this case the general solution is unknown, although the families of solutions of type II and D have been completely integrated in the case of constant $W_1 = 2\mu/3$ which are CSI spacetimes.

19.3.3 Kundt Metric Structure for $\mathbf{W_1(r)}$

The Kundt metric for $W_1(r)$ becomes

$$ds^2 = dr^2 + 2du\,dv + \left[v^2 F_2 + vF_1 + F_0\right] du^2 + 2\,v\,W_1\,du\,dr. \qquad (19.24a)$$

where the structural functions $W_1(r)$, and $F_i(u,r)$, $i = 0,1,2$, ought to satisfy the following equations:

$$\frac{d}{dr}\left(\frac{d}{dr}W_1 - \frac{1}{2}W_1^2 - 2\Lambda\right) + (\mu - \frac{3}{2}W_1)(\frac{d}{dr}W_1 - \frac{1}{2}W_1^2 - 2\Lambda) = 0, \qquad (19.24b)$$

$$F_2\,(r) = \frac{3}{4}\,W_1\,(r)^2 - \frac{d}{dr}\,W_1\,(r) + 3\,\Lambda, \qquad (19.24c)$$

$$\frac{\partial^2}{\partial r^2}F_1 + \left(\mu - \frac{1}{2}\,W_1\right)\frac{\partial}{\partial r}F_1 = 0, \qquad (19.24d)$$

$$\frac{\partial^3}{\partial r^3}F_0 + \left(\mu + \frac{3}{2}\,W_1\right)\frac{\partial^2}{\partial r^2}F_0 - \left(3\,\Lambda - \frac{7}{2}\,\frac{\partial}{\partial r}W_1 + \frac{1}{4}\,W_1{}^2 - W_1\mu\right)\frac{\partial}{\partial r}F_0$$
$$+ \left[4\,\Lambda\,\left(\mu - \frac{3}{2}\,W_1\right) - \left(\mu - \frac{9}{2}\,W_1\right)\frac{\partial}{\partial r}W_1 + W_1{}^2\left(\mu - \frac{3}{2}\,W_1\right)\right]F_0$$
$$= -\frac{\partial^2}{\partial u \partial r}F_1 - \frac{1}{2}\,F_1\,\frac{\partial}{\partial r}F_1. \qquad (19.24e)$$

The equation (19.24d) integrates as

$$F_1(r,u) = f_{11}(u) \int \exp(\int (W_1(r)/2 - \mu)dr)dr + f_{12}(u), \qquad (19.25)$$

which coincides with the the homogeneous solution of (19.23) for $W_1(r)$. By a coordinate transformation, one can send $f_{12}(u) \to 0$, as stated in Chow et al. (2010b).

19.4 Type II CSI Kundt Metric; $W_1 = 2\mu/3$

A particular type II metric, $S = -EVC/2 \neq 0$ arises for:

$$\text{type II: } W_1 = \frac{2}{3}\mu, \; S = \frac{2}{9}(\mu^2 + 9\Lambda) \neq 0. \qquad (19.26)$$

The corresponding type II metric gives rise to CSI solutions. Explicitly one has,

$$ds^2 = dr^2 + 2du\, dv + \left[v^2 F_2(r) + vF_1(u,r) + F_0(u,r)\right] du^2 + \frac{4\mu}{3} v\, du\, dr, \qquad (19.27a)$$

see Chow et al. (2010b), Eqn. (A.32), or more compactly using for this kind of citation the notation CPSb(A.32). The structural functions ought to satisfy the following equations

$$F_2(r) = \frac{1}{9}\mu^2 + 3\Lambda, \qquad (19.27b)$$

$$\frac{\partial^2}{\partial r^2} F_1 + \frac{2}{3}\mu \frac{\partial}{\partial r} F_1 = 0, \qquad (19.27c)$$

$$\frac{\partial^3}{\partial r^3} F_0 + 2\mu \frac{\partial^2}{\partial r^2} F_0 + \left(\frac{5}{9}\mu - 3\Lambda\right) \frac{\partial}{\partial r} F_0 = -\frac{\partial^2}{\partial u \partial r} F_1 - \frac{1}{2} F_1 \frac{\partial}{\partial r} F_1. \qquad (19.27d)$$

The solution of (19.27c) is

$$F_1(u,r) = f_{11}(u) \mathrm{e}^{-2\mu\, r/3} + f_{12}(u), \qquad (19.28)$$

which can be set equal to zero, $F_1(u,r) = 0$, via coordinate transformations:

$$v = \tilde{v} - \frac{1}{2(\mu^2/3 + 3\Lambda)} f_{11}(u) \mathrm{e}^{-2\mu\, r/3}.$$

Dropping tilde, the Petrov type II SCI Kundt metric amounts to CPSb(A.32)

$$ds^2 = \left(d\,r + \frac{2}{3}\mu\, v\, du\right)^2 + 2du\, dv - \frac{1}{9}\left(\mu^2 - 27\Lambda\right) v^2 du^2 + F_0(u,r)\, du^2 \qquad (19.29)$$

subjected to the equation for $F_0(u,r)$

$$\frac{\partial^3}{\partial r^3} F_0 + 2\mu \frac{\partial^2}{\partial r^2} F_0 + \left(\frac{5}{9}\mu - 3\Lambda\right) \frac{\partial}{\partial r} F_0 = 0, \qquad (19.30)$$

which possesses the general solution

$$F_0 = f_{01}(u)\,\mathrm{e}^{-\mu}\,\mathrm{e}^{\sqrt{4\,\mu^2+27\,\Lambda}\;r/3} + f_{02}(u)\,\mathrm{e}^{-\mu}\,\mathrm{e}^{-\sqrt{4\,\mu^2+27\,\Lambda}\;r/3} + f_{03}(u)\,, \quad (19.31)$$

allowing for a trigonometric representation in dependence of the relationship $\mu(\Lambda)$ and the sign of the radical $\sqrt{4\,\mu^2 + 27\,\Lambda}$; the corresponding sub-classes are reported in the forthcoming paragraphes.

Evaluating the traceless Ricci tensor one arrives at

$$S^\mu{}_\nu = S^v{}_v\,(\delta^\mu{}_v\delta^v{}_\nu + \delta^\mu{}_u\delta^u{}_\nu - 2\delta^\mu{}_r\delta^r{}_\nu) + S^v{}_u\delta^\mu{}_v\delta^u{}_\nu + S^r{}_u\delta^\mu{}_r\delta^u{}_\nu, \quad (19.32a)$$

$$S^v{}_v = \frac{1}{9}\left(\mu^2 + 9\,\Lambda\right) \neq 0,\; S^r{}_u = -2\,v\mu\,\frac{1}{9}\left(\mu^2 + 9\,\Lambda\right),$$

$$S^v{}_u = -\frac{1}{2}\left(\frac{\partial^2}{\partial r^2}F_0 + \frac{2}{3}\,\mu\,\frac{\partial}{\partial r}F_0\right). \quad (19.32b)$$

In general, within this branch of solutions with constant $W_1 = 2\mu/3$, the spacetime is of type II for $S^v{}_u \neq 0$,

$$(\mathcal{J}_{II}) = \begin{bmatrix} S^v{}_v & 1 & 0 \\ 0 & S^v{}_v & 0 \\ 0 & 0 & -2S^v{}_v \end{bmatrix},\quad F_0\,(u,r) \neq 0.$$

For $S^v{}_u = 0$, which occurs for $F_0(u,r) = 0$, the solution is of the type D. Returning to type II solutions, their explicit expressions depend on the relationship between $\mu(\Lambda)$ and also on the sign of Λ. Correspondingly, one may distinguish the following branches:

19.4.1 Negative Cosmological Constant; $\Lambda = -m^2$

For this negative cosmological constant, the metric is given by

$$ds^2 = \left(d\,r + \frac{2}{3}\mu\,v\,du\right)^2 + 2du\,dv - \frac{1}{9}\left(\mu^2 + 27\,m^2\right)v^2du^2 + F_0(u,r)\,du^2. \quad (19.33)$$

with structural function $F_0(u,r)$, in correspondence with CPSb(3.5)–(3.7), of the form:

$$\mu^2 > \frac{27}{4}m^2, \gamma = \sqrt{\frac{4\,\mu^2}{9\,m^2} - 3}:\; f_{0i} := f_{0i}(u),\; i = 1,2,3,$$

$$F_0 = \mathrm{e}^{-\mu\,r}\cosh(\gamma\,m\,r)f_{01} + \mathrm{e}^{-\mu\,r}\sinh(\gamma\,m\,r)f_{02} + f_{03}, \quad (19.34a)$$

$$\mu^2 < \frac{27}{4}m^2,\; \gamma = \sqrt{3 - \frac{4\,\mu^2}{9\,m^2}}:$$

$$F_0 = \mathrm{e}^{-\mu\,r}\cos(\gamma\,m\,r)f_{01} + \mathrm{e}^{-\mu\,r}\sin(\gamma\,m\,r)f_{02} + f_{03}, \quad (19.34b)$$

$$\mu^2 = \frac{27}{4}m^2:\; F_0 = r\,\mathrm{e}^{-\mu\,r}f_{01} + \mathrm{e}^{-\mu\,r}f_{02} + f_{03}, \quad (19.34c)$$

where the constants of integration $f_{0i}(u)$ depending on the variable u are denoted by $f_{0i} := f_{0i}(u)$, $i = 1, 2, 3$. Via a coordinate transformation the arbitrary function $f_{03}(u)$ can be set zero, $f_{03}(u) = 0$.

19.4.2 Positive Cosmological Constant; $\Lambda = m^2$

The metric in this case amounts to

$$ds^2 = \left(d\,r + \frac{2}{3}\mu\, v\, du\right)^2 + 2du\, dv - \frac{1}{9}\left(\mu^2 - 27m^2\right) v^2 du^2 + F_0(u, r)\, du^2. \quad (19.35)$$

with a single structural function $F_0(u, r)$:

$$\mu^2 > \frac{27}{4} m^2,\ \gamma = \sqrt{\frac{4}{9}\frac{\mu^2}{m^2} + 3} :$$
$$F_0 = \mathrm{e}^{-\mu\, r} \cosh(\gamma\, m\, r) f_{01}(u) + \mathrm{e}^{-\mu\, r} \sinh(\gamma\, m\, r) f_{02}(u) + f_{03}(u), \quad (19.36)$$

the function $f_{03}(u)$ can be equated to zero, $f_{03}(u) \rightarrow 0$, by coordinate transformations. This solution corresponds to the CPSb(3.12) one.

19.4.3 Zero Cosmological Constant; $\Lambda = 0$

The simplest case is determined by metric

$$ds^2 = \left(d\,r + \frac{2}{3}\mu\, v\, du\right)^2 + 2du\, dv - \frac{1}{9}\mu^2\, v^2 du^2 + F_0(u, r)\, du^2, \quad (19.37)$$

with the structural function $F_0(u, r)$ of the form:

$$F_0 = \mathrm{e}^{-\mu\, r/3} f_{01} + \mathrm{e}^{-5\,\mu\, r/3} f_{02} + f_{03}. \quad (19.38)$$

where $f_{0i} = f_{0i}(u)$. The function f_{03} can be equated to zero, $f_{03} \rightarrow 0$, by coordinate transformations. This solution is just the CPSb(3.11) one.

19.5 Type D CSI Kundt Solutions; $W_1 = 2\mu/3, F_0 = 0$

The type D CSI Kundt solution is given by the metric

$$ds^2 = (dr + \frac{2}{3}\mu\, v\, du)^2 + 2du\, dv - \frac{1}{9}\left(\mu^2 - 27\,\Lambda\right) v^2\, du^2, \quad (19.39a)$$

and it is characterized by the type D traceless Ricci–Cotton tensors, $S^{\mu}{}_{\nu} + C^{\mu}{}_{\nu}/\mu = 0$,

$$S^{\mu}{}_{\nu} = S^{v}{}_{v}\left(\delta^{\mu}{}_{v}\delta^{v}{}_{\nu} + \delta^{\mu}{}_{u}\delta^{u}{}_{\nu} - 2\delta^{\mu}{}_{r}\delta^{r}{}_{\nu}\right) + S^{r}{}_{u}\delta^{\mu}{}_{r}\delta^{u}{}_{\nu}, \quad (19.39b)$$

Table 19.5.1 *Type II and Type D SCI Kundt solutions*

$S = \partial_r W_1 - W_1^2/2 - 2\Lambda \neq 0$ and $W_1(u,r) = 2\mu/3$					
II	$W_1(u,r)$	$\mu(m)$	Eqs.	Rel. CPS	$\Lambda = \pm m^2$
II general	$W_1(u,r)$ unknown	μ	(19.12b)	(2.4–15)	Λ
II	$W_1 = 2\mu/3$	μ	(19.29)	(3.4)	Λ
II	$W_1 = 2\mu/3$	$\mu^2 > 27m^2/4$	(19.34a)	(3.5)	< 0
II	$W_1 = 2\mu/3$	$\mu^2 < 27m^2/4$	(19.34b)	(3.6)	< 0
II	$W_1 = 2\mu/3$	$\mu^2 = 27m^2/4$	(19.34c)	(3.7)	< 0
II	$W_1 = 2\mu/3$	$\mu^2 > 27m^2/4$	(19.35)	(3.12)	> 0
II	$W_1 = 2\mu/3$	μ	(19.37)	(3.11)	0
D	$W_1 = 2\mu/3$	μ	(19.39a)	(3.17)	< 0

$$S^v{}_v = \frac{1}{9}\left(\mu^2 + 9\,\Lambda\right) \neq 0,\; S^r{}_u = -\frac{2}{9}\,v\mu\,\left(\mu^2 + 9\,\Lambda\right),$$

$$\mathcal{J}_D = \begin{bmatrix} S^v{}_v & 0 & 0 \\ 0 & S^v{}_v & 0 \\ 0 & 0 & -2S^v{}_v \end{bmatrix}. \tag{19.39c}$$

Making the coordinate transformation

$$u = \frac{18}{(\mu^2 - 27\,\Lambda)}\tilde{u},\; v = \frac{1}{\tilde{v} - \tilde{u}},$$

one brings (19.39a) to the form

$$ds^2 = -\frac{36}{\mu^2 - 27\,\Lambda}\frac{d\tilde{v}\,d\tilde{u}}{(\tilde{u} - \tilde{v})^2} + \left(dr - \frac{12\mu}{\mu^2 - 27\,\Lambda}\,\frac{d\tilde{u}}{\tilde{u} - \tilde{v}}\right)^2, \tag{19.39d}$$

which, by the subsequent transformations

$$t = \frac{1}{2}\tilde{u} + \tilde{v},\; x = \frac{1}{2}\tilde{v} - \tilde{u},\; z = \frac{\mu^2 - 27\,\Lambda}{6\mu}r - \ln(\tilde{v} - \tilde{u}), \tag{19.39e}$$

becomes the: *biaxially squashed spacelike limit*

$$ds^2 = \frac{9}{\mu^2 - 27\,\Lambda}\left[\frac{dx^2 - dt^2}{x^2} + \frac{4\mu^2}{\mu^2 - 27\,\Lambda}\left(dz + \frac{dt}{x}\right)^2\right], \tag{19.39f}$$

where the $2D$ sector AdS_2 is written using Poincaré coordinates (t, x).

In a concise way all these solutions can be visualized in the table above.

19.6 Petrov Type III Kundt Metrics

If $S^1{}_1 \equiv S^v{}_v = S = -EVC/2 = 0$, the traceless Ricci $S^\alpha{}_\beta$ and Cotton $C^\alpha{}_\beta$ tensors are Petrov type III; the triple eigenvalue is equal to zero, $S^3 = 0$. Thus $W_1\,(u, r)$ ought to fulfill

$$\frac{\partial}{\partial r} W_1(u,r) - \frac{1}{2}(W_1)^2 - 2\Lambda = 0, \tag{19.40}$$

which guarantees the fulfillment of the key equation (19.24b). The solutions of this equation (19.40) depend on $r - f(u)$, thus, by a shift of the coordinate $r \to r + f(u)$ one brings the dependence to r. One distinguishes the following branches of solution depending on the sign of the cosmological constant Λ:

Case $\Lambda = 0$. For zero cosmological constant $\Lambda = 0$, there are two cases

$$\Lambda = 0: \; W_1(u,r) = 0, \tag{19.41a}$$

$$\Lambda = 0: \; W_1(u,r) = -\frac{2}{r}. \tag{19.41b}$$

Case $\Lambda = -m^2$. For negative cosmological constant $\Lambda = -m^2$, one distinguishes three different classes

$$\Lambda = -m^2: \; W_1(r) = -2m, \tag{19.41c}$$

$$\Lambda = -m^2: \; W_1(u,r) = -2m \coth(mr), \tag{19.41d}$$

$$\Lambda = -m^2: \; W_1(u,r) = -2m \tanh(mr). \tag{19.41e}$$

Case $\Lambda = m^2$. For positive cosmological constant $\Lambda = m^2$ there is a single family of solutions

$$\Lambda = m^2: \; W_1(r) = -2m \cot(mr). \tag{19.41f}$$

These structural functions W_1 coincide, modulo the identification of σ with the cosmological constant Λ, with the functions W_x^1, fulfilling (28) of Coley et al. (2008) (equivalent to (19.40)), listed in the paragraph 3.2.2. Segré types $\{3\}$, $\{(21)\}$, and $\{(1,11)\}$ of Coley et al. (2008).

Each of these cases will be treated in detail in the forthcoming sections; first one evaluates the function $F_2(u,r)$, next one proceeds to integrate the function $F_1(u,r)$ to end with the integration of the function $F_0(u,r)$.

19.7 Type III Kundt Solution; $\Lambda = 0, W_1 = 0$

We start the derivation of the Petrov type III Kundt solutions handling the easiest case of $\Lambda = 0$ and W_1 from (19.41a): evaluating F_2 from equation (19.24c), one gets

$$W_1(u,r) = 0 \to F_2 = 0. \tag{19.42}$$

Replacing these functions into the F_1 – equation (19.24d) – one arrives at

$$\frac{\partial^2}{\partial r^2} F_1(u,r) + \mu \frac{\partial}{\partial r} F_1 = 0 \tag{19.43}$$

with solution

$$F_1(u,r) = f_{12}(u) + f_{11}(u)\,\mathrm{e}^{-\mu r} \to F_1(u,r) = f_{11}(u)\,\mathrm{e}^{-\mu r}, \tag{19.44}$$

by coordinate transformations is achieved $f_{12}(u) \to 0$. Next, the equation (19.24e) for $F_0(u,r)$ becomes

$$\frac{\partial}{\partial r}\left(\mathrm{e}^{\mu r}\frac{\partial^2}{\partial r^2}F_0\right) = \mu\frac{df_{11}}{du} + \frac{1}{2}\mu\,\mathrm{e}^{-\mu r}f_{11}{}^2. \tag{19.45}$$

Taking into account that

$$\int \mathrm{e}^{-\mu r} r\,dr = -\frac{(1+\mu r)\,\mathrm{e}^{-\mu r}}{\mu^2},\quad \int^r\int^x z\mathrm{e}^{-\mu z}dz\,dx = \frac{(2+\mu r)\,\mathrm{e}^{-\mu r}}{\mu^3}$$

one arrives at

$$\begin{aligned} F_0(u,r) = {} & \frac{1}{\mu^2}\left(2\frac{df_{11}}{du} + f_{01}(u)\right)\mathrm{e}^{-\mu r} + f_{02}(u)\,r + f_{03}(u) \\ & -\frac{1}{8\mu^2}f_{11}{}^2\mathrm{e}^{-2\mu r} + \frac{1}{\mu}\mathrm{e}^{-\mu r}r\frac{df_{11}}{du}, \end{aligned} \tag{19.46}$$

which, by means of coordinate transformations, can be brought to the form

$$F_0(u,r) = f_{01}(u)\,\mathrm{e}^{-\mu r} - \frac{1}{8\mu^2}f_{11}{}^2\mathrm{e}^{-2\mu r} + \frac{1}{\mu}\mathrm{e}^{-\mu r}r\frac{df_{11}}{du}. \tag{19.47}$$

Finally, the resulting metric can be given as

$$ds^2 = dr^2 + 2du\,dv + \left[v\,f_{11}(u)\,\mathrm{e}^{-\mu r} + F_0(u,r)\,(19.47)\right]du^2. \tag{19.48}$$

This solution is characterized by a Cotton tensor $C_{\alpha\beta}$ and a traceless Ricci tensor $S_{\alpha\beta}$ fulfilling the relation $S_{\alpha\beta} + \frac{1}{\mu}C_{\alpha\beta} = 0$, expressible as

$$C^\alpha{}_\beta = C^v{}_u\delta^\alpha{}_v\delta^u{}_\beta + C^v{}_r\delta^\alpha{}_v\delta^r{}_\beta + C^r{}_u\delta^\alpha{}_r\delta^u{}_\beta, \tag{19.49a}$$

$$C^v{}_u = -\frac{\mu}{4}\,\mathrm{e}^{-\mu r}\left(f_{11}{}^2\mathrm{e}^{-\mu r} - 2\mu^2 v\,f_{11} - 2\mu^2 f_{01} + 4\frac{df_{11}}{du} - 2\mu r\frac{df_{11}}{du}\right), \tag{19.49b}$$

$$C^v{}_r = C^r{}_u = \frac{\mu^2}{2}\,f_{11}\mathrm{e}^{-\mu r}. \tag{19.49c}$$

19.7.1 Type N pp-Wave Limit

For vanishing function f_{11} one gets a type N metric

$$ds^2 = dr^2 + 2du\,dv + f_{01}(u)\,\mathrm{e}^{-\mu r}du^2 \tag{19.50a}$$

characterized by a Petrov type N Cotton tensor

$$C^\alpha{}_\beta = C^v{}_u k^\alpha k_\beta, C^v{}_u = \frac{\mu^3}{2}\,\mathrm{e}^{-\mu r}f_{01},\; k^\alpha = \delta^\alpha{}_v. \tag{19.50b}$$

19.8 Type III Kundt Solution; $\Lambda = 0, W_1 = -2/r$

Evaluating the function F_2, (19.24c), for W_1 given by (19.41b), one gets

$$W_1\,(u,r) = -\frac{2}{r} \to F_2 = \frac{1}{r^2}. \tag{19.51}$$

Substituting this function into the $F_1\,(u,r)$–equation (19.24d) one gets

$$\frac{\partial}{\partial r}\left(e^{\mu\,r}\,r\,\frac{\partial F_1}{\partial r}\right) = 0, \tag{19.52}$$

therefore, integrating

$$F_1\,(u,r) = -\int \frac{e^{-\mu\,r}}{r} dr\, f_{11}\,(u) =: Ei\,(1,\mu\,r)\; f_{11}\,(u), \tag{19.53}$$

where Ei(x) is the exponential integral,

$$Ei\,(1,\mu\,r) = -\int^{r} \frac{e^{-\mu\,x}}{x} dx,\; \frac{d}{dr} Ei\,(1,\mu\,r) = -\frac{e^{-\mu\,r}}{r}. \tag{19.54}$$

For the determined structural functions, the $F_0\,(u,r)$–equation (19.24e) becomes

$$\frac{\partial^3}{\partial r^3} F_0 + \left(\mu - \frac{3}{r}\right)\left[\frac{\partial^2}{\partial r^2} F_0 - \frac{2}{r}\frac{\partial}{\partial r} F_0 + \frac{2}{r^2} F_0\right] = Rhs(u,r),$$
$$Rhs(u,r) = \frac{1}{2\,r} {f_{11}}^2\, Ei\,(1,\mu\,r) e^{-\mu\,r} + \frac{1}{r}\frac{d f_{11}}{du} e^{-\mu\,r}. \tag{19.55}$$

Since the solution of the homogeneous equation is proportional to r, explicitly

$$F_{0\,H}\,(u,r) = r\,e^{-\mu\,r} f_{01}\,(u) + r^2 f_{02}\,(u) + r\,f_{03}\,(u) \tag{19.56}$$

then, searching for the solution to the non-homogeneous equation, it is more convenient to introduce an auxiliary function $\Phi(u,r)$ through

$$F_0\,(u,r) = \Phi\,(u,r)\; r. \tag{19.57}$$

In terms of the function Φ the equation (19.55) becomes

$$\frac{\partial}{\partial r}\left(e^{\mu\,r}\frac{\partial^2}{\partial r^2}\Phi\right) = Rhs\,\frac{e^{\mu\,r}}{r}, \tag{19.58}$$

consequently

$$\Phi_{NH}\,(u,r) = \int^{r}\!\!\int^{s}\left(\int^{z} Rhs\,(u,t)\,\frac{e^{\mu\,t}}{t} dt\right) e^{-\mu\,z} dz\, ds,$$
$$\Phi_H = e^{-\mu\,r} f_{01}\,(u) + r\,f_{02}\,(u) + f_{03}\,(u)\,, \tag{19.59}$$

accomplishing a partial integration one obtains

$$\Phi = \frac{1}{4}\,{f_{11}}^2 \int (Ei\,(1,\mu\,r))^2\, dr - \frac{1}{2}\,(f_{11})^2\; r\, Ei\,(1, 2\,\mu\,r)$$
$$-\frac{\mu}{2}\,{f_{11}}^2 \int^{r}\!\!\int^{s} Ei\,(1,\mu\,z) e^{-\mu\,z} dz\, ds + \frac{d f_{11}}{du}\; r\, Ei\,(1,\mu\,r)$$
$$+\frac{1}{4}\,(f_{11})^2\,\frac{e^{-2\,\mu\,r}}{\mu} - \frac{d f_{11}}{du}\,\frac{e^{-\mu\,r}}{\mu}. \tag{19.60}$$

Finally the corresponding $F_0(u,r)$ function can be given as

$$F_0(u,r) = r\,\Phi(19.60) + r\mathrm{e}^{-\mu r} f_{01}\,(u) + r^2 f_{02}\,(u) + r\,f_{03}\,(u)\,. \tag{19.61}$$

Solution via the Variation of Parameters (VP)

Consider an ordinary non-homogeneous third-order equation of the form

$$\frac{d^3X(r)}{dr^3} + p_2(r)\frac{d^2X(r)}{dr^2} + p_1(r)\frac{dX(r)}{dr} + p_0(r)X(r) = Rhs(r) \tag{19.62}$$

and assume that a set of three independent solutions $\{X_1(r), X_2(r), X_3(r)\}$ of the homogeneous equation has been determined; then the non-homogeneous (NH) solution is sought as a linear combination of them,

$$X_{NH} = C_1\,X_1 + C_2\,X_2 + C_3\,X_3$$

where the coefficients $C_i = C_i(r)$, $i = 1,2,3$ are considered at this stage as functions of the variable r and that their derivatives $\frac{d}{dr}C_i(r) = C_i{}'$ fulfill the set of equations:

$$C_1{}'\,X_1{}'' + C_2{}'\,X_2{}'' + C_3{}'\,X_3{}'' = Rhs, C_3{}' = \frac{Rhs}{W}\left(X_1X_2{}' - X_2X_1{}'\right), \tag{19.63a}$$

$$C_1{}'\,X_1 + C_2{}'\,X_2 + C_3{}'\,X_3 = 0,\, C_1{}' = \frac{Rhs}{W}\left(X_2X_3{}' - X_3X_2{}'\right), \tag{19.63b}$$

$$C_1{}'\,X_1{}' + C_2{}'\,X_2{}' + C_3{}'\,X_3{}' = 0,\, C_2{}' = \frac{Rhs}{W}\left(X_3X_1{}' - X_1X_3{}'\right), \tag{19.63c}$$

where $W = \det\left([X_i], [X_i{}'], [X_i{}'']\right)$ is the Wronskian of the system. Integrating the derivatives $C_i{}'$, one gets the functions $C_i(r)$ and consequently one determines the non-homogeneous solution, which, added to the homogeneous one, leads to the general solution.

One may use the variation of parameters procedure to derive the solution of

$$\begin{aligned}
&\frac{\partial^3}{\partial r^3}\Phi\,(u,r) + \mu\,\frac{\partial^2}{\partial r^2}\Phi\,(u,r) = rhs_\Phi,\\
&rhs_\Phi = \frac{1}{r^2}\mathrm{e}^{-\mu\,r}\,\frac{df_{11}}{du} + \frac{1}{2r^2}\,Ei(1,\mu\,r)\,f_{11}{}^2\mathrm{e}^{-\mu\,r}.
\end{aligned} \tag{19.64}$$

The independent solutions to the homogeneous equation for Φ are given by

$$\left\{\Phi_1 = 1, \Phi_2 = r, \Phi_3 = \mathrm{e}^{-\mu\,r}\right\}. \tag{19.65}$$

The solution for Φ to the non-homogeneous equation is sought in the form

$$\Phi_{NH}(u,r) = \Phi_1\,C_1(u,r) + \Phi_2\,C_2(u,r) + \Phi_3\,C_3(u,r). \tag{19.66}$$

Via the variation of the parameters $C_i(u,r)$, their derivatives are

$$\frac{\partial C_1}{\partial r} = -\frac{rhs_\Phi\,(1+\mu\,r)}{\mu^2},\ \frac{\partial C_2}{\partial r} = \frac{rhs_\Phi}{\mu},\ \frac{\partial C_3}{\partial r} = \frac{rhs_\Phi\,\mathrm{e}^{\mu\,r}}{\mu^2}.$$

Integrating these equations, one arrives at

$$C_1(u,r) = \frac{e^{-\mu r}}{\mu^2 r}\frac{df_{11}}{du} - \frac{{f_{11}}^2}{2}\int \frac{Ei(1,\mu r)\,e^{-\mu r}}{\mu^2 r^2} dr + \frac{{f_{11}}^2}{4\mu}\left(Ei(1,\mu r)\right)^2,$$
$$C_2(u,r) = \frac{df_{11}}{du}\left(-\frac{e^{-\mu r}}{\mu r} + Ei(1,\mu r)\right) + \frac{1}{2}{f_{11}}^2 \int \frac{Ei(1,\mu r)\,e^{-\mu r}}{\mu r^2} dr,$$
$$C_3(u,r) = -\frac{1}{\mu^2 r}\frac{df_{11}}{du} + \frac{1}{2\mu^2 r}{f_{11}}^2\left(e^{-\mu r} - Ei(1,\mu r)(\mu r + 1)\right).$$

Taking into account that the integral

$$\int \frac{e^{-\mu r} Ei(1,\mu r)}{r^2} dr = -\frac{e^{-\mu r} Ei(1,\mu r)}{r} + \frac{1}{2}\left(Ei(1,\mu r)\right)^2 \mu$$
$$-2\,Ei(1,2\mu r)\,\mu + \frac{e^{-2\mu r}}{r}$$

one obtains the final expression of $\Phi(u,r)$, namely

$$\begin{aligned}\Phi = {} & \frac{1}{\mu}\left[-e^{-\mu r} Ei(1,\mu r) + \frac{\mu r}{4}\left(Ei(1,\mu r)\right)^2 - Ei(1,2\mu r)(\mu r - 1)\right]{f_{11}}^2 \\ & + \frac{df_{11}}{du} Ei(1,\mu r)\,r + \frac{1}{2\mu} e^{-2\mu r}{f_{11}}^2 - \frac{1}{\mu}\frac{df_{11}}{du} e^{-\mu r} \\ & + f_{01}(u) e^{-\mu r} + f_{02}(u)\,r + f_{03}(u), \\ F_0(u,r) = {} & r\,\Phi(u,r), \end{aligned} \tag{19.67}$$

the arbitrary integration functions $f_{02}(u)$ and $f_{03}(u)$ of the variable u can be canceled by coordinate transformations. The metric is given by

$$ds^2 = dr^2 + 2du\,dv - \frac{4v}{r} du\,dr + \frac{v^2}{r^2} du^2 + \left[v Ei(1,\mu r) f_{11} + r\,\Phi(u,r)(19.67)\right] du^2. \tag{19.68}$$

This solution is characterized by the Cotton and traceless Ricci tensors, fulfilling $S_{\alpha\beta} + \frac{1}{\mu}C_{\alpha\beta} = 0$, of the form

$$\begin{aligned} C^\alpha{}_\beta &= C^v{}_u \delta^\alpha{}_v \delta^u{}_\beta + C^r{}_u \delta^\alpha{}_r \delta^u{}_\beta + C^v{}_r \delta^\alpha{}_v \delta^r{}_\beta, \\ C^v{}_u &= -\frac{\mu}{4}(1 + \mu r)\,{f_{11}}^2\,Ei(1,\mu r)\,e^{-\mu r} \\ & \quad + \frac{v}{2}\frac{\mu\,(\mu r + 1)}{r^2} f_{11} e^{-\mu r} + \frac{\mu}{4}{f_{11}}^2 e^{-2\mu r} - \frac{\mu}{2}\frac{df_{11}}{du} e^{-\mu r} + \frac{\mu^3}{2} r f_{01} e^{-\mu r}, \\ C^v{}_r &= C^r{}_u = \frac{\mu}{2r} f_{11} e^{-\mu r}. \end{aligned} \tag{19.69}$$

Generically, this metric is Petrov type III ($f_{11} \neq 0$).

19.8.1 Type N Limit

The type N branch arises for $f_{11} = 0$: its metric is given by

$$ds^2 = dr^2 + 2du\,dv - \frac{4v}{r} du\,dr + \frac{v^2}{r^2} du^2 + r\,f_{01}(u)\,e^{-\mu r}\,du^2, \tag{19.70a}$$

with Cotton tensor components

$$C^{\alpha}{}_{\beta} = \frac{\mu^3}{2} r f_{01} e^{-\mu r} k^{\alpha} k_{\beta},\ k^{\alpha} = \delta^{\alpha}{}_{v},\ k^{\alpha} k_{\alpha} = 0. \tag{19.70b}$$

Flat spacetime occurs for $f_{01} = 0 (= f_{11})$:

$$ds^2 = dr^2 + 2du\, dv - \frac{4v}{r} du\, dr + \frac{v^2}{r^2} du^2 \to ds^2 = dr^2 + 4\frac{r^2}{(u-\tilde{v})^2} du\, d\tilde{v},$$

where it has been used the transformation $v = 2r^2/(u - \tilde{v})$.

19.9 Type III Kundt Solution; $\Lambda = -m^2, W_1 = -2m$

In the presence of Λ and constant structural function W_1, the only possible case is the one with negative Λ, $\Lambda = -m^2$; the equation (19.40) is solved by (19.41c), on the other hand, the equation for F_2, (19.24c), gives $F_2 = 0$, thus

$$W_1 = -2m \to F_2 = 0, \tag{19.71a}$$

while the equation (19.24d) for F_1 amounts to

$$\frac{\partial^2 F_1}{\partial r^2} + (\mu + m) F_1 = 0. \tag{19.71b}$$

The solutions of this equation depend on the relationship between μ and m, and determine the features of the whole corresponding metrics; in the forthcoming paragraphs each possible branch is determined on its own account. In the derivation of the various branches of solutions one may adopt the following pattern: first, integrate the general case $\mu \neq \pm m$, next particularize the derived results and key equations to the sub-cases $\mu = \pm m$ and proceed to integrate them.

19.9.1 *Type III Kundt Solution;* $\Lambda = -m^2, W_1 = -2m, \mu \neq \pm m$

For this class of solutions, the function F_1 fulfilling (19.71b) is

$$F_1(u, r) = e^{-(\mu+m)r} f_{11}(u) + f_{12}(u),\ f_{12}(u) \to 0, \tag{19.72a}$$

which, substituted in the equation (19.24e) for F_0, yields

$$e^{r(-\mu+m)} \frac{\partial}{\partial r}\left[e^{(m+\mu)r} \frac{\partial}{\partial r}\left(e^{-2mr} \frac{\partial F_0}{\partial r}\right)\right]$$
$$= \frac{1}{2}(\mu + m)\left(2\frac{df_{11}}{du} e^{-(\mu+m)r} + f_{11}{}^2 e^{-2(\mu+m)r}\right), \tag{19.72b}$$

with a general solution

$$F_0(u, r) = e^{(m-\mu)r} f_{01}(u) + e^{2mr} f_{02}(u) + f_{03}(u) - \frac{e^{-(m+\mu)r}}{2m(\mu + 3m)} \frac{df_{11}}{du}$$

$$-\frac{e^{-2(m+\mu)r}}{8(\mu+2m)(\mu+3m)}\, {f_{11}}^2. \tag{19.72c}$$

Finally one may write the corresponding Kundt solution as

$$\begin{aligned} ds^2 &= dr^2 + 2du dv - 4m\, v du\, dr + \left[v\, e^{-(\mu+m)r} f_{11} + F_0(u,r)\right] du^2, \\ F_0 &= e^{(m-\mu)r} f_{01} - \frac{e^{-(m+\mu)r}}{2m(\mu+3m)} \frac{df_{11}}{du} - \frac{e^{-2(m+\mu)r} {f_{11}}^2}{8(\mu+2m)(\mu+3m)}, \end{aligned} \tag{19.72d}$$

where f_{02} and f_{03} have been canceled by means of coordinate transformations.

This solution is characterized by a Petrov type III Cotton tensor

$$\begin{aligned} C^\alpha{}_\beta &= C^v{}_u \delta^\alpha{}_v \delta^u{}_\beta + C^r{}_u \delta^\alpha{}_r \delta^u{}_\beta + C^v{}_r \delta^\alpha{}_v \delta^r{}_\beta, \\ C^v{}_u &= -\frac{1}{4}\frac{\mu\,(\mu+m)}{3m+\mu} e^{-2(\mu+m)r} {f_{11}}^2 + \frac{v}{2}\mu\,(\mu+m)^2 f_{11} e^{-(\mu+m)r} \\ &\quad -\frac{1}{4}\frac{\mu\,(\mu+m)}{m} e^{-(\mu+m)r}\frac{df_{11}}{du} + \frac{1}{2}\mu\,(\mu^2-m^2)\, f_{01} e^{(-\mu+m)r} \\ C^v{}_r &= C^r{}_u = \frac{1}{2}\mu\,(\mu+m)\, f_{11}\, e^{-(\mu+m)r}. \end{aligned} \tag{19.72e}$$

Type N Limit

For $f_{11} = 0$ this solution becomes a Petrov type N spacetime with metric

$$ds^2 = dr^2 + 2du dv - 4m\, v du\, dr + e^{(m-\mu)r} f_{01}(u)\, du^2, \tag{19.73a}$$

and Cotton tensor

$$C^\alpha{}_\beta = \frac{1}{2}\mu\,(\mu^2-m^2)\, f_{01} e^{(-\mu+m)r} k^\alpha k_\beta,\ k^\alpha = \delta^\alpha{}_v. \tag{19.73b}$$

19.9.2 Type III Kundt Solution; $\Lambda = -m^2, W_1 = -2m, \mu = m$

For this class of solutions, the function F_1 fulfilling (19.71b) is

$$F_1(u,r) = e^{-2mr} f_{11}(u),\ f_{12}(u) \to 0, \tag{19.74a}$$

which, substituted in the equation (19.24e) for F_0, yields

$$\frac{\partial}{\partial r}\left[e^{2mr}\frac{\partial}{\partial r}\left(e^{-2mr}\frac{\partial F_0}{\partial r}\right)\right] = 2m\frac{df_{11}}{du}e^{-4mr} + m{f_{11}}^2 e^{-6mr} \tag{19.74b}$$

with general solution

$$F_0(u,r) = r\, f_{01}(u) + e^{2mr} f_{02}(u) + f_{03}(u) - \frac{e^{-2mr}}{8m^2}\frac{df_{11}}{du} - \frac{e^{-4mr}}{96m^2}{f_{11}}^2. \tag{19.74c}$$

Finally one may write this branch of Kundt solutions as

$$\begin{aligned} ds^2 &= dr^2 + 2du dv - 4m\, v du\, dr + \left[v\, e^{-2mr} f_{11}(u) + F_0(u,r)\right] du^2, \\ F_0 &= r\, f_{01}(u) - \frac{e^{-2mr}}{8m^2}\frac{df_{11}}{du} - \frac{e^{-4mr}}{96m^2}{f_{11}}^2, \end{aligned} \tag{19.74d}$$

where f_{02} and f_{03} have been canceled by coordinate transformations. This solution is characterized by the type III Cotton and traceless Ricci tensor of the form

$$\begin{aligned} C^{\alpha}{}_{\beta} &= C^{v}{}_{u}\delta^{\alpha}{}_{v}\delta^{u}{}_{\beta} + C^{r}{}_{u}\delta^{\alpha}{}_{r}\delta^{u}{}_{\beta} + C^{v}{}_{r}\delta^{\alpha}{}_{v}\delta^{r}{}_{\beta}, \\ C^{v}{}_{u} &= -\frac{m}{8}\,\mathrm{e}^{-4\,m\,r}\,{f_{11}}^{2} + 2\,v\,m^{3}\,f_{11}\mathrm{e}^{-2\,m\,r} - \frac{m}{2}\,\mathrm{e}^{-2\,m\,r}\frac{df_{11}}{du} \\ &\quad -m^{2}\,(f_{01}+f_{11})\,,\ C^{v}{}_{r} = C^{r}{}_{u} = m^{2}\,f_{11}\,\mathrm{e}^{-2\,m\,r}. \end{aligned} \tag{19.74e}$$

Type N Limit

Nevertheless, for $f_{11} = 0$ this spacetime becomes a type N solution given by

$$ds^2 = dr^2 + 2dudv - 4m\,vdu\,dr + r\,f_{01}(u)du^2, \tag{19.75a}$$

with a type N Cotton tensor

$$C^{\alpha}{}_{\beta} = -m^2 f_{01} k^{\alpha} k_{\beta},\ k^{\alpha} = \delta^{\alpha}{}_{v},\ k^{\alpha}\,k_{\alpha} = 0. \tag{19.75b}$$

Moreover, for $f_{01} = 0$ the metric describes an AdS spacetime – a type O metric field – given by

$$ds^2 = dr^2 + 2dudv - 4m\,vdu\,dr = dr^2 + 2v\,du\,d(\ln v - 2mr) = dr^2 + 2v\,du\,dV,$$

which can be reached via the coordinate transformation $V = \ln v - 2mr,\ v = \mathrm{e}^{V+2mr}$.

19.9.3 Type III Kundt Solution; $\boldsymbol{\Lambda = -m^2, W_1 = -2m, \mu = -m}$

For this class of metrics with $\mu = -m$, the function F_1 fulfilling (19.71b), $\frac{\partial^2}{\partial r^2} F_1\,(u,r) = 0$, is given by

$$F_1(u,r) = r\,f_{11}(u). \tag{19.76a}$$

The evaluation of F_2, (19.24c), yields $F_2 = 0$, therefore, using these results in F_0, (19.24e), leads to

$$\frac{\partial^3 F_0}{\partial r^3} + 4\,m^2\frac{\partial F_0}{\partial r} - 4\,m\frac{\partial^2 F_0}{\partial r^2} = -\frac{1}{2}\,(f_{11})^2\,r - \frac{df_{11}}{du}, \tag{19.76b}$$

with general solution

$$F_0 = (r\,f_{01}(u) + f_{02}(u))\,\mathrm{e}^{2\,m\,r} + f_{03}(u) - \frac{r}{4m^2}\frac{df_{11}}{du} - \frac{r(2+mr)}{16\,m^3}\,{f_{11}}^2. \tag{19.76c}$$

Finally this Petrov type III Kundt solution can be written as

$$\begin{aligned} ds^2 &= dr^2 + 2dudv - 4m\,vdu\,dr + [v\,r\,f_{11}(u) + F_0(u,r)]\,du^2, \\ F_0 &= r\mathrm{e}^{2mr}\,f_{01}(u) - \frac{r}{4m^2}\frac{df_{11}}{du} - \frac{r(2+mr)}{16\,m^3}\,{f_{11}}^2, \end{aligned} \tag{19.76d}$$

where the functions $f_{02}(u)$ and $f_{03}(u)$ have been eliminated via coordinate transformations.

This solution is characterized by the Cotton and traceless Ricci tensor of the form

$$\begin{aligned} C^{\alpha}{}_{\beta} &= C^{v}{}_{u}\delta^{\alpha}{}_{v}\delta^{u}{}_{\beta} + C^{r}{}_{u}\delta^{\alpha}{}_{r}\delta^{u}{}_{\beta} + C^{v}{}_{r}\delta^{\alpha}{}_{v}\delta^{r}{}_{\beta}, \\ C^{v}{}_{u} &= -m^2 \mathrm{e}^{2\,m\,r} f_{01} - \frac{1}{4}\frac{df_{11}}{du} - \frac{1}{16}\frac{1+2\,m\,r}{m}\left(f_{11}\right)^2, \\ C^{v}{}_{r} &= C^{r}{}_{u} = \frac{1}{2} m f_{11}. \end{aligned} \tag{19.76e}$$

Type N Limit

Again, as in the previous branch of solutions, one finds the type N subcase arising for $f_{11} = 0$, namely

$$ds^2 = dr^2 + 2dudv - 4m\, vdu\, dr + re^{2mr} f_{01}(u) du^2, \tag{19.77a}$$

with a type N Cotton tensor

$$C^{\alpha}{}_{\beta} = -m^2 \mathrm{e}^{2\,m\,r} f_{01} k^{\alpha} k_{\beta},\ k^{\alpha} = \delta^{\alpha}{}_{v}. \tag{19.77b}$$

Moreover, for $f_{01} = 0$ the metric describes an AdS Type O spacetime.

19.10 Type III Kundt Metric; $\Lambda = -m^2$, $W_1 = -2\,m\,\coth(m\,r)$

The equation for vanishing eigenvalue S, (19.24b), which guarantees the Petrov type III of the Cotton tensor, for negative cosmological constant $\Lambda = -m^2$, allows for a solution of the form (19.41d), and evaluating the equation for F_2 (19.24c) one gets

$$W_1(u,r) = -2\,m\,\coth(m\,r) \rightarrow F_2 = \frac{m^2}{\sinh^2(m\,r)}. \tag{19.78}$$

Replacing these functions into the F_1 equation (19.24d) one arrives at

$$\frac{\partial^2}{\partial r^2} F_1(u,r) + [m\,\coth(m\,r) + \mu]\frac{\partial}{\partial r} F_1(u,r) = 0 \tag{19.79}$$

with solution

$$F_1(u,r) = f_{11}(u) \int \frac{m\,\mathrm{e}^{-\mu\,r}}{\sinh(m\,r)} dr = f_{11}\,\xi(r), \tag{19.80}$$

where it has been introduced the auxiliary function $\xi(r)$,

$$\frac{d\xi}{dr} = \frac{m\mathrm{e}^{-\mu\,r}}{\sinh(m\,r)} \rightarrow \xi(r) = \int \frac{m\,\mathrm{e}^{-\mu\,r}}{\sinh(m\,r)} dr. \tag{19.81}$$

The key equation (19.24e) for $F_0(u,r)$ becomes

$$\begin{aligned}
&\frac{\partial^3}{\partial r^3}F_0 + [\mu - 3m\,\coth(m\,r)]\,\frac{\partial^2}{\partial r^2}F_0 \\
&-2\,m\,\left[\mu\,\coth(m\,r) + 2\,m - 3\,m\,\coth^2(m\,r)\right]\frac{\partial}{\partial r}F_0 \\
&+2\,m^2\,\left[\mu\coth^2(m\,r) + 3\,m\,\coth(m\,r) - 3\,m\,\coth^3(m\,r) - \mu\right]F_0 = RHS, \\
&RHS := -\frac{1}{2}\left[2\,\frac{df_{11}}{du} + {f_{11}}^2\,\xi(r)\right]\frac{d}{dr}\xi(r).
\end{aligned} \tag{19.82}$$

Following the standard procedure of solving non-homogeneous linear ordinary differential equations by means of the variation of parameters of the determined general homogeneous solution, which in the present case amounts to

$$F_0 = \sinh(m\,r)\,\left(C_1(u)\,\mathrm{e}^{-\mu\,r} + C_2(u)\,\mathrm{e}^{-m\,r} + C_3(u)\,\mathrm{e}^{m\,r}\right), \tag{19.83}$$

one should arrive at an algebraic system of equations for the first-order derivatives with respect to the variable r of $C_i(u,r)$ considered at this stage of function of r too, beside of u. The integration of these derivatives, which depend on the RHS function, allows one to construct the non-homogeneous solution which added up to the homogeneous one, gives the sought general solution. Nevertheless, the structure of the homogeneous solution above points out to a better function choice, namely of an auxiliary function , $\Phi(u,r)$, in terms of which

$$F_0(u,r) = \sinh(m\,r)\;\Phi(u,r). \tag{19.84}$$

The key equation (19.82) in terms of Φ becomes

$$\frac{\partial^3}{\partial r^3}\Phi + \mu\,\frac{\partial^2}{\partial r^2}\Phi - m^2\frac{\partial}{\partial r}\Phi - m^2\mu\,\Phi = \frac{RHS}{\sinh(m\,r)} =: Rhs. \tag{19.85}$$

In what follows we proceed to determine the three different branches of solutions in the specific cases of $\mu \neq \pm m$, $\mu = -m$, and $\mu = m$ separately.

19.10.1 Type III Kundt Solution; $\Lambda = -m^2$, $W_1 = -2m\,\coth(m\,r)$, $\mu \neq \pm m$

Solution Through the Variation of Parameters

The homogeneous solution to the homogeneous equation (19.85) ($Rhs \to 0$), one searches in the form of $\Phi_H \simeq \mathrm{e}^{k\,r}$ arriving at the characteristic equation

$$k^3 + \mu\,k^2 - m^2\,k - m^2\mu = 0 = (k+\mu)(k-m)(k+m)$$

with solutions $k = -\mu$, $-m$, m, hence Φ_H is given by

$$\Phi(r) = C_1\,\mathrm{e}^{-\mu\,r} + C_2\,\mathrm{e}^{m\,r} + C_3\,\mathrm{e}^{-m\,r}. \tag{19.86a}$$

Thus the set of independent solutions amounts to

$$\left\{\Phi_1 = \mathrm{e}^{-\mu\,r}, \Phi_2 = \mathrm{e}^{m\,r}, \Phi_3 = \mathrm{e}^{-m\,r}\right\} \tag{19.86b}$$

Looking for the solution to the non-homogeneous equation ($Rhs \neq 0$) – the non-homogeneous solution – one assumes it to be of the form

$$\Phi\,(r) = C_1(u,r)\,\Phi_1 + C_2(u,r)\,\Phi_2 + C_3(u,r)\,\Phi_3, \tag{19.86c}$$

where u enters here as a parameter.

The variation of the parameters $C_i(u,r)$, within the VP procedure, yields

$$\begin{aligned}
\frac{\partial C_1}{\partial r} &= \frac{Rhs\,e^{\mu\,r}}{\mu^2 - m^2},\; C_1 = \frac{m}{2}\int \frac{\left[f_{11}{}^2\xi\,(r) + 2\,\frac{d}{du}\,(f_{11})\right]}{(m^2-\mu^2)\sinh^2(mr)}dr,\\
\frac{\partial C_2}{\partial r} &= \frac{1}{2}\frac{Rhs\,e^{-m\,r}}{m\,(m+\mu)},\; C_2 = -\frac{1}{4}\int e^{-(\mu+m)\,r}\frac{\left[f_{11}{}^2\xi\,(r) + 2\,\frac{d}{du}\,(f_{11})\right]}{(m+\mu)\sinh^2(m\,r)}dr,\\
\frac{\partial C_3}{\partial r} &= \frac{1}{2}\frac{Rhs\,e^{m\,r}}{m\,(m-\mu)},\; C_3 = -\frac{1}{4}\int \frac{e^{(m-\mu)\,r}\left[f_{11}{}^2\xi\,(r) + 2\,\frac{d}{du}\,(f_{11})\right]}{(m-\mu)\sinh^2(m\,r)}dr,
\end{aligned}$$

where it has been used $d\xi/dr$ from (19.81).

Notice the presence of $\mu \pm m$ in the denominators of the above expressions, which clearly point on the degeneracy of these quantities when dealing with the sub-branches $\mu = \pm m$; this is the reason how the quoted cases has to be integrated separately for the Φ-equation (19.85). Continuing with the general case $\mu \neq \pm m$, one may factorize the terms in the form

$$\begin{aligned}
\Phi_{NH} = \frac{df_{11}}{du}&\left[\frac{e^{-\mu\,r}}{m^2-\mu^2}\int\frac{dr}{\sinh^2(mr)}\right.\\
&\left.-\frac{1}{2}\frac{e^{mr}}{(m+\mu)}\int\frac{e^{-(m+\mu)r}}{\sinh^2(mr)}dr - \frac{1}{2}\frac{e^{-mr}}{(m-\mu)}\int\frac{e^{(m-\mu)r}}{\sinh^2(mr)}dr\right]\\
+f_{11}{}^2&\left[\frac{1}{2}\frac{e^{-\mu r}}{m^2-\mu^2}\int\frac{\xi\,(r)}{\sinh^2(mr)}dr - \frac{1}{4}\frac{e^{-m\,r}}{(m-\mu)}\int\frac{e^{(m-\mu)r}\xi\,(r)}{\sinh^2(mr)}dr\right.\\
&\left.-\frac{1}{4}\frac{e^{mr}}{m+\mu}\int\frac{e^{-(m+\mu)r}\xi\,(r)}{\sinh^2(m\,r)}dr\right],
\end{aligned}\tag{19.86d}$$

whereas the homogeneous solution is given by

$$\Phi_H(u,r) = f_{01}(u)\,e^{-r\mu} + f_{02}(u)\,e^{m\,r} + f_{03}(u)\,e^{-m\,r}, \tag{19.86e}$$

which together determine the general metric

$$\begin{aligned}
ds^2 &= dr^2 + 2du\,dv + (v^2F_2 + vF_1 + F_0)du^2 - 4m\,v\coth(m\,r)\,du\,dr,\\
F_2(u,r) &= \frac{m^2}{\sinh^2(m\,r)},\\
F_1(u,r) &= f_{11}(u)\xi(r),\; \xi(r) = \int\frac{m\,e^{-\mu\,r}}{\sinh(m\,r)}dr,\\
F_0(u,r) &= \sinh(m\,r)\,\left[\Phi_H(u,r) + \Phi_{NH}(u,r)(19.86d)\right].
\end{aligned}\tag{19.86f}$$

This solution is characterized by the Cotton:

$$
\begin{aligned}
C^{\alpha}{}_{\beta} &= C^{v}{}_{u}\delta^{\alpha}{}_{v}\delta^{u}{}_{\beta} + C^{v}{}_{r}\delta^{\alpha}{}_{v}\delta^{r}{}_{\beta} + C^{r}{}_{u}\delta^{\alpha}{}_{r}\delta^{u}{}_{\beta},\\
C^{v}{}_{u} &= \frac{\mu}{2}\left(\mu^2 - m^2\right) f_{03}(u)\,\mathrm{e}^{-\mu r}\sinh(m\,r)\\
&\quad -\frac{\mu\,m^2}{4} f_{11}{}^2\sinh(m\,r)\,\mathrm{e}^{-\mu r}\int^r\left[\frac{1}{\sinh^2(m\,y)}\int^y\left(\frac{\mathrm{e}^{-\mu\,z}}{\sinh(m\,z)}\right)dz\right]dy\\
&\quad -\frac{\mu m v}{2} f_{11}\,\mathrm{e}^{-\mu\,r}\frac{m\cosh(m\,r) + \mu\sinh(m\,r)}{\sinh^2(m\,r)} + \frac{\mu}{2}\frac{df_{11}}{du}\,\mathrm{e}^{-\mu\,r}\cosh(m\,r)\,,\\
C^{r}{}_{u} &= C^{v}{}_{r} = -\frac{m\mu}{2} f_{11}\frac{\mathrm{e}^{-\mu\,r}}{\sinh(m\,r)}.
\end{aligned}
\tag{19.86g}
$$

It should be pointed out that via the variations of parameters, the general solution is determined by a linear combinations of linear (single) integrals.

Type N Limit

The solution presented above (19.86f) allows for a type N Kundt subcase with metric

$$
\begin{aligned}
ds^2 &= dr^2 + 2du\,dv + (v^2F_2 + F_0)du^2 - 4\,m\,v\,\coth(m\,r)\,du\,dr,\\
F_2(u,r) &= \frac{m^2}{\sinh^2(m\,r)},\quad F_0(u,r) = \sinh(m\,r)\,f_{03}(u)\,\mathrm{e}^{-rm},
\end{aligned}
\tag{19.87a}
$$

with a type N Cotton tensor, $\mathcal{J}_{CN}$,

$$
C^{\alpha}{}_{\beta} = C^{v}{}_{u}k^{\alpha}k_{\beta},\,,\ C^{v}{}_{u} = \frac{\mu}{2}\left(\mu^2 - m^2\right) f_{03}(u)\,\mathrm{e}^{-\mu\,r}\sinh(m\,r)\,,\ k^{\alpha} = \delta^{\alpha}{}_{v}.
\tag{19.87b}
$$

Multi Exponent–Integral Representation of Φ; General Procedure

For a constant coefficients' equation involving up to n, $i = 1,\cdots,n$, order derivatives

$$
\sum_i^n a_i\frac{d^i}{dr^i}\phi(r) = f(r),\ i = 0,1,\cdots,n
\tag{19.88}
$$

where without loss of generality one may assume the highest coefficient $a_n = 1$, beside the VP procedure based on the determination of the roots of the characteristic equation, one can search for an expression of the equation of the form

$$
\frac{d}{dr}\left(\mathrm{e}^{\alpha_n r}\frac{d}{dr}\left(\mathrm{e}^{\alpha_{n-1} r}\frac{d}{dr}\cdots\frac{d}{dr}\left(\mathrm{e}^{\alpha_1 r}\phi(r)\right)\right)\right) = f(r)\mathrm{e}^{(\alpha_1+\alpha_2+\cdots+\alpha_n)r},
\tag{19.89}
$$

with n in number derivative operators $\frac{d}{dr}$. Comparing the coefficients a_i of (19.88) with those of (19.89) multiplying derivatives $\frac{d^i}{dr^i}\phi$ of the same order in both equations, one obtains a system of equations for α_i in terms of the known a_i, of the form

$$
\begin{aligned}
&1 = 1,\\
&n\alpha_1 + (n-1)\alpha_2 + \cdots + 1\,\alpha_n = a_{n-1},\\
&\cdots\cdots\cdots\cdots\cdots\cdots\cdots\cdots\cdots\cdots\\
&\alpha_1(\alpha_1+\alpha_2)\cdots(\alpha_1+\alpha_2+\cdots+\alpha_{n-1}+\alpha_n) = a_0,
\end{aligned}
\tag{19.90}
$$

whose solutions determine the exponents. Finally, integrating (19.89), one arrives at a compact expression for the main part of the non-homogeneous solution

$$
\phi_{NH}(r) = \mathrm{e}^{-\alpha_1 r}\int^r \left(\mathrm{e}^{-\alpha_2\,t}\cdots\left(\int^y f(z)\mathrm{e}^{(\alpha_1+\alpha_2+\cdots+\alpha_n)\,z}dz\right)dy\cdots\right)dt.
\tag{19.91}
$$

The general solution for the equation (19.88) with constant coefficients is just the sum of $\phi_{NH}(r)$ and the n-independent solutions to the homogeneous equation, which are of the exponential form $\mathrm{e}^{\alpha_i r}$ multiplied by powers of r in the case of degeneracy.

This procedure will be used extensively in the derivation of the solutions to the non-homogeneous equations, thus we shall refer to this subsection as 19.10.1.

Multi Exponent–Integral Representation of $\Phi(u,r)$

Returning to the search of the non-homogeneous solution to the equation (19.85) written in multi-exponent form

$$
\frac{\partial}{\partial r}\left(\mathrm{e}^{\gamma r}\frac{\partial}{\partial r}\left(\mathrm{e}^{\beta r}\frac{\partial}{\partial r}\left(\mathrm{e}^{\alpha r}\Phi(u,r)\right)\right)\right) = Rhs(u,r)\,\mathrm{e}^{(\alpha+\beta+\gamma)r}
\tag{19.92a}
$$

or accomplishing the differentiations

$$
\begin{aligned}
&\frac{\partial^3}{\partial r^3}\Phi + (\gamma + 3\,\alpha + 2\,\beta)\,\frac{\partial^2}{\partial r^2}\Phi + \left(2\,\gamma\,\alpha + \beta^2 + 3\,\alpha^2 + 4\,\beta\,\alpha + \gamma\,\beta\right)\frac{\partial}{\partial r}\Phi\\
&\qquad + \left(\gamma\,\beta\,\alpha + \gamma\,\alpha^2 + 2\,\beta\,\alpha^2 + \beta^2\alpha + \alpha^3\right)\Phi = Rhs\,(u,r)\,.
\end{aligned}
\tag{19.92b}
$$

Comparing the coefficients multiplying derivatives of the same order in the expanded equation (19.92b) with the corresponding of (19.85), one gets the system

$$
\begin{aligned}
&2\,\beta + 3\,\alpha + \gamma = \mu,\\
&\gamma\,\beta + 4\,\beta\,\alpha + 3\,\alpha^2 + 2\,\gamma\,\alpha + \beta^2 = -m^2,\\
&\gamma\,\beta\,\alpha + \gamma\,\alpha^2 + \beta^2\alpha + \alpha^3 + 2\,\beta\,\alpha^2 = -m^2\mu,
\end{aligned}
\tag{19.92c}
$$

with the following set of solutions

$$
\begin{aligned}
&\{\alpha = \mp m, \beta = \pm 2\,m, \gamma = \mp m + \mu\}\,,\ sum := \alpha + \beta + \gamma = \mu,\\
&\{\alpha = \mu, \beta = -\mu \mp m, \gamma = \pm 2\,m\}\,,\ sum = \pm\,m,\\
&\{\alpha = \mp m, \beta = \mu \pm m, \gamma = \pm\,m - \mu\}\,,\ sum = \pm\,m,
\end{aligned}
\tag{19.92d}
$$

where the upper (lower) signs match with the upper (lower). The substitution of the found exponents in the expression

$$\Phi(u,r) = \mathrm{e}^{-\alpha r} \int^{r} (\mathrm{e}^{-\beta t} \int^{t} (\mathrm{e}^{-\gamma y} \int^{y} (Rsh(u,z)\, \mathrm{e}^{z(\gamma+\beta+\alpha)})dz)dy)dt, \tag{19.92e}$$

determines an specific representation of the sought non-homogeneous solution. For instance, for the first set of exponents, one has

$$\begin{aligned}\Phi = & -\frac{df_{11}}{du}\mathrm{e}^{m\,r} \int^{r} \left(\mathrm{e}^{-2m\,t} \int^{t} \left(\mathrm{e}^{-(\mu-m)\,y} \int^{y} \left(\frac{\mathrm{e}^{\mu\,z}}{\sinh{(m\,z)}}\frac{d\xi}{dz}\right)dz\right)dy\right)dt \\ & -\frac{{f_{11}}^2}{2}\,\mathrm{e}^{m\,r} \int^{r} \left(\mathrm{e}^{-2m\,t} \int^{t} \left(\mathrm{e}^{-(\mu-m)\,y} \int^{y} \left(\frac{\mathrm{e}^{\mu\,z}\xi\,(z)}{\sinh{(m\,z)}}\frac{d\xi}{dz}\right)dz\right)dy\right)dt,\end{aligned} \tag{19.92f}$$

which determines in turn the non-homogeneous $F_0(u,r)$ function through

$$F_{0NH}(u,r) = \sinh(mr)\,\Phi(u,r). \tag{19.92g}$$

Consequently, the general solution

$$F_0(u,r) = \sinh(m\,r)\,\left(\mathrm{e}^{-\mu\,r}\,f_{01} + \mathrm{e}^{m\,r}\,f_{02} + \mathrm{e}^{-m\,r}\,f_{03} + \Phi(u,r)\right). \tag{19.92h}$$

Of course, this expression can be simplified in various manners; for instance one can partially integrate the first term in $\Phi(u,r)$ using the derivative $\frac{d\xi}{dr}$ from (19.81).

Finally, the metric becomes

$$\begin{aligned}ds^2 = & \, dr^2 + 2du\,dv + m^2v^2\mathrm{cosech}^2(mr)du^2 - 4\,m\,v\,\coth(m\,r)\,du\,dv \\ & + [v\xi(r)f_{11}(u) + F_0(u,r)(19.92h)]du^2.\end{aligned} \tag{19.92i}$$

VP of the CPS Representation of $F_0(u,r)$ through ϕ

The Chow et al. (2010b) representation of the solution for $F_0(u,r)$ is achieved via an alternative definition of the function $F_0(u,r)$ through a new auxiliary function $\phi(u,r)$ according to

$$\phi(u,r) = \frac{F_0(u,r)}{1-\mathrm{e}^{-2mr}},\ \Phi(u,r) = \frac{F_0(u,r)}{\sinh(mr)} = 2\,\mathrm{e}^{-mr}\phi(u,r), \tag{19.93a}$$

which brings the equation (19.85), (19.24e), to the form

$$\begin{aligned}\frac{\partial^3}{\partial r^3}\phi\ + (\mu - 3\,m)\,\frac{\partial^2}{\partial r^2}\phi + 2\,m\,(m-\mu)\,\frac{\partial}{\partial r}\phi = & \frac{RHS}{(1-\mathrm{e}^{-2\,mr})} = Rhs(u,r) \\ = -\frac{m\mathrm{e}^{-(\mu+m)r}\left({f_{11}}^2\xi\,(r) + 2\,\frac{d}{du}f_{11}\right)}{(1-\mathrm{e}^{-2\,mr})^2}.\end{aligned} \tag{19.93b}$$

The solution to the homogeneous equation is given by

$$\begin{aligned}\phi_H\,(u,r) = & \, C_1(u) + C_2(u)\,\mathrm{e}^{2\,mr} + C_3(u)\,\mathrm{e}^{(-\mu+m)r} \\ \equiv & \, C_1(u)\phi_1 + C_2(u)\,\phi_2 + C_3(u)\,\phi_3.\end{aligned} \tag{19.93c}$$

The non-homogeneous solution is searched in the form of

$$\phi_{NH}\,(r) = C_1(u,r)\phi_1 + C_2(u,r)\,\phi_2 + C_3(u,r)\,\phi_3 \tag{19.93d}$$

via the variation of the parameters $C_i(u,r)$, where u plays the role of a parameter. Thus their first derivatives are

$$\frac{\partial C_1}{\partial r} = \frac{Rhs}{2m(m-\mu)},\ \frac{\partial C_2}{\partial r} = \frac{e^{-2mr}}{2m(m+\mu)}Rhs,\ \frac{\partial C_3}{\partial r} = -\frac{e^{-(m-\mu)r}}{(m^2-\mu^2)}Rhs. \tag{19.93e}$$

Therefore the general $\phi(u,r)$ solution to the key equation (19.93b) via the variation of parameters method is

$$\begin{aligned}\phi_{NH} = {} & \frac{1}{4}e^{r(m-\mu)}\int \frac{m e^{r(m-\mu)}\left[2\frac{d}{du}f_{11} + {f_{11}}^2\xi\,(r)\right]e^{-r(m-\mu)}}{\sinh^2(m\,r)\,(m^2-\mu^2)}dr \\ & -\frac{1}{8}e^{2\,m\,r}\int \frac{e^{r(m-\mu)}\left[2\frac{d}{du}f_{11} + {f_{11}}^2\xi\,(r)\right]e^{-2\,m\,r}}{\sinh^2(m\,r)\,(\mu+m)}dr \\ & -\frac{1}{8}\int \frac{e^{r(m-\mu)}\left[2\frac{d}{du}f_{11} + {f_{11}}^2\xi\,(r)\right]}{\sinh^2(m\,r)\,(m-\mu)}dr,\end{aligned} \tag{19.93f}$$

which is a linear (single) integral representation of the solution under consideration. Of course, this representation is far from the triple integral CPS Chow et al. (2010b) representation (19.94j) of the next paragraph 19.10.2.

19.10.2 Multi Exponent–Integral Representation of ϕ

To obtain a CPS Chow et al. (2010b) representation of the solution to the key equation (19.93b) in terms of the function ϕ one has to follow the multi-exponent–integral procedure previously detailed in 19.10.1. Introducing an auxiliary function $H = \frac{\partial}{\partial r}\phi$ we rewrite the equation (19.93b) as

$$\frac{\partial}{\partial r}\left(e^{\beta\,r}\frac{\partial}{\partial r}\left(e^{\alpha\,r}H(u,r)\right)\right) = Rhs(u,r)\,e^{(\alpha+\beta)\,r},$$

$$H(u,r) = e^{-\alpha\,r}\int^r e^{-\beta\,t}\int^t Rhs(u,z)\,e^{(\alpha+\beta)\,z}dzdt, \tag{19.94a}$$

accomplishing the derivation one gets

$$\frac{\partial^2}{\partial r^2}H + (\beta + 2\,\alpha)\frac{\partial}{\partial r}H + \alpha\,(\beta+\alpha)\,H = Rhs. \tag{19.94b}$$

Comparing the coefficients of (19.94b) with those of (19.93b) one gets the system

$$\{\alpha\,(\beta+\alpha) = -2\,m\,(\mu-m)\,, \beta + 2\,\alpha = \mu - 3\,m\} \tag{19.94c}$$

with solutions

$$[[\alpha = -2\,m, \beta = m+\mu], [\alpha = \mu - m, \beta = -m-\mu]]. \tag{19.94d}$$

For the first set of solutions one obtains

$$H(u,r) = \mathrm{e}^{2m\,r} \int^r \mathrm{e}^{-(\mu+m)\,t} \int^t Rhs(u,z)\mathrm{e}^{(\mu-m)\,z} dzdt. \tag{19.94e}$$

The integral for ϕ is

$$\phi(u,r) = -\frac{1}{4} \int^r \mathrm{e}^{2\,m\,t} \int^t \mathrm{e}^{-(m+\mu)y} \int^y \frac{m\left[2\,\frac{df_{11}}{du} + {f_{11}}^2 \xi\,(z)\right]}{\sinh^2(mz)} dzdydt, \tag{19.94f}$$

which leads to the CPS $F_0(u,r)$ from (19.93a),

$$F_0(u,r) = 2\,\mathrm{e}^{-\,m\,r}\,\sinh(mr)\;\phi(u,r). \tag{19.94g}$$

Incidentally an integration of $1/\sinh^2(m\,r)$ in the term with factor $\frac{d}{du} f_{11}$ can be carried out, hence one may write

$$\begin{aligned}\phi =& \frac{1}{2}\frac{df_{11}}{du} \int^r \mathrm{e}^{2\,m\,t} \int^t \frac{\mathrm{e}^{-(\mu+m)\,z}\,\cosh(m\,z)}{\sinh(m\,z)} dzdt \\ &- \frac{m}{4} {f_{11}}^2 \int^r \mathrm{e}^{2\,m\,t} \int^t \mathrm{e}^{-(\mu+m)y} \int^y \frac{\xi\,(z)}{\sinh^2(m\,z)} dzdydt;\end{aligned} \tag{19.94h}$$

moreover, one has the relation

$$\frac{\mathrm{e}^{-r(\mu+m)}\cosh(m\,r)}{\sinh(m\,r)} = \frac{\mathrm{e}^{-\mu\,r}}{\sinh(rm)} - \mathrm{e}^{-(\mu+m)\,r}$$

therefore, the integration of this term in the factor of $\frac{d}{du} f_{11}$ yields

$$\int \frac{\mathrm{e}^{-r(\mu+m)}\cosh(m\,r)}{\sinh(m\,r)} dr = \int \frac{\mathrm{e}^{-\mu\,r}}{\sinh(m\,r)} dr + \frac{\mathrm{e}^{-\mu\,r - m\,r}}{\mu+m}$$

which leads to the sought solution ϕ in the form

$$\begin{aligned}\phi = \phi_H + \phi_{NH} =& f_{03}(u) + f_{02}(u)\mathrm{e}^{2mr} + f_{01}(u)\mathrm{e}^{(m-\mu)r} \\ &+ \frac{1}{2\,m}\frac{df_{11}}{du} \int^r dt' \mathrm{e}^{2\,m\,t'} \xi(t') \\ &- \frac{m}{4} {f_{11}}^2 \int^r \mathrm{e}^{2mt} dt \int^t dt' \mathrm{e}^{-(m+\mu)t'} \int^{t'} dt'' \frac{\xi(t'')}{\sinh^2(mt'')},\end{aligned} \tag{19.94i}$$

which coincides, modulo the identification $\xi \to F$, with the triple integral CPS solution representation given by the expression

$$\begin{aligned}\frac{F_0}{1-\mathrm{e}^{-2mr}} =& \mathrm{e}^{(m-\mu)r} f_{01} + \mathrm{e}^{2mr} f_{02} + f_{03} + \frac{1}{2m}\frac{df_{11}}{du} \int_\infty^r \mathrm{e}^{2mt} F(t) dt \\ &- \frac{m}{4} {f_{11}}^2 \int_\infty^r dt \int_\infty^t dt' \int_\infty^{t'} dt'' \frac{\mathrm{e}^{2m(t-t')}\mathrm{e}^{(m-\mu)t'}}{\sinh^2(mt'')} F(t''),\end{aligned} \tag{19.94j}$$

which determines the metric CPSb(A.32),

$$\begin{aligned}
&ds^2 = dr^2 + 2du\,dv + (v^2 F_2 + vF_1 + F_0)du^2 - 4\,m\,v\,\coth(m\,r)\,du\,dr,\\
&F_2(u,r) = \frac{m^2}{\sinh^2(m\,r)},\, F_1(u,r) = f_{11}(u)\xi(r),\ \xi(r) = \int \frac{m\,e^{-\mu\,r}}{\sinh(m\,r)}dr,\\
&F_0(u,r) = (1 - e^{-2mr})\,\phi(19.94i).
\end{aligned} \tag{19.94k}$$

19.10.3 Type III Kundt Solution; $\Lambda = -m^2$, $W_1 = -2m\coth(m\,r)$, $\mu = -m$

Solution F_0 through Φ via Variation of Parameters

The equation (19.85) for the function $\Phi(u,r)$ in the case $\mu = -m$ becomes

$$\begin{aligned}
&\frac{\partial^3}{\partial r^3}\Phi - m\frac{\partial^2}{\partial r^2}\Phi - m^2\frac{\partial}{\partial r}\Phi + m^3\Phi = \frac{RHS}{\sinh(m\,r)} = Rhs\\
&= -\frac{1}{2}\,\frac{me^{m\,r}\left[{f_{11}}^2\xi(r) + 2\frac{d}{du}f_{11}\right]}{\sinh^2(m\,r)}.
\end{aligned} \tag{19.95a}$$

Since the characteristic equation $(k+m)(k-m)^2 = 0$ allows for a double root, then the homogeneous solution is

$$\Phi_(H)\,(u\,r) = C_1\,e^{-m\,r} + C_2 e^{m\,r} + C_3\,r\,e^{m\,r}. \tag{19.95b}$$

Using the VP procedure for the non-homogeneous proposal

$$\Phi_{NH}\,(r) = C_1(u,r)\,e^{-m\,r} + C_2(u,r)\,e^{m\,r} + C_3(u,r)\,r\,e^{m\,r} \tag{19.95c}$$

one arrives at first-order equations for $C_i(u,r)$

$$\frac{\partial}{\partial r}C_1 = \frac{Rhs\,e^{m\,r}}{4m^2},\ \frac{\partial}{\partial r}C_2 = -\frac{Rhs\,(2\,m\,r+1)\,e^{-m\,r}}{4m^2},\ \frac{\partial}{\partial r}C_3 = \frac{e^{-m\,r}Rhs}{2m}.$$

Hence, factorizing terms with $(\frac{d}{du}f_{11})$ and ${f_{11}}^2$ one may write the non-homogeneous solution as

$$\begin{aligned}
\Phi_{NH}(u,r) = &-\frac{1}{4}\frac{df_{11}}{du}\left[e^{-m\,r}\int\frac{e^{2m\,r}}{m\sinh^2(m\,r)}dr - e^{m\,r}\int\frac{2\,m\,r+1}{m\sinh^2(m\,r)}dr\right.\\
&\left.+2\,r\,e^{m\,r}\int\frac{1}{\sinh^2(m\,r)}dr\right]\\
&-\frac{1}{8}{f_{11}}^2\left[e^{-m\,r}\int\frac{e^{2\,m\,r}\,\xi(r)}{m\sinh^2(mr)}dr - e^{m\,r}\int\frac{(2\,mr+1)\,\xi(r)}{m\,\sinh^2(mr)}dr\right.\\
&\left.+2\,r\,e^{m\,r}\int\frac{\xi(r)}{\sinh^2(mr)}dr\right],\\
\xi(r) = \ln&\left(1 - e^{-2\,m\,r}\right) = -m\,r + \ln(\sinh(m\,r)).
\end{aligned} \tag{19.95d}$$

Multi Exponent–Integral Representation of Φ

Let us search for a multi-exponent–integral representation, see 19.10.1, of the solution to the equation (19.95a). Comparing the coefficients of the derivatives of the same order one arrives at the system

$$\begin{aligned} &\gamma + 3\,\alpha + 2\,\beta = -m, \\ &2\gamma\,\alpha + \beta^2 + 3\,\alpha^2 + 4\,\beta\,\alpha + \gamma\,\beta = -m^2, \\ &\gamma\,\beta\,\alpha + \gamma\,\alpha^2 + 2\,\beta\,\alpha^2 + \beta^2\alpha + \alpha^3 = m^3 \end{aligned} \tag{19.96a}$$

with solutions

$$\begin{aligned} &\{\alpha = m, \beta = -2\,m, \gamma = 0\}\,, \{\alpha = -m, \beta = 2\,m, \gamma = -2\,m,\}\,, \\ &\{\alpha = -m, \beta = 0, \gamma = 2\,m\}\,. \end{aligned} \tag{19.96b}$$

The function Φ in this approach is expressible as

$$\Phi = \mathrm{e}^{-\alpha\,r} \int^r \mathrm{e}^{-\beta\,t} \int^t \mathrm{e}^{-\gamma\,y} \int^y \mathrm{e}^{(\gamma+\beta+\alpha)z}\, Rhs\,(u,z)\, dzdydt, \tag{19.96c}$$

thus, for each set of exponential constants one gets its corresponding exponent–integral representation.

VP of the CPS Representation of ϕ

The CPS representation of the solution for $F_0(u,r)$ is achieved via the definition of the function $F_0(u,r)$ in terms of $\phi(u,r)$ as

$$\phi(u,r) = \frac{F_0(u,r)}{1-\mathrm{e}^{-2mr}} \to \phi(u,r) = \frac{1}{2}\mathrm{e}^{mr}\,\Phi(u,r). \tag{19.97a}$$

The equation (19.95a), (19.24e), for ϕ becomes

$$\frac{\partial^3}{\partial r^3}\phi - 4m\frac{\partial^2}{\partial r^2}\phi + 4\,m^2\frac{\partial}{\partial r}\phi = -m\,\frac{\left[{f_{11}}^2\xi\,(r) + 2\,\frac{d}{du}f_{11}\right]}{(1-\mathrm{e}^{-2\,mr})^2} = Rhs. \tag{19.97b}$$

The solution to the homogeneous equation is given by

$$\begin{aligned} \phi_H\,(u,r) &= C_1(u) + C_2(u)\,\mathrm{e}^{2\,mr} + C_3(u)\,\mathrm{e}^{2\,mr}\,r \\ &\equiv C_1(u)\phi_1 + C_2(u)\,\phi_2 + C_3(u)\,\phi_3. \end{aligned} \tag{19.97c}$$

The non-homogeneous solution is sought in the form of

$$\phi_{NH}\,(u,r) = C_1(u,r)\phi_1 + C_2(u,r)\,\phi_2 + C_3(u,r)\,\phi_3 \tag{19.97d}$$

via the variation of parameters $C_i(u,r)$, where u plays the role of a parameter.

The first derivatives of $C_i(u,r)$ are

$$\frac{\partial C_1}{\partial r} = \frac{1}{4}\frac{Rhs}{m^2}, \frac{\partial C_2}{\partial r} = -\frac{1}{4}\frac{(2\,m\,r+1)\,\mathrm{e}^{-2\,m\,r}}{m^2}Rhs, \frac{\partial C_3}{\partial r} = \frac{1}{2}\frac{\mathrm{e}^{-2\,m\,r}}{m}Rhs.$$

Therefore the general non-homogeneous $\phi_{NH}(u,r)$ solution to the key equation (19.97b) via the variation of parameters is

$$\begin{aligned}\phi_{NH} = & \frac{1}{16\,m^2}\,{f_{11}}^2\Big\{\left(e^{2\,m\,r}-1\right)\left[\left(\ln\left(e^{2\,m\,r}-1\right)+1\right)^2-1\right] \\ & -2\,\mathrm{dilog}\left(e^{2\,m\,r}\right)e^{2\,m\,r}-4\,\ln\left(e^{2\,m\,r}-1\right)m\,r e^{2\,m\,r}-4\,m\,r e^{2\,m\,r}-2\Big\} \\ & +\frac{1}{4m^2}\,\frac{df_{11}}{du}\,\left[\left(e^{2\,m\,r}-1\right)\ln\left(e^{2\,m\,r}-1\right)-m\,r\,e^{2\,m\,r}\right] \qquad (19.97\mathrm{e})\end{aligned}$$

where it has been used in the relation

$$\int 2\,m\ln\left(e^{2\,mr}-1\right)dr = \mathrm{dilog}\left(e^{2\,mr}\right)+2\,m\,r\ln\left(e^{2\,mr}-1\right)$$

for the dilog function, which is presented in details in the next equations (19.99). The above-integrated solution (19.97e) can be given in the CPS form,

$$\begin{aligned}\frac{F_0(u,r)}{1-e^{2\,rm}} = & \frac{(f_{11})^2}{16m^2}\left[\left(e^{2\,rm}-1\right)\left(\left(1+\ln\left(e^{2\,rm}-1\right)\right)^2-1\right)\right. \\ & \left.+2e^{2m\,r}\,L_2\left(e^{2m\,r}\right)\right]+\frac{1}{4m^2}\left(\frac{df_{11}}{du}\right)\left(e^{2\,rm}-1\right)\ln\left(e^{2\,rm}-1\right).\end{aligned}$$
(19.97f)

Finally, this family $\mu = -m$ of Petrov type III CSI spacetimes is given by the metric

$$\begin{aligned} & ds^2 = dr^2+2du\,dv+(v^2F_2+vF_1+F_0)du^2-4\,v\,m\coth(m\,r)\,du\,dr, \\ & F_2(u,r) = \frac{m^2}{\sinh^2(m\,r)}, \\ & F_1(u,r) = f_{11}(u)\xi(r),\ \xi(r) = \ln\left(e^{2\,m\,r}-1\right) = m\,r+\ln\left(\sinh\left(m\,r\right)\right), \\ & F_0(u,r) = \left(1-e^{-2\,rm}\right)\left[\phi_H(u,r)+\phi_{NH}(u,r)(19.97\mathrm{e})\right]. \qquad (19.97\mathrm{g})\end{aligned}$$

This solution is characterized by the Cotton:

$$\begin{aligned} & C^{\alpha}{}_{\beta} = C^{v}{}_{u}\delta^{\alpha}{}_{v}\delta^{u}{}_{\beta}+C^{v}{}_{r}\delta^{\alpha}{}_{v}\delta^{r}{}_{\beta}+C^{r}{}_{u}\delta^{\alpha}{}_{r}\delta^{u}{}_{\beta}, \\ & C^{v}{}_{u} = -m^2\left(e^{2\,mr}-1\right)f_{03}+2\,v\,f_{11}\,\frac{m^3e^{2\,mr}}{\left(e^{2\,mr}-1\right)^2}-\frac{1}{2}\,me^{2\,mr}\frac{df_{11}}{du} \\ & \qquad -\frac{1}{4}\,m\,{f_{11}}^2\left[\ln\left(e^{2\,mr}-1\right)+e^{2\,mr}\right], \\ & C^{r}{}_{u} = C^{v}{}_{r} = m^2\frac{e^{2\,mr}}{e^{2\,mr}-1}f_{11}. \qquad (19.97\mathrm{h})\end{aligned}$$

Type N Limit

It allows for a type N Kundt solution with metric

$$\begin{aligned} & ds^2 = dr^2+2du\,dv+(v^2F_2+F_0)du^2-4\,v\,m\coth(m,r)\,du\,dr, \\ & F_2 = \frac{m^2}{\sinh^2(m\,r)},\ F_0 = 2f_{03}\,r\,\sinh(m\,r)\,e^{-rm} \qquad (19.98\mathrm{a})\end{aligned}$$

with a type N Cotton tensor, $\mathcal{J}_{CN}$,

$$C^{\alpha}{}_{\beta} = C^{v}{}_{u}\delta^{\alpha}{}_{v}\delta^{u}{}_{\beta},\; C^{v}{}_{u} = -m^2\left(\mathrm{e}^{2\,mr} - 1\right) f_{03}\left(u\right). \tag{19.98b}$$

The Dilogarithm dilog(x) and the Logarithmic Integral Li(x) Functions
By definition, the dilogarithm function dilog(x) is

$$\mathrm{dilog}\left(x\right) = \int_1^x \frac{\ln\left(t\right)}{1-t}dt,\; \frac{d}{dx}\mathrm{dilog}\left(x\right) = \frac{\ln\left(x\right)}{1-x}, \tag{19.99a}$$

hence

$$\begin{aligned}\mathrm{dilog}\left(1+\mathrm{e}^{\pm 2\,m\,r}\right) &= \mp 2m\int^{r}\ln\left(1+\mathrm{e}^{\pm 2\,mt}\right)dt,\\ \frac{d}{dr}\mathrm{dilog}\left(\mathrm{e}^{\pm 2\,m\,r}+1\right) &= \mp 2\,m\ln\left(\mathrm{e}^{\pm 2\,m\,r}+1\right),\end{aligned} \tag{19.99b}$$

$$\begin{aligned}\mathrm{dilog}\left(\mathrm{e}^{2\,m\,r}\right) + \mathrm{dilog}\left(\mathrm{e}^{-2\,m\,r}\right) + 2\,m^2r^2 &= 0,\\ \mathrm{dilog}\left(1+\mathrm{e}^{-2\,m\,r}\right) + \mathrm{dilog}\left(1+\mathrm{e}^{2\,m\,r}\right) + 2\,m^2\,r^2 &= 0,\end{aligned} \tag{19.99c}$$

$$\int 2\,m\ln\left(\mathrm{e}^{2\,mr}-1\right)dr = \mathrm{dilog}\left(\mathrm{e}^{2\,mr}\right) + 2\,m\,r\ln\left(\mathrm{e}^{2\,mr}-1\right). \tag{19.99d}$$

The logarithmic integral Li(x), denoted in CPS by $\mathrm{Li}_2(x)$, is defined as

$$\mathrm{Li}(x) = -\int \frac{\ln\left(|1-x|\right)}{x}dx \tag{19.99e}$$

hence depending on the range of value assumed by the variable of integration, one would have different functional relations with the dilog(x) and ln(x) functions:

- for $x \leq 1$

$$\begin{aligned}\mathrm{Li}(x) = \mathrm{dilog}(1-x),\; x\leq 1;\; \mathrm{Li}(\mathrm{e}^{-2\,m\,r}) &= \mathrm{dilog}\left(1-\mathrm{e}^{-2\,mr}\right),\\ \mathrm{Li}(-\mathrm{e}^{2\,m\,r}) = \mathrm{dilog}\left(1+\mathrm{e}^{2\,mr}\right),\end{aligned} \tag{19.99f}$$

- while for $x > 1$

$$\begin{aligned}\mathrm{Li}(x) &= -\,\mathrm{dilog}(x) - \ln(x-1)\,ln(x) + \pi^2/6,\; x>1;\\ \mathrm{Li}(1+\mathrm{e}^{-2\,mr}) &= -\mathrm{dilog}\left(1+\mathrm{e}^{-2\,mr}\right) + 2\,mr\ln\left(1+\mathrm{e}^{-2\,mr}\right) + \pi^2/6\\ &= -\mathrm{dilog}\left(1+\mathrm{e}^{-2\,mr}\right) - 4\,m^2\,r^2 + 2\,mr\ln(\mathrm{e}^{2\,mr}+1) + \pi^2/6,\end{aligned} \tag{19.99g}$$

see Abramowitz and Stegun (1965).

Multi Exponent–Integral Representation of ϕ

To obtain a CPS Chow et al. (2010b) representation of the solution ϕ to (19.97b) one has to follow the exponent–multiple integral procedure previously detailed in 19.10.1 or simple rewrite the equation (19.97b) as

$$\frac{\partial^2}{\partial r^2}\left(\mathrm{e}^{-2m\,r}\frac{\partial}{\partial r}\phi(u,r)\right) = \mathrm{e}^{-2m\,r}Rhs \tag{19.100a}$$

The integral for ϕ is

$$\phi\,(u,r) = \int^r \mathrm{e}^{2\,m\,t}\int^t\!\!\int^y \mathrm{e}^{-2\,m\,z}Rhs\,(u,z)\,dz\,dy\,dt$$

which arises from (19.94f) for $\mu = -m$, namely

$$\phi(u,r) = -\frac{1}{4}\int^r \mathrm{e}^{2\,m\,t}\int^t\int^y \frac{m\left(2\,\frac{d}{du}f_{11} + {f_{11}}^2\xi\,(z)\right)}{\sinh^2(mz)}dzdydt \tag{19.100b}$$

which yields to an expression comparable with CPS $\phi(u,r)$ giving rise to

$$\frac{F_0}{1-\mathrm{e}^{-2mr}} = f_{01} + \mathrm{e}^{2mr}f_{02} + r\mathrm{e}^{2mr}f_{03} + \frac{1}{2m}\frac{df_{11}}{du}\int_\infty^r \mathrm{e}^{2mt}F(t)dt$$
$$-\frac{m}{4}{f_{11}}^2\int_\infty^r dt\int_\infty^t dt'\int_\infty^{t'}dt''\frac{\mathrm{e}^{2m\,t}}{\sinh^2(mt'')}F(t''). \tag{19.100c}$$

19.10.4 Type III Kundt Solution; $\Lambda = -m^2, W_1 = -2m\coth(mr), \mu = m$

Solution $\Phi(u,r)$ through the Variation of Parameters

The equation (19.85) for the function $\Phi(u,r)$ in the case $\mu = m$ becomes

$$\frac{\partial^3}{\partial r^3}\Phi + m\frac{\partial^2}{\partial r^2}\Phi - m^2\frac{\partial}{\partial r}\Phi - m^3\Phi = \frac{RHS}{\sinh\,(m\,r)} = Rhs$$
$$= -\frac{m}{2}\frac{\mathrm{e}^{-m\,r}}{\sinh^2(m\,r)}\left[{f_{11}}^2\,\xi\,(r) + 2\,\frac{df_{11}}{du}\right], \tag{19.101a}$$

where

$$\frac{d}{dr}\xi(r) = \frac{2\,m}{\mathrm{e}^{2\,m\,r}-1},\ \xi(r) = \ln(1-\mathrm{e}^{-2\,m\,r}). \tag{19.101b}$$

Since the homogeneous solution is given as a linear combination of the set of independent functions $\{\mathrm{e}^{m\,r}, \mathrm{e}^{-m\,r}, r\,\mathrm{e}^{m\,r}\}$, then the non-homogeneous solution is of the form

$$\Phi_{NH} = C_1(u,r)\mathrm{e}^{m\,r} + C_2(u,r)\mathrm{e}^{-m\,r} + C_3(u,r)\mathrm{e}^{m\,r}\,r, \tag{19.101c}$$

Using the VP procedure one arrives at

$$\frac{\partial}{\partial r}C_1 = \frac{\mathrm{e}^{-m\,r}}{4m^2}\frac{RHS}{\sinh\,(m\,r)},\ \frac{\partial}{\partial r}C_2 = \frac{\mathrm{e}^{m\,r}\,(2\,m\,r-1)}{4\,m^2}\frac{RHS}{\sinh\,(m\,r)},$$
$$\frac{\partial}{\partial r}C_3 = -\frac{\mathrm{e}^{m\,r}}{2\,m}\frac{RHS}{\sinh\,(m\,r)}.$$

Consequently, integrating $\frac{\partial}{\partial r}C_i$, the non-homogeneous Φ solution amounts to

$$\Phi_{NH} = -\frac{1}{8m^2}\,{f_{11}}^2\left[-2\,\mathrm{dilog}\left(\mathrm{e}^{-2\,rm}\right) + 4\,rm\ln\left(1-\mathrm{e}^{-2\,rm}\right)\right.$$
$$\left. -\left(\mathrm{e}^{2\,rm}-1\right)\left(\left(1+\ln\left(1-\mathrm{e}^{-2\,rm}\right)\right)^2-1\right)\right]\mathrm{e}^{-rm}$$
$$+\frac{\mathrm{e}^{-rm}}{2m^2}\left(\frac{df_{11}}{du}\right)\left[\left(\mathrm{e}^{2\,rm}-1\right)\ln\left(\mathrm{e}^{2\,rm}-1\right) - 2\,m\,r\,\mathrm{e}^{2\,rm}\right]. \quad (19.101\mathrm{d})$$

Multi Exponent–Integral Representation of $\Phi(u,r)$

To determine the structural function Φ subjected to the equation

$$\frac{\partial^3}{\partial r^3}\Phi + m\,\frac{\partial^2}{\partial r^2}\Phi - m^2\frac{\partial}{\partial r}\Phi - m^3\Phi = Rhs, \quad (19.102\mathrm{a})$$

one follows the procedure given in 19.10.1.

Comparing the coefficients of the derivatives of the corresponding differential equations one arrives at an algebraic system for the exponents

$$\gamma + 2\,\beta + 3\,\alpha = m,$$
$$\gamma\,\beta + 2\,\gamma\,\alpha + \beta^2 + 4\,\beta\,\alpha + 3\,\alpha^2 = -m^2,$$
$$\gamma\,\beta\,\alpha + \gamma\,\alpha^2 + \beta^2\alpha + 2\,\beta\,\alpha^2 + \alpha^3 = -m^3 \quad (19.102\mathrm{b})$$

with the sets of solutions

$$\{\alpha = -m, \beta = 2\,m, \gamma = 0\},$$
$$\{\alpha = m, \beta = 0, \gamma = -2\,m\},$$
$$\{\alpha = m, \beta = -2\,m, \gamma = 2\,m\}. \quad (19.102\mathrm{c})$$

Therefore, one has various solution representations at one's disposal.

VP of the CPS Representation of $F_0(u,r)$ through ϕ

If one were using the variations of parameters method to determine the solution to the non-homogeneous equation one generalize the parameters C_i of the homogeneous solution

$$\phi_H\,(r) = C_1(u) + C_2(u)r + C_3(u)\,\mathrm{e}^{2\,m\,r} \equiv C_1(u)\phi_1 + C_2(u)\,\phi_2 + C_3(u)\,\phi_3 \quad (19.103\mathrm{a})$$

to be dependent on the variable r in such a way that the non-homogeneous solution is searched in the form of

$$\phi_{NH}\,(r) = C_1(u,r)\phi_1 + C_2(u,r)\,\phi_2 + C_3(u,r)\,\phi_3 \quad (19.103\mathrm{b})$$

via the variation of parameters $C_i(u,r)$, where u plays the role of a parameter. The derivatives of the C_i are

$$\frac{\partial C_1}{\partial r} = \frac{1}{8}\frac{\mathrm{e}^{m\,r}Rhs\,(2\,m\,r-1)}{m^2}, \quad \frac{\partial C_2}{\partial r} = -\frac{1}{4}\frac{\mathrm{e}^{m\,r}Rhs}{m}, \quad \frac{\partial C_3}{\partial r} = \frac{1}{8}\frac{Rhs}{\mathrm{e}^{m\,r}m^2},$$

hence the non-homogeneous ϕ_{NH} solution can be given as

$$\phi_{NH} = \frac{1}{8m^2}\int e^{m\,r}\,Rhs\,(u,r)\,(2\,m\,r-1)\,dr - \frac{r}{4m}\int e^{m\,r}\,Rhs\,(u,r)\,dr$$
$$+\frac{1}{8m^2}e^{2\,m\,r}\int Rhs\,(u,r)\,e^{-m\,r}dr, \tag{19.103c}$$

where the function $Rhs(u,r)$ entering in ϕ, using ξ from (19.101b), is given by

$$Rhs = -2\,\frac{\left(2\,\frac{d}{du}f_{11} + {f_{11}}^2\ln\left(1-e^{-2\,m\,r}\right)\right)me^{-3\,m\,r}}{\left(1-e^{-2\,m\,r}\right)^2}, \tag{19.103d}$$

therefore, factorizing $\frac{df_{11}}{du}$ and ${f_{11}}^2$ in the expression of ϕ, accomplishing the integration and using a relation between the dilog functions (19.99) in ϕ_{NH}, one gets

$$\phi_{NH} = \frac{{f_{11}}^2}{16m^2}\left\{-2\,\mathrm{dilog}\left(e^{2\,m\,r}\right) - 4\,m\,r\,\ln\left(e^{2\,mr}-1\right) + 4\,m^2\,r^2\right.$$
$$\left.+\left(e^{2\,m\,r}-1\right)\left[\left(1+\ln\left(1-e^{-2\,m\,r}\right)\right)^2-1\right]\right\}$$
$$+\frac{1}{4m^2}\left(\frac{df_{11}}{du}\right)\left(e^{2m\,r}-1\right)\ln\left(1-e^{-2\,m\,r}\right). \tag{19.103e}$$

In this expression, terms proportional to the independent homogeneous solutions $\{1, r, e^{2m\,r}\}$ have been eliminated. On the other hand, using $L_2(e^{2\,mr})$ from (19.99), one rewrites (19.103e) as

$$\phi_{NH} = \frac{{f_{11}}^2}{16m^2}\left\{\left(e^{2\,m\,r}-1\right)\left[\left(1+\ln\left(1-e^{-2\,m\,r}\right)\right)^2-1\right]\right.$$
$$\left.+2L_2(e^{2\,m\,r}) + 4\,m^2\,r^2\right\}$$
$$+\frac{1}{4m^2}\left(\frac{df_{11}}{du}\right)\left(e^{2m\,r}-1\right)\ln\left(1-e^{-2\,m\,r}\right), \tag{19.103f}$$

which coincides with the CPS Chow et al. (2010b) form of the solution:

$$ds^2 = dr^2 + 2du\,dv + (v^2F_2 + vF_1 + F_0)du^2 - 4\,v\,m\coth(m\,r)\,du\,dr,$$
$$F_2(u,r) = \frac{m^2}{\sinh^2(m\,r)},\quad F_1(u,r) = f_{11}(u)\xi(r),$$
$$\xi(r) = \ln(1-e^{-2\,m\,r}) = -2\,m\,r + \ln(e^{2\,m\,r}-1),$$
$$F_0(u,r) = \left(1-e^{-2\,m\,r}\right)\left[\phi_H(u,r) + \phi_{NH}(u,r)(19.103e)\right],$$
$$\phi_H = e^{2\,mr}f_{03}\,(u) + r\,f_{02}\,(u) + f_{01}\,(u)\,. \tag{19.103g}$$

This solution is characterized by the Cotton:

$$C^\alpha{}_\beta = C^v{}_u\delta^\alpha{}_v\delta^u{}_\beta + C^v{}_r\delta^\alpha{}_v\delta^r{}_\beta + C^r{}_u\delta^\alpha{}_r\delta^u{}_\beta,$$
$$C^v{}_u = -m^2 f_{02}\,(u)\,e^{-2\,mr}\left(e^{2\,mr}-1\right) - 2\,v f_{11}\,\frac{m^3e^{2\,mr}}{\left(e^{2\,mr}-1\right)^2}$$

$$+\frac{1}{2}me^{-2\,mr}\frac{df_{11}}{du}+\frac{1}{4}m\,f_{11}{}^{2}\left[\ln\left(1-\mathrm{e}^{-2\,mr}\right)+\mathrm{e}^{-2\,mr}\right],$$

$$C^{r}{}_{u}=C^{v}{}_{r}=-\frac{m^{2}}{\mathrm{e}^{2\,mr}-1}f_{11}. \tag{19.103h}$$

Type N Limit

It allows for a type N Kundt solution with metric

$$ds^2=dr^2+2du\,dv+(v^2F_2+F_0)du^2-4\,v\,m\coth(m,r)\,du\,dr,$$

$$F_2(u,r)=\frac{m^2}{\sinh^2(m\,r)},\ F_0(u,r)=2\,f_{02}\,r\,\sinh(m\,r)\,\mathrm{e}^{-rm}, \tag{19.104a}$$

with a type N Cotton tensor, $\mathcal{J}_{CN}$,

$$C^{\alpha}{}_{\beta}=C^{v}{}_{u}k^{\alpha}k_{\beta},\ C^{v}{}_{u}=-m^2f_{02}\,(u)\left(1-\ \mathrm{e}^{-2\,mr}\right),\ k^{\alpha}=\delta^{\alpha}{}_{v}. \tag{19.104b}$$

Multi Exponent–Integral Representation of ϕ

The Chow–Pope–Sezgin (Chow et al., 2010b) representation of the solution for $F_0(u,r)$ is achieved via an alternative definition of the function $F_0(u,r)$ through $\phi(r)$, namely

$$\phi(u,r)=\frac{1}{2}\mathrm{e}^{mr}\,\Phi(u,r)=\frac{F_0(u,r)}{1-\mathrm{e}^{-2mr}}, \tag{19.105a}$$

thus, replacing $\Phi\,(r)=2\,\mathrm{e}^{-rm}\phi\,(r)$ in (19.102a), one gets

$$\frac{\partial}{\partial r}\left(\mathrm{e}^{-2m\,r}\frac{\partial^2}{\partial r^2}\phi\right)=\frac{1}{2}\,\mathrm{e}^{-m\,r}Rhs,\ Rhs=\frac{RHS}{2\sinh(m\,r)} \tag{19.105b}$$

Incidentally, it arises from the equation (19.93b) for $\mu=m$. Integrating the equation (19.105b) one obtains the general solution for $\phi\,(r)$ as

$$\phi=\frac{1}{2}\int^{r}\!\!\int^{t}\mathrm{e}^{2\,m\,y}\int^{y}\mathrm{e}^{-m\,z}\,Rhs\,(z)\,dzdydt+C_3\mathrm{e}^{2\,m\,r}+C_2r+C_1.$$

Since the function $\xi(r)$ can explicitly be integrated in terms of ln, (19.101b), one can rewrite $\phi\,(u,r)$ as

$$\phi=-m\int^{r}\!\!\int^{t}\mathrm{e}^{2\,m\,y}\int^{y}\left(2\frac{df_{11}}{du}+f_{11}{}^{2}\ln\left(1-\mathrm{e}^{-2\,m\,z}\right)\right)\frac{dz\,dy\,dt}{(\mathrm{e}^{2\,m\,z}-1)^2}$$
$$+C_3(u)\,\mathrm{e}^{2\,m\,r}+C_2(u)\,r+C_1(u). \tag{19.105c}$$

19.11 Type III Kundt Metric; $\Lambda=-m^2$; $W_1=-2\,m\,\tanh(m\,r)$

Deriving the various possible branches of solutions, one has the case (19.41e),

$$W_1(r)=-2m\tanh(mr).$$

We follow here the same pattern used in the previous sections: first the general case of $\mu \neq \pm m$ is treated; next, the derived results and key equations are specialized to handle the sub-cases $\mu = \pm m$.

Evaluating the equation (19.24c) for F_2 one gets

$$F_2 = -\frac{m^2}{\cosh^2(m\,r)}. \tag{19.106}$$

Replacing these functions into the equation (19.24d) for F_1, one arrives at

$$\frac{\partial^2}{\partial r^2} F_1(u,r) + [\mu + m\tanh(m\,r)]\frac{\partial}{\partial r} F_1(u,r) = 0, \tag{19.107}$$

with solution

$$F_1(u,r) = f_{11}(u)\int \frac{m\,\mathrm{e}^{-\mu\,r}}{\cosh(m\,r)}dr = f_{11}\,\xi(r), \tag{19.108}$$

where it has been introduced the auxiliary function $\xi(r)$,

$$\frac{d\xi}{dr} = \frac{m\mathrm{e}^{-\mu\,r}}{\cosh(m\,r)} \rightarrow \xi(r) = m\int\frac{\mathrm{e}^{-\mu\,r}}{\cosh(m\,r)}dr. \tag{19.109}$$

The equation (19.24e) for $F_0(u,r)$ becomes

$$\begin{aligned}
&\frac{\partial^3}{\partial r^3}F_0 + [\mu - 3\,m\tanh(m\,r)]\frac{\partial^2}{\partial r^2}F_0\\
&+2\,m\left[-2\,m + 3\,m\tanh^2(m\,r) - \mu\tanh(m\,r)\right]\frac{\partial}{\partial r}F_0\\
&+2\,m^2\left(\tanh^2(m\,r) - 1\right)\left(\mu - 3\,m\tanh(m\,r)\right)F_0\\
&= -\frac{1}{2}\left(2\frac{d}{du}f_{11} + {f_{11}}^2\xi(r)\right)\frac{d}{dr}\xi(r) =: RHS.
\end{aligned} \tag{19.110}$$

Following the standard procedure for solving non-homogeneous linear ordinary differential equations by means of the variation of parameters of the homogeneous solution

$$F_{0H} = \cosh(m\,r)\left(C_1(u)\,\mathrm{e}^{-\mu\,r} + C_2(u)\,\mathrm{e}^{m\,r} + C_3(u)\,\mathrm{e}^{-m\,r}\right), \tag{19.111}$$

one arrives at an algebraic system of equations for the first-order derivatives of $C_i(u,r)$ with respect to r. Nevertheless, the structure of the homogeneous solution above points out to a better function choice, namely

$$F_0(u,r) = \Phi(u,r)\cosh(m\,r),\ \Phi(u,r) = \frac{F_0(u,r)}{\cosh(m\,r)}, \tag{19.112}$$

where u enters as a parameter. The equation (19.110) in terms of Φ becomes

$$\frac{\partial^3}{\partial r^3}\Phi + \mu\frac{\partial^2}{\partial r^2}\Phi - m^2\frac{\partial}{\partial r}\Phi - m^2\mu\,\Phi = \frac{RHS}{\cosh(m\,r)} = Rhs, \tag{19.113}$$

$$Rhs = -\frac{1}{2}\frac{\left[{f_{11}}^2\xi + 2\frac{df_{11}}{du}\right]}{\cosh(m\,r)}\frac{d\xi}{dr},\ \frac{d}{dr}\xi(r) = \frac{me^{-\mu\,r}}{\cosh(m\,r)}. \tag{19.114}$$

In what follows we proceed to determine the three different branches of solutions in the specific cases of $\mu \neq \pm m$, $\mu = -m$, and $\mu = m$ separately.

19.11.1 Type III Kundt Solution; $\Lambda = -m^2$, $W_1 = -2m\tanh(mr)$, $\mu \neq \pm m$

First, the derivation of the solution via the variation of parameters is given. Next, the determination of the solution in a multi-exponent–integral representation is done. Finally, the CPS representation of the solution through a key function introduced in Chow et al. (2010b) is given.

Solution Φ Through the Variation of Parameters

The homogeneous solution to the homogeneous equation (19.113) ($Rhs \to 0$), one searches in the form of $\Phi_H \simeq e^{k\,r}$ arriving at the characteristic equation

$$k^3 + \mu\,k^2 - m^2\,k - m^2\mu = 0 = (k+\mu)(k-m)(k+m)$$

with solutions $k = \{-\mu,\ -m,\ m\}$, hence Φ_H is given by $\Phi(r) = C_1\,e^{-\mu\,r} + C_2\,e^{m\,r} + C_3\,e^{-m\,r}$. Searching for the solution to the non-homogeneous equation ($Rhs \neq 0$) one assumes it to be of the form

$$\Phi(r) = C_1(u,r)\,\Phi_1 + C_2(u,r)\,\Phi_2 + C_3(u,r)\,\Phi_3, \tag{19.115a}$$

where u enters here as a parameter. The variation of the constants $C_i(u,r)$, within the variation of parameters procedure, yields

$$\frac{\partial}{\partial r}C_1 = \frac{e^{\mu\,r}}{(\mu^2 - m^2)}\frac{RHS}{\cosh(rm)},\ \frac{\partial}{\partial r}C_2 = \frac{1}{2}\frac{1}{m(\mu+m)}\frac{RHS}{\cosh(m\,r)}e^{-m\,r},$$
$$\frac{\partial}{\partial r}C_3 = -\frac{1}{2}\frac{1}{m(\mu-m)}\frac{RHS}{\cosh(m\,r)}e^{m\,r}.$$

Notice the presence of $\mu \pm m$ in the denominators of the above expressions, which clearly point to the degeneracy of these quantities when dealing with the sub-branches $\mu = \pm m$; this is the reason why the quoted cases have to be integrated separately for the equation (19.113).

Thus, the non-homogeneous solution amounts to

$$\Phi_{NH}(u,r) = \frac{df_{11}}{du}\left[\frac{e^{-\mu\,r}}{m^2-\mu^2}\int\frac{m}{\cosh^2(mr)}dr\right.$$
$$\left.-\frac{1}{2}\frac{e^{mr}}{(m+\mu)}\int\frac{e^{-(m+\mu)r}}{\cosh^2(mr)}dr - \frac{1}{2}\frac{e^{-mr}}{(m-\mu)}\int\frac{e^{(m-\mu)r}}{\cosh^2(mr)}dr\right]$$

$$+ f_{11}{}^2 \left[\frac{1}{2} \frac{m\,\mathrm{e}^{-\mu r}}{m^2 - \mu^2} \int \frac{\xi\,(r)}{\cosh^2(mr)} dr \right.$$
$$\left. - \frac{1}{4} \frac{\mathrm{e}^{mr}}{(m+\mu)} \int \frac{\mathrm{e}^{-(m+\mu)r}\xi\,(r)}{\cosh^2(mr)} dr - \frac{1}{4} \frac{\mathrm{e}^{-m\,r}}{(m-\mu)} \int \frac{\mathrm{e}^{(m-\mu)r}\xi\,(r)}{\cosh^2(mr)} dr \right], \qquad (19.115b)$$

whereas the homogeneous solution is given by

$$\Phi_H(u,r) = f_{01}(u)\,\mathrm{e}^{-r\mu} + f_{02}(u)\,\mathrm{e}^{m\,r} + f_{03}(u)\,\mathrm{e}^{-m\,r}, \qquad (19.115c)$$

which together determine the general metric of this spacetime:

$$\begin{aligned}
ds^2 &= dr^2 + 2du\,dv + (v^2F_2 + vF_1 + F_0)du^2 - 4\,v\,m\tanh(m\,r)\,du\,dr,\\
F_2(u,r) &= -\frac{m^2}{\cosh^2(m\,r)},\\
F_1(u,r) &= f_{11}(u)\ \xi(r),\ \xi(r) = \int \frac{m\,\mathrm{e}^{-\mu\,r}}{\cosh(m\,r)} dr,\\
F_0(u,r) &= \cosh(m\,r)\,(\Phi_H(u,r) + \Phi_{NH}(u,r)(19.115b)).
\end{aligned} \qquad (19.115d)$$

This spacetime is characterized by the Cotton tensor

$$\begin{aligned}
C^\alpha{}_\beta &= C^v{}_u\delta^\alpha{}_v\delta^u{}_\beta + C^v{}_r\delta^\alpha{}_v\delta^r{}_\beta + C^r{}_u\delta^\alpha{}_r\delta^u{}_\beta,\\
C^v{}_u &= \frac{\mu}{2}\left(\mu^2 - m^2\right)\mathrm{e}^{-\mu\,r} f_{01}(u)\cosh(mr) - \frac{\mu}{2}\frac{df_{11}}{du}\mathrm{e}^{-\mu\,r}\sinh(mr)\\
&\quad - \frac{m\,\mu\,v}{2} f_{11}\,\mathrm{e}^{-\mu\,r}\frac{[\mu\cosh(mr) + m\sinh(mr)]}{\cosh^2(mr)}\\
&\quad - \frac{\mu\,m^2}{4}\mathrm{e}^{-\mu\,r} f_{11}{}^2\cosh(mr) \int \frac{1}{\cosh^2(mr)} \int^r \frac{e^{-\mu\,z}}{\cosh(mz)} dz dr,\\
C^r{}_u &= C^v{}_r = -\frac{m\,\mu}{2} f_{11}\frac{\mathrm{e}^{-\mu\,r}}{\cosh(mr)}.
\end{aligned} \qquad (19.115e)$$

Type N Limit

The type N limit Kundt solution is given by the metric

$$\begin{aligned}
ds^2 &= dr^2 + 2du\,dv + (v^2F_2 + F_0)du^2 - 4\,v\,m\,\tanh(m\,r)du\,dr,\\
F_2 &= -\frac{m^2}{\cosh^2(m\,r)},\ F_0 = \cosh(m\,r)\,f_{01}(u)\,\mathrm{e}^{-\mu\,r},
\end{aligned} \qquad (19.116a)$$

with a type N Cotton tensor, $\mathcal{J}_{CN}$,

$$C^\alpha{}_\beta = C^v{}_u k^\alpha k_\beta,\ C^v{}_u = \frac{\mu}{2}\left(\mu^2 - m^2\right)\mathrm{e}^{-\mu\,r} f_{01}(u)\cosh(mr),\ k^\alpha = \delta^\alpha{}_v. \qquad (19.116b)$$

Multi Exponent–Integral Representation of Φ

In this approach, see details in 19.10.1, the non-homogeneous solution to the equation (19.113), which is assumed to be expressible as

$$\frac{\partial}{\partial r}\left(e^{\gamma r}\frac{\partial}{\partial r}\left(e^{\beta r}\frac{\partial}{\partial r}\left(e^{\alpha r}\Phi(u,r)\right)\right)\right)=\frac{RHS}{\cosh\left(m\,r\right)} \qquad (19.117a)$$

is searched in the multi-exponent–integral form.

Comparing the coefficients multiplying derivatives of the same order in equations (19.113) and (19.117a) one gets the algebraic system

$$\begin{aligned}&2\,\beta+3\,\alpha+\gamma=\mu,\\&\gamma\,\beta+4\,\beta\,\alpha+3\,\alpha^2+2\,\gamma\,\alpha+\beta^2=-m^2,\\&\gamma\,\beta\,\alpha+\gamma\,\alpha^2+\beta^2\alpha+\alpha^3+2\,\beta\,\alpha^2=-m^2\mu,\end{aligned} \qquad (19.117b)$$

with the following set of solutions

$$\begin{aligned}&\{\alpha=\mp m,\beta=\pm 2\,m,\gamma=\mp m+\mu\}\,,\ sum=\mu,\\&\{\alpha=\mu,\beta=-\mu\mp m,\gamma=\pm 2\,m\}\,,\ sum=\pm m,\\&\{\alpha=\mp m,\beta=\mu\pm m,\gamma=\pm m-\mu\}\,,\ sum=\pm m.\end{aligned} \qquad (19.117c)$$

The substitution of the found exponents in the multi-exponent–integral expression for $\Phi_{NH}(u,r)$

$$\Phi=-\frac{m}{2}e^{-\alpha\,r}\int^{r}e^{-\beta\,t}\int^{t}e^{-\gamma\,y}\int^{y}\frac{\left[{f_{11}}^2\,\xi\left(z\right)+2\,\frac{df_{11}}{du}\right]}{\cosh^2(m\,z)}\,e^{(\alpha+\beta+\gamma-\mu)\,z}dzdydt, \qquad (19.117d)$$

determines a specific representation of the sought inhomogeneous solution. For instance, for the first set of exponents, one has

$$\begin{aligned}\Phi_{NH}=&-\frac{df_{11}}{du}e^{m\,r}\int^{r}\left(e^{-2m\,t}\int^{t}\left(e^{-(\mu-m)\,y}\int^{y}\left(\frac{1}{\cosh^2\left(m\,z\right)}\right)dz\right)dy\right)dt\\&-\frac{{f_{11}}^2}{2}\,e^{m\,r}\int^{r}\left(e^{-2m\,t}\int^{t}\left(e^{-(\mu-m)\,y}\int^{y}\left(\frac{\xi\left(z\right)}{\cosh^2\left(m\,z\right)}\right)dz\right)dy\right)dt,\end{aligned} \qquad (19.117e)$$

which gives, in turn, the general $F_0(u,r)$ function through

$$F_0(u,r)=\cosh(mr)\,\left[e^{-\mu\,r}f_{01}(u)+e^{m\,r}f_{02}(u)+e^{-m\,r}f_{03}(u)+\Phi(u,r)\right]. \qquad (19.117f)$$

Consequently one gets the general type III Kundt metric

$$\begin{aligned}ds^2=&\,dr^2+2du\,dv+m^2v^2\mathrm{cosech}^2(mr)du^2-4mv\,\mathrm{cosech}(mr)du\,dv\\&+[v\xi(r)F_{11}(u)+F_0(u,r)]du^2.\end{aligned} \qquad (19.117g)$$

Of course this expression can be simplified in various manners; for instance, one can partially integrate the first integral in $\Phi(u,r)$.

VP of the CPS Representation of F_0 through ϕ

The Chow–Pope–Sezgin representation of the solution, in terms of new auxiliary function $\phi(u,r)$, is achieved via an alternative definition of the function $F_0(u,r)$ through a

$$\Phi = \frac{F_0\,(u,r)}{\cosh\,(m\,r)},\ \Phi = 2\,\phi\,\mathrm{e}^{-m\,r},\ \phi\,(u,r) = \frac{F_0\,(u,r)}{1+\mathrm{e}^{-2\,m\,r}} \tag{19.118a}$$

which transforms the equation (19.113), (19.24e), to

$$\frac{\partial^3}{\partial r^3}\phi + (\mu - 3\,m)\,\frac{\partial^2}{\partial r^2}\phi + 2\,m\,(m-\mu)\,\frac{\partial}{\partial r}\phi = \frac{1}{2}\,\frac{\mathrm{e}^{mr}RHS}{\cosh\,(mr)} = Rhs. \tag{19.118b}$$

The homogeneous solution, with characteristic polynomial $k(k-2m)(k+\mu-m) = 0$, reads

$$\begin{aligned}\phi_H\,(r) &= C_1(u) + C_2(u)\,\mathrm{e}^{2\,mr} + C_3(u)\,\mathrm{e}^{(-\mu+m)r}\\ &\equiv C_1(u)\phi_1 + C_2(u)\,\phi_2 + C_3(u)\,\phi_3\end{aligned} \tag{19.118c}$$

The non-homogeneous solution is searched in the form of

$$\phi_{NH}\,(r) = C_1(u,r)\phi_1 + C_2(u,r)\,\phi_2 + C_3(u,r)\,\phi_3 \tag{19.118d}$$

via the variation of parameters $C_i(u,r)$, where u plays the role of a parameter. The derivatives of C_i are given by

$$\frac{\partial}{\partial r}C_1 = \frac{1}{4\,(m-\mu)\,m}\,\frac{\mathrm{e}^{mr}RHS}{\cosh\,(mr)},\ \frac{\partial}{\partial r}C_2 = \frac{\mathrm{e}^{-2\,m\,r}}{4\,(m+\mu)\,m}\,\frac{\mathrm{e}^{mr}RHS}{\cosh\,(mr)},$$
$$\frac{\partial}{\partial r}C_3 = -\frac{\mathrm{e}^{-r(m-\mu)}}{2(m^2-\mu^2)}\,\frac{\mathrm{e}^{mr}RHS}{\cosh\,(mr)}.$$

Therefore the general $\phi_{NH}(u,r)$ solution to the key equation (19.118b) via VP is

$$\begin{aligned}\phi(u,r) &= \frac{F_0(u,r)}{(1+\mathrm{e}^{-2\,mr})} = C_1(u) + C_2(u)\,\mathrm{e}^{2\,mr} + C_3(u)\,\mathrm{e}^{(m-\mu)r}\\
&+{f_{11}}^2\left[-\frac{1}{2}\int\frac{\xi}{(1+\mathrm{e}^{-2m\,r})^2}\,\frac{\mathrm{e}^{-(m+\mu)r}}{m-\mu}\,dr - \frac{\mathrm{e}^{2mr}}{2(m+\mu)}\int\frac{\xi\,\mathrm{e}^{-(\mu+3m)r}}{(1+\mathrm{e}^{-2mr})^2}\,dr\right.\\
&\left.+\mathrm{e}^{(m-\mu)r}\int\frac{\xi}{(1+\mathrm{e}^{-2\,mr})^2}\,\frac{m\mathrm{e}^{-2\,mr}}{m^2-\mu^2}\,dr\right]\\
&+\frac{df_{11}}{du}\left[-\int\frac{1}{(1+\mathrm{e}^{-2m\,r})^2}\,\frac{\mathrm{e}^{-(m+\mu)\,r}}{m-\mu}\,dr - \frac{\mathrm{e}^{2mr}}{m+\mu}\int\frac{\mathrm{e}^{-(\mu+3\,m)r}}{(1+\mathrm{e}^{-2\,mr})^2}\,dr\right.\\
&\left.+2\mathrm{e}^{(m-\mu)r}\int\frac{1}{(1+\mathrm{e}^{-2\,mr})^2}\,\frac{m\,\mathrm{e}^{-2\,mr}}{m^2-\mu^2}\,dr\right],\end{aligned} \tag{19.118e}$$

which is a single integral representation of the solution while the CPSb(4.54) representation is multiple integral.

Multi Exponent–Integral Representation of ϕ

To obtain the representation CPSb(4.54) of the solution to the key equation (19.118b) one has to follow the multi-exponent–integral procedure previously detailed in 19.10.1. Introducing an auxiliary function $H = \frac{\partial}{\partial r}\phi$ we rewrite the equation (19.118b) as

$$EQH := \frac{\partial}{\partial r}\left(e^{\beta r}\frac{\partial}{\partial r}\left(e^{\alpha r}H(r)\right)\right) = Rhs\, e^{(\alpha+\beta)\, r},$$

$$H(r) = e^{-\alpha r}\int e^{-\beta r}\int Rhs(u,r)e^{(\alpha+\beta)\, r}drdr. \tag{19.119a}$$

Accomplishing the derivation one gets

$$\frac{\partial^2}{\partial r^2}H\,(r) + (\beta + 2\,\alpha)\,\frac{\partial}{\partial r}H\,(r) + \alpha\,(\beta+\alpha)\,H\,(r) = Rhs(u,r). \tag{19.119b}$$

Comparing the coefficients of (19.119b) with those of (19.118b) one obtains the system

$$\{\alpha\,(\beta+\alpha) = -2\,m\,(\mu - m)\,, \beta + 2\,\alpha = \mu - 3\,m\}\,, \tag{19.119c}$$

with solutions

$$[[\alpha = -2\,m, \beta = m + \mu], [\alpha = \mu - m, \beta = -m - \mu]]. \tag{19.119d}$$

For the first set of solutions one obtains

$$H(r) = e^{2m\, r}\int^{r} e^{-(\mu+m)\, y}\int^{y} Rhs(u,z)e^{(\mu-m)\, z}dzdy. \tag{19.119e}$$

Consequently, ϕ integrates as

$$\phi = \int^{r} e^{-\alpha\, t}\int^{t} e^{-\beta\, y}\int^{y} Rhs\,(z)\, e^{(\alpha+\beta)z}dzdydt,$$

and, for the case under consideration, ϕ is

$$\begin{aligned}\phi &= \int^{r} e^{2\,m\,t}\int^{t} e^{-(m+\mu)y}\int^{y} Rhs(z)e^{(\mu-m)\, z}dzdydt\\ &= -\frac{m}{4}\int^{r} e^{2\,m\,t}\int^{t} e^{-(m+\mu)y}\int^{y}\frac{\left[2\,\frac{df_{11}}{du} + {f_{11}}^2\xi\,(z)\right]}{\cosh^2(mz)}dzdydt.\end{aligned}$$

Integrating once, one gets

$$\begin{aligned}\phi_{NH} &= -\frac{1}{2}\frac{df_{11}}{du}\int^{r} e^{2\,m\,t}\int^{t}\frac{e^{-(m+\mu)z}\sinh\,(m\,z)}{\cosh\,(m\,z)}dzdt\\ &\quad -\frac{m}{4}{f_{11}}^2\int^{r} e^{2\,m\,t}\int^{t} e^{-(m+\mu)y}\int^{y}\frac{\xi\,(z)}{\cosh^2(m\,z)}dzdydt.\end{aligned} \tag{19.119f}$$

To obtain the CPS Chow et al. (2010b) representation, one accomplishes the substitution

$$e^{-(\mu+m)r}\frac{\sinh(m\,r)}{\cosh(m\,r)} = e^{-\mu\,r}\frac{[2-(1+e^{(-2m)r})]}{2\cosh(m\,r)} = \frac{e^{-\mu\,r}}{\cosh(m\,r)} - e^{-(\mu+m)r},$$

thus

$$\int e^{2\,m\,r}\int \frac{e^{-r(m+\mu)}\sinh(m\,r)}{\cosh(m\,r)}drdr = \frac{1}{m}\int e^{2\,m\,r}\xi(r)\,dr - \frac{e^{(m-\mu)\,r}}{\mu^2 - m^2},$$

the last term is proportional to one of the homogeneous solutions, hence, it can be dropped out. Consequently, one arrives at the solution coincident with the CPSb (4.54), see Chow et al. (2010b), by doing the identification $\xi \to F$, $r \to \rho$,

$$\phi_{CPS} = \frac{F_0(u,\rho)}{1+e^{-2m\rho}} = e^{(m-\mu)\rho}f_{01}(u) - \frac{1}{2m}\frac{df_{11}}{du}\int_\infty^\rho e^{2mt}F(t)dt$$
$$-\frac{1}{4m}{f_{11}}^2\int_\infty^\rho dt\int_\infty^t dt'\int_\infty^{t'} dt''\frac{e^{2m(t-t')}e^{(m-\mu)t'}}{\cosh^2(mt'')}F(t''). \qquad (19.119g)$$

19.11.2 Type III Kundt Solution; $\Lambda = -m^2, W_1 = -2m\tanh(mr), \mu = -m$

Solution Φ Through the Variation of Parameters

The key equation (19.113) for the function $\Phi(u,r)$ in the case $\mu = -m$ becomes

$$\frac{\partial^3}{\partial r^3}\Phi - m\frac{\partial^2}{\partial r^2}\Phi - m^2\frac{\partial}{\partial r}\Phi + m^3\Phi = \frac{RHS}{\cosh(m\,r)} = Rhs$$
$$= -\frac{m}{2}\frac{\left[{f_{11}}^2\,\xi(r) + 2\,\frac{df_{11}}{du}\right]e^{m\,r}}{\cosh^2(m\,r)}, \xi(r) = \ln(e^{2\,m\,r}+1). \qquad (19.120a)$$

Since the characteristic equation $(k+m)(k-m)^2 = 0$, allows for a double root, then the homogeneous solution is $\Phi_H(r) = C_1\,e^{-m\,r} + C_2\,e^{m\,r} + C_3\,r\,e^{m\,r}$.

Using the VP procedure for the non–homogeneous function

$$\Phi_{NH}(r) = C_1(u,r)\,e^{-m\,r} + C_2(u,r)\,e^{m\,r} + C_3(u,r)\,r\,e^{m\,r} \qquad (19.120b)$$

one arrives at first-order equations for $C_i(u,r)$

$$\frac{\partial C_1}{\partial r} = \frac{1}{4}\frac{Rhs\,e^{m\,r}}{m^2}, \quad \frac{\partial C_2}{\partial r} = -\frac{1}{4}\frac{Rhs\,(2\,m\,r+1)\,e^{-m\,r}}{m^2}, \quad \frac{\partial C_3}{\partial r} = \frac{1}{2}\frac{e^{-m\,r}Rhs}{m}.$$

Substituting the expression of Rhs, factorizing the terms with $(\frac{d}{du}f_{11})$ and ${f_{11}}^2$ and integrating one arrives at the non-homogeneous solution Φ_{NH}

$$\Phi_{NH} = -\frac{e^{-m\,r}}{2\,m^2}\frac{df_{11}}{du}\left(e^{2m\,r}+1\right)\ln\left(e^{2\,m\,r}+1\right)$$
$$-\frac{e^{-m\,r}}{8\,m^2}{f_{11}}^2\left\{\left(e^{2\,m\,r}+1\right)\left[\left(1+\ln\left(e^{2\,m\,r}+1\right)\right)^2 - 1\right]\right.$$
$$\left.+2e^{2\,m\,r}\mathrm{dilog}\left(e^{2\,m\,r}+1\right)\right\}. \qquad (19.120c)$$

Finally, the metric of this solution can be written as

$$\begin{aligned} ds^2 &= dr^2 + 2du\,dv + (v^2 F_2 + vF_1 + F_0)du^2 - 4\,v\,m\,\tanh(m\,r)\,du\,dr, \\ F_2(u,r) &= -\frac{m^2}{\cosh^2(m\,r)}, \\ F_1(u,r) &= f_{11}(u)\,\xi(r),\ \xi(r) = \ln\left(e^{2\,mr}+1\right), \\ F_0(u,r) &= \cosh(m\,r)\ (\Phi_H + \Phi_{NH}(19.120c)), \\ \Phi_H &= f_{01}\,(u)\,e^{-mr} + f_{02}\,(u)\,e^{mr} + f_{03}\,(u)\,e^{mr} r. \end{aligned} \tag{19.120d}$$

This spacetime is characterized by the Cotton tensor

$$\begin{aligned} C^\alpha{}_\beta &= C^v{}_u \delta^\alpha{}_v \delta^u{}_\beta + C^v{}_r \delta^\alpha{}_v \delta^r{}_\beta + C^r{}_u \delta^\alpha{}_r \delta^u{}_\beta, \\ C^v{}_u &= \frac{1}{2} m^2 \left(e^{2\,mr}+1\right) f_{03} + 2\,v\,f_{11}\,\frac{m^3 e^{2\,mr}}{\left(e^{2\,mr}+1\right)^2} \\ &\quad + \frac{m}{4} f_{11}{}^2 \left[\ln\left(e^{2\,mr}+1\right)+1\right] + \frac{m}{2}\frac{df_{11}}{du}, \\ C^r{}_u &= C^v{}_r = \frac{m^2 e^{2\,mr}}{e^{2\,mr}+1} f_{11}. \end{aligned} \tag{19.120e}$$

Type N Limit

For vanishing function f_{11} one gets a type N metric

$$\begin{aligned} ds^2 &= dr^2 + 2du\,dv + (v^2 F_2 + F_0)du^2 - 4m\,v\,\tanh(m\,r)\,du\,dr, \\ F_0(u,r) &= \cosh(m\,r)\left(f_{01}(u)\,e^{-m\,r} + f_{02}(u)\,e^{m\,r} + f_{03}(u)\,r\,e^{m\,r}\right), \\ F_2(r) &= -\frac{m^2}{\cosh^2(m\,r)}, \end{aligned} \tag{19.121a}$$

characterized by a Petrov type N Cotton tensor

$$C^\alpha{}_\beta = C^v{}_u k^\alpha k_\beta, C^v{}_u = \frac{1}{2} m^2 \left(e^{2\,mr}+1\right) f_{03},\ k^\alpha = \delta^\alpha{}_v. \tag{19.121b}$$

Multi Exponent–Integral Representation of Φ

Let us search for a multi-exponent–integral representation, see 19.10.1, of the solution to the equation (19.120a). One gets the algebraic system

$$\begin{aligned} &\gamma + 3\,\alpha + 2\,\beta = -m, \\ &2\,\gamma\,\alpha + \beta^2 + 3\,\alpha^2 + 4\,\beta\,\alpha + \gamma\,\beta = -m^2, \\ &\gamma\,\beta\,\alpha + \gamma\,\alpha^2 + 2\,\beta\,\alpha^2 + \beta^2\alpha + \alpha^3 = m^3 \end{aligned} \tag{19.122a}$$

with solutions

$$\begin{aligned} &\{\alpha = m, \beta = -2\,m, \gamma = 0\}, \{\alpha = -m, \beta = 2\,m, \gamma = -2\,m,\}, \\ &\{\alpha = -m, \beta = 0, \gamma = 2\,m\}. \end{aligned} \tag{19.122b}$$

Since in this approach Φ is expressible as

$$\Phi_{NH} = \mathrm{e}^{-\alpha\,r} \int^r \mathrm{e}^{-\beta\,t} \int^t \mathrm{e}^{-\gamma\,y} \int^y \mathrm{e}^{(\gamma+\beta+\alpha)z} Rhs\,(u,z)\,dzdydt, \tag{19.122c}$$

for each set of exponents one gets its corresponding multi-exponent–integral solution representation.

VP of the CPS Representation of ϕ

The Chow–Pope–Setzgin representation of the solution for $F_0(u,r)$ is achieved via an alternative definition of the function $F_0(u,r)$; instead of

$$\Phi(u,r) = \frac{F_0(u,r)}{\cosh(mr)} = 2\frac{F_0(u,r)}{(1+\mathrm{e}^{-2mr})}\mathrm{e}^{-mr}$$

one uses $\phi(u,r)$ defined as

$$\phi(u,r) = \frac{F_0(u,r)}{1+\mathrm{e}^{-2mr}} \to \phi(u,r) = \frac{1}{2}\mathrm{e}^{mr}\,\Phi(u,r). \tag{19.123a}$$

In terms of ϕ the equation (19.120a), (19.24e), becomes

$$\frac{\partial^3}{\partial r^3}\phi - 4m\frac{\partial^2}{\partial r^2}\phi + 4\,m^2\frac{\partial}{\partial r}\phi = \frac{1}{2}\frac{\mathrm{e}^{mr}RHS}{\cosh\,(mr)} = Rhs. \tag{19.123b}$$

The solution to the homogeneous equation is given by

$$\begin{aligned} \phi_h\,(r) &= C_1(u) + C_2(u)\,\mathrm{e}^{2\,mr} + C_3(u)\,r\,\mathrm{e}^{2\,mr} \\ &\equiv C_1(u)\phi_1 + C_2(u)\,\phi_2 + C_3(u)\,\phi_3. \end{aligned} \tag{19.123c}$$

The non-homogeneous solution is sought in the form of

$$\phi_{NH}\,(r) = C_1(u,r)\phi_1 + C_2(u,r)\,\phi_2 + C_3(u,r)\,\phi_3 \tag{19.123d}$$

via the variation of parameters $C_i(u,r)$, where u plays the role of a parameter.
The first derivatives are

$$\begin{aligned} &\frac{\partial}{\partial r}C_1 = \frac{1}{4}\frac{Rhs}{m^2},\ \frac{\partial C_2}{\partial r} = -\frac{1}{4}\frac{(2\,m\,r+1)\,\mathrm{e}^{-2\,m\,r}Rhs}{m^2}, \\ &\frac{\partial}{\partial r}C_3 = \frac{1}{2}\frac{\mathrm{e}^{-2\,m\,r}Rhs}{m}. \end{aligned}$$

Therefore the general $\phi(u,r)$ solution to the equation (19.123b) via the variation of parameters method is

$$\begin{aligned} \phi(u,r) = \frac{F_0(u,r)}{(1+\mathrm{e}^{-2\,mr})} &= f_{01}(u) + f_{02}(u)\,\mathrm{e}^{2\,mr} + f_{03}(u)\,r\,\mathrm{e}^{2\,m\,r} \\ &+ C_1(u,r) + C_2(u,r)\,\mathrm{e}^{2\,mr} + C_3(u,r)\,r\,\mathrm{e}^{2\,m\,r}, \end{aligned} \tag{19.123e}$$

explicitly the non-homogeneous solution is given by

$$\phi_{NH} = -\frac{{f_{11}}^2}{16\,m^2}\Big\{\left(e^{2\,m\,r}+1\right)\left[\left(1+\ln\left(e^{2\,m\,r}+1\right)\right)^2-1\right]$$
$$+2\,e^{2\,m\,r}\,\mathrm{dilog}\left(e^{2\,m\,r}+1\right)\Big\}$$
$$-\frac{1}{4\,m^2}\frac{df_{11}}{du}\left[\left(e^{2\,m\,r}+1\right)\ln\left(e^{2\,m\,r}+1\right)\right], \tag{19.123f}$$

which coincides with the CPS solution Chow et al. (2010b)(4.58) if one properly identifies the Li_2 function with the dilog appearing here, $\mathrm{Li}_2(-e^{2\,m\,r}) = \mathrm{dilog}\left(1+e^{2\,mr}\right)$, (19.99f).

Multi Exponent–Integral Representation of ϕ

To obtain a multi-exponent–integral representation of equation (19.123b) for $\phi(u,r)$ one simply rewrites it as

$$e^{2\,mr}\frac{\partial^2}{\partial r^2}\left(e^{-2\,mr}\frac{\partial}{\partial r}\phi\right) = \frac{1}{2}\frac{e^{mr}RHS}{\cosh(mr)} = Rhs = -\frac{m\left[{f_{11}}^2\xi(r)+2\,\frac{df_{11}}{du}\right]}{\left(1+e^{-2\,m\,r}\right)^2}. \tag{19.124a}$$

Therefore, the integral for $\phi(u,r)$ is

$$\phi(u,r) = -\frac{1}{4}\,m\int^{r}e^{2\,m\,t}\int^{t}\!\!\int^{y}\frac{\left[{f_{11}}^2\,\xi(z)+2\,\frac{df_{11}}{du}\right]}{\cosh^2(m\,z)}dz\,dy\,dt. \tag{19.124b}$$

It can be given as

$$\phi(u,r) = -\frac{m}{4}{f_{11}}^2\int^{r}e^{2\,m\,t}\int^{t}\!\!\int^{y}\frac{\xi(z)}{\cosh^2(m\,z)}dz\,dydt$$
$$-\frac{1}{4}\frac{df_{11}}{du}\frac{\ln\left(e^{2\,m\,r}+1\right)\left(e^{2\,m\,r}+1\right)}{m^2}. \tag{19.124c}$$

19.11.3 Type III Kundt Solution; $\Lambda = -m^2, W_1 = -2m\tanh(mr), \mu = m$

Solution Φ via Variation of Parameters

The equation (19.113) for the function $\Phi(u,r)$ in the case $\mu = m$ becomes

$$\frac{\partial^3}{\partial r^3}\Phi + m\frac{\partial^2}{\partial r^2}\Phi - m^2\frac{\partial}{\partial r}\Phi - m^3\Phi = \frac{RHS}{\cosh(m\,r)} = Rhs$$
$$= -\frac{1}{2\cosh(m\,r)}\left[2\frac{d}{du}f_{11}+{f_{11}}^2\xi\right]\frac{d\xi}{dr},$$
$$\frac{d\xi}{dr} = m\frac{e^{-m\,r}}{\cosh(m\,r)},\quad \xi(r) = -\ln\left(1+e^{-2\,m\,r}\right). \tag{19.125a}$$

Its homogeneous solution of the form $e^{k\,r}$ fulfills $(k+m)^2\,(k-m)=0$, because of the repeated roots, it is

$$\Phi_H = C_1 e^{m\,r} + C_2 e^{-m\,r} + C_3\,r\,e^{-m\,r}, \tag{19.125b}$$

thus the non-homogeneous solution is sought in the form

$$\Phi_{NH} = C_1(u,r)\,\mathrm{e}^{m\,r} + C_2(u,r)\,\mathrm{e}^{-m\,r} + C_3(u,r)\,r\,\mathrm{e}^{-m\,r}. \tag{19.125c}$$

Using the VP procedure, one arrives at

$$\frac{\partial}{\partial r}C_1 = \frac{\mathrm{e}^{-m\,r}}{4m^2}\,\frac{RHS}{\cosh(m\,r)},\ \frac{\partial}{\partial r}C_2 = \frac{\mathrm{e}^{m\,r}(2mr-1))}{4m^2}\,\frac{RHS}{\cosh(m\,r)},$$
$$\frac{\partial}{\partial r}C_3 = -\frac{\mathrm{e}^{m\,r}}{2\,m}\,\frac{RHS}{\cosh(m\,r)}.$$

The use of

$$\frac{d}{dr}\xi(r) = \frac{m\mathrm{e}^{-m\,r}}{\cosh(m\,r)} = 2\,\frac{m\mathrm{e}^{-2\,m\,r}}{1+\mathrm{e}^{-2\,m\,r}},\ \xi(r) = -\ln\left(1+\mathrm{e}^{-2\,m\,r}\right),$$

in the RHS, its substitution into the C_is and their integrations lead the non-homogeneous Φ_{NH} in the form

$$\Phi_{NH} = -\frac{1}{8m^2}{f_{11}}^2\Big\{\Big[\left(1+\ln\left(1+\mathrm{e}^{-2\,m\,r}\right)\right)^2 - 1\Big]\left(1+\mathrm{e}^{-2\,m\,r}\right)\mathrm{e}^{m\,r}$$
$$+2\,\mathrm{e}^{-m\,r}\,dilog\left(1+\mathrm{e}^{-2\,m\,r}\right) + 4\,m\,r\,\mathrm{e}^{-m\,r} + 2\,\mathrm{e}^{m\,r}\Big\}$$
$$+\frac{1}{2m^2}\,\frac{df_{11}}{du}\left[\left(\mathrm{e}^{m\,r}+\mathrm{e}^{-m\,r}\right)\ln\left(\mathrm{e}^{2\,m\,r}+1\right) - 2\,m\mathrm{e}^{m\,r}r\right]. \tag{19.125d}$$

Using the relation (19.99c) between the dilog functions and Li_2 (19.99f), excluding the terms proportional to the homogeneous functions, one finally arrives at

$$\Phi_{NH} = -\frac{\mathrm{e}^{-m\,r}}{8\,m^2}\,(f_{11})^2\Big\{\Big[\left(1+\ln\left(1+\mathrm{e}^{-2\,m\,r}\right)\right)^2 - 1\Big]\left(1+\mathrm{e}^{2\,m\,r}\right)$$
$$-2\,Li_2(-\mathrm{e}^{2m\,r}) - 4\,m^2\,r^2\Big\}$$
$$+\frac{\mathrm{e}^{-m\,r}}{2\,m^2}\,\frac{df_{11}}{du}\left(\mathrm{e}^{2m\,r}+1\right)\ln\left(\mathrm{e}^{2\,m\,r}+1\right). \tag{19.125e}$$

which coincides, up to the factor $2\mathrm{e}^{-m\,r}$, with CPSb(4.56).

Multi-Exponent–Integral Representation of Φ

In this approach the search of the solution to the key equation (19.125a), following 19.10.1, gives rise to the algebraic system

$$\begin{aligned}&\gamma + 2\,\beta + 3\,\alpha = m,\\ &\gamma\,\beta + 2\,\gamma\,\alpha + \beta^2 + 4\,\beta\,\alpha + 3\,\alpha^2 = -m^2,\\ &\gamma\,\beta\,\alpha + \gamma\,\alpha^2 + \beta^2\alpha + 2\,\beta\,\alpha^2 + \alpha^3 = -m^3\end{aligned} \tag{19.126a}$$

with the following set of solutions

$$\begin{aligned}&\{\alpha = -m, \beta = 2\,m, \gamma = 0\},\\ &\{\alpha = m, \beta = 0, \gamma = -2\,m\},\\ &\{\alpha = m, \beta = -2\,m, \gamma = 2\,m\},\end{aligned} \tag{19.126b}$$

with the corresponding solution representations. For instance, for the first set of exponents one has

$$\Phi_1 = e^{m\,r} \int e^{-2\,m\,r} \iint e^{m\,r}\, Rhs\,(r)\, dr\, dr dr.$$

VP of the CPS Representation of $\phi(r)$

The CPSb representation of the solution for $F_0(u,r)$ is achieved via an alternative definition of the function $F_0(u,r)$ through $\phi(r)$, namely

$$\phi(u,r) = \frac{F_0(u,r)}{1+e^{-2mr}},\; \Phi = \frac{F_0(u,r)}{\cosh(m\,r)} = 2\,e^{-mr}\,\phi \tag{19.127a}$$

thus replacing $\Phi = 2\,e^{-m\,r}\phi$ in (19.125a) one gets

$$\frac{\partial^3}{\partial r^3}\phi - 2\,m\frac{\partial^2}{\partial r^2}\phi = \frac{1}{2}\,e^{m\,r}\,\frac{RHS}{\cosh(mr)}. \tag{19.127b}$$

If one were using the variations of parameters method to determine the solution to the non-homogeneous equation one generalizes the parameters C_i of the homogeneous solution

$$\phi_H\,(r) = C_1(u) + C_2(u)r + C_3(u)\,e^{2\,m\,r}, \tag{19.127c}$$

to be dependent on the variable r in such a way that the non-homogeneous solution is search in the form of

$$\phi_{NH}\,(r) = C_1(u,r)\phi_1 + C_2(u,r)\,\phi_2 + C_3(u,r)\,\phi_3 \tag{19.127d}$$

via the variation of parameters $C_i(u,r)$, where u plays the role of a parameter. The derivatives of the C_i are

$$\frac{\partial C_1}{\partial r} = \frac{e^{r\,m}\,(2\,m\,r-1)}{8\,m^2}\,\frac{RHS}{\cosh(m\,r)},\; \frac{\partial C_2}{\partial r} = -\frac{e^{r\,m}}{4\,m}\,\frac{RHS}{\cosh(m\,r)},$$
$$\frac{\partial C_3}{\partial r} = \frac{e^{-r\,m}}{8\,m^2}\,\frac{RHS}{\cosh(m\,r)},$$

hence, the non-homogeneous ϕ_{NH} solution can be given as

$$\phi_{NH} = \frac{1}{8\,m^2}\int \frac{e^{m\,r}RHS\,(2\,m\,r-1)}{\cosh\,(m\,r)}dr$$
$$-\frac{r}{4\,m}\int \frac{e^{m\,r}RHS}{\cosh\,(m\,r)}dr + \frac{1}{8\,m^2}e^{2\,m\,r}\int e^{-m\,r}\frac{RHS}{\cosh\,(m\,r)}dr. \tag{19.127e}$$

Replacing the structural functions

$$\xi\,(r) = -\ln\left(1+e^{-2\,m\,r}\right),\; \cosh\,(m\,r) = 1/2\,e^{m\,r}\left(1+e^{-2\,m\,r}\right)$$

in the above equation and integrating one obtains

$$\phi_{NH} = \frac{1}{4\,m^2}\left(\frac{df_{11}}{du}\right)\left[\left(e^{2\,mr}+1\right)\ln\left(1+e^{-2\,m\,r}\right) + 2\,mr + e^{2\,m\,r}\right]$$

$$-\frac{1}{16\,m^2} {f_{11}}^2 \left\{ \left(\mathrm{e}^{2\,m\,r}+1\right) \left[\left(1+\ln\left(1+\mathrm{e}^{-2\,m\,r}\right)\right)^2 - 1 \right] + 4\,m\,r + 2\,\mathrm{e}^{2\,m\,r} \right.$$
$$\left. +2\,\mathrm{dilog}\left(1+\mathrm{e}^{-2\,m\,r}\right) \right\}. \tag{19.127f}$$

One can always eliminate the terms proportional to the independent homogeneous solutions $\{1, r, \mathrm{e}^{2mr}\}$, and use the relation (19.99c) between the dilog and Li_2 (19.99f) and write, without loss of generality, the function $\phi_{NH}(u,r)$

$$\begin{aligned}\phi_{NH} = & -\frac{1}{16\,m^2} {f_{11}}^2 \left\{ \left(\mathrm{e}^{2\,m\,r}+1\right) \left[\left(1+\ln\left(1+\mathrm{e}^{-2\,m\,r}\right)\right)^2 - 1 \right] \right. \\ & \left. -2\,Li_2\left(-\mathrm{e}^{2m\,r}\right) - 4\,m^2\,r^2 \right\} \\ & +\frac{1}{4\,m^2} \left(\frac{df_{11}}{du}\right) \left(\mathrm{e}^{2\,mr}+1\right) \ln\left(1+\mathrm{e}^{-2\,m\,r}\right),\end{aligned} \tag{19.127g}$$

i.e., the formula CPSb(4.56). Notice that the relation $\Phi = 2\,\mathrm{e}^{-\,m\,r}\,\phi$ holds, as indeed it should.

The metric of this solution can be written as

$$\begin{aligned} & ds^2 = dr^2 + 2du\,dv + (v^2 F_2 + v F_1 + F_0) du^2 - 4\,m\,v\,\tanh(m\,r)\,du\,dr, \\ & F_2(u,r) = -\frac{m^2}{\cosh^2(m\,r)}, \\ & F_1(u,r) = f_{11}(u)\,\xi(r),\ \xi(r) = -\ln\left(1+\mathrm{e}^{-2\,m\,r}\right), \\ & F_0(u,r) = \left(1+\mathrm{e}^{2\,mr}\right)\left[\phi_H(u,r;C_i(u)\to f_{0i}) + \phi_{NH}(u,r)\right]. \end{aligned} \tag{19.127h}$$

This spacetime is characterized by the Cotton tensor

$$\begin{aligned} & C^\alpha{}_\beta = C^v{}_u \delta^\alpha{}_v \delta^u{}_\beta + C^v{}_r \delta^\alpha{}_v \delta^r{}_\beta + C^r{}_u \delta^\alpha{}_r \delta^u{}_\beta, \\ & C^v{}_u = -m^2\left(1+\mathrm{e}^{-2\,mr}\right) f_{02} - 2\,v\,f_{11}\,\frac{m^3 \mathrm{e}^{2\,mr}}{\left(\mathrm{e}^{2\,mr}+1\right)^2} + \frac{m}{2}\,\mathrm{e}^{-2\,mr}\,\frac{df_{11}}{du} \\ & \qquad + \frac{m}{4}\left[\ln\left(1+\mathrm{e}^{-2\,mr}\right) - \mathrm{e}^{-2\,mr}\right](f_{11})^2, \\ & C^r{}_u = C^v{}_r = -\frac{m^2}{\mathrm{e}^{2\,mr}+1} f_{11}. \end{aligned} \tag{19.127i}$$

Type N Limit

Its Petrov type N limit is given by the metric

$$\begin{aligned} & ds^2 = dr^2 + 2du\,dv + (v^2 F_2 + v F_1 + F_0) du^2 - 4\,m\,v\tanh(m\,r)\,du\,dr, \\ & F_2(u,r) = -\frac{m^2}{\cosh^2(m\,r)}, \\ & F_0(u,r) = r\,\left(1+\mathrm{e}^{2\,mr}\right) f_{02}, \end{aligned} \tag{19.128a}$$

and is characterized by the Cotton tensor

$$C^\alpha{}_\beta = C^v{}_u k^\alpha k_\beta,\ C^v{}_u = -m^2\left(1+\mathrm{e}^{-2\,mr}\right) f_{02},\ k^\alpha = \delta^\alpha{}_v.$$

Multi Exponent–Integral Representation of ϕ

The CPS representation of the solution for $F_0(u,r)$ is achieved via an alternative definition of the function $F_0(u,r)$ through $\phi(r)$, namely

$$\phi(u,r) = \frac{F_0(u,r)}{1+\mathrm{e}^{-2mr}},\ \Phi = \frac{F_0(u,r)}{\cosh(m\,r)} = 2\,\mathrm{e}^{-mr}\,\phi, \tag{19.129a}$$

thus, replacing $\Phi = 2\,\mathrm{e}^{-m\,r}\phi$ in (19.125a), or $F_0(\phi)$ in (19.24e), one gets

$$\mathrm{e}^{mr}\frac{\partial}{\partial r}\left(\mathrm{e}^{-2mr}\frac{\partial^2\phi}{\partial r^2}\right) = \frac{1}{2}\frac{RHS}{\cosh(mr)}. \tag{19.129b}$$

Incidentally, it arises from the key CPS equation (19.118b) for $\mu = m$.

The first integral of the equation (19.129b) yields

$$\frac{\partial^2}{\partial r^2}\phi = \frac{1}{2}\,\mathrm{e}^{2\,m\,r}\int \mathrm{e}^{-m\,r}\,\frac{RHS}{\cosh(mr)}dr + A0\,\mathrm{e}^{2\,m\,r},$$

therefore, integrating twice, one obtains the general solution for $\phi\,(r)$ in the form

$$\phi = -\frac{1}{4}\,m\int^r\!\!\int^t \mathrm{e}^{2\,m\,y}\int^y \frac{\left[{f_{11}}^2\,\xi\,(z) + 2\,\frac{df_{11}}{du}\right]\mathrm{e}^{-2\,m\,z}}{\cosh^2\,(m\,z)}dzdy\,dt$$
$$+f_{01}(u)\mathrm{e}^{2m\,r} + f_{02}(u)\,r + f_{03}(u). \tag{19.129c}$$

The function $\xi(r)$ can be integrated in terms of logarithm, namely

$$\frac{d}{dr}\xi\,(r) = 2\,\frac{m\mathrm{e}^{-2\,m\,r}}{1+\mathrm{e}^{-2\,m\,r}},\ \xi(r) = -\ln\left(1+\mathrm{e}^{-2\,m\,r}\right), \tag{19.129d}$$

which substituted in ϕ yields

$$\phi = -m\int^r\!\!\int^t \mathrm{e}^{2\,m\,y}\int^y \frac{\mathrm{e}^{-4\,m\,z}dz}{(1+\mathrm{e}^{-2\,m\,z})^2}\left[2\,\frac{df_{11}}{du} - {f_{11}}^2\ln\left(1+\mathrm{e}^{-2\,m\,z}\right)\right]dy\,dt$$
$$+C_3(u)\,\mathrm{e}^{2\,m\,r} + C_2(u)\,r + C_1(u) \tag{19.129e}$$

this is an integral representation of the solution, although one can integrate it forward.

19.12 Type III Kundt Metric; $\Lambda = m^2, W_1 = -2m\cot(m\,r), \mu$

The equation (19.40) for vanishing eigenvalue S, which guarantees the Petrov type III of the Cotton tensor, for positive cosmological constant $\Lambda = m^2$, leads to

$$W_1\,(r) = -2\,m\,\cot\,(m r)\,,\ \text{from (19.41f)}. \tag{19.130}$$

Evaluating the equation (19.24c) for F_2 one gets

$$F_2 = \frac{m^2}{\sin^2\,(m\,r)}. \tag{19.131}$$

Replacing these functions into the equation (19.24d) for F_1 one arrives at

$$\frac{\partial^2}{\partial r^2}F_1\,(u,r) + (\mu + m\,\cot\,(m\,r))\,\frac{\partial}{\partial r}F_1 = 0 \tag{19.132}$$

with solution

$$F_1\,(u,r) = f_{11}\,(u)\,\int m\,\frac{e^{-\mu\,r}}{\sin\,(m\,r)}dr = f_{11}\,\xi(r); \tag{19.133}$$

where it has been introduced the auxiliary function $\xi(r)$,

$$\frac{d\xi}{dr} = \frac{m e^{-\mu\,r}}{\sin\,(m\,r)} \rightarrow \xi\,(r) = m\int\frac{e^{-\mu\,r}}{\sin\,(m\,r)}dr. \tag{19.134}$$

The key equation (19.24e) for $F_0(u,r)$ becomes

$$\begin{aligned}&\frac{\partial^3}{\partial r^3}F_0\,(u,r) + (\mu - 3m\,\cot\,(m\,r))\,\frac{\partial^2}{\partial r^2}F_0\\ &+2\,m\,(2\,m - \mu\,\cot\,(m\,r) + 3\,m\cot^2\,(rm))\,\frac{\partial}{\partial r}F_0\\ &+2\,m^2\,F_0\,(1+\cot^2\,(m\,r))\,(\mu - 3m\,\cot\,(m\,r))\\ &= -\frac{1}{2}\left[2\,\frac{d}{du}f_{11} + (f_{11})^2\,\xi\,(r)\right]\frac{d}{dr}\xi\,(r) = RHS.\end{aligned} \tag{19.135}$$

Following the standard VP procedure for solving non-homogeneous linear ordinary differential equations the solution $F_{0\,NH}$ is searched as

$$F_0 = \left[C_1\,(u,\,r)\,e^{-\mu\,r} + C_2\,(u,\,r)\sin\,(m\,r) + C_3\,(u,\,r)\cos\,(m\,r)\right]\sin\,(m\,r)\,. \tag{19.136}$$

Nevertheless, the structure of the solution above points to a better function choice, namely

$$\Phi\,(u,\,r) = F_0\,(u,r)/\sin\,(m\,r),\; F_0\,(u,r) = \Phi\,(u,\,r)\sin\,(m\,r)\,. \tag{19.137}$$

The key equation (19.135) in terms of Φ becomes

$$\frac{\partial^3}{\partial r^3}\Phi + \mu\,\frac{\partial^2}{\partial r^2}\Phi + m^2\frac{\partial}{\partial r}\Phi + \Phi m^2\mu = \frac{RHS}{\sin\,(m\,r)}. \tag{19.138}$$

In what follows we proceed to determine the single solution of this class. First the derivation of the solution via the variation of parameters procedure is done. Next, the solution in a multi-exponent–integral representation is curried out.

19.12.1 Solution Φ Through VP

The homogeneous solution to the homogeneous equation (19.138) ($Rhs \rightarrow 0$), one searches in the form of $\Phi_h \simeq e^{k\,r}$ arriving at the characteristic equation

$$k^3 + \mu\,k^2 + m^2\,k + m^2\mu = 0 = (k^2 + m^2)(k + \mu)$$

with solutions $k = \{-\mu,\, im,\, -i\,m\}$, hence Φ_H can be given as

$$\Phi_H\,(r) = C_1\,\mathrm{e}^{-\mu\,r} + C_2\,\sin\,(m\,r) + C_3\,\cos\,(m\,r) \equiv C_1\,\Phi_l + C_2\,\Phi_2 + C_3\,\Phi_3. \tag{19.139a}$$

Searching for the solution to the non-homogeneous equation ($Rhs \neq 0$) one assumes the solution to be of the form

$$\Phi_{NH}\,(u,r) = C_1(u,r)\,\Phi_l + C_2(u,r)\,\Phi_2 + C_3(u,r)\,\Phi_3. \tag{19.139b}$$

The variation of the parameters $C_i(u,r)$, within the VP procedure, yields

$$\begin{aligned}
\frac{\partial}{\partial r}C_1 &= \frac{\mathrm{e}^{\mu\,r}}{\mu^2+m^2}\frac{RHS}{\sin\,(m\,r)},\\
\frac{\partial}{\partial r}C_2 &= -\frac{(m\,\sin\,(m\,r) - \mu\,\cos\,(m\,r))}{m\,(\mu^2+m^2)}\frac{RHS}{\sin\,(m\,r)},\\
\frac{\partial}{\partial r}C_3 &= -\frac{(\mu\,\sin\,(m\,r) + m\,\cos\,(m\,r))}{m\,(\mu^2+m^2)}\frac{RHS}{\sin\,(m\,r)}.
\end{aligned} \tag{19.139c}$$

Some simplifications arise by replacing $\xi(r)$, or more precisely, its derivative from (19.134), to get the non-homogeneous solution as

$$\begin{aligned}
\Phi_{NH} = {} & \frac{1}{2}\sin\,(m\,r)\int\frac{\mathrm{e}^{-\mu\,r}\,(m\,\sin\,(m\,r) - \mu\cos\,(m\,r))\left({f_{11}}^2\xi + 2\,\frac{df_{11}}{du}\right)}{(\mu^2+m^2)\sin^2\,(m\,r)}dr\\
& + \frac{1}{2}\cos\,(m\,r)\int\frac{\mathrm{e}^{-\mu\,r}\,(\mu\,\sin\,(m\,r) + m\,\cos\,(m\,r))\left({f_{11}}^2\xi + 2\,\frac{df_{11}}{du}\right)}{\sin^2\,(m\,r)\,(\mu^2+m^2)}dr\\
& - \frac{1}{2}\mathrm{e}^{-\mu\,r}\int\frac{m\left({f_{11}}^2\xi + 2\,\frac{df_{11}}{du}\right)}{\sin^2\,(m\,r)\,(\mu^2+m^2)}dr,
\end{aligned} \tag{19.139d}$$

or, taking into account that the evaluation of the integrals multiplying $\frac{d}{du}f_{11}$ gives

$$\frac{d\,f_{11}}{du}\frac{\sin\,(m\,r)}{m^2}\,\xi\,(r)\,,$$

factorizing, one obtains a linear single integral solution for Φ_{NH}

$$\begin{aligned}
\Phi_{NH} = {} & \frac{d\,f_{11}}{du}\frac{\sin\,(m\,r)}{m^2}\,\xi\,(r) + \frac{{f_{11}}^2}{2\,(\mu^2+m^2)}\left[-\mathrm{e}^{-\mu\,r}\int\frac{\xi\,(r)\,m}{\sin^2\,(m\,r)}dr\right.\\
& + \sin\,(m\,r)\int\mathrm{e}^{-\mu\,r}\frac{(m\sin\,(m\,r) - \mu\,\cos\,(m\,r))\,\xi\,(r)}{\sin^2\,(m\,r)}dr\\
& \left. + \cos\,(m\,r)\int\frac{\mathrm{e}^{-\mu\,r}\xi\,(r)\,(\mu\,\sin\,(m\,r) + m\,\cos\,(m\,r))}{\sin^2\,(m\,r)}dr\right].
\end{aligned} \tag{19.139e}$$

The homogeneous solution can be written as

$$\Phi_H = f_{01}(u)\,\mathrm{e}^{-r\mu} + f_{02}(u)\,\sin(m\,r) + f_{03}(u)\,\cos(m\,r), \tag{19.139f}$$

which together determine the general solution. The metric of this solution can be written as

$$
\begin{aligned}
ds^2 &= dr^2 + 2du\,dv + (v^2F_2 + vF_1 + F_0)du^2 - 4\,m\,v\,\cot(m\,r)\,du\,dr, \\
F_2(u,r) &= \frac{m^2}{\sin^2(m\,r)},\ F_1(u,r) = f_{11}(u)\,\xi(r),\ \xi\,(r) = m\int \frac{e^{-\mu\,r}}{\sin\,(mr)}dr, \\
F_0(u,r) &= \sin(m\,r)\,\left[\phi_H(u,r) + \phi_{NH}(u,r)\right]. \qquad (19.139g)
\end{aligned}
$$

This spacetime is characterized by the Cotton tensor

$$
\begin{aligned}
C^\alpha{}_\beta &= C^v{}_u\delta^\alpha{}_v\delta^u{}_\beta + C^v{}_r\delta^\alpha{}_v\delta^r{}_\beta + C^r{}_u\delta^\alpha{}_r\delta^u{}_\beta, \\
C^v{}_u &= \frac{1}{2}\,\mu\,\left(\mu^2+m^2\right)\,f_{01}\,e^{-\mu\,r}\sin\,(mr) + \frac{1}{2}\,\mu\,\frac{df_{11}}{du}\,e^{-\mu\,r}\,\cos\,(mr) \\
&\quad -\frac{1}{4}\,\mu m\,\left(f_{11}\right)^2 e^{-\mu\,r}\sin\,(mr)\int\frac{\xi\,(r)}{\sin^2\,(mr)}dr \\
&\quad -\frac{1}{2}m\,\mu\,v\,f_{11}\,e^{-\mu\,r}\,\frac{\left[\mu\,\sin\,(mr) + m\,\cos\,(mr)\right]}{\sin^2\,(mr)}, \\
C^r{}_u &= C^v{}_r = -\frac{1}{2}\,m\mu\,f_{11}\,\frac{e^{-\mu\,r}}{\sin\,(mr)}. \qquad (19.139h)
\end{aligned}
$$

Type N Limit

Its Petrov type N limit is given by the metric

$$
\begin{aligned}
ds^2 &= dr^2 + 2du\,dv + (v^2F_2 + vF_1 + F_0)du^2 - 4\,m\,v\cot(m\,r)\,du\,dr, \\
F_2(u,r) &= \frac{m^2}{\sin^2(m\,r)},\ F_0(u,r) = f_{01}\,(u)\,e^{-\mu\,r}\,\sin(m\,r). \qquad (19.140a)
\end{aligned}
$$

and is characterized by the type N Cotton tensor

$$
C^\alpha{}_\beta = C^v{}_u k^\alpha k_\beta,\ C^v{}_u = \frac{1}{2}\,\mu\,\left(\mu^2+m^2\right)\,f_{01}\,e^{-\mu\,r}\sin\,(mr)\,,\ k^\alpha = \delta^\alpha{}_v. \qquad (19.140b)
$$

19.12.2 Multi Exponent–Integral Representation of Φ

Let us now search the non-homogeneous solution to the third-order equation (19.138) by guessing that this equation can be written in the form

$$
\frac{\partial}{\partial r}\left(e^{\gamma r}\frac{\partial}{\partial r}\left(e^{\beta r}\frac{\partial}{\partial r}\left(e^{\alpha r}\Phi(u,r)\right)\right)\right) = \frac{RHS}{\sin\,(m\,r)}\,e^{(\alpha+\beta+\gamma)r}, \qquad (19.141a)
$$

hence the solution will be

$$
\Phi(u,r) = e^{-\alpha r}\int^r(e^{-\beta t}\int^t(e^{-\gamma y}\int^y\frac{RHS(u,z)}{\sin\,(m\,z)}\,e^{(\alpha+\beta+\gamma)\,z}dz)dy)dt. \qquad (19.141b)
$$

Comparing the coefficients multiplying derivatives of the same order in equation (19.138) and with the expanded (19.141a) one gets the system

$$\begin{aligned}
&2\,\beta+3\,\alpha+\gamma=\mu,\\
&\gamma\,\beta+4\,\beta\,\alpha+3\,\alpha^2+2\,\gamma\,\alpha+\beta^2=m^2,\\
&\gamma\,\beta\,\alpha+\gamma\,\alpha^2+\beta^2\alpha+\alpha^3+2\,\beta\,\alpha^2=m^2\mu,
\end{aligned}$$

with the following set of complex solutions

$$\begin{aligned}
&[\alpha=\mu,\beta=-\mu+im,\gamma=-2\,im],[\alpha=\mu,\beta=-\mu-im,\gamma=2\,im],\\
&[\alpha=im,\beta=\mu-im,\gamma=-\mu-im],\\
&[\alpha=-im,\beta=\mu+im,\gamma=-\mu+im],\\
&[\alpha=-i\,m,\beta=2\,im,\gamma=\mu-im],\ [\alpha=im,\beta=-2\,im,\gamma=\mu+im].
\end{aligned}$$

The substitution of the found exponents in the expression (19.141b) determines a specific representation of the solution we are looking for. In particular, for the first set of constant exponents one gets a complex solution representation Φ_C of the CPS Chow et al. (2010b) solution:

$$\Phi_C=\mathrm{e}^{-\mu\,r}\int^r \mathrm{e}^{\mu\,t}\mathrm{e}^{-im\,t}\int^t \mathrm{e}^{2\,imy}\int^y \frac{RHS(z)}{\sin(m\,z)}\mathrm{e}^{-im\,z}dzdydt, \tag{19.141c}$$

or in term of the trigonometric functions

$$\begin{aligned}
\Phi_C=\mathrm{e}^{-\mu\,r}&\int^r \mathrm{e}^{\mu\,t}\left(\cos\left(m\,t\right)-i\sin\left(m\,t\right)\right)\int^t\left(\cos\left(2\,m\,y\right)+i\sin\left(2\,m\,y\right)\right)\\
&\times\int^y \frac{RHS\left(z\right)\left(\cos\left(m\,z\right)-i\sin\left(m\,z\right)\right)}{\sin\left(m\,z\right)}dzdydt,
\end{aligned}\tag{19.141d}$$

with real part

$$\begin{aligned}
\Phi_R=\mathrm{e}^{-\mu\,r}&\int^r \mathrm{e}^{\mu\,t}\sin\left(m\,t\right)\int^t\sin\left(2\,m\,y\right)\int^y\frac{RHS\left(z\right)\cos\left(m\,z\right)}{\sin\left(m\,z\right)}dzdydt\\
&+\mathrm{e}^{-\mu\,r}\int^r \mathrm{e}^{\mu\,t}\cos\left(m\,t\right)\int^t\sin\left(2\,m\,y\right)\int^y RHS\left(z\right)dzdydt\\
&-\mathrm{e}^{-\mu\,r}\int^r \mathrm{e}^{\mu\,t}\sin\left(m\,t\right)\int^t\cos\left(2\,m\,y\right)\int^y RHS\left(z\right)dzdydt\\
&+\mathrm{e}^{-\mu\,r}\int^r \mathrm{e}^{\mu\,t}\cos\left(m\,t\right)\int^t\cos\left(2\,m\,y\right)\int^y\frac{RHS\left(z\right)\cos\left(m\,z\right)}{\sin\left(m\,z\right)}dzdydt,
\end{aligned}\tag{19.141e}$$

and an imaginary part, which does not contribute to the solution since it fits the homogeneous equation. Therefore, we work out the real part, arriving at the compact "cosine" CPS Chow et al. (2010b) representation

$$\Phi_R=-\frac{m}{2}\int^r dt\int^t dy\int^y \mathrm{e}^{\mu(y-z-r)}\frac{\cos[m\,(t-2\,y+z)]}{\sin^2(m\,z)}\left[{f_{11}}^2\xi+2\frac{df_{11}}{du}\right]dz; \tag{19.141f}$$

interestingly, the factor of $\frac{df_{11}}{du}$ can be simplified up to

$$\frac{\sin(m\,r)\,\xi\,(r)}{m^2}\,\frac{df_{11}}{du},$$

as was done in the previous VP paragraph, coinciding with the expression of CPS Chow et al. (2010b), which determines in turn the non-homogeneous $F_0(u,r)$ function. Finally, the metric becomes

$$\begin{aligned}
ds^2 &= dr^2 + 2du\,dv + m^2v^2\mathrm{cosec}^2(m\,r)du^2 - 4mv\cot(m\,r)du\,dr \\
&\quad + [v\xi(r)f_{11}(u) + F_0(u,r)]du^2, \\
\frac{F_0(u,r)}{\sin(m\,r)} &= -\frac{m}{2}\,{f_{11}}^2\int^r dt\int^t dy\int^y \mathrm{e}^{\mu\,(t-z-r)}\frac{\cos[m\,(t-2\,y+z)]}{\sin^2(m\,z)}\xi dz \\
&\quad + \frac{\sin(mr)\,\xi\,(r)}{m^2}\frac{df_{11}}{du} + \mathrm{e}^{-\mu r}f_{01}(u) + \sin(m\,r)f_{02}(u) + \cos(m\,r)f_{03}(u), \\
\xi\,(r) &= m\int\frac{\mathrm{e}^{-\mu\,r}}{\sin(m\,r)}dr. \qquad (19.141g)
\end{aligned}$$

For other choice of exponents, one arrives at other representations of the sought inhomogeneous solution.

All the references of the solutions of the type III and N Kundt families can be represented schematically by the table:

Table 12.1.1 *Type III and type N Kundt solutions*

		$S = \partial_r W_1 - {W_1}^2/2 - 2\Lambda = 0$			
$W_1(u,r)$	$\mu(m)$	T III Secs.	Rel. to CPSb	T N Eqns.	$\Lambda = \pm m^2$, 0
0	μ	19.7	(4.15)	–	0
$-2/r$	μ	19.8	(4.39)	(19.70)	0
	$\mu \neq \pm m$	19.9.1	(4.3)	(19.73)	< 0
-2m	$\mu = m$	19.9.2	(4.9)	(19.75)	< 0
	$\mu = -m$	19.9.3	(4.13)	(19.77)	< 0
	$\mu \neq \pm m$	19.10.1	(4.26)	(19.87)	< 0
$-2m\coth(2mr)$	$\mu = -m$	19.10.3	(4.34)	(19.98)	< 0
	$\mu = m$	19.10.4	(4.30)	(19.104)	< 0
	$\mu \neq \pm m$	19.11.1	(4.53)	(19.116)	< 0
$-2m\tanh(2mr)$	$\mu = -m$	19.11.2	(4.57)	(19.121)	< 0
	$\mu = m$	19.11.3	(4.55)	(19.128)	< 0
$-2m\cot(2mr)$	μ	19.12	(4.44)	(19.140)	> 0

20

Cotton Tensor in Riemannian Spacetimes

In this chapter a systematic derivation of the Cotton tensor is presented following García et al. (2004). In $3D$ spaces the Cotton tensor is prominent as the substitute for the Weyl tensor. It is conformally invariant and its vanishing is equivalent to conformal flatness. However, the Cotton tensor arises in the context of the Bianchi identities and is present in *any* dimension n. In this text, its irreducible decomposition is performed and the number of independent components as $n(n^2-4)/3$ is determined. Subsequently, its characteristic properties are exhibited. An algebraic classification of the Cotton tensor in three dimensions is accomplished. This classification is used throughout this text; for each class of the derived here solutions the Cotton tensor is evaluated and classified.

The nonlinear coupling of gravity to matter in general relativity presents difficult technical problems in attempts to understand the gravitational interaction of elementary particles and strings or to investigate details of the gravitational collapse; see Witten (1991). Progress in the former area has come mainly from treating quantum fields as propagating on fixed background geometries: Polchinski (1998), whereas much of the progress in the latter has come from detailed numerical works: Choptuik (1993); Abrahams and Evans (1993); Gundlach (1999).

Exact solutions of the relevant matter–gravity equations can play an important role by shedding light on questions of interest in both general relativity and string theory. One is often interested in certain classes of solutions with specified asymptotic properties; the most common of them are the asymptotically flat spacetimes. Recent work in string theory has, via the AdS/CFT conjecture, highlighted the importance of the asymptotically AdS spacetimes, see Maldacena (1998). The AdS/CFT correspondence relates a quantum field theory in d dimensions to a theory in $d+1$ dimensions that includes gravity Gubser et al. (1998), and Witten (1998). This is the motivation for looking at the *conformally flat* spaces and at the spaces of *constant curvature*. For this reason we decided to review the subject and to collect some old and new results that are nowadays important in the context of anti-de Sitter spacetimes and to present them in modern language. These results seem presently not to be too well known in the community.

In the theory of conformal spaces the main geometrical objects to be analyzed are the Weyl (1918) and the Cotton (1899) tensors. It is well known that for conformally flat spaces the Weyl tensor has to vanish. However, the Cotton tensor is only conformally invariant in three dimensions. Any $3D$ space is conformally flat if the Cotton tensor vanishes. If matter is present, the Ricci tensor is related to the energy–momentum tensor of matter by means of the Einstein equations. Then the vanishing of the Cotton tensor imposes severe restrictions on the energy–momentum tensor. The Cotton tensor also plays a role in the context of the Hamiltonian formulation of general relativity; see Arnowitt et al. (1962).

First the Cotton 2-form in the context of the Bianchi identities is derived. Subsequently its characteristic properties are described and an irreducible decomposition with respect to the (pseudo-)orthogonal group is performed. This allows ones to determine the number of irreducible components in any dimension. Moreover, in four dimensions, the relation of the Cotton to the Bach tensor is exhibited. After that it is shown how to derive the Cotton 2-form in three dimensions by means of a variational procedure. The classification of the Cotton 2-form in three dimensions is accomplished by means of its eigenvalues.

Our notation and conventions for this chapter are taken from Schouten (1954) and for exterior calculus we refer to Hehl et al. (1995). For quick and easy reference, we display our conventions for index positions and signs of the Christoffel symbol, the Riemann tensor, and the Ricci tensor (holonomic indices $i, j, \dots = 0, 1, 2, 3$). The sign of the Ricci tensor is the same as those of the L_{ij} tensor and the Cotton tensor. In particular, the Ric_{ij} sign introduces a relative sign between the L_{ij} tensor and the Weyl tensor in the decomposition of the curvature:

$$\nabla_i T_j{}^k = \partial_i T_j{}^k - \Gamma_{ij}{}^\ell T_\ell{}^k + \Gamma_{i\ell}{}^k T_j{}^\ell \,, \tag{20.1}$$

$$+R_{ijk}{}^\ell = \partial_i \Gamma_{jk}{}^\ell - \partial_j \Gamma_{ik}{}^\ell + \Gamma_{im}{}^\ell \Gamma_{jk}{}^m - \Gamma_{jm}{}^\ell \Gamma_{ik}{}^m \,, \tag{20.2}$$

$$+\mathrm{Ric}_{jk} = R_{ijk}{}^i \,, \tag{20.3}$$

$$+R = R_{ij}{}^{ji} \,, \tag{20.4}$$

$$\mathrm{Weyl}_{ijk\ell} = R_{ijk\ell} + \frac{4}{n-2}\, g_{[i|[k} L_{\ell]|j]} \,. \tag{20.5}$$

An extensive comparison between the various conventions can be found in Misner et al. (1973).

20.1 Bianchi Identities and the Irreducible Decomposition of the Curvature

Let V_n be a Riemannian space of n dimensions. We have the *coframe* ϑ^α and its dual *frame* e_α according to

$$\vartheta^\alpha = e_i{}^\alpha \, dx^i \,, \quad e_\alpha = e^i{}_\alpha \partial_i \,, \quad e_\alpha \rfloor \vartheta^\beta = e^i{}_\alpha \, e_i{}^\beta = \delta^\beta_\alpha \,, \tag{20.6}$$

where x^i are local coordinates and the $e^i{}_\alpha$ are the n-bein coefficients. Greek letters α, β, $\dots = 1, \dots, n$ denote anholonomic, Latin letters $i, j, \dots = 1, \dots, n$ holonomic indices. The symbol $\rfloor$ stands for the interior product.

We introduce a metric g according to

$$g = g_{\alpha\beta} \, \vartheta^\alpha \otimes \vartheta^\beta = g_{ij} \, dx^i \otimes dx^j \,. \tag{20.7}$$

With the metric at our disposal we also have the Hodge dual operator. Then we define the η-basis according to

$$\eta := {}^\star 1 \,, \quad \eta_\alpha := {}^\star \vartheta_\alpha \,, \quad \eta_{\alpha\beta} := {}^\star(\vartheta_\alpha \wedge \vartheta_\beta) \,, \quad \eta_{\alpha\beta\gamma} := {}^\star(\vartheta_\alpha \wedge \vartheta_\beta \wedge \vartheta_\gamma). \tag{20.8}$$

Furthermore, we equip our manifold with a symmetric connection (u is an arbitrary vector),

$$\nabla_u \, e_\alpha = \Gamma_\alpha{}^\beta(u) \, e_\beta \,, \quad \Gamma_\alpha{}^\beta(\partial_i) = \Gamma_{i\alpha}{}^\beta \,, \quad \Gamma_\alpha{}^\beta = \Gamma_{i\alpha}{}^\beta \, dx^i \,, \tag{20.9}$$

with $d\vartheta^\beta = -\Gamma_\alpha{}^\beta \wedge \vartheta^\alpha$. In a Riemannian space the connection 1-form can be expressed in terms of the metric and the object of anholonomity $\Omega^\alpha := d\vartheta^\alpha$ and $\Omega_\alpha := g_{\alpha\beta} \, \Omega^\beta$,

$$\Gamma_{\alpha\beta} = \Gamma_{i\alpha\beta} \, dx^i = \frac{1}{2} \, dg_{\alpha\beta} + \left(e_{[\alpha} \rfloor dg_{\beta]\gamma}\right) \vartheta^\gamma + e_{[\alpha} \rfloor \Omega_{\beta]} - \frac{1}{2} \left(e_\alpha \rfloor e_\beta \rfloor \Omega_\gamma\right) \vartheta^\gamma \,. \tag{20.10}$$

The curvature 2-form reads

$$R_\alpha{}^\beta := d\Gamma_\alpha{}^\beta - \Gamma_\alpha{}^\gamma \wedge \Gamma_\gamma{}^\beta = \frac{1}{2} \, R_{\mu\nu\alpha}{}^\beta \, \vartheta^\mu \wedge \vartheta^\nu \,. \tag{20.11}$$

Since a metric is given, we can lower the second index. Then the curvature 2-form is antisymmetric $R_{\alpha\beta} = -R_{\alpha\beta}$. This can be seen if we choose an orthonormal frame

$$g_{\alpha\beta} = \mathrm{diag}(\underbrace{-1, -1, \dots}_{\text{ind}}, \underbrace{1, 1, \dots}_{n\text{-ind}}) \,, \tag{20.12}$$

where *ind*, the number of negative eigenvalues, denotes the index of the metric. Then $dg = 0$ and, according to (20.10), the connection is antisymmetric. In turn, from (20.11) we can infer the antisymmetry of $R_{\alpha\beta}$.

The exterior covariant derivative of a p-form $X_\alpha{}^\beta$ is given by

$$DX_\alpha{}^\beta = dX_\alpha{}^\beta - \Gamma_\alpha{}^\gamma \wedge X_\gamma{}^\beta + \Gamma_\gamma{}^\beta \wedge X_\alpha{}^\gamma \,. \tag{20.13}$$

The Bianchi identities can be formulated concisely with the help of this definition. In a Riemannian space the torsion $T^\alpha = D\vartheta^\alpha$ vanishes. Thus, the first Bianchi identity reads

$$0 = DT^\alpha = DD\vartheta^\alpha = R_\beta{}^\alpha \wedge \vartheta^\beta \,, \tag{20.14}$$

or, in components,

$$R_{[\alpha\beta\gamma]}{}^\delta = 0 \,. \tag{20.15}$$

The first Bianchi identity is a (co-)vector valued 3-form with

$$n \binom{n}{3} = \frac{n^2(n-1)(n-2)}{3!} \tag{20.16}$$

independent components that imposes the same number of constraint equations on the components of the curvature. Accordingly, in n-dimensions, the curvature 2-form has

$$\binom{n}{2}\binom{n}{2} - n\binom{n}{3} = \frac{n^2(n-1)(n+1)}{12} \tag{20.17}$$

independent components. For $n = 3$, we have six independent components and for $n = 4$ (the case of GR) 20 independent components.

The second Bianchi identity is

$$DR_\alpha{}^\beta = 0\,, \qquad \nabla_{[\lambda} R_{\mu\nu]\alpha}{}^\beta = 0\,. \tag{20.18}$$

We now perform the irreducible decomposition of the curvature with respect to the pseudo-orthogonal group (Hehl et al., 1995):

$$\begin{array}{llcl} n=1 & R_{\alpha\beta} & = & 0 \\ n=2 & R_{\alpha\beta} & = & \mathrm{Scalar}_{\alpha\beta} \\ n=3 & R_{\alpha\beta} & = & \mathrm{Scalar}_{\alpha\beta} + \mathrm{Ri\not{c}ci}_{\alpha\beta} \\ n\geq 4 & R_{\alpha\beta} & = & \mathrm{Scalar}_{\alpha\beta} + \mathrm{Ri\not{c}ci}_{\alpha\beta} + \mathrm{Weyl}_{\alpha\beta} \end{array} \tag{20.19}$$

- The $\mathrm{Scalar}_{\alpha\beta}$-piece is given by

$$\mathrm{Scalar}_{\alpha\beta} := -\frac{1}{n(n-1)} R\,\vartheta_\alpha \wedge \vartheta_\beta\,, \quad R := e_\alpha \rfloor \mathrm{Ric}^\alpha\,, \quad \mathrm{Ric}_\alpha := e_\beta \rfloor R_\alpha{}^\beta\,, \tag{20.20}$$

where R is the curvature scalar and Ric_α the Ricci 1-form. This piece has 1 independent component and is present in any dimension $n > 1$. In components we have

$$\mathrm{Ric}_\alpha = \mathrm{Ric}_{\mu\alpha}\,\vartheta^\mu\,, \quad \mathrm{Ric}_{\mu\alpha} = R_{\lambda\mu\alpha}{}^\lambda\,, \quad R = R_{\lambda\mu}{}^{\mu\lambda}\,, \tag{20.21}$$

and

$$\mathrm{Scalar}_{\mu\nu\alpha\beta} = -\frac{2}{n(n-1)} R\, g_{\mu[\alpha} g_{\beta]\nu}\,. \tag{20.22}$$

The Scalar piece enjoys the obvious symmetry

$$\mathrm{Scalar}_{\alpha\beta} \wedge \vartheta^\beta = 0\,, \qquad \mathrm{Scalar}_{[\mu\nu\alpha]\beta} = 0\,. \tag{20.23}$$

- From dimension 3 onwards the tracefree Ricci piece comes into play,

$$\mathrm{Ri\not{c}ci}_{\alpha\beta} := -\frac{2}{n-2}\,\vartheta_{[\alpha} \wedge \mathrm{Ri\not{c}}_{\beta]}\,, \quad \mathrm{Ri\not{c}}_\beta := \mathrm{Ric}_\beta - \frac{1}{n} R\,\vartheta_\beta\,. \tag{20.24}$$

It has $\frac{1}{2}(n+2)(n-1)$ independent components. In index notation this corresponds to

$$\mathrm{Ri\not{c}ci}_{\mu\nu\alpha\beta} = -\frac{4}{n-2}\, g_{[\mu|[\alpha}\, \mathrm{Ri\not{c}}_{\beta]|\nu]}\,, \quad \mathrm{Ri\not{c}}_{\alpha\beta} = \mathrm{Ric}_{\alpha\beta} - \frac{1}{n} R\, g_{\alpha\beta}\,. \tag{20.25}$$

If we contract the first Bianchi identity (20.14), we find

$$0 = e_\beta \rfloor (R_\alpha{}^\beta \wedge \vartheta^\alpha) = \mathrm{Ric}_\alpha \wedge \vartheta^\alpha \,, \tag{20.26}$$

since $R_\alpha{}^\alpha = 0$ in a Riemannian space. Thus, $\mathrm{Ric}_{\mu\alpha}\,\vartheta^\mu \wedge \vartheta^\alpha = 0$ or

$$\mathrm{Ric}_{\alpha\beta} = \mathrm{Ric}_{\beta\alpha} \,, \tag{20.27}$$

that is, the Ricci tensor is symmetric. This also implies

$$\mathrm{Ri}\not{\mathrm{c}}\mathrm{ci}_{\alpha\beta} \wedge \vartheta^\alpha = 0 \,. \tag{20.28}$$

- Finally, in dimension greater than three, the Weyl 2-form emerges according to

$$\mathrm{Weyl}_{\alpha\beta} := R_{\alpha\beta} - \mathrm{Scalar}_{\alpha\beta} - \mathrm{Ri}\not{\mathrm{c}}\mathrm{ci}_{\alpha\beta} \,. \tag{20.29}$$

From the construction it is clear that the Weyl 2-form is totally traceless, i. e.,

$$e_\alpha \rfloor \mathrm{Weyl}^{\alpha\beta} = -e_\beta \rfloor \mathrm{Weyl}^{\alpha\beta} = 0 \,, \quad e_\alpha \rfloor e_\beta \rfloor \mathrm{Weyl}^{\alpha\beta} = 0 \,. \tag{20.30}$$

This property also explains the vanishing of the Weyl 2-form in three dimensions. An arbitrary antisymmetric tensor-valued 2-form $A_{\alpha\beta} = -A_{\beta\alpha} = A_{\mu\nu\alpha\beta}\,\vartheta^\mu \wedge \vartheta^\nu/2$ in three dimensions has nine independent components. The condition $e_\alpha \rfloor A^{\alpha\beta} = 0$ results in three 1-forms, i.e., nine constraint equations that eventually yield the vanishing of all components.

According to Heinicke (2001), we can combine $\mathrm{Scalar}_{\alpha\beta}$ and $\mathrm{Ri}\not{\mathrm{c}}\mathrm{ci}_{\alpha\beta}$,

$$\mathrm{Scalar}_{\alpha\beta} + \mathrm{Ri}\not{\mathrm{c}}\mathrm{ci}_{\alpha\beta} = -\frac{2}{n-2}\,\vartheta_{[\alpha} \wedge L_{\beta]} \,, \tag{20.31}$$

with

$$L_\alpha := e_\beta \rfloor R_\alpha{}^\beta - \frac{1}{2(n-1)}\,R\,\vartheta_\alpha \,, \tag{20.32}$$

i.e., this sum can be expressed in a coherent way in terms of the 1-form L_α. From $\mathrm{Scalar}_{\alpha\beta}$ and $\mathrm{Ri}\not{\mathrm{c}}\mathrm{ci}_{\alpha\beta}$ it inherits the property

$$L_\alpha \wedge \vartheta^\alpha = 0 \,. \tag{20.33}$$

We may expand L_α in components as

$$L_{\alpha\beta} = L_{\beta\alpha} = \mathrm{Ric}_{\alpha\beta} - \frac{1}{2(n-1)}\,R\,g_{\alpha\beta} \,. \tag{20.34}$$

This tensor is sometimes called *Schouten tensor*. Also the names *rho tensor* or $P_{\alpha\beta}$ can be found in the literature. Then the curvature 2-form can be expressed as

$$R_{\alpha\beta} = \mathrm{Weyl}_{\alpha\beta} - \frac{2}{n-2}\,\vartheta_{[\alpha} \wedge L_{\beta]} \tag{20.35}$$

or, in components,

$$\mathrm{Weyl}_{\alpha\beta\gamma\delta} = R_{\alpha\beta\gamma\delta} + \frac{4}{n-2}\,g_{[\alpha|[\gamma}L_{\delta]|\beta]} \,. \tag{20.36}$$

20.2 Cotton 2-Form

By applying the exterior covariant derivative to (20.35), we obtain the following representation of the second Bianchi identity,

$$0 = DR_{\alpha\beta} = D\mathrm{Weyl}_{\alpha\beta} + \frac{2}{n-2}\vartheta_{[\alpha} \wedge C_{\beta]}\,, \tag{20.37}$$

where we encounter the *Cotton 2-form*

$$C_\alpha := DL_\alpha = \frac{1}{2}\, C_{\mu\nu\alpha}\, \vartheta^\mu \wedge \vartheta^\nu \tag{20.38}$$

or, in components,

$$C_{\alpha\beta\gamma} = 2\left(\nabla_{[\alpha}\mathrm{Ric}_{\beta]\gamma} - \frac{1}{2(n-1)}\,\nabla_{[\alpha} R g_{\beta]\gamma}\right). \tag{20.39}$$

We perform an irreducible decomposition of the Cotton 2-form with respect to the Lorentz group. We can use the decomposition for the torsion, since this is also a vector-valued 2-form. Then we have

$$\begin{aligned} C^\alpha &= {}^{(1)}C^\alpha + {}^{(2)}C^\alpha + {}^{(3)}C^\alpha \\ &= \mathrm{TENCOT} + \mathrm{TRACOT} + \mathrm{AXICOT}\,, \end{aligned} \tag{20.40}$$

with the following counting of their independent components

$$\frac{1}{2}\,n^2(n-1) = \frac{1}{3}\,n(n^2-4) \;+ n + \;\frac{1}{6}\,n(n-1)(n-2)\,, \tag{20.41}$$

where

$${}^{(2)}C^\alpha := \frac{1}{n-1}\,\vartheta^\alpha \wedge (e_\beta \rfloor C^\beta)\,, \tag{20.42}$$

$${}^{(3)}C^\alpha := \frac{1}{3}\,e^\alpha \rfloor (C_\beta \wedge \vartheta^\beta)\,, \tag{20.43}$$

$${}^{(1)}C^\alpha := C^\alpha - {}^{(2)}C^\alpha - {}^{(3)}C^\alpha\,, \tag{20.44}$$

or, in components,

$${}^{(2)}C_{\mu\nu}{}^\alpha = -\frac{2}{n-1}\,\delta^\alpha_{[\mu}\, C_{\nu]\beta}{}^\beta\,, \tag{20.45}$$

$${}^{(3)}C_{\mu\nu}{}^\alpha = \frac{1}{3!}\,C_{[\mu\nu\beta]}\, g^{\alpha\beta}\,, \tag{20.46}$$

$${}^{(1)}C_{\mu\nu}{}^\alpha = C_{\mu\nu}{}^\alpha - {}^{(2)}C_{\mu\nu}{}^\alpha - {}^{(3)}C_{\mu\nu}{}^\alpha\,. \tag{20.47}$$

TENCOT, TRACOT, and AXICOT are the computer algebra names of the pieces of the Cotton 2-form, denoting the tensor, the trace, and the axial pieces, respectively. The number of independent components of these pieces is given in (20.41). They arise as follows: TRACOT corresponds to a scalar-valued 1-form $C := e_\alpha \rfloor C^\alpha$ with n independent components. In general, a (co-)vector-valued 2-form has

$$n\binom{n}{2} = \frac{n^2(n-1)}{2} \tag{20.48}$$

independent components. AXICOT corresponds to a scalar valued 3-form $(C^\alpha \wedge \vartheta_\alpha)$ and thus has

$$\binom{n}{3} = \frac{n(n-1)(n-2)}{6} \tag{20.49}$$

independent components. Thus, TENCOT is left with

$$\frac{n^2(n-1)}{2} - \frac{n(n-1)(n-2)}{6} - n = \frac{n}{3}(n-2)(n+2) \tag{20.50}$$

independent components.

We now show that in a Riemannian space the trace piece (TRACOT) and the axial piece (AXICOT) vanish. Hence, only the tensor piece (TENCOT) with its $n(n^2-4)/3$ independent components survives. This insight seems to be new. For $n=3$, we have 5 and for $n=4$ (the case of GR) 16 independent components.

In order to see the vanishing of AXICOT, we contract the Cotton 2-form with the coframe and use (20.33):

$$\vartheta^\alpha \wedge C_\alpha = \vartheta^\alpha \wedge DL_\alpha = -D(\vartheta^\alpha \wedge L_\alpha) = 0\,, \tag{20.51}$$

or

$$C_{[\mu\nu\alpha]} = \frac{2}{3!}\,\nabla_{[\mu}L_{\nu\alpha]} = 0\,. \tag{20.52}$$

The second Bianchi identity leads to a vanishing trace of the Cotton 2-form (TRACOT), $C = e_\alpha \rfloor C^\alpha = 0$. In order to see this, we contract (20.37) twice:

$$0 = e_\beta \rfloor DR^{\alpha\beta} = e_\beta \rfloor D\mathrm{Weyl}^{\alpha\beta} - \frac{n-3}{n-2}\,C^\alpha - \frac{1}{n-2}\vartheta^\alpha \wedge C\,, \tag{20.53}$$

$$0 = e_\alpha \rfloor e_\beta \rfloor DR^{\alpha\beta} = e_\alpha \rfloor e_\beta \rfloor D\mathrm{Weyl}^{\alpha\beta} - 2C = -2C\,, \tag{20.54}$$

or

$$C_{\alpha\beta}{}^{\alpha} = \nabla_\alpha \left(\mathrm{Ric}_\beta{}^\alpha - \frac{1}{2}\,R\,\delta^\alpha_\beta \right) = 0. \tag{20.55}$$

As we see, the second Bianchi identity relates the derivative of the Weyl 2-form to the Cotton 2-form,

$$e_\beta \rfloor D\mathrm{Weyl}_\alpha{}^\beta = \frac{n-3}{n-2}\,C_\alpha\,. \tag{20.56}$$

This formula allows us to rewrite the Einstein equation as a Maxwell-like equation for the Weyl tensor, see Bini et al. (2001), e. g., the Ricci identity intertwines the derivative of the Cotton 2-form with the Weyl 2-form,

$$\begin{aligned} DC_\alpha = DDL_\alpha = -R_\alpha{}^\beta \wedge L_\beta &= -\mathrm{Weyl}_\alpha{}^\beta \wedge L_\beta + \frac{2}{n-2}\,\vartheta_{[\alpha} \wedge L_{\beta]} \wedge L^\beta \\ &= -\mathrm{Weyl}_\alpha{}^\beta \wedge L_\beta\,. \end{aligned} \tag{20.57}$$

Consequently, in three dimensions, C_α is a covariantly conserved 2-form, with $DC_\alpha = 0$. Thus it is a candidate for a conserved current that can be derived by means of a variational procedure. The properties of the Cotton tensor are summarized in Table 20.2.1.

Something similar emerges in four dimensions. By means of the Ricci identity and the decomposition of the curvature we have

$$\begin{aligned} DD^\star C_\alpha &= -R_\alpha{}^\beta \wedge {}^\star C_\beta \\ &= -\mathrm{Weyl}_\alpha{}^\beta \wedge {}^\star C_\beta - \mathrm{Ri}\not{\mathrm{c}}\mathrm{ci}\,{}_\alpha{}^\beta \wedge {}^\star C_\beta - \mathrm{Scalar}_\alpha{}^\beta \wedge {}^\star C_\beta \,. \end{aligned} \tag{20.58}$$

For p-forms ϕ, ψ of the same degree, there holds ${}^\star\phi \wedge \psi = {}^\star\psi \wedge \phi$. By means of $\vartheta^\alpha \wedge {}^\star\phi = (-1)^{p-1}\,{}^\star(e^\alpha \rfloor \phi)$ we can prove that $\mathrm{Scalar}_{\alpha\beta} \wedge {}^\star C^\alpha = 0$. Performing a "partial integration" we arrive at

$$DD^\star C_\alpha = -D\left({}^\star\mathrm{Weyl}_\alpha{}^\beta \wedge L_\beta\right) + \left(D^\star\mathrm{Weyl}_\alpha{}^\beta\right) \wedge L_\beta - {}^\star\mathrm{Ri}\not{\mathrm{c}}\mathrm{ci}\,{}_\alpha{}^\beta \wedge C_\beta \,. \tag{20.59}$$

Next, we use the "double duality relations" for the irreducible pieces of the curvature,

$${}^\star\mathrm{Weyl}_{\alpha\beta} = \mathrm{Weyl}_{\mu\nu}\,\frac{1}{2}\,\eta^{\mu\nu}{}_{\alpha\beta}\,, \tag{20.60}$$

$${}^\star\mathrm{Ri}\not{\mathrm{c}}\mathrm{ci}\,{}_{\alpha\beta} = -\mathrm{Ri}\not{\mathrm{c}}\mathrm{ci}\,{}_{\mu\nu}\,\frac{1}{2}\,\eta^{\mu\nu}{}_{\alpha\beta}\,, \tag{20.61}$$

$${}^\star\mathrm{Scalar}_{\alpha\beta} = \mathrm{Scalar}_{\mu\nu}\,\frac{1}{2}\,\eta^{\mu\nu}{}_{\alpha\beta}\,. \tag{20.62}$$

Together with eqs.(20.37) and (20.31), we obtain

$$\begin{aligned} \left(D^\star\mathrm{Weyl}_\alpha{}^\beta\right) \wedge L_\beta &= \frac{1}{2}\,\eta^{\mu\nu}{}_\alpha{}^\beta\, D\mathrm{Weyl}_{\mu\nu} \wedge L_\beta = -\frac{1}{2}\,\eta^{\mu\nu}{}_\alpha{}^\beta\,\vartheta_{[\mu} \wedge C_{\nu]} \wedge L_\beta \\ &= -\frac{1}{2}\,\eta^{\mu\beta\nu}{}_\alpha\,\vartheta_\mu \wedge L_\beta \wedge C_\nu \\ &= -{}^\star\mathrm{Ri}\not{\mathrm{c}}\mathrm{ci}\,{}^\nu{}_\alpha \wedge C_\nu + {}^\star\mathrm{Scalar}^\nu{}_\alpha \wedge C_\nu \\ &= {}^\star\mathrm{Ri}\not{\mathrm{c}}\mathrm{ci}\,{}_\alpha{}^\nu \wedge C_\nu \,. \end{aligned} \tag{20.63}$$

Substituting this into (20.59) one gets

$$DD^\star C_\alpha = -D\left({}^\star\mathrm{Weyl}_\alpha{}^\beta \wedge L_\beta\right)\,. \tag{20.64}$$

Thus,

$$B_\alpha := D^\star C_\alpha + {}^\star\mathrm{Weyl}_\alpha{}^\beta \wedge L_\beta =: B_\alpha{}^\beta\,\eta_\beta \tag{20.65}$$

or, in components,

$$B_{\alpha\beta} = \nabla^\mu C_{\alpha\mu\beta} + L^{\mu\nu}\,\mathrm{Weyl}_{\alpha\mu\beta\nu}\,, \tag{20.66}$$

is a covariantly conserved 3-form:

$$DB_\alpha = 0 \qquad \left(\nabla_\beta B_\alpha{}^\beta = 0\right). \tag{20.67}$$

One recognizes the *Bach tensor* $B_{\alpha\beta}$, Bach (1921); Schouten (1954); Penrose and Rindler (1986); Tsantilis et al. (1996). From the symmetry properties of C_α, L_α, and $\mathrm{Weyl}_{\alpha\beta}$ it follows that

$$B_\alpha \wedge \vartheta^\alpha = 0 \quad (B_\alpha{}^\alpha = 0)\,, \qquad e_\alpha \rfloor B^\alpha = 0 \quad (B_{[\alpha\beta]} = 0)\,. \tag{20.68}$$

Moreover, it transforms as a conformal density and can be derived from a variational principle. Since in a conformally flat space the Weyl and the Cotton tensors vanish, the vanishing of the Bach tensor is also a necessary (but not sufficient) condition for a $4D$ space to be conformally flat.

C_α *as a Variational Derivative*

It is well known Deser et al. (1984); Baekler et al. (1992) that C_α can be obtained by means of varying the $3D$ Chern–Simons action

$$C_{\mathrm{RR}} = -\frac{1}{2}\left(\Gamma_\alpha{}^\beta \wedge d\Gamma_\beta{}^\alpha - \frac{2}{3}\Gamma_\alpha{}^\beta \wedge \Gamma_\beta{}^\gamma \wedge \Gamma_\gamma{}^\alpha\right) \tag{20.69}$$

with respect to the metric keeping the connection fixed. In order to enforce vanishing torsion

$$T^\alpha := D\vartheta^\alpha = d\vartheta^\alpha - \Gamma_\beta{}^\alpha \wedge \vartheta^\beta \tag{20.70}$$

and vanishing non-metricity

$$Q_{\alpha\beta} := -Dg_{\alpha\beta} = -dg_{\alpha\beta} + \Gamma_\alpha{}^\gamma\, g_{\gamma\beta} + \Gamma_\beta{}^\gamma\, g_{\alpha\gamma}\,, \tag{20.71}$$

we have to apply Lagrange multipliers. Then the total Lagrangian reads

$$L = C_{\mathrm{RR}} + \lambda_\alpha \wedge T^\alpha + \lambda^{\alpha\beta} \wedge Q_{\alpha\beta}\,, \tag{20.72}$$

where λ_α is a 1-form and $\lambda^{\alpha\beta} = \lambda^{\beta\alpha}$ a symmetric 2-form.

The variation of this Chern–Simons Lagrangian (20.69), which only depends on the connection, turns out to be

$$\delta C_{\mathrm{RR}} = -\delta\Gamma_\alpha{}^\beta \wedge R_\beta{}^\alpha + \frac{1}{2}\, d\left(\Gamma_\alpha{}^\beta \wedge \delta\Gamma_\beta{}^\alpha\right)\,. \tag{20.73}$$

In the next step, we enforce vanishing torsion and non-metricity by means of respective Lagrange multiplier terms:

$$L = C_{\mathrm{RR}} + \lambda_\alpha \wedge T^\alpha + \lambda^{\alpha\beta} \wedge Q_{\alpha\beta}\,. \tag{20.74}$$

The variation then yields

$$\begin{aligned}
\delta L &= \delta C_{\mathrm{RR}} + \delta\lambda_\alpha \wedge T^\alpha + \lambda_\alpha \wedge \delta T^\alpha + \delta\lambda^{\alpha\beta} \wedge Q_{\alpha\beta} + \lambda^{\alpha\beta} \wedge \delta Q_{\alpha\beta} \\
&= -\delta\Gamma_\alpha{}^\beta \wedge R_\beta{}^\alpha + \delta\lambda_\alpha \wedge T^\alpha + \lambda_\alpha \wedge \left(d\delta\vartheta^\alpha + \delta\Gamma_\beta{}^\alpha \wedge \vartheta^\beta + \Gamma_\beta{}^\alpha \wedge \delta\vartheta^\beta\right) \\
&\quad + \frac{1}{2}\, d\left(\Gamma_\alpha{}^\beta \wedge \delta\Gamma_\beta{}^\alpha\right) + \delta\lambda^{\alpha\beta} \wedge Q_{\alpha\beta} \\
&\quad + \lambda^{\alpha\beta} \wedge \left(-d\delta g_{\alpha\beta} + \delta\Gamma_\alpha{}^\gamma\, g_{\gamma\beta} + \Gamma_\alpha{}^\gamma\, \delta g_{\gamma\beta} + \delta\Gamma_\beta{}^\gamma\, g_{\alpha\gamma} + \Gamma_\beta{}^\gamma\, \delta g_{\alpha\gamma}\right)
\end{aligned}$$

$$= -\delta\Gamma_\alpha{}^\beta \wedge R_\beta{}^\alpha + \delta\lambda_\alpha \wedge T^\alpha + \delta\lambda^{\alpha\beta} \wedge Q_{\alpha\beta} + \frac{1}{2} d\left(\Gamma_\alpha{}^\beta \wedge \delta\Gamma_\beta{}^\alpha\right)$$
$$+\lambda_\alpha \wedge D\delta\vartheta^\alpha - \delta\Gamma_\beta{}^\alpha \wedge \lambda_\alpha \wedge \vartheta^\beta - \lambda^{\alpha\beta} \wedge D\delta g_{\alpha\beta} + \delta\Gamma_\alpha{}^\beta \wedge (\lambda^\alpha{}_\beta + \lambda_\beta{}^\alpha)$$
$$= \delta\lambda_\alpha \wedge T^\alpha + \delta\lambda^{\alpha\beta} \wedge Q_{\alpha\beta} + \delta\vartheta^\alpha \wedge D\lambda_\alpha + \delta g_{\alpha\beta}\, D\lambda^{\alpha\beta}$$
$$-\delta\Gamma_\alpha{}^\beta \wedge (R_\beta{}^\alpha + \lambda_\beta \wedge \vartheta^\alpha - \lambda^\alpha{}_\beta - \lambda_\beta{}^\alpha)$$
$$-d\left(-\lambda_\alpha \wedge \delta\vartheta^\alpha + \frac{1}{2}\Gamma_\alpha{}^\beta \wedge \delta\Gamma_\beta{}^\alpha - \lambda^{\alpha\beta}\,\delta g_{\alpha\beta}\right). \tag{20.75}$$

The corresponding field equations read

$$\frac{\delta L}{\delta\lambda_\alpha} = T^\alpha = 0\,, \tag{20.76}$$
$$\frac{\delta L}{\delta\lambda^{\alpha\beta}} = Q_{\alpha\beta} = 0\,, \tag{20.77}$$
$$\frac{\delta L}{\delta\Gamma_\alpha{}^\beta} = -R_\beta{}^\alpha - \lambda_\beta \wedge \vartheta^\alpha + 2\lambda_\beta{}^\alpha = 0\,, \tag{20.78}$$
$$\frac{\delta L}{\delta\vartheta^\alpha} = D\lambda_\alpha = 0\,, \tag{20.79}$$
$$\frac{\delta L}{\delta g_{\alpha\beta}} = D\lambda^{\alpha\beta} = 0\,. \tag{20.80}$$

We can solve (20.78) for its symmetric and its antisymmetric parts,

$$R_{[\alpha\beta]} + \vartheta_{[\alpha} \wedge \lambda_{\beta]} = 0\,, \tag{20.81}$$
$$-R_{(\alpha\beta)} + \vartheta_{(\alpha} \wedge \lambda_{\beta)} + 2\lambda_{\alpha\beta} = 0\,. \tag{20.82}$$

Because of (20.77), the symmetric part of the curvature vanishes,

$$0 = DQ_{\alpha\beta} = -DDg_{\alpha\beta} = R_\alpha{}^\gamma\, g_{\gamma\beta} + R_\beta{}^\gamma g_{\alpha\gamma} = 2R_{(\alpha\beta)}\,. \tag{20.83}$$

Thus, by means of (20.82)

$$\lambda_{\alpha\beta} = -\frac{1}{2}\,\vartheta_{(\alpha} \wedge \lambda_{\beta)}\,. \tag{20.84}$$

According to (20.35), in three dimensions, $R_{\alpha\beta} = -2\,\vartheta_{[\alpha} \wedge L_{\beta]}$. We substitute this into (20.81) and find

$$\lambda_\beta = 2\,L_\beta\,, \qquad \lambda_{\alpha\beta} = -\vartheta_{(\alpha} \wedge L_{\beta)}\,. \tag{20.85}$$

Eventually,

$$\frac{1}{2}\frac{\delta L}{\delta\vartheta^\alpha} = C_\alpha\,, \qquad \frac{\delta L}{\delta g_{\alpha\beta}} = -\vartheta^{(\alpha} \wedge C^{\beta)}\,, \qquad -\frac{2}{n-1}\,e_\beta\rfloor\frac{\delta L}{\delta g_{\alpha\beta}} = C^\alpha\,. \tag{20.86}$$

In the presence of matter, the gravitational field equation is given by $\delta(L + L_{\rm mat})/\delta\vartheta^\alpha = 0$. Hence, the Cotton 2-form can be coupled to the energy–momentum 2-form of matter

$$\Sigma_\alpha := \frac{\delta L_{\rm mat}}{\delta\vartheta^\alpha}\,. \tag{20.87}$$

This is carried out in the topologically massive gravity model of Deser, Jackiw and Templeton (DJT); see Deser et al. (1982a, 1982b), where the Lagrangian (20.72) is enriched by a Hilbert–Einstein term and a cosmological term (ℓ is the gravitational constant and θ a dimensionless coupling constant):

$$\begin{aligned} L_{DJT} &= \theta\, C_{\rm RR} + V_{\rm HE} + V_\Lambda + \lambda_\alpha \wedge T^\alpha + \lambda^{\alpha\beta}\, Q_{\alpha\beta} + L_{\rm mat} \\ &= \theta\, C_{\rm RR} - \frac{1}{2\ell}\, R^{\alpha\beta} \wedge \eta_{\alpha\beta} - \frac{\Lambda}{\ell}\, \eta + \lambda_\alpha \wedge T^\alpha + \lambda^{\alpha\beta}\, Q_{\alpha\beta} + L_{\rm mat}\,. \end{aligned} \tag{20.88}$$

Then the field equation (20.79) reads

$$G_\alpha + \Lambda \eta_\alpha + \frac{1}{\mu}\, C_\alpha = \ell \Sigma_\alpha\,, \tag{20.89}$$

where $G_\alpha = \frac{1}{2}\,\eta_{\alpha\beta\gamma} \wedge R^{\beta\gamma}$ is the Einstein 2-form and the DJT coupling constant $1/\mu = -2\theta\ell$. The model of topologically massive gravity in Riemann–Cartan space by Baekler et al. (1992) considers additionally to the Chern–Simons term for the curvature also a corresponding term for the torsion. This model includes Einstein, Einstein–Cartan, and the DJT field equations as limiting cases.

In three dimensions the Cotton tensor arises from the variation of the topological Chern–Simons term. Recently, a similar procedure was proposed by Jackiw et al. Jackiw and Pi (2003); Guralnik et al. (2003); Jackiw (2006) for the case $n = 4$ by starting from the corresponding four-dimensional topological Lagrangian $(1/2)\,\theta\, R_\alpha{}^\beta \wedge R_\beta{}^\alpha = \theta\, dC_{RR}$, where θ is an external, prescribed field. The variation with respect to the metric yields a four-dimensional Cotton type tensor that differs from the one in our definition (20.38) with (20.32).

The Einstein $(n-1)$-form G_α is equivalent to the 1-form L_α according to

$$G_\alpha = L^\beta \wedge \eta_{\beta\alpha}; \tag{20.90}$$

see Heinicke (2001). Hence, we may rewrite the DJT-field equation as a differential equation for L_α,

$$DL_\alpha + \mu\, L^\beta \wedge \eta_{\beta\alpha} = \ell\mu\, \Sigma_\alpha - \mu\Lambda\eta_\alpha\,. \tag{20.91}$$

The Bianchi identities imply full integrability of this system.

In the case of Einstein gravity, the equivalence of G_α and L_α implies a relation between the Cotton 2-form and the energy–momentum $(n-1)$-form in any dimension. We can solve

$$G_\alpha + \Lambda\, \eta_\alpha = \ell \Sigma_\alpha \tag{20.92}$$

for L_α and obtain by covariant exterior differentiation

$$C_\alpha = (-1)^{n-1+\rm ind}\, \ell\, D^\star \big[\Sigma_\alpha - \frac{1}{n-1}\, e_\alpha \rfloor (\vartheta^\beta \wedge \Sigma_\beta)\big]\,. \tag{20.93}$$

Note that the cosmological constant Λ, which induces a constant curvature term, drops out.

Table 20.2.1 *Properties of the Cotton 2-form C_α in arbitrary dimensions*

$C_\alpha := DL_\alpha\,,\quad L_\alpha := e_\beta \rfloor R_\alpha{}^\beta - \frac{1}{2(n-1)}\, R\,\vartheta_\alpha$	Cotton 2-form
$\vartheta^\alpha \wedge C_\alpha = 0$ (axialfree)	1st Bianchi identity
$e_\alpha \rfloor C^\alpha = 0$ (tracefree)	contracted 2nd Bianchi identity
$D\mathrm{Weyl}_{\alpha\beta} = -\frac{2}{n-2}\,\vartheta_{[\alpha} \wedge C_{\beta]}$	2nd Bianchi identity
$DC_\alpha = -\mathrm{Weyl}_\alpha{}^\beta \wedge L_\beta$	Ricci identity
$\widehat{C}_\alpha = C_\alpha + (n-2)\,\sigma_{,\beta}\,\mathrm{Weyl}_\alpha{}^\beta$	conformal transformation
$C_\alpha = \ell\, D^\star[\Sigma_\alpha - \frac{1}{n-1}\, e_\alpha \rfloor(\Sigma_\beta \wedge \vartheta^\beta)]$	Einstein equation differentiated

We recognize that all vacuum solutions of Einstein's theory have a vanishing Cotton 2-form. Therefore, via the Bianchi identity, the Weyl 2-form is divergenceless, see (20.56). This property considerably simplifies the classification of the Weyl tensor. Petrov type D spacetimes with vanishing Cotton 2-form have been classified in Ferrando and Sáez (2002).

20.3 Conformal Correspondence

The conformal correspondence between two n-dimensional manifolds V_n and $\widehat{V}_n$ is achieved by means of a conformal transformation of the form Eisenhart (1966); Schouten (1954)

$$\hat{g}_{\alpha\beta} = \exp(2\sigma) g_{\alpha\beta}\,, \qquad \hat{g}^{\alpha\beta} = \exp(-2\sigma) g^{\alpha\beta}\,, \tag{20.94}$$

where σ is an arbitrary function. In general, a conformal transformation (20.94) is *not* associated with a transformation of coordinates, i.e., with a diffeomorphism of V_n; both metrics in (20.94) are given in the same coordinate system and frame. Since these transformations preserve angles between corresponding directions, the causal structure of the manifold is preserved. As a rule, indices of quantities with *hat* , $\hat{A}$, are raised and lowered by means of $\hat{g}^{\alpha\beta}$ or $\hat{g}_{\alpha\beta}$, respectively, those of untransformed quantities by $g^{\alpha\beta}$ or $g_{\alpha\beta}$. The transformed connection reads

$$\widehat{\Gamma}_\alpha{}^\beta = \Gamma_\alpha{}^\beta + \left(\delta^\beta_\alpha\, d\sigma - \vartheta_\alpha\, \sigma^{,\beta} + \sigma_{,\alpha}\, \vartheta^\beta\right) =: \Gamma_\alpha{}^\beta + S_\alpha{}^\beta\,, \tag{20.95}$$

a comma denotes partial and a semicolon covariant differentiation. If $\hat{D} = d + \Gamma_\alpha{}^\beta + S_\alpha{}^\beta$ is the exterior covariant derivative with respect to $\widehat{\Gamma}_\alpha{}^\beta$, the transformed curvature is

$$\widehat{R}_\alpha{}^\beta = d\widehat{\Gamma}_\alpha{}^\beta - \widehat{\Gamma}_\alpha{}^\gamma \wedge \widehat{\Gamma}_\gamma{}^\beta = R_\alpha{}^\beta + 2\,\vartheta_{[\alpha} \wedge S_{\gamma]}\, g^{\gamma\beta}\,, \tag{20.96}$$

with

$$S_\gamma := D\sigma_{,\gamma} - \sigma_{,\gamma}\, d\sigma + \frac{1}{2}\,\sigma^{,\alpha}\sigma_{,\alpha}\,\vartheta_\gamma\,. \tag{20.97}$$

By contracting (20.96) with the frame $e_\beta\rfloor$, we infer

$$\widehat{L}_\alpha = L_\alpha - (n-2)\, S_\alpha\,, \tag{20.98}$$

$$\widehat{\mathrm{Weyl}}_\alpha{}^\beta = \mathrm{Weyl}_\alpha{}^\beta\,, \tag{20.99}$$

$$\widehat{R} = \exp(-2\sigma)\,[R - 2(n-1)\,\sigma^{,\alpha}{}_{;\alpha} - (n-1)(n-2)\sigma_{,\alpha}\sigma^{,\alpha}]\,. \tag{20.100}$$

The Weyl 2-form is conformally invariant since a conformal transformation does not act on the trace-free part of the curvature. Application of $\widehat{D}$ onto (20.98) yields the transformation behavior of the Cotton 2-form,

$$\widehat{C}_\alpha = C_\alpha + (n-2)\,\sigma_{,\beta}\,\mathrm{Weyl}_\alpha{}^\beta\,. \tag{20.101}$$

Thus, in $n = 3$, where the Weyl 2-form vanishes, the Cotton 2-form becomes conformally invariant.

20.4 Criteria for Conformal Flatness

In the following paragraphs we investigate the criteria for conformal flatness, i. e., the possibilities to transform the curvature to zero by means of a conformal transformation. We basically follow Schouten (1954). Since we have seen that the curvature 2-form in two, three, and more than three dimensions is built up rather differently, we have to investigate these cases separately.

n = 2

In $n = 2$ the only non-vanishing curvature piece is the curvature scalar R. Its behavior under conformal transformation is given by

$$\widehat{R} = \exp(-2\sigma)\,\left(R - 2\,\sigma^{,i}{}_{;i}\right) = 0\,. \tag{20.102}$$

Thus,

$$\widehat{R} = 0 \quad\Longleftrightarrow\quad \sigma^{,i}{}_{;i} = \frac{R}{2}\,. \tag{20.103}$$

This is a scalar wave equation for the conformal factor σ with R as source. Since the wave equation always has a solution, we conclude that all $2D$ spaces are conformally flat.

n ≥ 3

For more than two dimensions we start from (20.35), namely

$$R_{\alpha\beta} = \mathrm{Weyl}_{\alpha\beta} - \frac{2}{n-2}\,\vartheta_{[\alpha} \wedge L_{\beta]}\,. \tag{20.104}$$

Since the Weyl 2-form is conformally invariant it cannot be transformed to zero by means of a conformal transformation. Consequently, the vanishing of the Weyl 2-form is a necessary condition for conformal flatness.

The L_α 1-form transforms according to

$$\widehat{L}_\alpha = L_\alpha - (n-2)\,S_\alpha\,. \tag{20.105}$$

We can transform L_α to zero if there is a function σ such that

$$L_\alpha = (n-2)\,S_\alpha\,. \tag{20.106}$$

This will impose a differential restriction on L_{ij}. By means of (20.97), we rewrite the latter equation as a differential equation for $\sigma_{,i}$,

$$D\sigma_{,i} = \sigma_{,i}\,\sigma_{,j}\,\vartheta^j - \frac{1}{2}\,\sigma^{,j}\,\sigma_{,j}\,\vartheta_i + \frac{1}{n-2}\,L_i\,. \tag{20.107}$$

If we apply the covariant derivative to both sides of (20.107), we obtain a necessary condition for the integrability,

$$-\,R_i{}^j\,\sigma_{,j} = DD\sigma_{,i} = \sigma_{,j}\,D\sigma_{,i}\wedge\vartheta^j - \sigma^{,j}\,D\sigma_{,j}\wedge\vartheta_i + \frac{1}{n-2}\,C_i\,. \tag{20.108}$$

This becomes a necessary and sufficient condition of integrability if the dependence on $\sigma_{,i}$ can be eliminated; see Schouten (1954, 1958). Thus we substitute $D\sigma_{,i}$ from (20.107) into (20.108):

$$-\,R_i{}^j\,\sigma_{,j} = -\frac{2}{n-2}\,L_{[i}\wedge\vartheta_{j]}\,\sigma^{,j} + \frac{1}{n-2}\,C_i\,. \tag{20.109}$$

Using the decomposition (20.35) of the curvature, we finally arrive at

$$-\,(n-2)\,\mathrm{Weyl}_i{}^j\,\sigma_{,j} = C_i\,. \tag{20.110}$$

For $n = 3$, the Weyl 2-form is zero and $C_\alpha = 0$ is the integrability condition for the conformal factor. Thus, if the Cotton 2-form is zero, the space is conformally flat. Conversely, if the space is conformally flat, there is a conformal transformation such that $\widehat{R}_\alpha{}^\beta = 0 \Leftrightarrow \widehat{L}_\alpha = 0 \Rightarrow \widehat{C}_\alpha = 0$. Since the Cotton 2-form is conformally invariant in three dimensions, we find $C_\alpha = 0$. Hence, the vanishing of the Cotton 2-form is the necessary and sufficient condition for a V_3 to be conformally flat.

In more than three dimensions the vanishing of the Weyl 2-form is a necessary condition for conformal flatness. Thus, also in dimensions greater than three, $C_\alpha = 0$ is the integrability condition for the conformal factor. However, for $n > 3$, the contracted second Bianchi identity (20.53) implies the vanishing of the Cotton 2-form when the Weyl 2-form is zero. Hence, the vanishing of the Weyl 2-form is also the sufficient condition for conformal flatness.

20.5 Classification of the Cotton 2-Form in 3D

A vector-valued 2-form in three dimensions has nine independent components, the same as the number of components of a 3×3 matrix. A mapping between these two can be achieved by means of the Hodge dual. The Hodge dual of a vector-valued 2-form in three dimensions is a vector-valued 1-form with the same number of independent components. Its components form a second rank tensor ("matrix"),

$$C_{\alpha\beta} := e_\alpha\rfloor{}^\star C_\beta = {}^\star(C_\beta\wedge\vartheta_\alpha) \tag{20.111}$$

or, in components,

$$C_\alpha{}^\beta = \nabla_\mu \left(\mathrm{Ric}_{\nu\alpha} - \frac{1}{4} R g_{\nu\alpha} \right) \eta^{\mu\nu\beta} . \tag{20.112}$$

This alternative representation of the Cotton 2-form, often called *Cotton–York tensor* – see York Jr. (1971) (even though it was already discussed explicitly by ADM Arnowitt et al., 1962) – can only be defined in three dimensions. Sometimes it appears under the name Bach tensor in the literature; see Christodoulou and Klainerman (1993), for example This seems to be a misnomer.

The Cotton tensor is tracefree

$$C_\alpha{}^\alpha = e_\alpha \rfloor {}^\star C^\alpha = {}^\star(C^\alpha \wedge \vartheta_\alpha) = 0 . \tag{20.113}$$

In the three dimensions, the second Bianchi identity (20.37) amounts to $\vartheta_{[\alpha} \wedge C_{\beta]} = 0$. In view of the definition (20.111), we infer that the Cotton tensor is symmetric $C_{\alpha\beta} = C_{\beta\alpha}$. Introducing this symmetry explicitly into (20.112), we obtain the alternative representation

$$C^{\alpha\beta} = C^{\beta\alpha} = \eta^{\mu\nu(\alpha} \nabla_\mu \mathrm{Ric}_\nu{}^{\beta)} . \tag{20.114}$$

We now perform a classification of the Cotton tensor with respect to its eigenvalues. The corresponding generalized eigenvalue problem reads:

$$\left(C^{\alpha\beta} - \lambda\, g^{\alpha\beta}\right) V_\beta = 0 , \quad C^{[\alpha\beta]} = 0 , \quad C^{\alpha\beta}\, g_{\alpha\beta} = 0 . \tag{20.115}$$

By lowering one index, we can reformulate this as ordinary eigenvalue problem for the matrix $C_\alpha{}^\beta$. However, in that case, the symmetry $C^{\alpha\beta} = C^{\beta\alpha}$ is no longer manifest:

$$\left(C_\alpha{}^\beta - \lambda\, \delta_\alpha^\beta\right) V_\beta = 0 , \quad C_\alpha{}^\alpha = 0 . \tag{20.116}$$

20.5.1 Euclidean Signature

The case of Euclidean signature is simple: the generalized eigenvalue problem reduces to an ordinary one. As a real symmetric matrix $C^{\alpha\beta}$ possesses three real eigenvalues and the eigenvectors form a basis. With respect to this basis, $C^{\alpha\beta}$ takes a diagonal form. Since $C^{\alpha\beta}$ is tracefree, the sum of the eigenvalues is zero. Consequently, we can distinguish three classes:

- Class A
  Three distinct eigenvalues: $\lambda_1 \neq \lambda_2$ and $\lambda_3 = -(\lambda_1 + \lambda_2)$.
- Class B
  Two distinct eigenvalues: $\lambda_1 = \lambda_2 \neq 0$, $\lambda_3 = -2\lambda_1$.
- Class C
  One distinct eigenvalue: $\lambda_1 = \lambda_2 = \lambda_3 = 0$.

  In the present context of Euclidean signature, this implies $C_{\alpha\beta} = 0$.

20.5.2 Lorentzian Signature

In the case of an indefinite metric, the roots of the characteristic polynomial

$$\det\left(C^{\alpha\beta} - \lambda\, g^{\alpha\beta}\right) = 0 \tag{20.117}$$

may be complex. Accordingly, the matrix $C_\alpha{}^\beta$ is no longer symmetric and in the equivalent ordinary eigenvalue problem

$$\det\left(C_\alpha{}^\beta - \lambda\, \delta_\alpha^\beta\right) = 0 \tag{20.118}$$

complex eigenvalues occur, too. This point seems to have been overlooked by the authors of Barrow et al. (1986). Consequently, the classification will not be as simple as was the case for the Euclidean metric.

In the following, we will present a classification of $C_\alpha{}^\beta$. The tracefree condition $(20.116)_2$, in orthonormal coordinates, reads explicitly

$$C_1{}^1 + C_2{}^2 + C_3{}^3 = 0\,. \tag{20.119}$$

Accordingly, we can eliminate $C_3{}^3$, e.g., from $(20.116)_1$. Then the secular determinant reads

$$\det\begin{vmatrix} C_1{}^1 - \lambda & C_1{}^2 & C_1{}^3 \\ -C_1{}^2 & C_2{}^2 - \lambda & C_2{}^3 \\ -C_1{}^3 & C_2{}^3 & -C_1{}^2 - C_2{}^2 - \lambda \end{vmatrix} = 0\,, \tag{20.120}$$

with the five matrix elements $C_1{}^2, C_1{}^2, C_1{}^3, C_2{}^2, C_2{}^3$. We compute the determinant and order according to powers of λ,

$$\lambda^3 + b\,\lambda + c = 0\,. \tag{20.121}$$

A cubic polynomial with real coefficients has at least one real root and the complex roots have to be complex conjugates. Since one eigenvalue is real, types D and II with only one independent eigenvalue $\lambda_1 = \lambda_2 = -2\lambda_3$ are always real. For class I, besides the real eigenvalue, two complex eigenvalues may occur. In that case, they are complex conjugated. Therefore, class I can be subdivided into class I with three real eigenvalues, [111], and class I′ with one real and two complex conjugated eigenvalues, $[1z\bar{z}]$.

We can now specify simple criteria for deciding to which of these classes the Cotton tensor $C_\alpha{}^\beta$ belongs. First determine the eigenvalues.

1. Three different eigenvalues (2 independent)
   1. all real $\Rightarrow$ Class I
   2. one real, two complex $\Rightarrow$ Class I′
2. Two different eigenvalues (1 independent $\lambda_1 = \lambda_2 = -2\lambda_3$)
   1. $(C_\alpha{}^\beta - \lambda_1\, \delta_\alpha^\beta)(C_\beta{}^\gamma + \frac{1}{2}\lambda_1\, \delta_\beta^\gamma) = 0 \Rightarrow$ Class D
   2. else $\Rightarrow$ Class II

3. All eigenvalues zero
   1. $C_\alpha{}^\beta = 0 \Rightarrow 0$
   2. $C_\alpha{}^\beta C_\beta{}^\gamma = 0 \Rightarrow$ Class N
   3. else $\Rightarrow$ Class III

The Jordan normal forms of the Cotton tensor read:

"Petrov" type	Jordan form	Segré notation	eigenvalues $\lambda_1+\lambda_2+\lambda_3=0$
I	$\begin{pmatrix} \lambda_1 & 0 & 0 \\ 0 & \lambda_2 & 0 \\ 0 & 0 & -\lambda_1-\lambda_2 \end{pmatrix}$	[111]	Real; $\lambda_1 \neq \lambda_2$
I′	$\begin{pmatrix} \mathrm{Re}\, z & \mathrm{Im}\, z & 0 \\ -\mathrm{Im}\, z & \mathrm{Re}\, z & 0 \\ 0 & 0 & -2\,\mathrm{Re}\, z \end{pmatrix}$	$[1z\bar{z}]$	$\lambda_2 = z,\ \lambda_3 = \bar{z}$
D	$\begin{pmatrix} \lambda_1 & 0 & 0 \\ 0 & \lambda_1 & 0 \\ 0 & 0 & -2\lambda_1 \end{pmatrix}$	[(11)1]	$\lambda_1 = \lambda_2 \neq 0, \lambda_3 = -2\lambda_1$
II	$\begin{pmatrix} \lambda_1 & 1 & 0 \\ 0 & \lambda_1 & 0 \\ 0 & 0 & -2\lambda_1 \end{pmatrix}$	[21]	$\lambda_1 = \lambda_2 \neq 0, \lambda_3 = -2\lambda_1$
N	$\begin{pmatrix} 0 & 1 & 0 \\ 0 & 0 & 0 \\ 0 & 0 & 0 \end{pmatrix}$	[(21)]	$\lambda_1 = \lambda_2 = \lambda_3 = 0$
III	$\begin{pmatrix} 0 & 1 & 0 \\ 0 & 0 & 1 \\ 0 & 0 & 0 \end{pmatrix}$	[3]	$\lambda_1 = \lambda_2 = \lambda_3 = 0$
O	$\begin{pmatrix} 0 & 0 & 0 \\ 0 & 0 & 0 \\ 0 & 0 & 0 \end{pmatrix}$		

This parallels exactly the Petrov classification of the Weyl tensor in four dimensions; see Stephani et al. (2003). This comes about since the Weyl tensor in $4D$ is equivalent to a (complex) 3×3 tracefree matrix, as $C_\alpha{}^\beta$ in $3D$; for a similar classification of $C_{\alpha\beta}$, see Hall and Capocci (1999). A curvature classification in terms of spinors has been done by Torres del Castillo and Gómez-Ceballos (2003).

References

Anninos, D., Li, W., Padi, M., Song, W., and Strominger, A. (2009). "Warped AdS_3 black holes," *JHEP* 0903, 130, arXiv:0807.3040.

Aragone, C. (1987). "Topologically massive gravity in the dreibein light-front gauge," *Class. Quant. Grav.* **4**, L1.

Ayón-Beato, E., Cataldo, M., and Garcia, A. A. (2005). "Electromagnetic fields in stationary cyclic symmetric 2+1 gravity", In *Proceedings of the 10th. PASCOS04 and Pran Nath Fest,* Eds. G. Alverson, E. Barberi, P. Nath, M.T. Vaughn (World Scientific, 2005), p.359.

Ayón-Beato, E., Garcia, A. A., Macias, A., and Perez-Sanchez, J. M. (2000). "Note on scalar fields nonminimally coupled to (2+1) gravity," *Phys. Lett.* **B495**, 164 [gr-qc/0101079].

Ayón-Beato, E. and Hassaïne, M. (2005). "pp–waves of conformal gravity with self-interacting source," *Annals Phys.* **317**, 175. hep-th/0409150.

Ayón-Beato, E. and Hassaïne, M. (2006). "Exploring AdS waves via nonminimal coupling," *Phys. Rev.* **D 73**, 104001. hep-th/0512074.

Ayón-Beato, E., Martinez, C., and Zanelli, J. (2004). "Birkhoff's theorem for three-dimensional AdS gravity," *Phys. Rev.* **D70**, 044027. [hep-th/0403227].

Bañados, M., Henneaux, M. Teitelboim, C., and Zanelli, J. (1993). "Geometry of the (2+1) black hole," *Phys. Rev.* **D48**, 1506. [gr-qc/9302012].

Bañados, M., Teitelboim, C., and Zanelli, J. (1992). "The black hole in three-dimensional spacetime," *Phys. Rev. Lett.* **69**, 1849. [hep-th/9204099].

Barrow, J. D., Burd, A. B., and Lancaster, D. (1986). "Three-dimensional classical spacetimes," *Class. Quant. Grav.* **3**, 551–567.

Barrow, J. D., Shaw, D. J., and Tsagas, C. G. (2006). "Cosmology in three dimensions: Steps towards the general solution," *Class. Quant. Grav.* **23**, 5291. [gr-qc/0606025].

Bouchareb A. and Clément, G. (2007). "Black hole mass and angular momentum in topologically massive gravity," *Class. Quant. Grav.* **24**, 5581. arXiv:0706.0263.

Brown, J. D., Creighton, J., and Mann, R. B. (1994). "Temperature, energy and heat capacity of asymptotically anti-de Sitter black holes," *Phys. Rev.* **D50**, 6394. [gr-qc/9405007].

Brown, J. D. and York Jr., J. W. (1993). "Quasilocal energy and conserved charges derived from the gravitational action", *Phys. Rev.* **D47**, 1407.

Carlip, S. (1998). *Quantum Gravity in 2+1 Dimensions*, Cambridge, UK: Cambridge University Press.

Carlip, S., Deser, S., Waldron, A., and Wise, D. K. (2008). "Topologically massive AdS gravity," *Phys. Lett.* **B666**, 272. arXiv:0807.0486.

Carlip, S., Deser, S., Waldron, A., and Wise, D. K. (2009). "Cosmological topologically massive gravitons and photons," *Class. Quant. Grav.* **26**, 075008.

Cataldo, M. (2002). "Azimuthal electric field in a static rotationally symmetric (2+1)-dimensional space-time," *Phys. Lett.* **B529**, 143. [gr-qc/0201047].

Cataldo, M. (2004). "Rotating perfect fluids in (2+1)–dimensional Einstein gravity," *Phys. Rev.* **D69**, 064015.

Cataldo, M., Crisostomo, J., del Campo, S., and Salgado, P. (2004). "On magnetic solution to (2+1) Einstein–Maxwell gravity," hep-th/0401189.

Cataldo, M., Cruz, N., del Campo, S., and Garcia, A. (2000). "(2+1)–dimensional black hole with Coulomb-like field," *Phys. Lett.* **B484**, 154. [hep-th/0008138].

Cataldo, M., del Campo, S., and Garcia, A. (2001). "BTZ black hole from (3+1) gravity," *Gen. Rel. Grav.* **33**, 1245. [gr-qc/0004023].

Cataldo, M., and García, A. (1999). "Three dimensional black hole coupled to the Born–Infeld electrodynamics," *Phys. Lett.* **B456**, 28.

Cataldo, M. and Garcia, A. (2000). "Regular (2+1)-dimensional black holes within nonlinear electrodynamics," *Phys. Rev.* **D61**, 084003. [hep-th/0004177].

Cataldo, M. and Salgado, P. (1996). "Static Einstein–Maxwell solutions in (2+1)-dimensions," *Phys. Rev.* **D54**, 2971.

Cavaglia, M. (1999). "The Birkhoff theorem for topologically massive gravity," *Grav. Cosmol.* **5**, 101. gr-qc/9904047.

Chan, K. C. K. (1996). "Comment on the calculation of the angular momentum for the (anti)selfdual charged spinning BTZ black hole," *Phys. Lett.* **B373**, 296. [gr-qc/9509032].

Chan, K. C. K. (1997). "Modifications of the BTZ black hole by a dilaton and scalar," *Phys. Rev.* **D55**, 3564. [gr-qc/9603038].

Chan, K. C. K. and Mann, R. B. (1994). "Static charged black holes in (2+1)-dimensional dilaton gravity," *Phys. Rev.* **D50**, 6385, [Erratum: *Phys. Rev.* **D52**, 2600. (1995)]. [gr-qc/9404040].

Chan, K. C. K. and Mann, R. B. (1996). "Spinning black holes in (2+1)-dimensional string and dilaton gravity," *Phys. Lett.* **B371**, 199. [gr-qc/9510069].

Chow, D. D. K., Pope, C. N., and Sezgin, E. (2010a). "Classification of solutions in topologically massive gravity," *Class. Quant. Grav.* **27**, 105001.

Chow, D. D. K., Pope, C. N., and Sezgin, E. (2010b). "Kundt spacetimes as solutions of topologically massive gravity," *Class. Quant. Grav.* **27**, 105002.

Clément, G. (1976). "Field-theoretic extended particles in two space dimensions," *Nucl. Phys.* **B114**, 437.

Clément, G. (1985a). "Stationary solutions in three-dimensional general relativity," *Int. J. Theor. Phys.* **24**, 267.

Clément, G. (1992a). "Stationary rotationally symmetric solutions in topologically massive gravity, *Class. Quant. Grav.* **9**, 2615.

Clément, G. (1992b). "Localised solutions in topologically massive gravity," *Class. Quant. Grav.* **9**, S35.

Clément, G. (1993). "Classical solutions in three-dimensional Einstein–Maxwell cosmological gravity," *Class. Quant. Grav.* **10**, L49.

Clément, G. (1994). "Particle-like solutions to topologically massive gravity," *Class. Quant. Grav.* **11**, L115, gr-qc/9404004.

Clément, G. (1996). "Spinning charged BTZ black holes and selfdual particle-like solutions," *Phys. Lett.* **B367**, 70. [gr-qc/9510025].

Coley, A., Hervik, S., and Pelavas, N. (2006). "On spacetimes with constant scalar invariants," *Class. Quant. Grav.* **23**, 3053. gr-qc/0509113.

Coley, A., Hervik, S., and Pelavas, N. (2008). "Lorentzian spacetimes with constant curvature invariants in three dimensions," *Class. Quant. Grav.* **25**, 025008. arXiv:0710.3903.

Coley, A., Hervik, S., and Pelavas, N. (2009). "Lorentzian spacetimes with constant curvature invariants in four dimensions," *Class. Quant. Grav.* **26**, 125011. arXiv:0904.4877.

Coley, A., Hervik, S., Papadopoulos, G. O., and Pelavas, N. (2009). "Kundt spacetimes," *Class. Quant. Grav.* **26**, 105016. arXiv:0901.0394.

Collas, P. (1977). "General relativity in two-and three-dimensional space-times," *Am. J. Phys.* **45**, 833.

Cornish, N. J. and Frankel, N. E. (1991). "Gravitation in (2+1)–dimensions," *Phys. Rev.* **D43**, 2555.

Cornish, N. J. and Frankel, N. E. (1994). "Gravitation versus rotation in (2+1)–dimensions," *Class. Quant. Grav.* **11**, 723. [gr-qc/9306012].

Coussaert, O. and Henneaux, M. (1994b). "Self-dual solutions of the 2 + 1 Einstein gravity with a negative cosmological constant," [hep-th9407181].

Cruz, N. and Martínez, C. (2000). "Cosmological scaling solutions of minimally coupled scalar fields in three dimensions" *Class. Quant. Grav.* **17**, 2867.

Cruz, N. and Zanelli, J. (1995). "Stellar equilibrium in (2+1)-dimensions," *Class. Quant. Grav.* **12**, 975 [gr-qc/9411032].

Dereli, T. and Sarioğlu, O. (2001). "Supersymmetric solutions to topologically massive gravity and black holes in three dimensions," *Phys. Rev.* **D 64**, 027501. gr-qc/0009082.

Dereli, T. and Tucker, R. W. (1988). "Gravitational interactions in 2 + 1 dimensions," *Class. Quant. Grav.* **5**, 951.

Deser, S. (1984)."Cosmological topological supergravity," in *Quantum Theory of Gravity*, ed. S.M. Christensen, Adam Hilger, London.

Deser, S. and Jackiw, R. (1989). "String sources in (2+1)-dimensional gravity," *Ann. of Phys.* **192**, 352.

Deser, S., Jackiw, R., and Pi, S. Y. (2005). "Cotton blend gravity pp waves," *Acta Phys. Polon.* **B36**, 27. [gr-qc/0409011].

Deser, S., Jackiw, R. and Templeton, S. (1982a). "Three-Dimensional Massive Gauge Theories," *Phys. Rev. Lett.* **48**, 975.

Deser, S., Jackiw, R., and Templeton, S. (1982b). "Topological massive gauge theories," *Ann. of Phys.* **140**, 372.

Deser, S., Jackiw, R. and 't Hooft, G. (1984). "Three-dimensional Einstein gravity: dynamics of flat space," *Ann. of Phys.* **152**, 220–235.

Deser, S. and Mazur, P. O. (1985). "Static solutions in $D = 3$ Einstein–Maxwell theory," *Class. Quant. Grav.* **2** L51–L56.

Deser, S. and Steif, A.R. (1992). "Gravity theories with lightlike sources in $D = 3$," *Class. Quant. Grav.* **9**, L153. hep-th/9208018.

Dias, O. J. C. and Lemos, J. P. S. (2002b). "Rotating magnetic solution in three-dimensional Einstein gravity," *JHEP* 01(2002), 006. [hep-th/0201058].

Garbarz, A., Giribet, G., and Vásquez, Y. (2009). "Asymptotically AdS 3 solutions to topologically massive gravity at special values of the coupling constants," *Phys. Rev.* **D 79**, 044036. arXiv:0811.4464.

García, A. (1999). "On the rotating charged BTZ metric," hep-th/9909111.

García, A. A. (2004). "Stationary circularly symmetric 2+1 rigidly rotating perfect fluid," *Phys. Rev.* **D69**, 124024.

Garcia, A. A. (2009). "Three-dimensional stationary cyclic symmetric Einstein–Maxwell solutions; black holes," *Ann. of Phys.* **324**, 2004–2050.

Garcia-Diaz, A. (2014). "Hydrodynamic equilibrium of a static star in the presence of a cosmological constant in 2 + 1 dimensions," arXiv:1412.5620[gr-qc].

García, A. A. and Campuzano, C. (2003). "All static circularly symmetric perfect fluid solutions of 2+1 gravity," *Phys. Rev.* **D67**, 064014.

García, A., Cataldo, M., and del Campo, S. (2003). "Relationship between (2+1) and (3+1) Friedmann–Robertson–Walker cosmologies," *Phys. Rev.* **D68**, 124022. [hep-th/0309098].

García, A., Hehl, F. W., Heinicke, C., and Macias, A. (2004). "The Cotton tensor in Riemannian space-times," *Class. Quant. Grav.* **21**, 1099 [gr-qc/0309008].

Garcia-Diaz, A. A. (2013). "Three-dimensional stationary cyclic symmetric Einstein–Maxwell solutions; energy," [gr-qc/0309008].

Garcia-Diaz, A. and Gutierrez-Cano, G. (2014a). "Low energy 2+1 string gravity; black hole solutions," arXiv:1412.5618[gr-qc].

Garcia-Diaz, A. and Gutierrez-Cano, G. (2014b). "Dilaton minimally coupled to 2+1 Einstein–Maxwell fields: stationary cyclic symmetric black holes," arXiv:1412.5621 [gr-qc].

Gibbons, G. W., Pope, C. N., and Sezgin, E. (2008). "The general supersymmetric solution of topologically massive supergravity," *Class. Quant. Grav.* **25**, 205005, arXiv:0807.2613.

Giddings, S., Abbott, J., and Kuchar, K. (1984). "Einstein's theory in a three-dimensional space-time," *Gen. Rel. Grav.* **16**, 751.

Gott, J. R. and Alpert, M. (1984). "General relativity in a (2+1)-dimensional space-time," *Gen. Rel. Grav.* **16**, 243.

Gott, J. R., Simon, J. Z., and Alpert, M. (1986). "General relativity in a (2+1)-dimensional space-time: An electrically charged solution," *Gen. Rel. Grav.* **18**, 1019.

Gürses, M. (1994). "Perfect fluid sources in 2+1 dimensions," *Class. Quant. Grav.* **11**, 2585–2587.

Gürses, M. (2008). "Gödel type metrics in three dimensions," *Gen. Rel. Grav.*,**42**, 1413. arXiv:0812.2576.

Hall, G. S. and Capocci, M. S. (1999). "Classification and conformal symmetry in three-dimensional space-time," *J. Math. Phys.* **40**, 1466–1478.

Hall, G. S., Morgan, T., and Perjés, Z. (1987). "Three-dimensional space-times," *Gen. Rel. Grav.* **19**, 1137.

Hirschmann, E. W. and Welch, D. L. (1996). "Magnetic solutions to (2+1) gravity," *Phys. Rev.* **D53**, 5579 [hep-th/9510181].

Horne, J. H. and Horowitz, G. T. (1992). "Exact string solution in three dimensions," *Nucl. Phys.* **B368**, 444.

Horowitz, G. T. (1992). "The dark side of string theory; Black holes and black strings," [arXiv:hep-th/9210119].

Horowitz, G. T. (2005). "Creating naked singularities and negative energy," *Phys. Scripta* **T117**, 86. [hep-th/0312123].

Horowitz, G. T. and Welch, D. L. (1993). "Exact three-dimensional black holes in string theory," *Phys. Rev. Lett.* **71**, 328. [hep-th/9302126].

Jackiw, R. (2006). "4-dimensional Einstein gravity extended by a 3-dimensional gravitational Chern–Simons term," *Int. J. Theor. Phys.* **45**, 1431. [gr-qc/0310115].

Kamata, M. and Koikawa, T. (1995). "The electrically charged BTZ black hole with self (antiself) dual Maxwell field," *Phys. Lett.* **B353**, 196. [hep-th/9505037].

Kamata, M. and Koikawa, T. (1997). "2+1-dimensional charged black hole with (anti-) self dual Maxwell field," *Phys. Lett.* **B391**, 196.

Kogan, I. I. (1992). "About some exact solutions for (2+1) gravity coupled to gauge fields," *Mod. Phys. Lett.* **A7**, 2341. [hep-th/9205095].

Lubo, M., Rooman, M. and Spindel, P. (1999). "(2+1)-dimensional stars," *Phys. Rev.* **D59**, 044012 [gr-qc/9806104].

Macías, A. and Camacho, A. (2005). "Kerr–Schild metric in topological massive (2+1) gravity," *Gen. Rel. Grav.* **37**, 759.

Mann, R. B. and Ross, S. F. (1993). "Gravitationally collapsing dust in 2 + 1 dimensions", *Phys. Rev.* **D47**, 3319.

Martinez, E. A. and Shepley, L. (1986). preprint, University of Texas.

Martínez, C., Teitelboim, C. and Zanelli, J. (2000). "Charged rotating black hole in three space-time dimensions," *Phys. Rev.* **D61**, 104013 [hep-th/9912259].

Martínez, C. and Zanelli, J. (1996). "Conformally dressed black hole in (2+1)-dimensions," *Phys. Rev.* **D 54**, 3830. [gr-qc/9604021].

Matyjasek, J. and Zaslavskii, O. B. (2004). "Extremal limit for charged and rotating (2+1)–dimensional black holes and Bertotti-Robinson geometry," *Class. Quant. Grav.* **21**, 4283 [gr-qc/0404090].

Melvin, M. A. (1986). "Exterior solutions for electric and magnetic stars in 2+1 dimensions," *Class. Quant. Grav.* **3**, 117-131.

Moussa, K. A., Clément, G., and Leygnac, C. (2003). "The black holes of topologically massive gravity," *Class. Quant. Grav.* **20**, L277, gr-qc/0303042.

Nutku, Y. (1993). "Exact solutions of topologically massive gravity with a cosmological constant," *Class. Quant. Grav.* **10**, 2657.

Nutku, Y. and Baekler, P. (1989). "Homogeneous, anisotropic three-manifolds of topologically massive gravity," *Annals Phys.* **195**, 16.

Ortiz, M. E. (1990). "Homogeneous solutions to topologically massive gravity," *Annals Phys.* **200**, 345.

Obukhov, Y. N. (2003). "New solutions in 3-D gravity," *Phys. Rev.* **D68**, 124015. [gr-qc/0310069].

Ölmez, S., Sarioğlu, O., and Tekin, B. (2005). "Mass and angular momentum of asymptotically AdS or flat solutions in the topologically massive gravity," *Class. Quant. Grav.* **22**, 4355, gr-qc/0507003.

Peldan, P. (1993). "Unification of gravity and Yang-Mills theory in (2+1)-dimensions," *Nucl. Phys.* **B395**, 239 [gr-qc/9211014].

Percacci, R., Sodano, P., and Vuorio, I. (1987). "Topologically massive planar universes with constant twist," *Ann. of Phys.* **176**, 344.

Rooman, M. and Spindel, P. (1998). "Gödel metric as a squashed anti-de Sitter geometry," *Class. Quant. Grav.* **15**, 3241 [gr-qc/9804027].

Saslaw, W. C. (1977). "A relation between the homogeneity of the Universe and the dimensional of space" *Mon. Not. R. Astr. Soc.* **179**, 659.

Staruszkiewicz, A. (1963). "Gravitation theory in three-dimensional space," *Acta Phys. Polon.* **24**, 735.

Torres del Castillo, G. F. and Gómez-Ceballos, L. F. (2003). "Algebraic classification of the curvature of three-dimensional manifolds with indefinite metric," *J. Math. Phys.* **44**, 4374 .

Vuorio, I. (1985). "Topologically massive planar universe," *Phys. Lett.* **B163**, 91. *Gen. Rel. Grav.* **23**, 181–187.

Zaslavskii, O. B. (1994). "Thermodynamics of 2+1 black holes," *Class. Quant. Grav.* **11**, L33–L38.

Additional Literature

Abrahams, A. M. and Evans, C. R. (1993). "Critical behavior and scaling in vacuum axisymmetric gravitational collapse," *Phys. Rev. Lett.* **70**, 2980.

Abramowitz, M. and Stegun, I. (1965). *Handbook of Mathematical Functions.* New York: Dover Publications Inc.

Adler, R., Bazin, M., and Schiffer, M. (1965). *Introduction to General Relativity*, McGraw–Hill.

Arnowitt, R., Deser, S., and Misner, C. W. (1962). "The dynamics of General Relativity." In: *Gravitation: An Introduction to Current Research*, L. Witten (Ed.). Wiley, New York.

Ashtekar, A., and Krishnan, B. (2004). "Isolated and dynamical horizons and their applications," *Living Rev. Rel.* **7**, 10.

Ayón-Beato, E. and García, A. (1998). "Regular black hole in general relativity coupled to a nonlinear electrodynamics," *Phys. Rev. Lett.* **80**, 5056.

Bach, R. (1921). "Zur Weylschen Relativitätstheorie und der Weylschen Erweiterung des Krümmungsbegriffs," *Math. Zeitschr.* **9**, 110–135.

Baekler, P., Mielke, E. W., and Hehl, F. W. (1992). "Dynamical symmetries in topological 3D gravity with torsion," *Nuovo Cimento.* **B107**, 91 110.

Barrow, J. D., and Saich, P. (1993). "Scalar–field cosmologies," *Class. Quant. Grav.* **10**, 279.

Bergshoeff, E. A., Hohm, O., and Townsend, P. K. (2009) "Massive gravity in three dimensions," *Phys. Rev. Lett.***102**, 201301.

Bertotti, I. (1959). "Uniform electromagnetic field in the theory of general relativity," *Phys. Rev.* **116**, 1331–1333.

Bini, D., Jantzen, R. T., and Minutti, G. (2001). "The Cotton, Simon–Mars and Cotton–York tensors in stationary spacetimes," *Class. Quant. Grav.* **18**, 4969.

Borde, A. (1997). "Regular black holes an topological change," *Phys. Rev.* **D50**, 7615.

Born, M., and Infeld, L. (1934). "On the quantum theory of the electromagnmetic field," *Proc. Roy. Soc.* (London) **A144**, 425.

Brinkmann, M. W. (1923). "On Riemann spaces conformal to Euclidean spaces," *Proc. Natl. Acad. Sci. U.S.* **9**, 1.

Buchdahl, H. A. (1959). "General relativistic fluid spheres," *Phys. Rev.* **116**, 1027.

Buscher, T. (1993). "Path integral derivation of quantum duality in nonlinear sigma models," *Phy. Lett.***B201**, 466.

Carter, B. (1968). "Hamilton–Jacobi and Schrodinger separable solutions of Einstein's equations," *Commun. Math. Phys.* **10**, 280 (1968), (1968). "A new family of Einstein spaces," *Phys. Lett.* **A26**, 399.

Cotton, E. (1899). "Sur les variétés a trois dimensions," *Ann. Fac. d. Sc. Toulouse (II)* **1**, 385.

Choptuik, M. W. (1993). "Universality and scaling in gravitational collapse of a massless scalar field," *Phys. Rev. Lett.* **70**, 9.

Christodoulou, D. and Klainerman, S. (1993). *The Global Nonlinear Stability of the Minkowski Space.* Princeton University Press, Princeton, NJ.

de Rham, C. (2014). "Massive Gravity," Living Rev.Rel. **17**, 7. arXiv:1401.4173 [hep-th].

Deser, S. and Gibbons, G. W. (1998). "Born–Infeld–Einstein actions," *Class. Quant. Grav.* **15**, L35.

Ehlers, J. (1961). "Beiträge zur realativistischen Mechanik kontinuierlich Medien," *Abh. Mainzer Akad. Wiss., Math.–natuwiss,*" No, 11, English Translation: (1993) "Contributions to relativistic mechanics of continuous media," *Gen. Rel. Grav.* **25**,

1225–1266. See also, (2002). “Relativistic hydrodynamics,” *Gen. Rel. Grav.* **34**, 2171.

Eisenhart, L. P. (1966). *Riemannian Geometry,* Princeton University Press. Princeton, NJ.

Ellis, G. F. R. and Elst, H. (1999). *Theoretical and Observational Cosmology,* Ed. ML Lachieze-Rey (Dordrecht: Kluwer) p.1.

Ferrando, J. J. and Sáez, A. J. (2002). “On the classification of Type D spacetimes,” [gr-qc/0212086].

Fradkin, E. and Tseytlin, A. (1985). “Nonlinear electrodynamics from quantized strings,” *Phys. Lett.* **B163**, 123.

Frolov, V., Hendy, S., and Larsen, A. L. (1996). “Stationary strings and principal Killing triads in 2+1 gravity,” *Nucl. Phys. B* **468**, 336.

García, A. A. (1988). “Comments on a paper by Collinson,” *Gen. Rel. Grav.* **20**, 589.

Gibbons, G. W. and Rasheed, D. A. (1995). “Electric–magnetic duality rotations in non-linear electrodynamics,” *Nucl. Phys.* **B454**, 185.

Gubser, S. S., Klebanov, I. R., and Polyakov, A. M. (1998). “Gauge Theory Correlators from Non-Critical String Theory,” *Phys. Lett.* **B428**, 105 (arXiv: hep-th/9802109).

Gundlach, C. (1999). “Critical Phenomena in Gravitational Collapse,” *Living Rev. Rel.* **2**, 4 (arXiv: gr-qc/0001046).

Guralnik, G., Iorio, A., Jackiw, R., and Pi, S.-Y. (2003). “Dimensionally reduced gravitational Chern–Simons term and its kink,” *Ann. Phys.* (NY) **308**, 222–236. (arXiv: hep-th/0305117).

Gürses, M. and Gürsey, Y. (1975). “Conformal uniqueness and various forms of the Schwarzschild interior metric,” *Nuovo Cimento* **25 B**, 786.

Hayward, S. A. (2008). “Dynamics of black holes,” arXiv:0810.0923 [gr-qc].

Hehl, F. W., McCrea, J. D., Mielke, E. W., and Ne’eman, Y. (1995). “Metric–affine gauge theory of gravity: Field equations, Noether identities, world spinors, and breaking of dilation invariance,” *Phys. Repts.* **258**, 1.

Heinicke, C. (2001). “The Einstein 3-form G_α and its equivalent 1–form L_α in Riemann–Cartan spacetime,” *Gen. Relat. Grav.* **33**, 1115.

Israel, R. B. (1966). “Singular hypersurfaces and thin shells in General Relativity,” *Nuovo Cimento.* XLIV B, 1.

Infeld, L. and Plebański, J. F. (1960). *Motion and Relativity,* Pergamon Press, New York.

Jackiw, R. S. and Pi, Y. (2003). “Chern–Simons Modification of General Relativity,” arXiv: gr-qc/0308071.

Korn, G. A. and Korn, T. M. (1961). *Mathematics Handbook for Scientists and Engineers: Definitions, Theorems, and Formulas for References and Review,* McGraw–Hill, New York.

Kottler, F. (1918). “Über die physikalischen Grudlagen der Einsteischen Gravitationstheorie,” *Annalen Physik* **56**, 410.

Kramer, D., Stephani, H., MacCallum, M., and Hertl, E. (1980). *Exact Solutions of the Einstein’s Field Equations,* Deutsch. Ver. der Wiss., Berlin.

Landau, L. and Lifshitz, E. (1970). *Théorie des Champs,* Mir, Moscow,

Liddle, A. R. and Lyth, D. H. (2000). *Cosmological Inflation and Large-Scale Structure,* Cambridge University Press.

Lucchin, F. and Matarrese, S. (1985). “Power–law inflation,” *Phys. Rev.* **D32**, 1316.

Madsen, M. S. (1986). “An unusual cosmological solutions for $\Lambda\phi^4$ theory with broken symmetry,” *Gen. Rel. and Grav.* **18**, 879.

Maldacena, J. (1998). "The large N limit of superconformal field theories and supergravity," *Adv. Theor. Math. Phys.* **2**, 231 (arXiv: hep-th/9711200).

Mandal, G., Sengupta, A. M. and Wadia, S. R. (1991). "Classical solutions of 2–dimensional string theory," *Mod. Phys. Lett.* **A 6** 1685.

Mak, M. K. and Harko, T. (2000). "Maximum mass–radius ratio for compact general relativistc objects in Swarzschild–de Sitter geometry," *Mod. Phys. Lett.* **A15**, 2105.

Maki, T. and Shiraishi, K. (1993). "Multi-black hole solutions in cosmological Einstein–Maxwell dilaton theory," *Class. Quant. Grav.* **10**, 2171.

Misner, C. W., Thorne, K. S., and Wheeler, J. A. (1973). *Gravitation,* San Francisco, Freeman.

Ortín, T. (2004). *Gravity and Strings,* Cambridge University Press.

Oppenheimer, J. R. and Snyder, H. (1939). "On continued gravitational contraction," *Phys. Rev.* **56**, 455.

Oppenheimer, J. R. and Volkoff, G. (1939). "On massive neutron cores," *Phys. Rev.* **55**, 374.

Penrose, R. and Rindler, W. (1986). *Spinors and Space-Time.* 2 Vols. Cambridge University Press, Cambridge, UK.

Plebański, J. F. (1964). "The algebraic structure of the tensor of matter", Acta Phys. Pol. **26**, 963.

Plebański, J. F. (1967). *On Conformally Equivalent Riemannian Spaces,* Monograph of CINVESTAV-IPN, México.

Plebański, J. F. (1975). "A class of solutions of Einstein–Maxwell equations," *Ann. of Phys. (USA)* **90**,196.

Polchinski, J. (1998). *String Theory,* 2 vols. Cambridge University Press, Cambridge, UK.

Ponce de Leon, J. and Cruz, N. (2000). "Hydrostatic equilibrium of a perfect fluid sphere with exterior higher-dimensional Schwarzschild spacetime," *Gen. Rel. Grav.* **32**, 1207.

Robinson, I. (1959). "A solution of the Einstein–Maxwell equations," *Bull. Acad. Polon. Sci., Ser. Math. AStr. Phys.* **7**, 351.

Salazar, H., García, A., and Plebański, J. F. (1984). "Type D solutions of the Einstein and Born–Infeld nolinear electrodynamics equations," *Nuovo Cimento* **B84**, 65.

Salazar, H., García, A., and Plebański, J. F. (1987). "Duality rotations and type D solutions to Einstein equations with nonlinear electrodynamics sources," *J. Math. Phys.* **28**, 2171.

Schouten, J. A. (1954). *Ricci Calculus,* Springer–Verlag, Berlin.

Schouten, J. A. (1958). "Über die konforme Abbildung n-dimensionaler Mannigfaltigkeiten mit quadratischer Maßbestimmung auf eine Mannigfaltigkeit euklidischer Maßbestimmung," *Math. Zeit.* **11**, 58.

Stephani, H. (1990). *General Relativity–An Introduction to the Theory of Gravitational Field,* Cambridge University Press, Sec. Ed.

Stephani, H., Kramer, D., MacCallum, M., Honselaers, C. and Herlt, E. (2003). *Exact Solutions to Einstein's Field Equations,* Second Edn. Cambridge University Press.

Straumann, N. (1984). *General Relativity and Relativistic Astrophysics,* Springer–Verlag, Berlin.

Tsantilis, E., Puntigam, R. A., and Hehl, F. W. (1996). "A quadratic curvature Lagrangian of Pawłowski and Rączka: A finger exercise with MathTensor." In: *Relativity and Scientific Computing*, F. W. Hehl, R. A. Puntigam, H. Ruder (Eds.). Springer, Berlin (arXiv: gr-qc/9601002).

Tangherlini, F. R. (1963). "Schwarzschil field in n dimensions and the dimensionality of the space problem," *Il Nuovo Cimento.* **27**, 636.

Tonnelat, M. A. (1959). *Les principles de la théorie èlectromagnètique et de la relativltè,* Masson et Cie, Editeurs, 120 Boulevard Saint Germain, Paris.

Visser, M. (1992). "Dirty black holes: Thermodynamics and horizon structure," *Phys. Rev.* **D46**, 2445.

Wald, R. M. (1984). *General Relativity,* Chicago, University of Chicago Press.

Weinberg, S. (1972). *Gravitation and Cosmology: Principles and applications of the General Theory of Relativity,* Wiley, New York.

Witten, E. (1991). "String theory and black holes," *Phys. Rev.* **D44**, 314.

Witten, E. (1998). "Anti de Sitter space and holography," *Adv. Theor. Math. Phys.* **2**, 253. (arXiv: hep-th/9802150).

Weyl, H. (1918). "Reine Infinitesimalgeometrie," *Math. Zeitschr.* **2**, 384.

York Jr., J. W. (1971). "Gravitational degrees of freedom and the initial-value problem," *Phys. Rev. Lett.* **26**, 1656–1658.

Zarro, C. A. D. (2009). "Buchdahl limit for d-dimensional spherical solutions with a cosmological constant," *Gen. Rel. Grav.* **41**, 453.

Index